Lecture Notes in Chemistry

Volume 91

The Lecture Notes in Chemistry

The series Lecture Notes in Chemistry (LNC) reports new developments in chemistry and molecular science-quickly and informally, but with a high quality and the explicit aim to summarize and communicate current knowledge for teaching and training purposes. Books published in this series are conceived as bridging material between advanced graduate textbooks and the forefront of research. They will serve the following purposes:

- provide an accessible introduction to the field to postgraduate students and nonspecialist researchers from related areas,
- provide a source of advanced teaching material for specialized seminars, courses and schools, and
- be readily accessible in print and online.

The series covers all established fields of chemistry such as analytical chemistry, organic chemistry, inorganic chemistry, physical chemistry including electrochemistry, theoretical and computational chemistry, industrial chemistry, and catalysis. It is also a particularly suitable forum for volumes addressing the interfaces of chemistry with other disciplines, such as biology, medicine, physics, engineering, materials science including polymer and nanoscience, or earth and environmental science.

Both authored and edited volumes will be considered for publication. Edited volumes should however consist of a very limited number of contributions only. Proceedings will not be considered for LNC.

The year 2010 marks the relaunch of LNC.

More information about this series at http://www.springer.com/series/632

Yongfang Li
Editor

Organic Optoelectronic Materials

Springer

Editor
Yongfang Li
Institute of Chemistry
Chinese Academy of Sciences
Beijing
China

ISSN 0342-4901 ISSN 2192-6603 (electronic)
Lecture Notes in Chemistry
ISBN 978-3-319-16861-6 ISBN 978-3-319-16862-3 (eBook)
DOI 10.1007/978-3-319-16862-3

Library of Congress Control Number: 2015935407

Springer Cham Heidelberg New York Dordrecht London

Printed on acid-free paper

Springer International Publishing AG Switzerland is part of Springer Science+Business Media (www.springer.com)

Preface

Organic optoelectronic materials, including organic semiconductors, organic conductors, organic superconductors, conducting polymers, and conjugated polymers, have attracted great attentions since the discoveries of organic semiconductors in the 1950s and conducting polymers in the 1970s. Their novel physicochemical properties and promising applications in organic field-effect transistors (OFET), organic/polymer light-emitting diodes (OLED/PLED), and organic/polymer solar cells [OSC/PSC or OPV (organic photovoltaics)] stimulated and promoted broad research interests and development of new materials and new devices based on them.

As a book in the series *Lecture Notes in Chemistry*, this book is designed for graduate students and researchers who look for up-to-date knowledge on organic optoelectronic materials and their applications in OFETs, OLEDs/PLEDs, OPVs, and transparent conducting electrodes. This book can also be used as a reference book or text book for related researchers and graduate students. The molecular structures, synthetic methods, and physicochemical and optoelectronic properties of organic optoelectronic materials are introduced and described in detail. The structures and working mechanisms of organic optoelectronic devices are elucidated. The key scientific problems and future research directions of organic optoelectronic materials are also addressed. In more detail, Chaps. 1 and 2 cover the development history and physicochemical properties of organic semiconductors, organic conductors, organic superconductors, and conducting polymers. Chapter 3 introduces OFETs and the molecular structures and charge-carrier mobilities (hole and electron mobilities) of various *p*-type and *n*-type organic semiconductors. Chapters 4 and 5 describe photovoltaic materials and devices for OPVs based on organic small molecules and conjugated polymers, respectively. Chapters 6 and 7 elucidate electroluminescent materials and devices for OLEDs based on organic small molecules and PLEDs based on conjugated polymers, respectively. Chapter 8 outlines the knowledge of transparent conducting polymers for application in flexible transparent electrodes.

The research field of organic optoelectronic materials and devices has been developing quickly in recent years. The contents of this book may be limited by the knowledge and the understanding of the authors, and there may be some errors or mistakes. Any comments and suggestions or questions about the contents of this book are welcomed by me or by the contributing authors.

Beijing
January 2015

Yongfang Li

Contents

1 Organic Semiconductors, Conductors, and Superconductors 1
Yue Yue and Bin Zhang
1.1 Introduction 1
1.2 Crystal Engineering of Charge-Transfer Complexes 2
1.2.1 Charge Transfer Salts of AB Type 4
1.2.2 Charge Transfer Salts of A_2B Type 4
1.2.3 Charge Ordering in Organic ET Compounds 6
1.3 Magnetism in Charge Transfer Salt 11
1.4 Dual-Functional, Multifunctional Molecular Crystals 11
1.5 Relationship Between Organic Superconductors and Inorganic Superconductors: Resonating Valence-Bonding Solids and Jahn–Teller Distortion 14
1.6 Summary 15
References 16

2 Conducting Polymers 23
Yongfang Li
2.1 Molecular Structure of Conducting Polymers 24
2.1.1 Electronic Structure of Intrinsic Conjugated Polymers 25
2.1.2 Doping Structures of Conducting Polymers 27
2.1.3 Charge Carriers in Conducting Polymers 28
2.2 Doping Characteristics 29
2.2.1 Chemical Doping 29
2.2.2 Electrochemical Doping 30
2.3 Conductivity Characteristics 30
2.4 Absorption Spectra 31
2.5 Solubility 34
2.5.1 Effect of Substituents on Solubility of Conjugated Polymers 34

2.5.2 Effect of Substitution on the Conductivity of Conducting Polymers. 35
2.6 Electrochemical Properties 36
2.6.1 Electrochemical Properties of Conducting Polypyrrole 36
2.6.2 Electrochemical Properties of Conducting Polyaniline. 39
2.6.3 Electrochemical Properties of Polythiophene and Other Conjugated Polymers 39
2.6.4 Electrochemical Measurement of HOMO and LUMO Energy Levels of Conjugated Polymers 41
2.7 Optoelectronic Properties of Conjugated Polymers. 41
2.8 Synthesis of Conducting Polymers 42
2.8.1 Electrochemical Oxidation Polymerization of Conducting Polymers. 42
2.8.2 Chemical Polymerization of Conducting Polymers. . . 46
2.9 Summary 47
References. 49

3 Organic Semiconductors for Field-Effect Transistors 51
Weifeng Zhang and Gui Yu
3.1 Introduction 52
3.1.1 Overview. 52
3.1.2 History and Work Principle of OFETs. 53
3.1.3 Device Configuration and Processing Technique of OFETs 57
3.1.4 Factors Influencing the Performance of OFETs. 60
3.2 p-Type Semiconductors 67
3.2.1 Selected p-Type Small-Molecule Semiconductors . . . 67
3.2.2 Selected *p*-Type Polymer Semiconductors 93
3.3 n-Type Semiconductors 104
3.3.1 Selected n-Type Small-Molecule Semiconductors . . . 104
3.3.2 Selected n-Type Polymer Semiconductors 126
3.4 Ambipolar Semiconductors. 130
3.4.1 Selected Ambipolar Small-Molecule Semiconductors. 131
3.4.2 Selected Ambipolar Polymer Semiconductors 136
3.5 Outlook 141
References. 141

4 **Organic Semiconductor Photovoltaic Materials** 165
Zhi-Guo Zhang
4.1 Introduction 165
4.2 Organic Solar Cells by Vacuum Deposition 166
4.3 Organic Solar Cells by Solution Processing 169
4.3.1 Dyes 169
4.3.2 Triphenylamine Derivatives 173
4.3.3 Oligothiophenes 176
4.3.4 Linear D-A Oligothiophenes. 178
4.3.5 Organic Molecule Acceptors. 183
4.4 Conclusion and Future Perspectives. 188
References. 188

5 **Conjugated Polymer Photovoltaic Materials**. 195
Long Ye and Jianhui Hou
5.1 Introduction 195
5.1.1 Brief Summary of Photovoltaic Polymers. 196
5.1.2 Design Considerations of Conjugated Polymer Photovoltaic Materials. 196
5.2 Conjugated Polymer Donor Materials 200
5.2.1 Three Important Types of Homopolymer 200
5.2.2 Donor–Acceptor Copolymers 208
5.3 Conjugated Polymer Acceptor Materials. 226
5.4 Summary and Outlook. 229
References. 229

6 **Organic Semiconductor Electroluminescent Materials**. 241
Gufeng He
6.1 Introduction 241
6.2 Working Mechanism of OLEDs 242
6.2.1 Working Mechanism 242
6.2.2 Anode and Hole Injection Material 243
6.2.3 Cathode and Electron Injection Material. 245
6.2.4 Hole and Electron Transport Materials. 246
6.2.5 *p*- and *n*-Type Doping Materials 247
6.3 Fluorescent Electroluminescent Materials 250
6.3.1 Red Fluorescent Materials 251
6.3.2 Green Fluorescent Materials 252
6.3.3 Blue Fluorescent Materials 257
6.3.4 Advanced Delayed Fluorescent Materials 261
6.4 Phosphorescent Electroluminescent Materials 263
6.4.1 Red Phosphorescent Materials 264
6.4.2 Green Phosphorescent Materials 275
6.4.3 Blue Phosphorescent Materials 283
6.5 Summary and Outlook. 290
References. 291

7 **Conjugated Polymer Electroluminescent Materials** 303
Xing Guan, Shenjian Liu and Fei Huang
7.1 Introduction 303
7.1.1 Electroluminescence and PLEDs 304
7.2 Conjugated Electroluminescent Polymers and Performance Tuning 305
7.2.1 Early Efforts. 305
7.2.2 Performance Tuning 306
7.3 Luminescent Polymers Based on Dopant/Host System 321
7.3.1 Electrofluorescent Polymers 322
7.3.2 Electrophosphorescent Polymers 327
7.3.3 Single White Emitting Polymers 333
7.4 Hyperbranched Polymers 336
7.5 Supramolecular Luminescent Polymers 345
7.6 Conclusion. 347
References. 348

8 **Transparent Conducting Polymers** 359
Yijie Xia and Jianyong Ouyang
8.1 Electronic Structure and Optical Properties of Conducting Polymers. 360
8.2 Transparent Conducting Polymers 362
8.3 Preparation of PEDOTs by Electrochemical Polymerization 364
8.4 Preparation of PEDOTs by Chemical Synthesis. 364
8.5 Vapor-Phase Polymerization of EDOT. 366
8.6 Development of Highly Conductive PEDOT:PSS 367
8.6.1 Structure of PEDOT:PSS 367
8.6.2 Conductivity Enhancement by Adding Compounds to PEDOT:PSS Aqueous Solution 368
8.6.3 Conductivity Enhancement of PEDOT:PSS Through a Post-coating Treatment. 370
8.6.4 Mechanisms for the Conductivity Enhancements of PEDOT:PSS. 379
8.7 Application of PEDOT:PSS for Optoelectronic Devices 383
8.8 Outlook for Transparent Conducting Polymers 386
References. 387

Chapter 1
Organic Semiconductors, Conductors, and Superconductors

Yue Yue and Bin Zhang

1.1 Introduction

In general, organic solids are insulators. However, there have been extensive and intensive efforts in materials science and technology to make them conductive. The family of organic solids, starting from insulators, has widened to include organic semiconductors, organic conductors, and organic superconductors. The distinctions between them are based on the band structure of the materials as well as the electron occupancy of these bands. In 1954, the first organic semiconductor was discovered and the conductivity reached 10^{-3} S/cm [1]. This illustrates a new direction for the synthesis of organic conductors, when organic material was first doped with an electron donor or acceptor as a charge-transfer complex. In the 1960s, a conducting organic solid was first achieved with the charge-transfer complex of TCNQ [2]. The organic/metal product TTF-TCNQ was obtained in 1973 [3] and the first organic superconductor $TMTSF_2 \cdot PF_6$ was discovered in 1980 [4–6]. After that, the critical temperature of organic superconductors quickly increased from 0.6 to 18 K. In 1991, the electron-transfer superconductor A_3C_{60} was discovered with superconducting transition at 33 and 35 K [7, 8], respectively and eventually single-component molecular metals were synthesized in 2001 [9].

Organic conductors are critical for electronic applications as they are as efficient as metals but lighter and more flexible. Scientists working on organic electronics want to improve the conductivity, stability, and tailorability of highly conjugated organic semiconductors and conductors. The way to challenging high performance

Y. Yue (✉)
School of Mechanical Engineering and Automation, Beihang University,
Beijing 100191, People's Republic of China
e-mail: yueyue@buaa.edu.cn

B. Zhang
Institute of Chemistry, Chinese Academy of Sciences, Beijing 100190,
People's Republic of China

Y. Li (ed.), *Organic Optoelectronic Materials*,
Lecture Notes in Chemistry 91, DOI 10.1007/978-3-319-16862-3_1

optical and electronic organic devices is to understand the processes that determine charge transport of organic molecular and polymeric materials. Small molecules can also be grown as single crystals as model systems to demonstrate the intrinsic electronic properties. This chapter focuses on the charge transport of organic materials, and some prototype organic solids are also discussed.

1.2 Crystal Engineering of Charge-Transfer Complexes

One way to produce the organic conductors is to use charge-transfer reactions from donor to acceptor and the produced crystal is called a charge transfer complex (salt) [10]. The formation of the charge transfer complex is through hybridization between the HOMO (highest occupied molecular orbital) of the donor and the LUMO (lowest unoccupied molecular orbital) of the acceptor. Scientists' efforts from the 1960s led to the organic acceptor 7,7,8,8-tetracyanoquinodimethane (TCNQ) [11] and the donor tetrathiafulvalene (TTF) [12, 13] (Fig. 1.1a, b).The first stable organic conductor TTF-TCNQ was synthesized in 1973 [3]. In 1978, the derivative of TTF, combining a conjugated TTF unit and ethylene group, BEDT-TTF, (Fig. 1.1d) was synthesized,

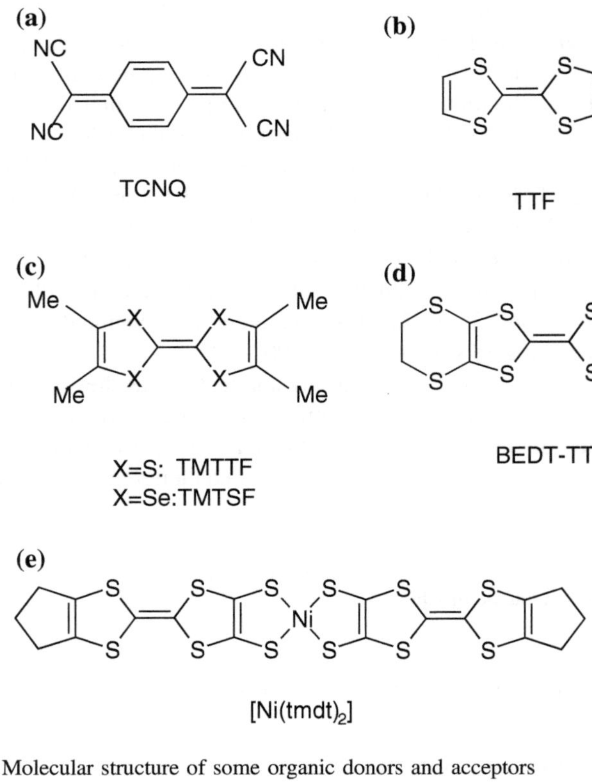

Fig. 1.1 Molecular structure of some organic donors and acceptors

showing a two-dimensional layer in the crystal and contributing most of the organic superconductor properties as κ-(BEDT-TTF)$_2$Cu[N(CN)$_2$]Cl, $T_c = 13.2$ K [14]. With regard to this, most organic conductors were synthesized by the charge transfer reaction until the new superconductor Ni(dmit)$_2$ (Fig. 1.1e) was synthesized in 2001 [9]. In this single molecular conductor, the gap between HOMO and LUMO is so small that it can form partially filled bands. The characterization of conducting organic material is carried out for a high-quality single crystal because the crystal defect traps the carrier inside the material. Electrocrystallization is a powerful method for obtaining high quality organic conductors and superconductors.

The charge carrier transport properties of organic solids have been investigated extensively and can be used to investigate and optimize the structure-property relations of the materials used in existing optoelectronic devices and to predict the ideal materials for the next generation of electronic and optoelectronic devices. The electronic properties are controlled by weak interactions between the π-units (donor: TTF, BEDT-TTF; acceptor: Ni(dmit)$_2$). The interaction between π-units and transition metal counterions as π–d interaction plays an important role in the physical properties. For example, when a π-unit was put in one column or two-dimensional layer, within the orbital overlap between neighbor π-unit as an S…S contact at distance less than 3.6 Å (sum of Van der Waals value of S), the channel for the conduction electron resulted. The crystal showed semiconductive metallic to superconductive behavior.

Polytypism and polymorphism are popular in charge-transfer complexes because of the assembly of molecular crystals in crystal engineering. For example, the charge-transfer complexes of BEDT-TTF and I_3^- with compositions of 2:1, 3:2, and 3:5 and the charge-transfer salts of BEDT-TTF and $FeCl_4^{n-}$ with composition of 2:1, 3:2, 1:1, and 1:2 are examples of polytypism. Depending on the donor arrangement of the BEDT-TTF molecule, more than ten arrangement modes known as α, β, γ, κ, λ, δ, …, etc., were observed [15], displaying different transport properties. Regarding polymorphism, in charge-transfer complexes, α-(BEDT-TTF)$_2$I$_3$ shows metal–insulator transition at 150 K, β-(BEDT-TTF)$_2$I$_3$ and γ-(BEDT-TTF)$_2$I$_3$ show superconductivity at 7 and 6 K, respectively. Mott insulator β′-(BEDT-TTF)$_3$(FeCl$_4$)$_2$ and metal δ-(BEDT-TTF)$_3$(FeCl$_4$)$_2$ have also been investigated.

The conductivity of crystal and charge-transfer complexes is controlled by the arrangement of π-units and crystal structures, respectively. For example, β-(BEDT-TTF)$_2$I$_3$ shows metal to superconductor transition at 6 K, β-(BEDT-TTF)$_3$[CrMn(C$_2$O$_4$)$_3$] shows as metallic to 2 K. α-(BEDT-TTF)$_2$I$_3$ shows metal to insulator transition at 150 K, and metal to insulator transition was observed at 150 K in α-(BEDT-TTF)$_3$[CrMn(C$_2$O$_4$)$_3$]. Conductivity could be influenced by counterions when the arrangement of π-units remained the same. For example, a metallic to insulator transition at 200 K is observed in θ^{21}-(BEDT-TTF)$_3$Ag$_{6.4}$I$_8$ with $\sigma_{rt} = 50$ S/cm, and θ^{21}-(BEDT-TTF)$_3$[Cu$_2$(C$_2$O$_4$)$_3$](CH$_3$OH)$_2$ is a semiconductor with $\sigma_{rt} = 4$ S/cm. Conductivity can be influenced by the guest solvent molecules. For example, in (BEDT-TTF)$_4$(H$_3$O)Fe(C$_2$O$_4$)$_3$ in solvent, $T_c = 7.0$ K is observed when the solvent is C_6H_5CN and 4.0 K when solvent is C_6H_5Br. As the donor arrangement remained the same as δ-phase with the counteranion of $GaCl_4^-$, room-temperature conductivity increased from 0.1 to 1 S/cm when solvent molecules C_6H_5Cl intercalated into an anion sheet. Some of the crucial factors relating to conducting molecular solids are as follows.

1.2.1 Charge Transfer Salts of AB Type

One of the highlights at this stage is the TTF-TCNQ, which is one-dimensional (1D) charge-transfer conducting salt with a Peierls transition at low temperature and synthesized between the π-electron molecules: the electron donor TTF and the acceptor TCNQ [3, 13]. The ratio of the TTF and TCNQ is 1:1. As a donor, TTF has four sulfur heteroatoms which can easily donate electrons when combining with the acceptor molecule. TCNQ, as an acceptor, can be easily reduced to form an anion radical $TCNQ^{-}$. The conductivity of this salts reaches $\sigma = 1.47 \times 10^4$ (S/cm) at around 60 K, where a metal to insulator phase transition was also observed [3] and the metallic behavior was confirmed by polarized reflection spectroscopy [16]. The divergent peak ($\sigma_{MAX} > 10^6$ S/cm) of conductivity at 58 K in a TTF-TCNQ crystal was reported [17] and the conductivity was found to originate from the fluctuations of Frohlich superconductivity, which is based on the coupled electron–phonon collective mode in a 1D system [18]. This metal to insulator phase transition is attributed to the fluctuation of charge density waves by impurities or lattice instability [19]. After this discovery, Scientists synthesized many types of derivatives of TTF and TCNQ such as TSeF-TCNQ [20], HMTSF-TCNQ [21], and TMTSF-DMTCNQ [22], which show metallic conductivity at very low temperatures. AB type charge transfer salts have generally demonstrated insulating ground states because of the instability of metallic states intrinsic for 1D systems.

1.2.2 Charge Transfer Salts of A_2B Type

More conductive states have been found in charge transfer salts of the A_2B type compared to the AB type. In 1980, the first superconductor $(TMTSF)_2PF_6$ at 0.9 K under 12 kbar was discovered [6, 23]. This transition originated from the spin density wave (SDW) and occurs at 12 K [24–26], an antiferromagnetic ordering being observed by using NMR [27] and static magnetic susceptibility measurements [14]. In the vast $(TM)_2X$ family (see Fig. 1.1c), scientists mainly found two isostructural groups: selenium TMTSF salts which are metals with a formally 3/4-filled conduction band and sulfur TMTTF salts which are close to the Mott–Hubbard insulating state because of the high anisotropy, dimerization, and on-site Coulomb repulsion [28]. X in $(TM)_2X$ can be several possible anions such as $(TMTSF)_2PF_6$, $(TMTSF)_2AsF_6$, $(TMTSF)_2SbF_6$, and $(TMTSF)_2TaF_6$ which show the metal–insulator transition at 11–17 K below that of the SDW state [24–26]. $(TMTTF)_2PF_6$ and $(TMTTF)_2SbF_6$ undergo superconducting transitions at 1.8 K under 54 kbar and

Fig. 1.2 Generalized phase diagram for: TM_2X [23]

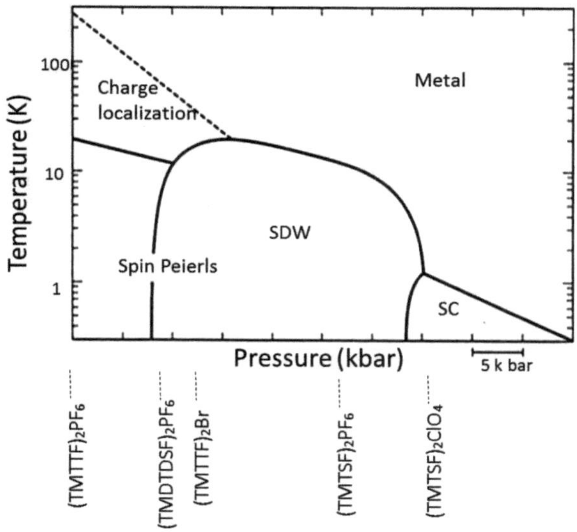

2.6 K under 61 kbar, respectively [29, 30]. Moreover, the superconducting phase transition of $(TMTSF)_2ClO_4$ was observed at ambient pressure down to 1 K [31]. Figure 1.2 shows the phase diagram of $(TM)_2X$ [23]. This diagram suggests various different phases such as normal metals, superconductors, spin-density-wave states, spin-Peierls state, and antiferromagnetic state as a function of decreasing pressure. Although the band structure of $(TMTSF)_2PF_6$ is calculated to have a quasi 1D Fermi surface, intermolecular Se…Se contact was observed between the TMTSF stacks [32]. Scientists found the way to synthesize 2D organic conductors from $(TMTSF)_2PF_6$ by increasing the bandwidth and dimensionality [33].

The one-dimensional A_2B systems may be unstable in the insulating state and the ideal 2D A_2B systems superconductor was first made from β-$(BEDT\text{-}TTF)_2ReO_4$ at 2 K under 4 kbar [34]. β-$(BEDT\text{-}TTF)_2I_3$ at 1.4 K at ambient pressure [27, 35, 36] and κ-$ET_2Cu(NCS)_2$ at 10.4 K [35], and recently β'-ET_2ICl_2 showed the highest T_c among organic superconductors at 14 K under 82 kbar [37, 38]. BEDT-TTF as a donor, was first synthesized in 1978 [14]. The π-electron orbitals of the donor aromatic rings overlap to form a conducting band. This BEDT-TTF molecule forms various phases with various anions. Figure 1.3 shows the four different donor planes of the BEDT-TTF compound. The β-type organic BEDT-TTF salts were known very early because of their superconducting state at ambient pressure—e.g., $(BEDT\text{-}TTF)_2IBr_2$ at 2.7 K and $(BEDT\text{-}TTF)_2AuI_2$ at 3.8 K [32, 39]. β′ and β″ types are similar to the β type whereas the molecular stackings are different. Figure 1.4 shows the phase diagram of the θ phase family θ—$(BEDT\text{-}TTF)_2MM'(SCN)_4$ (M = Rb, Tl, Cs, M′ = Co, Zn) concerning the charge ordering phenomenon [40, 41]. The electronic state, including insulators, superconductors, and metals, is parameterized by the dihedral angel between columns [40]. In the phase diagram, the metallic phase is reduced with increasing dihedral angle. All compounds become insulators at low temperature.

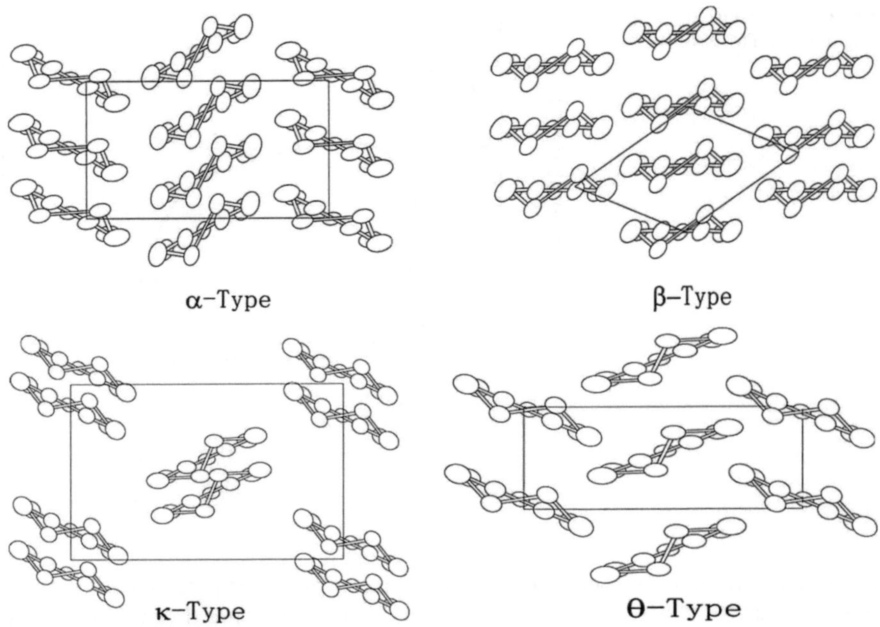

Fig. 1.3 Schematic view of some molecular configurations of the BEDT-TTF compound

The α-type BEDT-TTF salts are similar to the θ phase and show a weak dimerization. There are two different kinds of typical groups in α-type BEDT-TTF salts. One is the family of α-$(BEDT\text{-}TTF)_2MHg(SCN)_4$ (M = K, Rb, Tl, NH_4) in which K, Rb, and Tl compounds produce the SDW below 10 K [38] and NH_4 salt shows a superconductivity at 1.15 K [42]. Another group is α-$(BEDT\text{-}TTF)_2X$ (X = I_3, IBr_2, ICl_2, etc.). Material α-$(BEDT\text{-}TTF)_2I_3$ undergoes an MI transition at 136 K [41, 43, 44]. Charge-ordering phenomena were found in NMR experiments [45]. After the success of the 1D TMTSF and 2D BEDT-TTF salts, scientists made efforts to synthesize many new 3D molecular superconductors such as K_3–C_{60} with T_c = 18 K [46] and Cs_2RbC_{60} with the highest T_c = 33 K [47].

1.2.3 Charge Ordering in Organic ET Compounds

The family of 2D organic conductors $(ET)_2X$ is known to exhibit a variety of interesting electronic properties. The theoretical studies of Kino and Fukuyama developed a systematic way to understand the diversity in their ground state properties [48]. Another interesting conclusion of Kino and Fukuyama is that α-type compounds show an insulating state with charge transfer in their notation (charge ordering) [49]. Arising from a strong correlation, the charge ordered (CO) state is

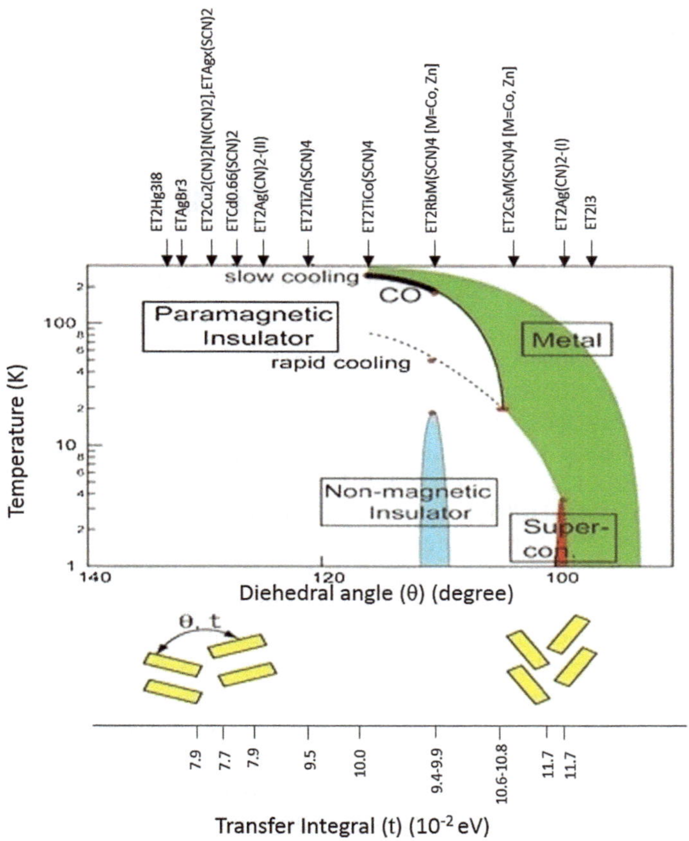

Fig. 1.4 Universal phase diagram of θ-type BEDT-TIF compounds [40, 41]

one of the typical ground states of molecular conductors. As to the electron correlation phenomenon, it draws growing attention to understanding the organic conductor's low temperature properties [49–52]. Charge ordering can be understood as self-organization of localized charge carriers. For example, in the charge-ordered state of a one-dimensional system with a quarter-filled conduction band, the localized charge carriers occupy or do not occupy the lattice site individually. If the conduction band is not filled completely, charge disproportionation can be observed. Charge order in organic conductors was first suggested in the 1D dimensional system $(DI\text{-}DCNQI)_2Ag$ [53]. It was shown that below 220 K, ^{13}C-NMR spectra are split. Nonequivalent differently charged molecules appear along the chain axis and the ratio is 3:1 below 130 K. U is the on-site Coulomb repulsion and V is the nearest neighbor interaction. The inter-site Coulomb repulsion V is the driving force for charge ordering to occur as well as the onsite Coulomb repulsion U [49, 50]. If V exceeds a certain value, the charges arrange themselves with a long enough distance

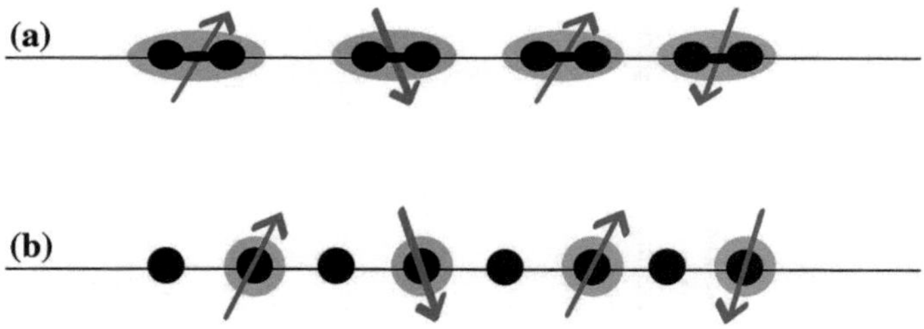

Fig. 1.5 **a** Dimer Mott–Hubbard insulator. **b** Wigner crystal type charge ordering [56]

to minimize the influence of the V. The extended Hubbard model is a good description of the relevant energies [49, 54–56].

Here we discuss the charge ordering state using quarter-filled systems. Figure 1.5 shows the two cases [56]: (1) dimer Mott–Hubbard insulator such as 1D MEM-$TCNQ_2$ and 2D κ-$(BEDT\text{-}TTF)_2X$, λ-$BETS_2X$ and (2) Wigner crystal type charge ordering such as DI-$DCNQI_2Ag$ and $TMTTF_2X$, 2D θ-$(BEDT\text{-}TTF)_2X$, and α-$(BEDT\text{-}TTF)_2X$ [57]. In the first case, because of the strong dimerization, the single electron occupies the bonding state of each dimer. The Mott insulating state is realized because of this strong effective Coulomb interaction within a dimer. In the second case, however, inter-site Coulomb interaction, V plays an important role, and the charge-ordered state called the Wigner crystal is realized on the lattice. In the absence of a dimerization structure of the 2D system such as α, θ, and β″ type compounds, several types of a CO state which is called stripe type CO state are found as a ground state [58]. Electrons stay apart from each other if the kinetic energy is rather small compared to the Coulomb interaction. Moreover, the anisotropy in the transfer integrals is also important for the arrangement of the localized charges. Figure 1.6 shows the different pattern of CO.

The charge-ordered state has been studied by means of NMR [59], XRD [60], and vibrational spectroscopy [61–65]. The NMR spectrum shows a splitting or broadening depending on the distribution of carrier density. The first CO was found in $(DIDCNQI)_2Ag$ by ^{13}C-NMR measurement [53]. The spin/charge configuration of $(TMTTF)_2X$ ($X = SCN, Br, PF_6, AsF_6$) was also confirmed by NMR experimentally [66–70] and theoretically [71]. $(TMTTF)_2PF_6$ and $(TMTTF)_2AsF_6$ undergo a spin-Peierls transition [72, 73], whereas $(TMTTF)_2$ SCN [66] and $(TMTTF)_2Br$ [67] have 1010 type ordering and CO was directly confirmed as the splitting of signals into charge-rich site and charge-poor sites at low temperature by ^{13}C-NMR [69]. In 2D systems, θ-$(BEDT\text{-}TTF)_2RbZn(SCN)_4$, θ-$(BEDT\text{-}TTF)_2CsZn(SCN)_4$, and α-$(BEDT\text{-}TTF)_2I_3$ were investigated and were found to be in CO states at low temperature and in CD state at high temperature by NMR [45, 59, 74–80]. In the case of α-$(BEDT\text{-}TTF)_2I_3$, the ratio of the effective charges are also estimated from the amplitude of the curves [45, 78], and the horizontal stripe CO pattern predicted theoretically [49] was confirmed from experimental results not only by ^{13}C-NMR but also by X-ray [81, 82] and IR/Raman spectroscopy [63, 64, 83]. Among the various

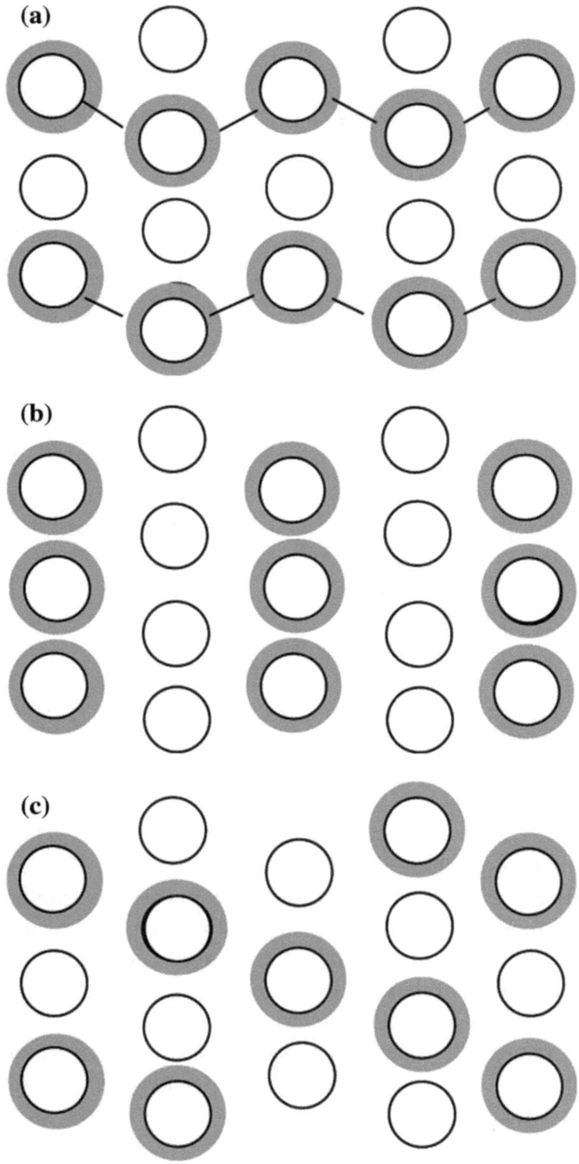

Fig. 1.6 **a** Horizontal stripe. **b** Vertical stripe. **c** Diagonal stripe

techniques for charge ordering research, vibrational spectroscopy such as IR/Raman can be one of the powerful methods [84, 85]. In vibrational spectroscopy, most charge-sensitive modes for BEDT-TTF molecule are the stretching modes ν_3, (Raman active), the in-phase ν_2 (Raman active), and out-of-phase ν_{27} (infrared active) (Fig. 1.7) [86]. The ν_2 and ν_3 modes include the stretching vibrations of the central

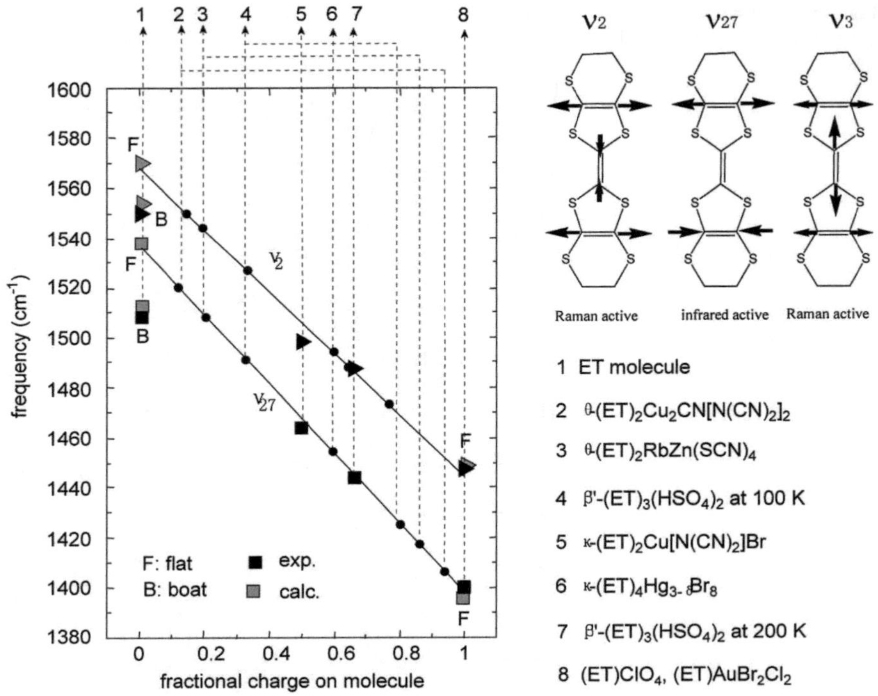

Fig. 1.7 Frequencies of the υ_2 and υ_{27} modes plotted as a function of the charge ρ on the BEDT-TTF molecule [83]

C=C bond and the symmetric ring C=C bond. The ν_{27} mode corresponds to the stretching vibration of the anti-symmetric ring C=C bond. In these three sensitive modes, ν_3 is more strongly perturbed by electron-molecular-vibration interaction than by molecular charge. Therefore, it is inappropriate to use υ_3 for estimating the fractional charge on molecules. ν_2 and ν_{27} are mainly perturbed by molecular charge, have a linear relationship between the frequency and the charge on the molecules, and can be used to calculate the fractional charge in charge ordering state at low temperature [83]. The linear relationship between the frequency and site charges is shown in Fig. 1.7: $\nu_2(\rho) = 1447 + 120(1 - \rho)$ and $\nu_{27}(\rho) = 1398 + 140(1 - \rho)$ [83]. Vibrational spectroscopy was first applied to the study of charge-ordering in θ-(BDT-TTP)$_2$(SCN)$_4$ [85]. θ-(BEDT-TTF)$_2$RbZn(SCN)$_4$ undergoes the CO–CD phase transition at 200 K. The assignments for ν_2 modes which split into two and ν_3 modes which split into four were performed based on the ^{13}C-substituted sample by IR/Raman spectroscopy [64]. Based on this assignment, the horizontal stripe was confirmed. The horizontal stripe of the CO pattern was also reported by analyzing the electronic transition in the infrared region [87]. The same IR/Raman method was applied to the study of charge ordering in α-(BEDT-TTF)$_2$I$_3$ below and above 136 K from ambient pressure to 3.6 GPa [63]. The splitting of ν_2 indicates the charge disproportionation caused by charge localization and it formed the horizontal CO stripe perpendicular to the stacks.

1.3 Magnetism in Charge Transfer Salt

Naturally, magnetism relates closely with conductivity. Classic magnetism is found in charge-transfer complexes such as the long-range ferromagnetic ordering at 4.5 K in insulator $(NH_4)_2[Ni(mnt)_2]\cdot H_2O$ [88]. Recently, quantum magnetism as spin liquid was observed in molecular insulators κ-$(BEDT\text{-}TTF)_2[Cu(CN)_3]$ and $EtMe_3Sb[Pt(dmit)_2]_2$ with spin on π-units [89–92]. The conductivity of a charge-transfer complex of TCNQ was studied before the discovery of the TTF series of organic superconductors, and the room-temperature conductivity of (5,8-dihydroxyquinolineH)$(TCNQ)_2$ reached 10^2 S/cm in 1971 [93]. When TCNE was used as ligand, the conducting magnet was produced. In 1991, the room-temperature ferrimagnet $V(TCNE)_2(CH_2Cl_2)_{0.5}$ was discovered [94]. It was a semiconductor with $\sigma_{rt} = 10^{-4}$ S/cm [95]. It is one of the best examples of combined magnetism and conductivity in a molecule-based conducting magnet. When TCNE, TCNQ, and its derivatives were used as coordination ligands, a large number of molecule-based conducting magnets, including dynamic conducting magnets, were obtained. No metal product was found [96–98].

There are two sources of magnetism in coordination compounds: one is the interaction between cation and anion through weak interactions such as antiferromagnetic ordering at 3.0 K in $(C_2H_5)_4NFeCl_4$, the other comes from magnetic interaction between metal ions in a counter-anion such as oxalate-bridged Cr^{3+} and Mn^{2+} ions in $(C_4H_9)_4N[CrMn(C_2O_4)_3]$. This shows ferromagnetic ordering at 6 K [99, 100]. Magnetism in charge-transfer salt was also influenced by the arrangement of donor and counter-anion in the crystal, such as β′-$(BEDT\text{-}TTF)_3(FeCl_4)_2$ and δ-$(BEDT\text{-}TTF)_3(FeCl_4)_2$. β′-$(BEDT\text{-}TTF)_3(FeCl_4)_2$ shows antiferromagnetic transition at 2.7 K and δ-$(BEDT\text{-}TTF)_3(FeCl_4)_2$ at 4.8 K [101].

1.4 Dual-Functional, Multifunctional Molecular Crystals

Endowing the molecular conductor with magnetism or certain optical properties produces dual-functional molecular crystals such as the magnetic conductor [102], the magnetochiral conductor [103], and the single-molecular magnet with luminescence [104]. Combining the magnetic or photonic building block with conducting π-unit is one of most popular way to approach the goal.

Supramolecular chemistry is the key to designing new dual-functional, multifunctional molecular conductors. The functional units are synthons, and the arrangement and weak interaction between π-units decide the conductivity.

The magnetic conductor is the hottest research area in dual-functional molecular crystals because of the close relationship between magnetism and conductivity in organic superconductors and between molecular conductors and molecular magnetism. In the phase diagram of the inorganic superconductor, the diamagnetic superconductor is close to the antiferromagnetic insulator. An antiferromagnetic insulator could become a diamagnetic superconductor after hole or charge doping.

The antiferromagnetic Mott insulator attracts attention because of their potential for conversion into a superconductor after carrier-doping.

Top-down is another way to obtain dual-function, multifunction material. The intercalation of alkali metal into layered compounds, such as intercalated graphene, can produce a superconductor with transition temperatures ranging from 0.14 [105] to 11.5 K [106]. When alkali metal was intercalated into an isomer of graphene–C_{60}, the superconducting transition temperature reached higher than 50 K. Intercalated compound of aromatic compounds have recently been studied, and new materials with superconducting transition temperatures of about 30 K (18 K [107]; 5 K [108]) have been obtained. More exciting results can be obtained when the crystal structures are confirmed. (One of the shortcomings of the top-down approach is that it is always difficult to obtain high-quality single crystals.)

When an electric field, magnetic field, ultrabright laser, or high-pressure is applied to a single crystal, the energy state may be modified. Thereby (electric) field-induced organic superconductor doping with hole or electron is obtained when gate voltage is changed in a field-effect-transistor. The electric-field-induced superconductor was observed in the inorganic layer compound MoS_2 [109, 110], the (magnetic) field-induced reaction being obtained when the intra-magnetic field inside the crystal from spin-orbital coupling as a π–d interaction was compensated by application of a magnetic field. Irradiation of the crystal under a laser could change the electronic structure of the crystal as an injection of energy, and laser-induced metallic reaction was observed in $(EDO\text{-}TTF)_2PF_6$ [111]. This indicates the possibility of modulating the conductivity state with photo-irradiation.

High-pressure was one of the most powerful and the earliest method used to increase interactions between molecular π-units; it could suppress the metal–insulator transition by Peierls transition, charge-ordering, charge-localization, or Fermi nesting in organic compounds when the temperature decreased. Now the pressure of 200 GPa can be achieved with a diamond cell. However, the crystal is sensitive to pressure, so the experiments should be carried out carefully and slowly, step by step [112]. Bottom-up is a powerful method to obtain material with controllable designed properties. Magnetic conductors were synthesized by combining conducting organic π-units with magnetic inorganic coordination anions as organic–inorganic hybrids. Zero-dimensional anions, such as $FeCl_4^-$, $MnCl_4^{2-}$, $CoCl_4^{2-}$, and $CuCl_4^{2-}$, could produce π–d interaction between donor and anion through S…Cl contact in charge-transfer salts. Charge-transfer salts with strong π–d interaction showed negative magnetoresistance around 4.2 K [113, 114], magnetic-field-induced superconductivity was observed in $\lambda\text{-BETS}_2FeCl_4$ with $J\pi d = 17.7$ K [115], and by diluting Fe with Ga as Fe/Ga alloy in $\lambda\text{-BETS}_2Fe_{0.40}Ga_{0.6}Cl_4$ with insulator metal superconductor modulation by an applied magnetic field [116]. The band engineering method succeeded on charge-transfer salts of β′-(BEDT-TTF)$_3$(FeCl$_4$)$_2$. A strong π–d interaction was observed in β′-(BEDT-TTF)$_3$(FeCl$_4$)$_2$ with $J\pi d = 25.82$ K, so it is a Mott insulator.

A one-dimensional anion, such as $[Fe(C_2O_4)Cl_2^-]_n$, was used as counter-anion for a magnetic conductor. In ammonium salts of $[Fe(C_2O_4)Cl_2^-]_n$, a broad maximum for low-dimensional antiferromagnetism was observed at around 20–50 K, some of them showing long-range magnetic ordering as spin canting. In TTF[Fe(C$_2$O$_4$)Cl$_2$], the strong π–d interaction between TTF dimer and $[Fe(C_2O_4)Cl_2^-]_n$ produced a three-dimensional antiferromagnetic ordering at 19.8 K [117]. The weak ferromagnetic conductor with metallic properties to 0.6 K was obtained with BETS stacks in a two-dimensional κ′-phase, hysteresis with a loop of 150 Oe being observed at 150 Oe. The bifurcation of ZFCM/FCM at 4.5 K suggested a long-range magnetic ordering [118]. In the charge-transfer salt (BEDT-TTF)[Fe(C$_2$O$_4$)Cl$_2$](CH$_2$Cl$_2$), BEDT-TTF dimer and CH$_2$Cl$_2$ coexisted in a donor layer, this being a semiconductor as is TTF[Fe(C$_2$O$_4$)Cl$_2$] [119].

The single molecular magnet (SMM) and single chain magnet (SCM) are of great interest as quantum magnets to chemists. They could be used as counterions to synthesize charge-transfer salts with TTF or dmit units [120] or to connect TTF units to coordination ligands to form coordination compounds [121]. An excellent way to obtain a magnetic conductor is to merge TTF and dmit units into one unit as a single-component compound. Antiferromagnetic transition was observed at 110 K produced by Fermi nesting [122].

Molecular magnets provide abundant magnetic units for dual-functional molecular crystals with magnetism and conductivity. In 1992, (Bu$_4$N)[CrMn(C$_2$O$_4$)$_3$] was reported to have ferromagnetic transition at 5.5 K [100]. It was not until 2001 that the first organic-inorganic hybrid dual-functional molecular crystal as charge-transfer salt of (BEDT-TTF)$_3$[CrMn(C$_2$O$_4$)$_3$] was reported with magnetism from layered anions and conductivity from donors as the β-phase in β-(BEDT-TTF)$_2$I$_3$, respectively [102]. The charge-transfer salt α-BETS$_3$[CrMn(C$_2$O$_4$)$_3$] shows ferromagnetic transition at 5.5 K and a metal-to-semiconductor transition at 150 K [123]. These two crystals have incommensurate structures, and the donor and anion structures were determined separately.

When homometallic honeycomb anion $[Cu_2(C_2O_4)_3^{2-}]_n$ was used as counter-anion, the high-quality single crystal (BEDT-TTF)$_3$[Cu$_2$(C$_2$O$_4$)$_3$](CH$_3$OH)$_2$ was obtained. By means of the Jahn–Teller distortion of Cu^{2+}, a distorted honeycomb anion was formed. The donor arrangement belongs to the θ^{21}-phase, and when BEDT-TTF was replaced with BETS the isostructural compound was obtained. This is different from charge-transfer salts of heterometallic honeycomb anions where high-quality crystal structures are obtained from single crystal X-ray diffraction experiments [124, 125]. The spin-orbital coupling of Cu^{2+} produces spin frustration in these crystals. The frustration factor f is larger than 60, at least when the conductivity and susceptibility were measured above 1.8 K. Experiments at lower temperatures may bring some exciting results [126].

1.5 Relationship Between Organic Superconductors and Inorganic Superconductors: Resonating Valence-Bonding Solids and Jahn–Teller Distortion

Organic superconductors are important in the study of superconductors, not only because of their conductivity but also their magnetism. Research on the organic superconductor covers a wide area including the conductivity of insulators, semiconductors, conductors, and superconductor, and magnetism from classic magnets to quantum magnets.

After the discovery of the superconductor, people were confused by the mechanism of superconductivity for decades. Designing new superconductor systems is still a challenge for chemists. In 1986, Muller discovered the first high-temperature superconductor $Ba_xLa_{5-x}Cu_5O_{5(3-y)}$ with an onset temperature of 30 K from his initial exploration of the Jahn–Teller effect in the presence of spin-orbital coupling in perovskite material [127, 128]. The Jahn–Teller polaron in superconductors was confirmed by the observation by scanning tunnel microscopy. After that, Jahn–Teller distortion could be used to interpret the superconductivity in new inorganic superconductors, such as octahedral Co^{2+} in $Na_xCoO_2(H_2O)_y$ and tetrahedral Fe^{2+} in La–O–Fe–As (iron pnictide) [129]. Because Jahn–Teller distortion could be observed from crystal structure with bond-length of coordination polyhedron, it could be treated as distorted octahedral or tetrahedral coordination environments in the crystal structure.

Another investigation of high-temperature superconductor was carried out by Anderson in 1987 who looked at resonating valance bonding in solids relating to the electron structure [130]. He introduced Pauling's valance bond theory from chemistry into condensed state material as valance bond solids (VBS) and proposed the spin frustration state in the triangular lattice in 1973 to be a resonating valance bond (RVB) state [131–135]. Then he developed his theory by the discovery of high-temperature cuprate superconductors and proposed the possibility that a copper pair was formed by coupling of spin in the spin liquid state. Carrier doping on parent antiferromagnetic La_2CuO_4, Na_2CoO_2, and LaFeAs produced a new superconductor intermediate by spin fluctuation. The two-dimensional antiferromagnetic correlation as spin frustration or antiferromagnetic ordering came from the resonant valence-band state [136]. So at first, an extended (infinite) coordination polymer is needed, such as the Cu–O plane in cuprate, Co–O plane in $Na_xCoO_2(H_2O)_y$, and Fe-As plane in iron pnictides. This guarantees the transportation channel for the carrier in the solid. Then the Jahn–Teller distorted transition metals with their variable valances should exist. From the structure point of view, a two-dimensional extended metal-O layer is the key to these materials. The conducting layer behaves as an acceptor, so the high-temperature superconductor acts as a charge-transfer salt. Because the superconductivity was first discovered in the ceramic phase in these systems, it always takes time to confirm the composition and growth of high-quality single crystals with the critical ratio of atoms. This is the main difference between organic and inorganic superconductors. In research on organic superconductors, the high quality and clear crystal structure is always the first step. For example, a series of candidates for quantum spin liquids was discovered in charge-transfer complexes with π-units in a κ-phase arrangement. In these compounds, interactions between donor pairs are isotropic and produce a resonating valence state. The first quantum spin liquid with triangular lattice was observed in a charge-transfer salt with κ-phase donor arrangement κ-(BEDT-TTF)$_2$Cu$_2$(CN)$_3$ and confirmed by specific heat and ESR experiment [92, 137]. It is a Mott insulator and could be superconductive at 3.9 K under 0.06 GPa [138]. Such a Mott insulator, β'-(BEDT-TTF)$_2$ICl$_2$, shows $T_c = 14.2$ K under 8.2 GPa [37]. The charge-transfer salt $(C_2H_5)(CH_3)_3Sb[Pd(dmit)_2]_2$ was found to be another quantum spin liquid with triangular lattice [91]. A third one was discovered in a single component compound with κ-phase arrangement, κ-H$_3$(cat(cat-EDT-TTF)$_2$ [139]. The Kagome lattice is another ideal model to spin liquid; when people were looking for coordination compounds with kagome lattice, the weak interactions between organic molecule formed a kagome lattice in [EDT-TTF-CONH$_2$]$_6$[Re$_6$Se$_8$(CN)$_6$] [140] and antiferromagnetic spin fluctuation was observed. This was earlier than the first coordination compound with kagome lattice $Zn_{0.33}Cu_{0.67}(OH)_6Cl_2$ [141, 142].

The single component organic conductor and superconductor is an important area in organic superconductor technology. Apart from the calculation of band structure, the theory of the resonating valance bonding state could be useful in explaining this system. The weak interaction between molecules from supramolecular chemistry forms a resonating valence bonding state. So the crystal could show semiconductor, conductor, and superconductor properties. The Jahn–Teller distortion originating from metal coordination compounds now extends to organic chemistry as representative of energy degeneracy. The International Symposium on the Jahn–Teller effect, which was initiated by Prof. Muller, is held every 2 years. He combined the RVB with the Jahn–Teller effect in inorganic superconductors, and it is also suitable for use with organic superconductors [129]. The research area can be expanded after the combination of inorganic superconductor units and molecular crystals [23, 143].

1.6 Summary

Organic materials have received considerably more attention than inorganic-based materials for use in modern optoelectronic devices, such as organic solar cells, light-emitting diodes, and field effect transistors, because of their low-cost, easy deposition, and wide variety and tailorable or tunable properties. This chapter focuses on the charge transport of organic materials, and some prototype organic solids are also discussed. Organic superconductors can be obtained from organic conductors, semiconductors, and insulators under suitable conditions at low temperatures. These materials therefore still represent a challenging and exciting research field for the near future.

References

1. Akamatu H, Inokuchi H, Matsunaga Y (1954) Electrical conductivity of the perylene-bromine complex. Nature 173:168–169
2. Kepler RG, Bierstedt PE, Merrifield RE (1960) Electronic conduction and exchange interaction in a new class of conductive organic solids. Phys Rev Lett 5:503
3. Ferraris J, Cowan DO, Walatka V, Perlstein JH (1973) Electron transfer in a new highly conducting donor-acceptor complex. J Am Chem Soc 95:948–949
4. Greene R, Engler E (1980) Pressure dependence of superconductivity in an organic superconductor. Phys Rev Lett 45:1587–1590
5. Andres K, Wudl F, McWhan D, Thomas G, Nalewajek D, Stevens A (1980) Observation of the Meissner effect in an organic superconductor. Phys Rev Lett 45:1449–1452
6. Jerome D, Mazaud A, Ribault M, Bechgaard K (1980) Superconductivity in a synthetic organic conductor (TMTSF) $_2PF_6$. J de Phys Lettres 41:95–98
7. Fleming RM, Ramirez AP, Rosseinsky MJ, Murphy DW, Haddon RC, Zahurak SM, Makhija AV (1991) Relation of structure and superconducting transition temperatures in A_3C_{60}. Nature 352:787–788
8. Ganin AY, Takabayashi Y, Jeglic P, Arcon D, Potocnik A, Baker PJ, Ohishi Y, McDonald MT, Tzirakis MD, McLennan A, Darling GR, Takata M, Rosseinsky MJ, Prassides K (2010) Polymorphism control of superconductivity and magnetism in Cs_3C_{60} close to the Mott transition. Nature 466:221–225
9. Tanaka H, Okano Y, Kobayashi H, Suzuki W, Kobayashi A (2001) A three-dimensional synthetic metallic crystal composed of single-component molecules. Science 291:285–287
10. Mulliken RS (1952) Molecular compounds and their spectra II. J Am Chem Soc 74:811–824
11. Acker DS, Harder RJ, Hertler WR, Mahler W, Melby LR, Benson RE, Mochel WE (1960) 7, 7, 8, 8-Tetracyanoquinodimethane and its electrically conducting anion-radical derivatives. J Am Chem Soc 82:6408–6409
12. Coffen DL, Chambers JQ, Williams DR, Garrett P, Canfield N (1971) Tetrathioethylenes. J Am Chem Soc 93:2258–2268
13. Wudl F, Smith GM, Hufnagel EJ (1970) Bis-1, 3-dithiolium chloride: an unusually stable organic radical cation. J Chem So D Chem Commun 1453–1454
14. Mizuno M, Garito AF, Cava MP (1978) 'Organic metals': alkylthio substitution effects in tetrathiafulvalene-tetracyanoquinodimethane charge-transfer complexes. J Chem Soc Chem Commun 18–19
15. Zhang B, Zhu D (2012) Molecular-based conducting magnet. Sci China Chem 55:883–892
16. Bright AA, Garito AF, Heeger AJ (1973) Optical properties of (TTF) (TCNQ) in the visible and infrared. Solid State Commun 13:943–948
17. Coleman LB, Cohen MJ, Sandman DJ, Yamagishi FG, Garito AF, Heeger AJ (1973) Superconducting fluctuations and the peierls instability in an organic solid. Solid State Commun 12:1125–1132
18. Bardeen J (1973) Superconducting fluctuations in one-dimensional organic solids. Solid State Commun 13:357–359
19. Lee PA, Rice TM, Anderson PW (1974) Conductivity from charge or spin density waves. Solid State Commun 14:703–709
20. McMillan WL (1975) Landau theory of charge-density waves in transition-metal dichalcogenides. Phys Rev B 12:1187
21. Greene RL, Mayerle JJ, Schumaker R, Castro G, Chaikin PM, Etemad S, LaPlaca SJ (1976) The structure, conductivity, and thermopower of HMTTF-TCNQ. Solid State Commun 20:943–946
22. Andersen JR, Bechgaard K, Jacobsen CS, Rindorf G, Soling H, Thorup N (1978) The crystal and molecular structure of the organic conductor 2,3,6,7-tetramethyl-1,4,5,8-tetraselenafulvalenium 2,5-dimethyl-7,7,8,8-tetracyano-p-quinodimethanide (TMTSF-DMTCNQ). Acta Crystallogr Sect B 34:1901–1905

23. Jerome D (1991) The physics of organic superconductors. Science 252:1509–1514
24. Walsh WM, Wudl F, Thomas GA, Nalewajek D, Hauser JJ, Lee PA, Poehler T (1980) Restoration of metallic behavior in organic conductors by small electric fields. Phys Rev Lett 45:829–832
25. Andrieux A, Jérome D, Bechgaard K (1981) Spin-density wave ground state in the one-dimensional conductor (TMTSF) $_2PF_6$: microscopic evidence from 77Se and 1H NMR experiments. J Phys Lett 42:87–90
26. Mortensen K, Tomkiewicz Y, Bechgaard K (1982) Antiferromagnetism in the organic conductor bis-tetramethyltetraselenafulvalene hexafluoroarsenate. Phys Rev B 25:3319–3325
27. Laukhin VN, Kostyuchenko E, Sushko YV, Shchegolev IF, Yagubskii É (1985) Effect of pressure on the superconductivity of β-(BEDT-TTF) $_2I_3$. JETP Lett 41:81–84
28. Jérome D (2004) Organic conductors: from charge density wave TTF-TCNQ to superconducting (TMTSF) $_2PF_6$. Chem Rev 104:5565–5592
29. Coulon C, Delhaes P, Amiell J, Manceau JP, Fabre JM, Giral L (1982) Effect of doping (TMTSF) $_2ClO_4$ with TMTTF-I. Ambient pressure results: a competition between the different possible ground states. J de Phys 43:1721–1729
30. Yu W, Zhang F, Zamborszky F, Alavi B, Baur A, Merlic CA, Brown SE (2004) Electron-lattice coupling and broken symmetries of the molecular salt $(TMTTF)_2$ SbF_6. Phys Rev B 70:121101
31. Bechgaard K, Carneiro K, Olsen M, Rasmussen FB, Jacobsen CS (1981) Zero-pressure organic superconductor: di-(Tetramethyltetraselenafulvalenium)-perchlorate $[(TMTSF)_2 ClO_4]$. Phys Rev Lett 46:852
32. Wang HH, Beno MA, Geiser U, Firestone MA, Webb KS, Nunez L, Crabtree GW, Carlson KD, Williams JM (1985) Ambient-pressure superconductivity at the highest temperature (5 K) observed in an organic system: beta.-(BEDT-TTF)2AuI2. Inorg Chem 24:2465–2466
33. Kobayashi A, Fujiwara E, Kobayashi H (2004) Single-component molecular metals with extended-TTF dithiolate ligands. Chem Rev 104:5243–5264
34. Parkin SSP, Engler EM, Schumaker RR, Lagier R, Lee VY, Scott JC, Greene RL (1983) Superconductivity in a new family of organic conductors. Phys Rev Lett 50:270
35. Yagubskii EB, Shchegolev IF, Laukhin VN, Kononovich PA, Karstovnik MV, Zvarykina AV, Buravov LI (1984) Normal-pressure, superconductivity in an organic metal (BEDT-TTF)$_2I_3$ [bis (ethylene dithiolo) tetrathiof ulvalene triiodide]. JETP Lett 39:12–16
36. Murata K, Tokumoto M, Anzai H, Bando H, Saito G, Kajimura K, Ishiguro T (1985) Pressure phase diagram of the organic superconductor β-(BEDT-TTF)$_2I_3$. J Phys Soc Jpn 54:2084–2087
37. Taniguchi H, Miyashita M, Uchiyama K, Satoh K, Môri N, Okamoto H, Miyagawa K, Kanoda K, Hedo M, Uwatoko Y (2003) Superconductivity at 14.2 K in layered organics under extreme pressure. J Phys Soc Jpn 72:468–471
38. Sasaki T, Sato H, Toyota N (1990) Magnetic breakdown effect in organic superconductor κ-(BEDT-TTF)$_2$Cu(NCS)$_2$. Solid State Commun 76:507–510
39. Williams JM, Wang HH, Beno MA, Emge TJ, Sowa LM, Copps PT, Behroozi F, Hall LN, Carlson KD, Crabtree GW (1984) Ambient-pressure superconductivity at 2.7 K and higher temperatures in derivatives of (BEDT-TTF)$_2IBr_2$: synthesis, structure, and detection of superconductivity. Inorg Chem 23:3839–3841
40. Mori H, Tanaka S, Mori T (1998) Systematic study of the electronic state in θ-type BEDT-TTF organic conductors by changing the electronic correlation. Phys Rev B 57:12023
41. Bender K, Hennig I, Schweitzer D, Dietz K, Endres H, Keller HJ (1984) Synthesis, Structure and Physical Properties of a Two-Dimensional Organic Metal, Di[bis(ethylenedithiolo) tetrathiofulvalene] triiodide, (BEDT-TTF)$_2^+I_3^-$. Mol Cryst Liq Cryst 108:359
42. Wang HH, Carlson KD, Geiser U, Kwok WK, Vashon MD, Thompson JE, Larsen NF, McCabe GD, Hulscher RS, Williams JM (1990) A new ambient-pressure organic superconductor: (BEDT-TTF)$_2(NH_4)Hg(SCN)_4$. Physica C 166:57–61

43. Rothaemel B, Forró L, Cooper JR, Schilling JS, Weger M, Bele P, Brunner H, Schweitzer D, Keller HJ (1986) Magnetic susceptibility of α and β phases of di [bis (ethylenediothiolo) tetrathiafulvalene] tri-iodide [$(BEDT\text{-}TTF)_2I_3$] under pressure. Phys Rev B 34:704
44. Pokhodnya KI, Sushko YV, Tanatar MA (1987) Singularities of the phase states and of the metal-insulator phase transition in the a-$(BEDT\text{-}TTF)_2I_3$ system. Sov Phys JETP 65:795
45. Takano Y, Hiraki K, Yamamoto HM, Nakamura T, Takahashi T (2001) Charge disproportionation in the organic conductor, α-$(BEDT\text{-}TTF)_2I_3$. J Phys Chem Solids 62:393–395
46. Hebard AF, Rosseinky MJ, Haddon RC, Murphy DW, Glarum SH, Palstra TTM, Ramirez AP, Karton AR (1991) Potassium-doped C_{60}. Nature 350:600–601
47. Tanigaki K, Ebbesen TW, Saito S, Mizuki J, Tsai JS, Kubo Y, Kuroshima S (1991) Superconductivity at 33 K in $Cs_xRb_yC_{60}$. Nature 352:222–223
48. Kino H, Fukuyama H (1995) Electronic states of conducting organic κ-$(BEDT\text{-}TTF)_2X$. J Phys Soc Jpn 64:2726–2729
49. Seo H (2000) Charge ordering in organic ET compounds. J Phys Soc Jpn 69:805–820
50. Kino H, Fukuyama H (1996) Phase diagram of two-dimensional organic conductors: $(BEDT\text{-}TTF)_2X$. J Phys Soc Jpn 65:2158–2169
51. McKenzie RH, Merino J, Marston JB, Sushkov OP (2001) Charge ordering and antiferromagnetic exchange in layered molecular crystals of the θ type. Phys Rev B 64:085109
52. Torsten Clay R, Mazumdar SK, Campbell D (2002) Charge ordering in θ-$(BEDT\text{-}TTF)_2X$ materials. J Phys Soc Jpn 71:1816–1819
53. Hiraki K, Kanoda K (1998) Wigner crystal type of charge ordering in an organic conductor with a quarter-filled band: $(DI\text{-}DCNQI)_2Ag$. Phys Rev Lett 80:4737
54. Hoffmann R (1963) An extended Hückel theory. I. hydrocarbons. J Chem Phys 39:1397–1412
55. Mori T, Kobayashi A, Sasaki Y, Kobayashi H, Saito G, Inokuchi H (1984) The Intermolecular interaction of tetrathiafulvalene and bis (ethylenedithio)-tetrathiafulvalene in organic metals. Calculation of orbital overlaps and models of energy-band structures. Bull Chem Soc Jpn 57:627–633
56. Seo H, Hotta C, Fukuyama H (2004) Toward systematic understanding of diversity of electronic properties in low-dimensional molecular solids. Chem Rev 104:5005–5036
57. Hubbard J (1978) Generalized Wigner lattices in one dimension and some applications to tetracyanoquinodimethane (TCNQ) salts. Phys Rev B 17:494
58. Mori T (1998) Structural genealogy of BEDT-TTF-based organic conductors I. Parallel molecules: BETA and BETA″ phases. Bull Chem Soc Jpn 71:2509–2526
59. Miyagawa K, Kawamoto A, Kanoda K (2000) Charge ordering in a quasi-two-dimensional organic conductor. Phys Rev B 62:R7679
60. Watanabe M, Noda Y, Nogami Y, Mori H (2004) Transfer integrals and the spatial pattern of charge ordering in θ-$(BEDT\text{-}TTF)_2RbZn(SCN)_4$ at 90 K. J Phys Soc Jpn 73:116–122
61. Mizoguchi T, Tanaka I, Yoshioka S, Kunisu M, Yamamoto T, Ching WY (2004) First-principles calculations of ELNES and XANES of selected wide-gap materials: dependence on crystal structure and orientation. Phys Rev B 70:045103
62. Ōnuki Y, Settai R, Sugiyama K, Takeuchi T, Kobayashi TC, Haga Y, Yamamoto E (2004) Recent advances in the magnetism and superconductivity of heavy fermion systems. J Phys Soc Jpn 73:769–787
63. Wojciechowski R, Yamamoto K, Yakushi K, Inokuchi M, Kawamoto A (2003) High-pressure Raman study of the charge ordering in α-$(BEDT\text{-}TTF)_2I_3$. Phys Rev B 67:224105
64. Yamamoto K, Yakushi K, Miyagawa K, Kanoda K, Kawamoto A (2002) Charge ordering in θ-$(BEDT\text{-}TTF)_2RbZn(SCN)_4$ studied by vibrational spectroscopy. Phys Rev B 65:085110
65. Suzuki K, Yamamoto K, Yakushi K (2004) Charge-ordering transition in orthorhombic and monoclinic single-crystals of θ-$(BEDT\text{-}TTF)_2TlZn(SCN)_4$ studied by vibrational spectroscopy. Phys Rev B 69:085114

66. Nakamura T, Kinami R, Takahashi T, Saito G (1997) 1H-NMR study of the magnetic structure in $(TMTTF)_2$ SCN. Synth Met 86:2053–2054
67. Nakamura T, Nobutoki T, Kobayashi Y, Takahashi T, Saito G (1995) 1H-NMR investigation of the SDW wave-number in (TMTTF) $_2$Br. Synth Met 70:1293–1294
68. Fujiyama S, Nakamura T (2004) Charge disproportionation in $(TMTTF)_2SCN$ observed by C^{13} NMR. Phys Rev B 70:045102
69. Chow DS, Zamborszky F, Alavi B, Tantillo DJ, Baur A, Merlic CA, Brown SE (2000) Charge ordering in the TMTTF family of molecular conductors. Phys Rev Lett 85:1698
70. Zamborszky F, Yu W, Raas W, Brown SE, Alavi B, Merlic CA, Baur A (2002) Competition and coexistence of bond and charge orders in $(TMTTF)_2AsF_6$. Phys Rev B 66:081103
71. Seo H, Fukuyama H (1997) Antiferromagnetic phases of one-dimensional quarter-filled organic conductors. J Phys Soc Jpn 66:1249–1252
72. Delhaes P, Coulon C, Amiell J, Flandrois S, Toreilles E, Fabre JM, Giral L (1979) Physical properties of one dimensional conductors. Mol Cryst Liq Cryst 50:43–58
73. Laversanne R, Coulon C, Gallois B, Pouget JP, Moret R (1984) Structural and electrical properties of $(TMTTF)_2MF_6$ salts (M = P, As, Sb). Rôle of the anions. J de Phys Lettres 45:393–399
74. Chiba R, Hiraki K, Takahashi T, Yamamoto HM, Nakamura T (2004) Extremely slow charge fluctuations in the metallic state of the two-dimensional molecular conductor θ-$(BEDT\text{-}TTF)_2RbZn(SCN)_4$. Phys Rev Lett 93:216405
75. Chiba R, Yamamoto HM, Hiraki K, Nakamura T, Takahashi T (2001) Charge ordering in θ-$(BEDT\text{-}TTF)_2RbZn(SCN)_4$. Synth Met 120:919–920
76. Chiba R, Yamamoto H, Hiraki K, Takahashi T, Nakamura T (2001) Charge disproportionation in $(BEDT\text{-}TTF)_2RbZn(SCN)_4$. J Phys Chem Solids 62:389–391
77. Watanabe M, Nogami Y, Oshima K, Mori H, Tanaka S (1999) Novel pressure-induced 2k F CDW state in organic low-dimensional compound θ-$(BEDT\text{-}TTF)_2CsCo(SCN)_4$. J Phys Soc Jpn 68:2654–2663
78. Takano Y, Hiraki K, Yamamoto HM, Nakamura T, Takahashi T (2001) Charge ordering in α-$(BEDT\text{-}TTF)_2I_3$. Synth Met 120:1081–1082
79. Kawamoto A, Miyagawa K, Kanoda K (1997) Deuterated κ-$(BEDT\text{-}TTF)_2Cu[N(CN)_2]Br$: A system on the borderof the superconductor–magnetic-insulator transition. Phys Rev B 55:14140
80. Moroto S, Hiraki KI, Takano Y, Kubo Y, Takahashi T, Yamamoto HM, Nakamura T (2004) Charge disproportionation in the metallic state of α-$(BEDT\text{-}TTF)_2I_3$. Science 114:399–340
81. Guionneau P, Kepert CJ, Bravic G, Chasseau D, Truter MR, Kurmoo M, Day P (1997) Determining the charge distribution in BEDT-TTF salts. Synth Met 86:1973–1974
82. Kobayashi H, Kato R, Mori T, Kobayashi A, Sasaki Y, Saito G, Enoki T, Inokuchi H (1984) Crystal Structures and Electrical Properties of BEDT-TTF Coeipounds. Mol Cryst Liq Cryst 107:33–43
83. Yamamoto T, Uruichi M, Yamamoto K, Yakushi K, Kawamoto A, Taniguchi H (2005) Examination of the charge-sensitive vibrational modes in bis (ethylenedithio) tetrathiafulvalene. J Phys Chem B 109:15226–15235
84. Kozlov ME, Pokhodnia KI, Yurchenko AA (1989) Electron molecular vibration coupling in vibrational spectra of BEDT-TTF based radical cation salts. Spectrochim Acta Part A 45:437–444
85. Ouyang J, Yakushi K, Misaki Y, Tanaka K (2001) Raman spectroscopic evidence for the charge disproportionation in a quasi-two-dimensional organic conductor θ-$(BDT\text{-}TTP)_2Cu(NCS)_2$. Phys Rev B 63:054301
86. Eldridge JE, Homes CC, Williams JM, Kini AM, Wang HH (1995) The assignment of the normal modes of the BEDT-TTF electron-donor molecule using the infrared and Raman spectra of several isotopic analogs. Spectrochim Acta Part A Mol Biomol Spectrosc 51:947–960

87. Tajima H, Kyoden S, Mori H, Tanaka S (2000) Estimation of charge-ordering patterns in θ-ET_2 MM′$(SCN)_4$ (MM′ = RbCo, RbZn, CsZn) by reflection spectroscopy. Phys Rev B 62:9378
88. Coomber AT, Beljonne D, Friend RH, Brédas J-L, Charlton A, Robertson N, Underbill AE, Kurmoo M, Day P (1996) Intermolecular interactions in the molecular ferromagnetic NH_4Ni $(mnt)_2{\cdot}H_2O$. Nature 380:144–146
89. Shimizu Y, Miyagawa K, Kanoda K, Maesato M, Saito G (2003) Spin liquid state in an organic Mott insulator with a triangular lattice. Phys Rev Lett 91:107001
90. Yamashita S, Nakazawa Y (2010) Heat capacities of antiferromagnetic dimer-Mott insulators in organic charge-transfer complexes. J Therm Anal Calorim 99:153–157
91. Yamashita M, Nakata N, Senshu Y, Nagata M, Yamamoto HM, Kato R, Shibauchi T, Matsuda Y (2010) Highly mobile gapless excitations in a two-dimensional candidate quantum spin liquid. Science 328:1246–1248
92. Pratt FL, Baker PJ, Blundell SJ, Lancaster T, Ohira-Kawamura S, Baines C, Shimizu Y, Kanoda K, Watanabe I, Saito G (2011) Magnetic and non-magnetic phases of a quantum spin liquid. Nature 471:612–616
93. Bonniface DW, Braithwaite MJ, Eley DD, Evans RG, Pethig R, Willis MR (1971) Factors affecting conduction in polymeric complex TCNQ salts. Discuss Faraday Soc 51:131–138
94. Manriquez JM, Yee GT, McLean RS, Epstein AJ, Miller JS (1991) A room-temperature molecular/organic-based magnet. Science 252:1415–1417
95. Carlegrim E, Kanciurzewska A, Nordblad P, Fahlman M (2008) Air-stable organic-based semiconducting room temperature thin film magnet for spintronics applications. Appl Phys Lett 92:163308
96. Lopez N, Zhao H, Ota A, Prosvirin AV, Reinheimer EW, Dunbar KR (2010) Unprecedented binary semiconductors based on TCNQ: Single-crystal X-ray studies and physical properties of $Cu(TCNQX_2)$ X = Cl. Br Adv Mater 22:986–989
97. Chifotides HT, Schottel BL, Dunbar KR (2010) The π-accepting arene $HAT(CN)_6$ as a halide receptor through charge transfer: multisite anion interactions and self-assembly in solution and the solid state. Ang Chem Int Ed 49:7202–7207
98. Miyasaka H, Motokawa N, Matsunaga S, Yamashita M, Sugimoto K, Mori T, Toyota N, Dunbar KR (2010) Control of charge transfer in a series of Ru2II, II/TCNQ two-dimensional networks by tuning the electron affinity of TCNQ units: a route to synergistic magnetic/ conducting materials. J Am Chem Soc 132:1532–1544
99. Edwards PR, Johnson CE (1968) Mössbauer hyperfine interactions in tetrahedral Fe(III) ions. J Chem Phys 49:211–216
100. Tamaki H, Zhong ZJ, Matsumoto N, Kida S, Koikawa M, Achiwa N, Hashimoto Y, Okawa H (1992) Design of metal-complex magnets. Syntheses and magnetic properties of mixed-metal assemblies $NBu_4[MCr(ox)_3]_x$ (NBu_4^+= tetra(n-butyl)ammonium ion; ox^{2-} = oxalate ion; M = Mn^{2+}, Fe^{2+}, Co^{2+}, Ni^{2+}, Cu^{2+}, Zn^{2+}). J Am Chem Soc 114:6974–6979
101. Zhang B, Kurmoo M, Mori T, Zhang Y, Pratt FL, Zhu D (2009) Polymorphism in Hybrid Organic-Inorganic Bilayered Magnetic Conductors $(BEDT\text{-}TTF)_3$ $(FeIIICl4)_2$, BEDT-TTF = bis (ethylenedithio) tetrathiafulvalene. Cryst Growth Des 10:782–789
102. Coronado E, Galán-Mascarós JR, Gómez-García CJ, Laukhin V (2000) Coexistence of ferromagnetism and metallic conductivity in a molecule-based layered compound. Nature 408:447–449
103. Pop F, Auban-Senzier P, Canadell E, Rikken GLJA, Avarvari N (2014) Electrical magnetochiral anisotropy in a bulk chiral molecular conductor. Nat Commun 5:3757
104. Pointillart F, Le Guennic B, Cauchy T, Sp Golhen, Cador O, Maury O, Ln Ouahab (2013) A Series of tetrathiafulvalene-based lanthanide complexes displaying either single molecule magnet or luminescence-direct magnetic and photo-physical correlations in the ytterbium analogue. Inorg Chem 52:5978–5990
105. Hannay NB, Geballe TH, Matthias BT, Andres K, Schmidt P, MacNair D (1965) Superconductivity in graphitic compounds. Phys Rev Lett 14:225

106. Emery N, Hérold C, d'Astuto M, Garcia V, Bellin C, Marêché JF, Lagrange P, Loupias G (2005) Superconductivity of bulk CaC_6. Phys Rev Lett 95:087003
107. Mitsuhashi R, Suzuki Y, Yamanari Y, Mitamura H, Kambe T, Ikeda N, Okamoto H, Fujiwara A, Yamaji M, Kawasaki N (2010) Superconductivity in alkali-metal-doped picene. Nature 464:76–79
108. Wang XF, Liu RH, Gui Z, Xie YL, Yan YJ, Ying JJ, Luo XG, Chen XH (2011) Superconductivity at 5 K in alkali-metal-doped phenanthrene. Nat Commun 2:507
109. Suzuki R, Sakano M, Zhang YJ, Akashi R, Morikawa D, Harasawa A, Yaji K, Kuroda K, Miyamoto K, Okuda T (2014) Valley-dependent spin polarization in bulk MoS_2 with broken inversion symmetry. Nat Nanotechnol 9:611–617
110. Kawasugi Y, Yamamoto HM, Tajima N, Fukunaga T, Tsukagoshi K, Kato R (2011) Electric-field-induced Mott transition in an organic molecular crystal. Phys Rev B 84:125129
111. Chollet M, Guerin L, Uchida N, Fukaya S, Shimoda H, Ishikawa T, Matsuda K, Hasegawa T, Ota A, Yamochi H (2005) Gigantic photoresponse in ¼-filled-band organic salt (EDO-TTF)$_2$PF$_6$. Science 307:86–89
112. Cui H, Kobayashi H, Ishibashi S, Sasa M, Iwase F, Kato R, Kobayashi A (2014) A single-component molecular superconductor. J Am Chem Soc 129:12618–12619
113. Xiao X, Hayashi T, Fujiwara H, Sugimoto T, Noguchi S, Weng Y, Yoshino H, Murata K, Aruga Katori H (2007) An antiferromagnetic molecular metal based on a new bent-donor molecule. J Am Chem Soc 129:12618–12619
114. Hayashi T, Xiao X, Fujiwara H, Sugimoto T, Nakazumi H, Noguchi S, Fujimoto T, Yasuzuka S, Yoshino H, Murata K (2006) A metallic (EDT-DSDTFVSDS)$_2$FeBr$_4$ salt: antiferromagnetic ordering of d spins of FeBr$_4$-ions and anomalous magnetoresistance due to preferential π-d interaction. J Am Chem Soc 128:11746–11747
115. Uji S, Shinagawa H, Terashima T, Yakabe T, Terai Y, Tokumoto M, Kobayashi A, Tanaka H, Kobayashi H (2001) Magnetic-field-induced superconductivity in a two-dimensional organic conductor. Nature 410:908–910
116. Zhang B, Tanaka H, Fujiwara H, Kobayashi H, Fujiwara E, Kobayashi A (2002) Dual-action molecular superconductors with magnetic anions. J Am Chem Soc 124:9982–9983
117. Zhang B, Wang Z, Fujiwara H, Kobayashi H, Kurmoo M, Inoue K, Mori T, Gao S, Zhang Y, Zhu D (2005) Tetrathiafulvalene [FeIII(C_2O_4)Cl_2]: an organic-inorganic hybrid exhibiting canted antiferromagnetism. Adv Mater 17:1988–1991
118. Zhang B, Wang Z, Zhang Y, Takahashi K, Okano Y, Cui H, Kobayashi H, Inoue K, Kurmoo M, Pratt FL (2006) Hybrid organic-inorganic conductor with a magnetic chain anion: κ-BETS$_2$[FeIII(C_2O_4)Cl_2][BETS = Bis(ethylenedithio)tetraselenafulvalene]. Inorg Chem 45:3275–3280
119. Zhang B, Zhang Y, Gao Z, Chang G, Su S, Wang D, Guo Y, Zhu D (2014) Synthesis, crystal structure, and characterization of the charge-transfer salt (BEDT-TTF)[Fe(C_2O_4)Cl_2] (CH_2Cl_2),{BEDT-TTF = Bis(ethylenedithio)tetrathiafulvalene}. Eur J Inorg Chem 24:4028–4032
120. Kubo K, Shiga T, Yamamoto T, Tajima A, Moriwaki T, Ikemoto Y, Yamashita M, Sessini E, Mercuri ML, Deplano P (2011) Electronic state of a conducting single molecule magnet based on Mn-salen type and Ni-dithiolene complexes. Inorg Chem 50:9337–9344
121. Ouahab L, Mori T (2014) Molecular conductors. Eur J Inorg Chem 24:3783–3784
122. Zhou B, Shimamura M, Fujiwara E, Kobayashi A, Higashi T, Nishibori E, Sakata M, Cui H, Takahashi K, Kobayashi H (2006) Magnetic transitions of single-component molecular metal [Au(tmdt)$_2$] and its alloy systems. J Am Chem Soc 128:3872–3873
123. Alberola A, Coronado E, Galán-Mascarós JR, Giménez-Saiz C, Gómez-García CJ (2003) A molecular metal ferromagnet from the organic donor bis (ethylenedithio) tetraselenafulvalene and bimetallic oxalate complexes. J Am Chem Soc 125:10774–10775
124. Zhang B, Zhang Y, Zhu D (2012) (BEDT-TTF)$_3$Cu$_2$(C_2O_4)$_3$(CH_3OH)$_2$: an organic–inorganic hybrid antiferromagnetic semiconductor. Chem Commun 48:197–199

125. Zhang B, Zhang Y, Wang Z, Gao S, Guo Y, Liu F, Zhu D (2013) $BETS_3[Cu_2(C_2O_4)_3](CH_3OH)_2$: an organic–inorganic hybrid antiferromagnetic metal (BETS = bisethylene (tetraselenfulvalene)). Cryst Eng Commun 15:3529–3535
126. Balents L (2010) Spin liquids in frustrated magnets. Nature 464:199–208
127. Bednorz JG, Müller KA (1986) Possible high T_c superconductivity in the Ba–La–Cu–O system. Zeitschrift für Physik B Condens Matter 64:189–193
128. Müller KA, Bednorz JG (1987) The discovery of a class of high-temperature superconductors. Science 237:1133–1139
129. Müller KA (2007) On the superconductivity in hole doped cuprates. J Phys Condens Matter 19:251002
130. Anderson PW (1987) The resonating valence bond state in La_2CuO_4 and superconductivity. Science 235:1196–1198
131. Pauling L (1933) The calculation of matrix elements for Lewis electronic structures of molecules. J Chem Phys 1:280–283
132. Pauling L, Wheland GW (1933) The nature of the chemical bond. V. The quantum-mechanical calculation of the resonance energy of benzene and naphthalene and the hydrocarbon free radicals. J Chem Phys 1:362–374
133. Pauling L (1938) The nature of the interatomic forces in metals. Phys Rev 54:899
134. Copper DL, Gerratt J, Raimondi M (1986) The electronic structure of the benzene molecule. Nature 325:396
135. Anderson PW (1973) Resonating valence bonds: A new kind of insulator? Mater Res Bull 8:153–160
136. Moriya T, Takahashi Y, Ueda K (1990) Antiferromagnetic spin fluctuations and superconductivity in two-dimensional metals-A possible model for high T_c oxides. J Phys Soc Jpn 59:2905–2915
137. Yamashita S, Nakazawa Y, Oguni M, Oshima Y, Nojiri H, Shimizu Y, Miyagawa K, Kanoda K (2008) Thermodynamic properties of a spin-1/2 spin-liquid state in a κ-type organic salt. Nat Phys 4:459–462
138. Komatsu T, Matsukawa N, Inoue T, Saito G (1996) Realization of superconductivity at ambient pressure by band-filling control in κ-$(BEDT\text{-}TTF)_2Cu_2(CN)_3$. J Phys Soc Jpn 65:1340–1354
139. Isono T, Kamo H, Ueda A, Takahashi K, Kimata M, Tajima H, Tsuchiya S, Terashima T, Uji S, Mori H (2014) Gapless quantum spin liquid in an organic spin-1/2 triangular-lattice κ–$H_3(Cat\text{-}EDT\text{-}TTF)_2$. Phys Rev Lett 112:177201
140. Baudron SA, Batail P, Coulon C, Clérac R, Canadell E, Laukhin V, Melzi R, Wzietek P, Jérome D, Auban-Senzier P (2005) $(EDT\text{-}TTF\text{-}CONH_2)_6[Re_6Se_8(CN)_6]$, a metallic kagome-type organic-inorganic hybrid compound: electronic instability, molecular motion, and charge localization. J Am Chem Soc 127:11785–11797
141. Shores MP, Nytko EA, Bartlett BM, Nocera DG (2005) A structurally perfect S = 1/2 kagome antiferromagnet. J Am Chem Soc 127:13462–13463
142. Han T-H, Helton JS, Chu S, Nocera DG, Rodriguez-Rivera JA, Broholm C, Lee YS (2012) Fractionalized excitations in the spin-liquid state of a kagome-lattice antiferromagnet. Nature 492:406–410
143. Williams JM, Schultz AJ, Geiser URS, Carlson KD, Kini AM, Wang HH, Kwok W-K, Whangbo M-H, Schirber JE (1991) Organic superconductors—new benchmarks. Science 252:1501–1508

Chapter 2
Conducting Polymers

Yongfang Li

In general, conducting polymers include electronically conducting polymers and ionically conducting polymers. Ionically conducting polymers are usually called polymer electrolytes. Electronically conducting polymers can also include conjugated conducting polymers and the insulating polymers blending with conducting materials. In this chapter, the conducting polymers are limited to conjugated conducting polymers, unless otherwise stated.

Traditionally, polymers are thought of as insulators. However, in 1977 a discovery by Alan G. MacDiarmid, Hideki Shirakawa, and Alan J. Heeger et al. changed the traditional concept. They found that conductivity of polyacetylene—after doping with electron-withdrawing AsF_5—increased ninefold, reaching the order of 10^3 S/cm [1, 2]. Soon after this discovery, a series of stable conducting polymers, including polypyrrole (PPy), polyaniline (PAn), and polythiophene (PTh), were reported from the end of the 1970s to the beginning of the 1980s, which greatly promoted the research on conducting polymers. Actually, the conductivity of almost all conjugated polymers can reach the order of 10^{-3}–10^3 S/cm after doping. Now we can expand the class of conducting polymers to include all the doped conjugated polymers.

In 1990, Friends et al. found the electroluminescent properties of poly(*p*-phenylene vinylene) (PPV) [3] and opened up a new field of polymer light-emitting diodes (PLEDs) with semiconducting intrinsic conjugated polymers as the active light-emitting layer. In 1995, Heeger et al. reported bulk-heterojunction polymer solar cells (PSCs) with a conjugated polymer MEH-PPV as donor and a fullerene derivative PCBM as acceptor [4], which further extended the research on conjugated polymers to the field of organic photovoltaics. Since then, conjugated polymer

Y. Li (✉)
Institute of Chemistry, Chinese Academy of Sciences, Beijing 100190, China
e-mail: liyf@iccas.ac.cn

Y. Li
College of Chemistry, Chemical Engineering and Materials Science,
Soochow University, Suzhou 215123, China

Y. Li (ed.), *Organic Optoelectronic Materials*,
Lecture Notes in Chemistry 91, DOI 10.1007/978-3-319-16862-3_2

optoelectronic materials and devices including PLEDs and PSCs have attracted great attention all over the world and have developed into hot research fields. Because of the importance of conjugated polymers, Heeger, MacDiarmid, and Shirakawa were awarded the Nobel Prize in Chemistry in 2000, in recognition of their great contributions to the discovery and developments of conducting polymers.

The main advantages of conducting polymers are that they possess not only the electronic and optical properties of metals and inorganic semiconductors, but also the flexible mechanics and processability of polymers. In addition, there is special electrochemical redox activity with conducting polymers. Obviously, conducting polymers, including doped conducting polymers and intrinsic semiconducting conjugated polymers, will play a key role in the future development of organic optoelectronic and electrochemical devices.

2.1 Molecular Structure of Conducting Polymers

The unique characteristic of conducting polymers is the conjugated molecular structure of the polymer main chain where the π-electrons delocalize over the whole polymer chain. Conjugated polymers become conducting polymers after doping. Figure 2.1 shows the main chain structures of representative conjugated polymers, including polyacetylene (PA), polypyrrole (PPy), polyaniline (PAn or PANi), polythiophene (PT or PTh), poly(*p*-phenylene vinylene) (PPV), poly(*p*-phenylene) (PPP), and polyfluorene (PF).

In the conjugated polymers, polyacetylene shows the simplest main chain structure composed of an alternate single bond and double bond carbon chain. According to the locations of the hydrogen atoms on the double bond carbons, there are two kinds of structures: *trans*-polyacetylene with the two hydrogen atoms on

Fig. 2.1 Main chain structures of several representative conjugated polymers

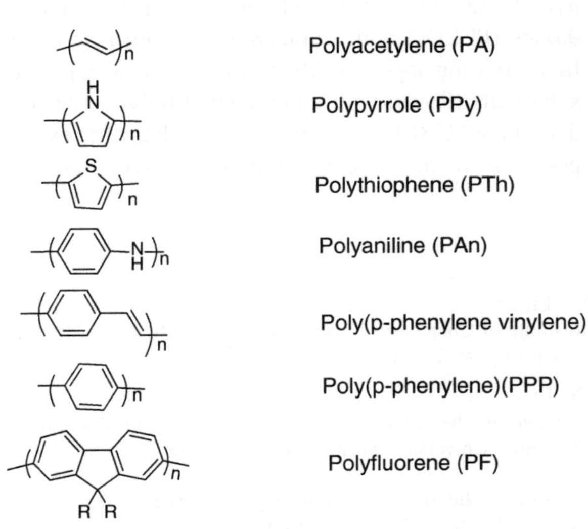

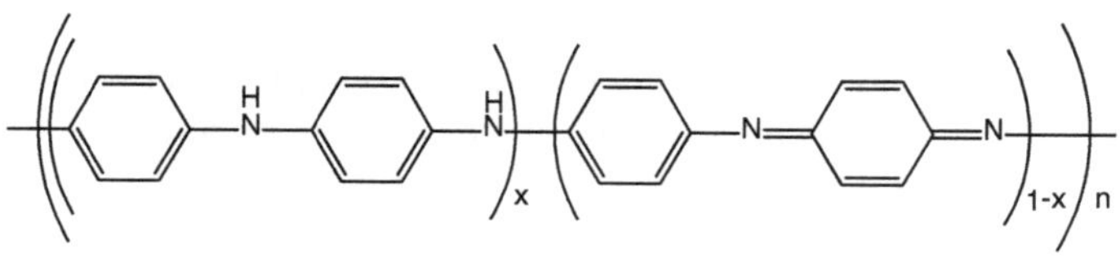

Fig. 2.2 Main chain structure of polyaniline

opposite sides of the double bond carbons and *cis*-polyacetylene with the two hydrogen atoms located on the same side of the double bond. *trans*-Polyacetylene is a degenerate conjugated polymer which possesses an equivalent structure after exchanging its double bond and single bond. *cis*-Polyacetylene and other conjugated polymers are nondegenerate conjugated polymers which have non-equivalent structures after exchanging their double and single bonds.

Among the various conjugated polymers, the main chain structure of polyaniline (PAn) is a little complicated. Figure 2.2 shows the conjugated main chain structure of PAn. Commonly, there exist three structure forms: leucoemeraldine (PAn-I) where $x = 1$, emeraldine base (EB, PAn-II) where $x = 0.5$, and pernigraniline (PAn-III) where $x = 0$. The molecular structure of PAn in Fig. 2.1 is leucoemeraldine. Actually, the structure of emeraldine base (EB) is the most important structure for conducting polyaniline, because proton-acid doping of the EB structure (Fig. 2.4) turns it into conducting PAn.

2.1.1 *Electronic Structure of Intrinsic Conjugated Polymers*

Conjugated polymers possess delocalized π-electron structures, including the band structure of π-valence band and π^*-conduction band. In the basic state of the intrinsic conjugated polymers, all the valence bands are filled by electrons and the conduction bands are all empty. The difference between the top of the valence band (the highest occupied molecular orbital, HOMO) and the bottom of the conduction band (the lowest unoccupied molecular orbital, LUMO) is called the bandgap (E_g) of the conjugated polymers. The E_g values of most conjugated polymers are in the range 1.5–3.0 eV. Therefore, the intrinsic conjugated polymers are organic semiconductors.

The E_g values of conjugated polymers can be measured by absorption spectroscopy of the conjugated polymer films. From the absorption edge wavelength (λ_{edge}) of the absorption spectra, E_g can be calculated according to the following equation:

$$E_g = \frac{1240}{\lambda_{edge}} \text{ (eV)}$$

where the unit of λ_{edge} is nm.

HOMO and LUMO energy levels of conjugated polymers can be estimated from onset oxidation and onset reduction potentials measured by electrochemical cyclic voltammetry [5]. The detailed measurement method is described in Sect. 2.6.4. The difference between the HOMO and LUMO energy levels of the conjugated polymers also corresponds to E_g values.

When the conjugated polymers were used as electroluminescent materials in PLEDs, the bandgap E_g of the conjugated polymer determines the emitted color of the PLEDs, and the HOMO and LUMO energy levels of the polymer influence the holes and electrons injection efficiency in the devices. Chapter 5 gives a more detailed discussion of the electroluminescent characteristics of conjugated polymers. For the use of conjugated polymers as donor materials in the active layer of PSCs, the E_g value of the conjugated polymer determines the absorption wavelength range of the devices, and the HOMO and LUMO energy levels influence the exciton dissociation efficiency at the donor/acceptor interface and the open circuit voltage of the PSCs. Therefore, it is very important to understand the effect of the molecular structure on the energy bandgap and electronic energy levels of the conjugated polymers.

The factors influencing the electronic structure and E_g values of the conjugated polymers are as follows:

1. For the degenerate *trans*-polyacetylene, E_g values decrease on decreasing the difference between the alternating single bond length and double bond length.
2. For the conjugated polymers formed by connecting the aromatic rings with single bonds, such as polypyrrole, polythiophene, and poly(*p*-phenylene), the deviation of the conjugated main chain from planar structure between the two neighboring aromatic rings will result in the increase of the E_g values. The larger the angle between the two neighboring aromatic rings in the main chain of the conjugated polymer, the smaller the overlap between the two molecular orbitals of the conjugated ring units and the higher the E_g values of the conjugated polymer.
3. The nature of the substituents on the main chain also influences the electronic structure of the conjugated polymers. The electron-donating substituents up-shift the LUMO and HOMO energy levels and reduce the E_g of the conjugated polymers (the up-shift of the HOMO is more than that of the LUMO). The electron-withdrawing substituents down-shift the LUMO and HOMO energy levels and also reduce the E_g of the conjugated polymers (the down-shift of the LUMO is more than that of the HOMO).
4. Copolymerization of conjugated electron-donating (D) unit and electron-accepting (A) unit results in lower bandgap conjugated D-A copolymers, and the HOMO and LUMO energy levels can be tuned by selecting suitable donor and acceptor units in the copolymers. The absorption spectra of the D-A copolymers are broadened and red-shifted because of the intramolecular charge transfer between the donor and acceptor units.
5. The existence of quinone structure in the polymer main chain can decrease the E_g values of the conjugated polymers.

6. The aggregation of the polymer main chains in the solid state also influences the E_g and electronic energy levels of the conjugated polymers. The strong intermolecular interaction in the planar main chain decreases the E_g of the conjugated polymers.

2.1.2 Doping Structures of Conducting Polymers

The distinguished characteristic of conducting polymers is the p-doped and n-doped states of the conjugated polymer main chains. In the p-doped state, the main chain of the conducting polymer is oxidized with counteranion doping for keeping the electron neutrality of the whole molecule. There are holes in the main chains (lost electrons) which make the conducting polymer p-type conducting. In the n-doped state, the main chain of the conducting polymer is reduced with countercation doping for keeping the electron neutrality of the whole molecule. There are electrons in the main chains which make the conducting polymer n-type conductive.

Figure 2.3 shows the p-doped structure of conducting PPy. The positive charge is delocalized on the PPy main chain. A^- represents counteranions such as NO_3^-, ClO_4^-, Cl^-, TsO^-.

The number of counteranions per monomer unit of the conducting polymer (or the concentration of the charge carrier in the conjugated main chain of the conducting polymer) is called doping degree of the conducting polymer. The maximum doping degree is related to the main chain structure of the conducting polymers. For example, the doping degree for polyacetylene is usually 0.1–0.2, it is 0.25–0.35 for polypyrrole, 0.4–0.5 for polyaniline, and ca. 0.3–0.4 for polythiophene. For the p-doped polypyrrole, the doping degree of 0.25–0.35 implies that the conjugated chain including 3–4 pyrrole units can be doped with 1 counteranion (or there is a hole within the polypyrrole main chain containing 3–4 pyrrole units), as shown in Fig. 2.3.

The doping processes of conducting polyaniline are related to its structures of conjugated polyaniline. For leucoemeraldine (PAn-I), the doping process is similar to polypyrrole, i.e., it can be p-doped by oxidation. The EB, PAn-II can be doped by protonation, as shown in Fig. 2.4. In the protonation process of the PAn-II, the positive charge on the proton delocalizes on the whole conjugated main chain of polyaniline, becoming the positive charge carrier (hole) of the conducting

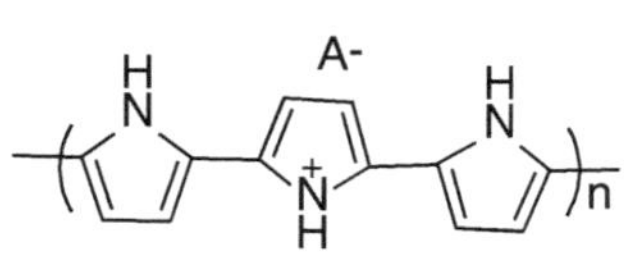

Fig. 2.3 p-Doped structure of conducting PPy

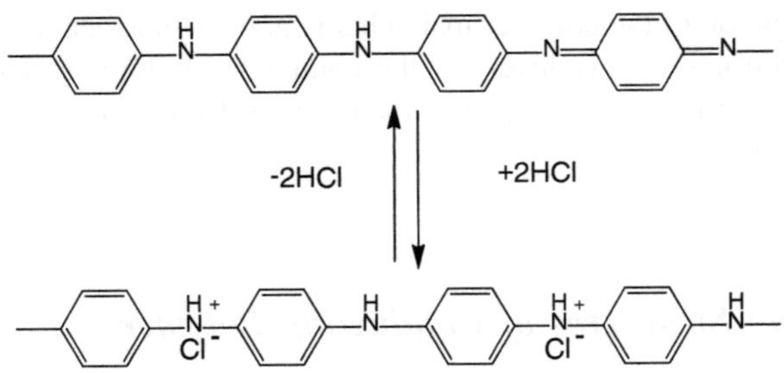

Fig. 2.4 Proton-acid doping of emeraldine base polyaniline (EB) and its proton acid doping structure

polyaniline. At the same time, the anion of the proton acid becomes the counter-anion of the conducting polyaniline.

It should be mentioned that the nature of the doping in conducting polymers is different from that of the doping in inorganic semiconductors. In inorganic semiconductors with crystalline structures, doping is realized by replacing some bulk atoms (such as Si) with outer shell electrons one more or one less than Si to achieve n-doping or p-doping. The doping concentration is very low. Although the doping of conjugated polymers with amorphous structure needs charge injection by oxidation or reduction of its conjugated main chain, counterions doping is required for keeping the charge neutrality. The doping degree is much higher in conducting polymers where the charge carrier concentration reaches $10^{21}/cm^3$, which is several orders higher than that of the inorganic semiconductors. In addition, the doping in conducting polymers also results in volume expansion and morphology changes because of the counteranion doping.

2.1.3 Charge Carriers in Conducting Polymers

Novel structures lead to novel charge carriers in conducting polymers. For *trans*-polyacetylene with the degenerate basic state, the charge carriers are solitons and polarons. However, for the basic state nondegenerate *cis*-polyacetylene, polypyrrole, polythiophene, polyaniline, etc. the charge carriers are polarons and bipolarons [6]. The *soliton* (**S**) is an unpaired π-electron resembling the charge on free radicals, which can be delocalized on a long conjugated polymer main chain (*trans*-polyacetylene main chain). The neutral soliton can be oxidized to lose an electron and form a positive soliton, or it can be reduced to gain an electron and become a negative soliton. The soliton possesses a spin of 1/2, whereas there is no spin for the positive and negative solitons. The electronic energy level of the soliton is located at the middle of the bandgap of the *trans*-polyacetylene. There is no electron or there are a couple of electrons on the soliton energy levels for the positive soliton and negative soliton, respectively.

Polarons are the major charge-carriers in conducting polymers including basic state degenerate *trans*-polyacetylene and the basic state non-degenerate conjugated polymers. The positive polaron with positive charge and the negative polaron with negative charge are denoted as $\mathbf{P^+}$ and $\mathbf{P^-}$, respectively. P^+ is formed after oxidation of the conjugated polymer main chain and P^- is formed after reduction of the conjugated polymer main chain. The appearance of the polarons produces two new polaron energy levels in the bandgap of the conjugated polymers. $\mathbf{P^+}$ and $\mathbf{P^-}$ possess spin of 1/2.

The *bipolaron* is the charge carrier that possesses double charges by coupling of two P^+ or two P^- on a conjugated polymer main chain. The bipolaron has no spin, and it can be formed when the concentration of polarons are high in the conjugated polymer main chains. The positive bipolaron and negative bipolaron correspond to the hole pair or the electron pair.

2.2 Doping Characteristics

As mentioned above, the doping of conducting polymers is natively different from that in inorganic semiconductors. Doping of conducting polymers can be realized chemically or electrochemically by oxidation or reduction of the conjugated polymers.

2.2.1 Chemical Doping

Conducting polyacetylene was discovered by chemical doping [1, 2]. The chemical doping includes p-type doping and n-type doping.

p-*Doping* is also called oxidation doping, which refers to the oxidation process of the conjugated polymer main chain to form polarons. The oxidants I_2, Br_2, AsF_5, etc. can be used as p-dopants. After p-doping, the conjugated polymer is oxidized and loses electron to form p-doped conjugated polymer chain, and the dopant gains an electron to become the counteranion. The following reaction is an example of the p-doping process:

$$\mathrm{CP} + (3/2)\mathrm{I_2} \rightarrow \mathrm{CP^+(I_3^-)} \tag{2.1}$$

where CP denotes conducting polymers.

n-Doping is also called reduction doping, which refers to the reduction process of the conjugated polymer main chain to form negative charge carriers. Some strong reductants, such as alkali metal vapor, $Na^+(C_{10}H_8)^-$, etc., can be used as n-type dopants. After n-doping, the conjugated polymer is reduced and gains electrons to form an n-doped conjugated polymer main chain, whereas the dopant losses an electron to become the countercation. The following reaction is an example of the n-doping process:

$$CP + Na^{+}(C_{10}H_{8})^{-} \rightarrow CP^{-}(Na^{+}) + C_{10}H_{8} \quad (2.2)$$

Proton acid doping of polyaniline (see Fig. 2.4) is one kind of chemical doping.

2.2.2 Electrochemical Doping

Electrochemical doping is realized by electrochemical oxidation or reduction of the conjugated polymers on an electrode.

For electrochemical p-doping, the conjugated polymer main chain is oxidized to lose an electron (gain a hole) accompanying the doping of counteranions from electrolyte solution:

$$CP - e^{-} + A^{-} \rightleftharpoons CP^{+}(A^{-}) \quad (2.3)$$

where A^{-} denotes the solution anion, $CP^{+}(A^{-})$ represents the conducting polymer with the main chain oxidized and counteranion doped.

For electrochemical n-doping, the conjugated polymer main chain is reduced to gain an electron accompanying the doping of countercations from electrolyte solution:

$$CP + e^{-} + M^{+} \rightleftharpoons CP^{-}(M^{+}) \quad (2.4)$$

2.3 Conductivity Characteristics

Conductivity is the most important property of conducting polymers. The conductivity of common doped conducting polymers is in the range of 10^{-3}–10^{3} S/cm, whereas that of the intrinsic conjugated polymers without doping is in the range of 10^{-9}–10^{-6} S/cm. After doping, conductivity of conjugated polymers increases by six to ninefold. The highest conductivity reported in the literature is 10^{5} S/cm for drawing-extended ordering conducting polyacetylene film [7].

Conducting polymers usually have an amorphous structure, in some cases with ordered domains. The charge-transporting mechanism in conducting polymers is different from that in the crystalline conducting materials where there exist conduction bands and valence bands and the charge carriers can move freely in the energy bands. In conducting polymers the charge carriers are located in the local doping energy levels (limited length of conjugated polymer chain) or in a very narrow doping energy band in the case of ordered domains. The charge carriers can move easily on the conjugated polymer main chain, but the charges have to hop for the transportation between the conjugated polymer chains. The activation energy for the hopping of the charge carriers is much higher than that of the charge transportation within the conjugated polymer main chains. Obviously, the charge transportation in conducting polymers is limited by the hopping between the conjugated polymer chains. Therefore, the conductivity of conducting polymers shows characteristics of hopping transportation.

The conductivity of conducting polymers shows a temperature dependence similar to that of semiconductors, and it obeys the Mott Variable Range Hopping (VRH) model:

$$\sigma(T) = \sigma_0 \exp[-(T_0/T)^{1/(n+1)}] \tag{2.5}$$

where σ_0 is a factor weakly related to temperature, n is the dimension number, $n = 1, 2, 3$ indicate that it is one-dimension, two–dimension, and three-dimension VRH transportation. For the common three dimension system, the conductivity equation is [8]

$$\sigma(T) = \sigma_0^{3d} \exp[-(T_0^{3d}/T)^{1/4}] \tag{2.6}$$

where

$$T_0^{3d} = c/[k_B N(E_F) L^3] \tag{2.7}$$

In (2.7), c is a constant, k_B is the Boltzmann constant, L is the localization length (effective conjugated chain length), and $N(E_F)$ is the state density at the Fermi energy level.

The conductivity of conducting polymers is closely related to the doping degree and the degree of ordering of the polymer main chain in the solid film. The doping degree relates to the charge carrier concentration on the conjugated polymer main chain. In the low doping degree region (far lower than the saturated doping degree), the conductivity of the conjugated polymers increases linearly with increasing the doping degree of the conjugated polymers.

2.4 Absorption Spectra

Intrinsic conjugated polymers and doped conducting polymers can be distinguished by absorption spectra. There is a strong absorption peak in the near-infrared (NIR) region for doped conducting polymers, caused by the existence of the polaron and bipolaron energy levels within the bandgap of the conjugated polymers. The NIR absorption disappears after dedoping for the intrinsic conjugated polymers. Figure 2.5 shows the absorption spectra of doped and dedoped polypyrrole films,

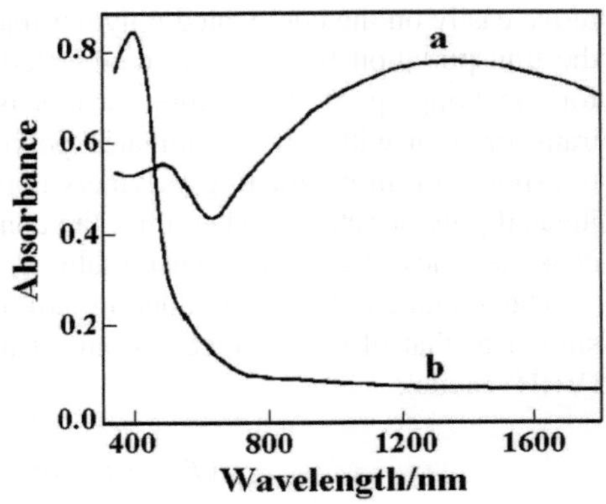

Fig. 2.5 Absorption spectra of polypyrrole. **a** Doped state. **b** Intrinsic state

which represents the typical characteristic absorption spectra of the doped conducting polymer and the dedoped intrinsic conjugated polymer. There is an absorption peak at ca. 400 nm for the dedoped intrinsic polypyrrole film, which corresponds to the π-π* transition absorption of the conjugated polypyrrole main chain. After doping, a strong and broad absorption peak appears in the NIR region from 700 nm to ca. 2,000 nm, which corresponds to the polaron and bipolaron energy levels.

The absorption spectra of doped and intrinsic polyaniline are more complicated in comparison with those of polypyrrole because of the three structure change of polyaniline. Figure 2.6 shows the absorption spectra of polyaniline in different doping states. The absorption spectrum of doped (proton-acid doping or electrochemical doping) conducting polyaniline is similar to that of the doped conducting polypyrrole: there is a strong and broad absorption peak in the NIR region (peaked at ca. 950 nm) which corresponds to the polaron and bipolaron absorption, and an absorption peak at ca. 320 nm corresponding to the π-π* transition absorption of the

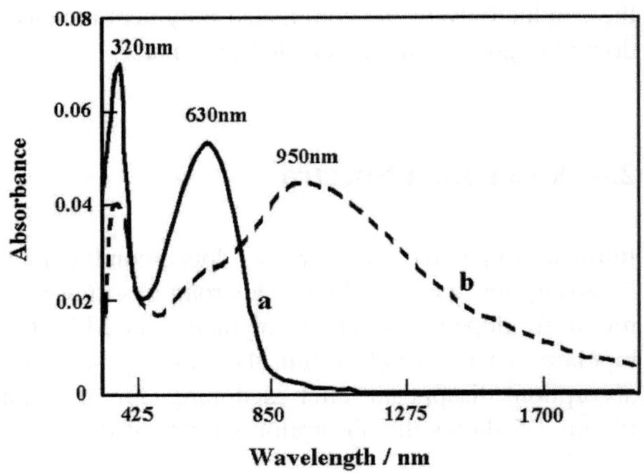

Fig. 2.6 Absorption spectra of polyaniline. **a** Pernigraniline (PAn-III). **b** Proton-acid-doped conducting polyaniline

conjugated polyaniline main chain. With the reduction (dedoping) of the conducting polyaniline, the NIR absorption is weakened and finally disappears when it is reduced to its intrinsic conjugated state. Conducting polyaniline can be proton-acid dedoped to become a pernigraniline (PAn-III) which shows an absorption spectra with two strong absorption peaks at ca. 320 and 630 nm, respectively. The absorption peak at ca. 320 nm corresponds to the π-π* absorption of the conjugated polyaniline main chain, whereas the absorption peak at ca. 630 nm can be ascribed to the electron transition between the HOMO of the benzo-structure and the LUMO of the quinone-structure of PAn-III.

Figure 2.7 shows the absorption spectra of several representative intrinsic conjugated polymers including polyfluorene (PFO), MEH-PPV, poly(3-hexylthiophene) (P3HT), and poly(3-hexylthienylene-vinylene) (P3HTV). All the spectra correspond to the π-π* absorption of the conjugated polymer main chains. The bandgap (E_g) of the conjugated polymers can be calculated from its absorption edge. It can be seen from Fig. 2.7 that the absorption edges of PFO, MEH-PPV, P3HT, and P3HTV are 443, 568, 669, and 749 nm, respectively. From the absorption edge wavelength, the bandgaps of PFO, MEH-PPV, P3HT, and P3HTV can be calculated to be 2.8, 2.18, 1.85, and 1.66 eV, respectively. The results indicate that inserting a carbon–carbon double bond between the benzene or thiophene rings in the polymer main chain can reduce the bandgap and red-shift the absorption spectra of the conjugated polymers significantly.

The characteristics of the absorption spectra of conjugated polymers play a crucial role in the applications of PSCs. Broad and strong absorption in the visible and NIR regions is pursued for the high performance polymer photovoltaic materials to harvest solar light efficiently.

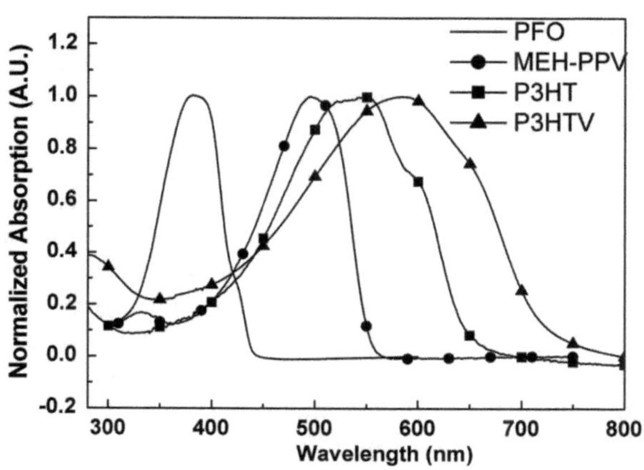

Fig. 2.7 Absorption spectra of the intrinsic PFO, MEH-PPV, P3HT, and P3HTV films

2.5 Solubility

Conducting polymers are usually insoluble and infusible because of their rigid conjugated main chain, which limits their application. There are several strategies to solve the problems as discussed in the following. In 1992, Cao et al. [9] from UNIAX company prepared soluble conducting polyaniline (by counteranion induced solubility) and solved the difficulty of processing of conducting polymers, which paved the way for large-scale application of conducting polymers.

2.5.1 Effect of Substituents on Solubility of Conjugated Polymers

Conjugated polymers without substituents are all insoluble. Attaching appropriate flexible side chains (substituents) can make the conjugated polymers soluble in organic solvents. For example, polythiophene without substituents is insoluble in any solvent, whereas the hexyl-substituted polythiophene derivative P3HT is soluble in organic solvents such as toluene, chlorobenzene, dichlorobenzene, etc. In addition, for application as electroluminescent polymers or photovoltaic polymers, the side chains can also tune the bandgap and electronic energy levels (HOMO and LUMO energy levels) of the conjugated polymers. The bandgap determines the color of the PLEDs with the conjugated polymer as active layer, and it influences the photovoltaic properties of the conjugated polymers in PSCs. The HOMO and LUMO energy levels are very important for improving the optoelectronic performance of conjugated polymers.

Bredas et al. [10] studied the effect of electron-donating or electron-accepting ability of the substituents on the electronic energy levels of PPV derivatives by quantum chemistry calculation with the VEH method. They calculated the energy bandgap (E_g), ionization potential (IP) (which corresponds to the HOMO energy level with IP = −HOMO), and electron affinity (EA) (which corresponds to the LUMO energy level with EA = −LUMO) of the PPV derivatives. The calculation results are listed in Table 2.1. The E_g values calculated are in good agreement with those obtained from the absorption edges of their absorption spectra, indicating that the theoretical calculation results are quite reliable. It can be seen from Table 2.1 that IP and EA values of the PPV derivatives substituted by electron-donating alkoxy groups decreased in comparison with those of PPV without substituents, with more decrease of IP than EA. IP and EA values of the PPV derivatives substituted by electron-withdrawing cyano groups increased in comparison with those of PPV without substituents, with more increase of EA than IP. Irrespective of the electron-donating or electron-withdrawing substitution, the E_g values of the PPV derivatives are reduced to some extent.

Table 2.1 Electronic properties of PPV and its derivatives calculated by the VEH method

PPV and its derivatives	E_g/eV	IP/eV	EA/eV
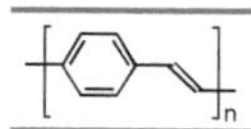	2.32	5.05	2.73
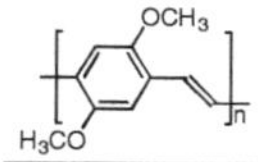	2.07	4.72	2.65
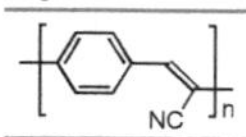	2.17	5.27	3.10
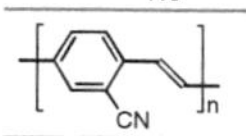	2.24	5.15	2.91
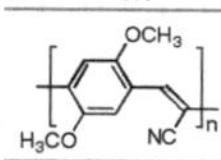	1.97	5.12	3.15
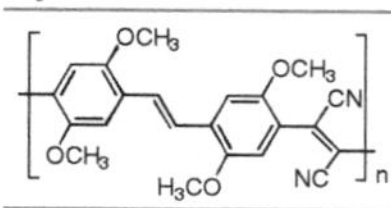	1.74	5.08	3.34

2.5.2 *Effect of Substitution on the Conductivity of Conducting Polymers*

Actually, for doped conducting polymers, substitution by flexible side chains is not a good method for solving their solubility, because it usually results in a conductivity decrease in the conducting polymers. For example, the conducting polypyrrole with a long alkyl substituent on the 3- or 4-position or the *N*-position of its pyrrole ring is soluble in organic solvents, but conductivity of the conducting polypyrrole is decreased significantly to ca. 0.001 S/cm. The reason for the conductivity decrease is that introducing a substituent on the conducting polymer main chain results in distortion of the conjugated main chain, thereby decreasing the conjugation degree and the conductivity of the conducting polymers.

The best way to make conducting polymers soluble is the counteranion induced method proposed by Yong Cao et al. [9]—using proton acid and acid containing anions with flexible side chain [such as dodecyl-benzene sulfonic acid (DBSA)] to make the doped polyaniline soluble. This is the most successful method, and polyaniline film prepared from conducting polyaniline solution possesses high conductivity. Yong Cao et al. prepared conducting polyaniline film from polyaniline solution, and the conductivity of the conducting polyaniline film reached 10^2 S/cm [9], which is higher than that of conducting polyaniline films prepared by electrochemical polymerization or other methods.

2.6 Electrochemical Properties

Electrochemical doping/dedoping property is one of the most important properties of conducting polymers, because many applications of conducting polymers, including electrode materials for batteries, electrochromic materials, modified electrodes, and enzyme electrodes, etc., are based on the electrochemical properties. Studies on the electrochemical properties of conducting polymers are mainly focused on the electrochemical redox potentials, reversibility, and reaction mechanism of the doping/dedoping processes.

Electrochemical studies of conducting polymers are commonly carried out for the conducting polymer films on the working electrode. The electrochemical redox processes of conducting (or conjugated) polymers are quite complicated in comparison with common organic and inorganic molecules. Electrochemical oxidation of conjugated polymers is accompanied by the intercalation (or doping) of counteranions from electrolyte solution, so that the oxidation process of the conjugated polymers is often called oxidation doping (or p-doping). The electrochemical reduction of conjugated polymers is accompanied by the intercalation (or doping) of countercations from electrolyte solution, so the reduction process of the conjugated polymers is often called reduction doping (or n-doping). In addition to the common electron transfer on the electrode/electrolyte interface, there are both diffusion of counterions in the conjugated polymer films and expansion of the conjugated polymer films because of the intercalation of the counterions.

2.6.1 Electrochemical Properties of Conducting Polypyrrole

Polypyrrole (PPy) is a typical p-type conjugated polymer with a very low oxidation potential in the range of ca.−0.6 to 0.3 V versus SCE, which makes the p-doped conducting PPy stable, and the neutral PPy is very easily oxidized into its p-doped state. The electrochemical properties of conducting PPy are therefore usually inferred from the reduction (dedoping)/re-oxidation (doping) of the p-doped PPy. Generally, the electrochemical reaction of conducting PPy can be expressed as follows:

$$PPy^{+}(A^{-}) + e^{-} \rightleftharpoons PPy^{0} + A^{-} \tag{2.8}$$

where $PPy^{+}(A^{-})$ denotes the oxidation-doped conducting PPy doped with counteranion A^{-} (p-doped PPy), PPy^{0} denotes the neutral (intrinsic) PPy.

The electrochemical reduction/re-oxidation (dedoping/doping) processes of conducting PPy (p-doped PPy) in aqueous solution are closely related to counteranions of the electrolyte and the pH values of the aqueous solution. Figure 2.8 shows cyclic voltammograms of nitrate-doped conducting PPy film in neutral and weakly acidic $NaNO_3$ aqueous solutions [11]. In weak acidic pH 3 solution, there are

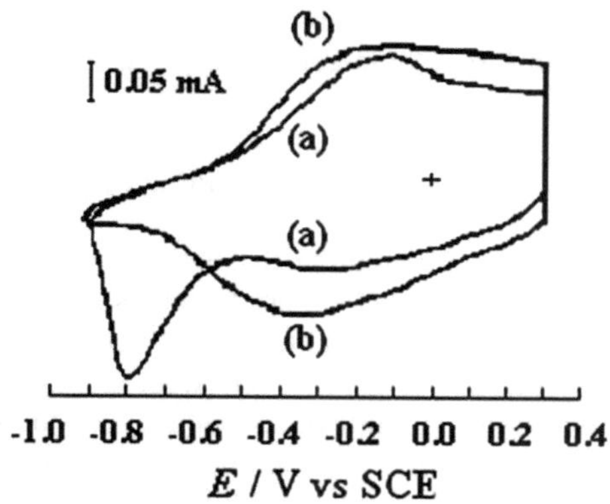

Fig. 2.8 Electrochemical cyclic voltammograms of PPy(NO_3^-) in 0.5 mol/L $NaNO_3$ aqueous solution at potential scan rate of 20 mV/s: **a** neutral (pH 7) aqueous solution; **b** acidic (pH 3) aqueous solution

a couple of reversible reduction/re-oxidation peaks in the potential range of 0.3 to −0.8 V versus SCE (Fig. 2.8b). The reaction processes can be expressed as that in reaction (2.8). In neutral (pH 7) aqueous solution there are two reduction peaks in the potential range of 0.3 to −0.8 V versus SCE (Fig. 2.8a), which corresponds to the two doping structures (oxidation doping structure and proton-acid doping structure) of conducting PPy [12]. Actually, the electrochemical reduction of conducting PPy in acidic solutions also involves two reduction processes, the two reduction peaks in the acidic solutions being mixed together to show a broad reduction peak in the cyclic voltammogram. The in situ absorption spectra at different reduction potentials of conducting PPy in a pH 3 $NaNO_3$ aqueous solution clearly indicate the two reduction processes in the potential ranges of 0.3 to −0.3 V and −0.3 to −0.8 V versus SCE [11]. For the re-oxidation of the reduced PPy (neutral PPy), it is reversible if the upper-limited potential is lower than 0.3 V versus SCE. However, if the oxidation potential is higher than 0.5 V versus SCE, some overoxidation of the p-doped PPy takes place. The potential value where the overoxidation starts to occur is closely related to the pH value of the aqueous electrolyte solution—the higher the pH value, the lower the potential [13]. That is, PPy is more easily overoxidized in an alkaline aqueous solution than in an acidic solution.

The electrochemical reduction and re-oxidation processes of conducting PPy are closely related to the nature of the counteranions in the polymer films and the anions in the electrolyte solutions [14]. When the anions are small and spherically shaped such as NO_3^- and Cl^- etc., the reversible reduction/re-oxidation peaks can be observed in the cyclic voltammograms of the conducting PPy, as mentioned above. However, if the counteranions in the polymer film are large surfactant anions such as TsO^-, the reduction peak appears at a much lower potential because of the difficulty of dedoping the large counteranions, and the solution cations dope into the conducting PPy instead of the anions dedoping [11]. The reaction mechanism of the conducting PPy film with large counteranions such as TsO^- can be expressed as follows:

$$PPy^+(A^-) + e^- + M^+ \rightleftharpoons PPy^0(M^+A^-) \quad (2.9)$$

On the other hand, if the cyclic voltammetry is performed in an aqueous solution containing the electrolyte salt with large anions such as TsO^-, the first reduction of $PPy(NO_3^-)$ is the dedoping of NO_3^- counteranions, but the re-oxidation process is irreversible because of the difficulty of TsO^- doping [14]. In a weak alkaline aqueous solution, the original counteranions in the conducting polypyrrole are exchanged with strong nucleophilic OH^- anions [15]. Then the reduction and re-oxidation of the PPy film are accompanied with the dedoping and redoping of OH^- [15, 16]. In a strongly alkaline aqueous solution, the doping structure of conducting PPy is unstable and the conjugated polymer chain of PPy can be degraded and destroyed, which results in the loss of conductivity and electrochemical properties of PPy [13].

For the PPy film prepared from an organic electrolyte, an abnormal cyclic voltammogram of the PPy film can be observed in an organic electrolyte solution. There is a high overpotential for the first reduction process of the PPy, and no reduction current until −0.6 V versus SCE, as shown in Fig. 2.9. The reoxidation and the redox processes from the second cycle resumes to normal cyclic voltammograms of PPy in weakly acidic aqueous solutions [17]. This phenomenon can be explained as follows. The diffusion coefficient of the counteranions during the first reduction of PPy in organic electrolyte solution is very small and the solvation energy of the counteranions with organic solvent molecules is also very small, so it is very difficult for the counteranions to dedope from PPy into the organic solution. Then the solvated cations with organic solvent molecules dope into PPy when the potential reaches a very low value (negative potential), which increases the diffusion coefficient of the counteranions in PPy and makes the following redox reaction reversible [17].

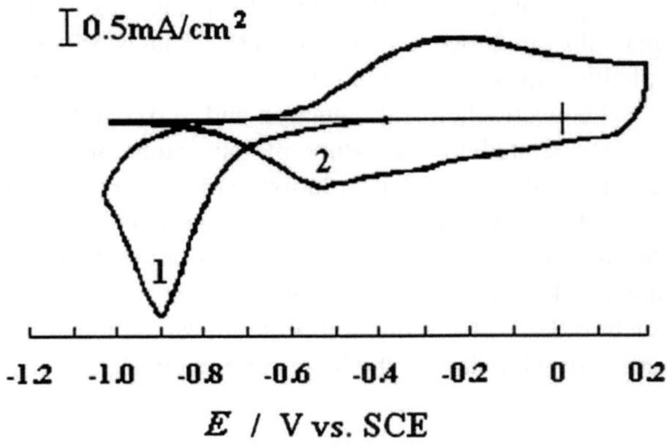

Fig. 2.9 Cyclic voltammograms of PPy (ClO_4^-) in 0.5 mol/L $NaClO_4$ PC solution. The numbers in the figure indicate the sequence of the cyclic potential scan

2.6.2 Electrochemical Properties of Conducting Polyaniline

Polyaniline (PAn) is another important and well studied conducting polymer. There are many potential applications (such as electrode materials for batteries and supercapacitors, anti-corrosion materials, modified electrodes and biosensors, etc.) for PAn based on its electrochemical properties. Therefore, understanding the electrochemical properties is of great importance for the applications of PAn.

The most important characteristic of PAn, in comparison with other conducting polymers such as PPy, etc., is its proton-acid doping. The proton-acid doping benefits the preparation of conducting PAn solutions by counteranions induced proton-acid doping [9], but it also makes the doped PAn unstable in basic and neutral aqueous solutions. The doped PAn is easily dedoped by removing proton-acid in the basic and neutral solutions. PAn becomes an insulator and loses its electrochemical activity after the dedoping. Therefore, the doped PAn is stable only in acidic solutions and the electrochemical properties of PAn are mainly studied in acidic electrolyte solution.

There are two reversible redox processes for PAn in an acidic aqueous solution in the potential range of −0.5 to 0.7 V versus SCE. Figure 2.10 shows cyclic voltammograms of polyaniline doped with NO_3^- counteranions ($PAn(NO_3^-)$) in an acidic (pH 1.5) 1 mol/L $NaNO_3$ aqueous solution [18]. Based on the in situ absorption spectra at different redox potentials, the redox processes in the potential range of 0.3 to −0.5 V versus SCE are similar to that for PPy in acidic solution [see reaction (2.8)], the reduction peak and reoxidation peak corresponding to the dedoping and redoping of conducting polyaniline. From 0.3 to 0.7 V versus SCE, there is another unique redox processes for PAn; the doped PAn is further oxidized into the completely oxidized polyaniline pernigraniline (PAn-III) (see Fig. 2.2) which loses conductivity. In addition, weak redox peaks often appear in the potential range of 0.3–0.5 V versus SCE, as shown in Fig. 2.10. These weak middle redox peaks are related to an overoxidation structure of PAn. If the electropolymerization potential during the preparation of PAn is higher than 0.8 V, or the upper-limit potential during cyclic voltammetry is higher than 0.8 V versus SCE, the middle redox peaks appear in the cyclic voltammograms [19].

2.6.3 Electrochemical Properties of Polythiophene and Other Conjugated Polymers

In comparison with polypyrrole and polyaniline, polythiophene (PTh) possesses a narrower bandgap and a higher oxidation doping potential, which indicates that polythiophene can be both p-doped by oxidation and n-doped by reduction. The higher oxidation doping potential results in lower stability of the p-doped conducting polythiophene in comparison with conducting polypyrrole and polyaniline [20]. Nevertheless, it makes the neutral polythiophene stable which guarantees the optoelectronic applications of polythiophene as semiconducting polymers.

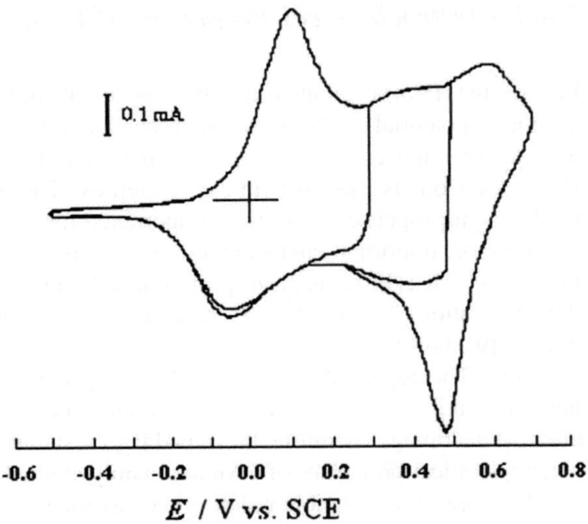

Fig. 2.10 Cyclic voltammograms of PAn (NO_3^-) in pH 1.5, 1 mol/L $NaNO_3$ aqueous solution with a potential scan rate of 40 mV/s

Organic solutions are used in the measurement of the electrochemical properties of polythiophene, because of the high oxidation potential for p-doping and lower reduction potential for n-doping of polythiophene. Figure 2.11 shows the cyclic voltammogram of poly(3-hexylthiophene) (P3HT) in 0.1 mol/L Bu_4NPF_6 acetonitrile solution. It can be seen that there are a couple of redox peaks in the higher potential range (0–0.6 V vs. Ag/Ag^+) corresponding to the p-doping/dedoping of polythiophene, and a couple of redox peaks in the negative potential range (−1.8 to −2.6 V vs. Ag/Ag^+) corresponding to the n-doping/dedoping of polythiophene.

For other semiconducting conjugated polymers used in optoelectronic devices, such as PLEDs or PSCs, their electrochemical properties are similar to those of polythiophene mentioned above. Usually, there are a couple of redox peaks in a

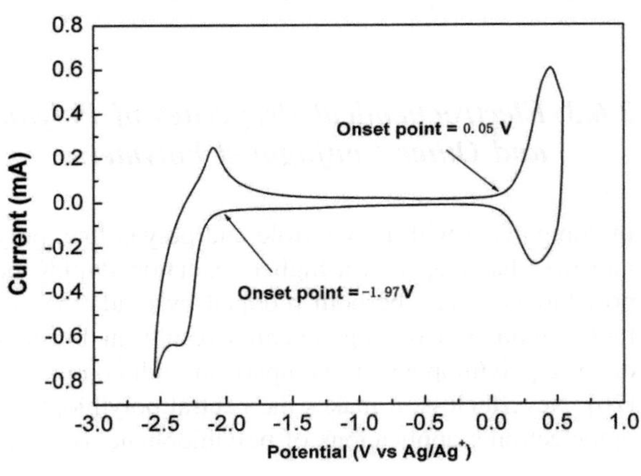

Fig. 2.11 Cyclic voltammogram of poly(3-hexylthiophene) (P3HT) in 0.1 mol/L Bu_4NPF_6 acetonitrile solution

higher positive potential range and a couple of redox peaks in the negative lower potential range, the only difference being their different onset oxidation/reduction potentials or their redox peak potentials in their cyclic voltammograms.

2.6.4 Electrochemical Measurement of HOMO and LUMO Energy Levels of Conjugated Polymers

As mentioned above, in the cyclic voltammograms of polythiophene and other optoelectronic conjugated polymers, there are a couple of redox peaks in the higher potential range corresponding to their p-doping/dedoping and a couple of redox peaks in the lower (negative) potential range corresponding to their n-doping/dedoping (see Fig. 2.11). Actually, the onset p-doping (oxidation) potential corresponds to the HOMO energy level and the onset n-doping (reduction) potential corresponds to the LUMO energy level of the conjugated polymers. Therefore, cyclic voltammetry has commonly been used to measure the HOMO and LUMO energy levels of the conjugated polymers [5, 21, 22]. The HOMO and LUMO energy levels of the conjugated polymers can be calculated from the following equations:

$$\mathrm{HOMO} = -\mathrm{e}\ (E_{\mathrm{ox}} + \mathrm{C})(\mathrm{eV});\quad \mathrm{LUMO} = -\mathrm{e}\ (E_{\mathrm{red}} + \mathrm{C})(\mathrm{eV})$$

where E_{ox} and E_{red} denote the onset oxidation and onset reduction potentials with the unit of V respectively, and C is a constant related to the reference electrode (RE) used in the measurement of onset redox potentials. C usually takes the value of 4.4 for the RE of SCE, 4.8 for the RE of Fc/Fc^+ (Fc denotes ferrocene), and 4.71 for the RE of Ag/Ag^+ [21, 22]. If silver wire was used as the RE, C could be 4.39 [5] (it should be calibrated with the ferrocene).

2.7 Optoelectronic Properties of Conjugated Polymers

Since the discovery of PLEDs by Friend et al. in 1990, [3] the optoelectronic properties of conjugated polymers have drawn great attention. There are detailed descriptions for the optoelectronic properties and their applications in PLEDs, PSCs, and organic field effect transistors for various conjugated polymers and conjugated organic molecules in Chaps. 3, 5, and 7 in this book.

2.8 Synthesis of Conducting Polymers

Conducting polymers can be prepared by chemical or electrochemical oxidation polymerization or by chemical catalytic synthesis.

2.8.1 Electrochemical Oxidation Polymerization of Conducting Polymers

The electrochemical preparation of conducting polymers is usually carried out through oxidative polymerization of their corresponding monomers by constant current, constant potential, or cyclic voltammetry in a potential range. The electropolymerization is performed in an electrolyte solution which contains solvent, electrolyte salt, and the monomer. There are many factors influencing the electropolymerization processes, such as solvents, supporting electrolyte salts, concentration of the monomers, and pH value of the electrolyte solutions, as well as polymerization potential, current, temperature, etc. Among these factors, the polymerization potential of the monomers is the most important. Table 2.2 lists the oxidation polymerization potentials of the most important monomers pyrrole, aniline, and thiophene. The lower oxidation polymerization potentials of pyrrole and aniline make the electropolymerization of polypyrrole and polyaniline easier, and it can be performed in aqueous solutions.

2.8.1.1 Electrochemical Preparation of Conducting Polypyrrole

Polypyrrole (PPy) is one of the most stable and environmentally-friendly conducting polymers. In 1979, Diaz et al. [23] first reported the preparation of a PPy film with conductivity (σ) of ca. 100 S/cm by electrochemical polymerization on Pt electrode in acetonitrile solution. Pyrrole can be electropolymerized both in organic and in aqueous solutions benefitting from its lower polymerization potential. Of course, aqueous solutions are the first selection for the preparation of conducting PPy.

The anions of the salts in the aqueous solutions influence conductivity of the as-prepared PPy films by electrochemical polymerization [24]. Flexible PPy films with

Table 2.2 Oxidative polymerization potential of monomers and conductivity of the as-prepared conducting polymers [39]

Monomer	Polymerization potential (V vs. SCE)	Conductivity of the as-prepared polymers (S/cm)
Pyrrole	0.7	30–100
Aniline	0.8	1–20
Thiophene	1.6	10–100

σ higher than 100 S/cm can be produced with the surfactant anions such as tosylate, benzene sulfonate, etc. Warrant et al. found that the acidity of the anions plays an important role in the conductivity of PPy films, the stronger the conjugated acids of the anions the higher the conductivity of the PPy films prepared from the anion-containing solution [25]. The concentration of the electrolyte anions also plays an important role. The concentration of the anions should be no lower than 0.1 M; too low an anion concentration leads to poor PPy films. The concentration of pyrrole monomer is usually 0.1 M.

The effect of solvent on the electropolymerization depends on the donor number (DN) of the solvent [26]. The low DN solvents, such as acidic water, propylene carbonate (PC), and CH_3NO_2, are very important for getting high conductivity of the as-prepared PPy films. The solvent effect can be explained from the cation radical coupling mechanism of the electropolymerization. High DN value of the solvent means high nucleophilicity or strong basicity. The solvent molecules with high DN value attack the cation radical formed by oxidation of pyrrole, which hampers the electropolymerization [26]. Water is a special solvent with changeable acidity by changing pH values. The optimum pH value of the aqueous solutions for pyrrole electropolymerization is between pH 2 and pH 5.5 [27].

The quality of the conducting PPy films can be improved by using a small amount of additive in the electrolyte solutions for the electropolymerization. By adding surfactant additives in the electrolyte solution, the smoothness, mechanical property, and conductivity of the as-prepared PPy films improves significantly [28, 29]. By using nonionic surfactant nonylphenol polyethyleneoxy (10) as an additive in the TsONa aqueous solution, the tensile strength of the PPy film produced from the solution reached 127 MPa, which is five times higher than that of the PPy film prepared without the surfactant additive [28].

As mentioned above, the electropolymerization can be performed with a potential-controlled method (constant potential or cyclic voltammetry) or a current-controlled method (constant current). In the potential-controlled method, the potential should be controlled no higher than 0.75 V versus SCE (usually at 0.65–0.70 V vs. SCE) for the electropolymerization of PPy. For the electropolymerization with constant current, Maddison et al. [30]. studied the effect of current and found that the PPy film with highest conductivity was obtained at 2.8 mA/cm^2. Good PPy films can usually be obtained at the current density of 1–2 mA/cm^2.

Temperature is another factor that influences electropolymerization. High quality PPy films can usually be obtained at lower temperatures (lower than 20 °C). At higher temperature, defect structures of PPy are easily formed, which results in lower conductivity.

For the mechanism of the electrochemical oxidation polymerization of PPy, Genies et al. [31] proposed a cation-radical polymerization in 1983. According to this mechanism, after applying the oxidation polymerization potential, pyrrole monomers are first oxidized into cation radicals on the anode, then two cation-radicals couple together to form a dimer with losing two protons. The dimer is oxidized into its cation radical more easily than monomer because of its lower oxidation potential. The cation radical of the dimer couples with other cation radicals to make a trimer or longer chain of polymer, and so on. The PPy films produced by electropolymerization should be in its p-doping (oxidized) state, because the p-doping potential of the polymer is much lower than the oxidation potential of the monomers. The cation radical mechanism can successfully explain the effect of solvent and solution anions on the electropolymerization. However, it fails to explain the effect of pH values of the electrolyte solutions. Qian et al. [32] modified the cation radical mechanism and proposed a pre-protonation cation-radical polymerization mechanism to elucidate the pH value effect on the electropolymerization. They think pyrrole monomer is first protonated on its β-carbon, which benefits the formation of the pyrrole cation radical. Then the polymerization follows the cation-radical mechanism.

The cation-radical mechanism and the modified mechanism with pre-protonation doesn't consider the effect of solution anions on the electropolymerization. Actually, the concentration and nature of the solution anions influence the oxidative polymerization rate and the quality of the as-prepared PPy films significantly. Therefore, the solution anions should take part in the electropolymerization processes. In addition, the coupling process of two cation-radicals in the cation-radical mechanism is unreasonable because the two cation radicals with positive charge should expel each other when one approaches to another for coupling. Based on this consideration, Li proposed an anion-participated cation-radical polymerization mechanism [33]. This mechanism proposes that pyrrole monomers and solution anions competitively adsorb on the anode under the oxidation polymerization potentials, and the cation-radicals formed on the anode should combine with solution anions to form neutral cation-anion pairs. Then the cation radical-anion pairs couple together to form a dimer with losing two anions and two protons. The experimental results of competition doping of two kinds of anions into PPy during pyrrole polymerization support the anion-participated mechanism [34].

2.8.1.2 Electrochemical Preparation of Polyaniline

Polyaniline (PAn) can be prepared electrochemically in a strongly acidic aqueous solution at ca. 0.8 V versus SCE. However, the PAn product prepared by electrochemical polymerization is powder attached to the electrode, which cannot form flexible films as do electropolymerized PPy and polythiophene (PTh). Therefore, the electrochemical preparation of PAn is mainly for deposition of a PAn modified electrode for electrochemical studies and for some applications in electrocatalysis, sensors, and electrode materials in lithium batteries.

Common electrolyte solution for the electrochemical preparation of PAn is 0.1 M aniline in 1 M H_2SO_4, HCl, $HClO_4$, or HBF_4 aqueous solutions. Conductivity of the electropolymerized PAn is usually of the order of $10^{-1} \sim 10^{1}$ S/cm. The electropolymerization of aniline in the acidic aqueous solutions at a constant potential (such as 0.8 V vs. SCE) often produces PAn with partial overoxidation [35]. Therefore, cyclic voltammetry in a potential range (such as −0.15 to 0.78 V vs. SCE [36]) is usually preferred for the preparation of a high quality PAn product.

2.8.1.3 Electrochemical Preparation of Polythiophene

The most important characteristic of thiophene electrochemical polymerization is its high polymerization potential of 1.6 V versus SCE (see Table 2.2), which requires the electrochemical polymerization of thiophene to be performed in organic solution and the conducting polythiophene (PTh) produced is easily overoxidized during the electropolymerization. Therefore, much effort has been devoted to decreasing the polymerization potential with various strategies. Diaz et al. [37] and Garnier et al. [38] studied the effect of substituents of thiophene on the polymerization potential (see Table 2.3). The oxidative polymerization potentials of bithiophene (dimer of thiophene) and the thiophene derivatives with the electron-donating substituents are obviously decreased, which benefits the electropolymerization.

Tourillon and Garnier [39] prepared polythiophene (PTh) electrochemically at 1.6 V versus SCE on a Pt anode in a solution of CH_3CN + 0.1 M $(Bu)_4NClO_4$ + 0.01 M thiophene (containing ca. 0.01 M water) with Ar pretreatment (bubbling) for 15 min. They obtained PTh film with a conductivity of 10–100 S/cm. They found that the pretreatment of the polymerization solution by Ar bubbling is very important. Without the Ar pretreatment to remove oxygen in the electrolyte solution, the conductivity of the as-prepared PTh films is only ca. 0.1 S/cm. The PTh films deposited on the anode are easily overoxidized in the presence of oxygen and water at the high electropolymerization potential. Sato et al. [40] performed the electropolymerization of thiophene in a dry solution (removed water carefully) under an Ar atmosphere, and they obtained the PTh film with the high conductivity of 190 S/cm.

Thiophene derivatives with electron-donating substituents, such as alkyl or alkoxy groups, show lower oxidative polymerization potentials, which is beneficial for the electropolymerization. For example, the oxidative polymerization potential

Table 2.3 Oxidative polymerization potential of thiophene and it derivatives [38]

Monomers	Oxidative polymerization potential (V vs. SCE)
Thiophene (T)	1.65
2,2′-Bithiophene (2,2′-bT)	1.20
3-Methyl thiophene (3-MeT)	1.35
3-Bromothiophene (3-BrT)	1.85
3,4-Dibromo thiophene (3,4-BrT)	2.00
3,4-Dimethyl thiophene (3,4-MeT)	1.25
3,4-Methyl ethyl thiophene (3,4-MeEtT)	1.26
3,4-Diethyl thiophene (3,4-EtT)	1.23
3-Thiomethyl thiophene (3-SCH_3T)	1.30

of 3-methyl thiophene (3-MeT) dropped to 1.35 V in comparison with 1.65 V for thiophene (see Table 2.3). The conductivity of poly(3-methylthiophene) [P(3-MeT)] film, prepared in a solution of 0.2 M 3-methylthiophene, 0.03 M Et_4NPF_6, PC (removing water carefully) at 5 °C under an Ar atmosphere, reached 450–510 S/cm [40]. 3-Methoxythiophene can be electropolymerized in an aqueous solution thanks to its lower polymerization potential with the substitution of strong electron-donating methoxy group [41].

Poly(3,4-ethylenedioxythiophene) (PEDOT) has drawn much attention recently because of its broad applications in transparent electrode materials, antistatic painting, solid state capacitors, and the electrode buffer layer material on ITO electrodes for PLEDs and PSCs. PEDOT can also be prepared by the electropolymerization of its monomer EDOT. The polymerization potential of EDOT is 1.49 V versus SCE which is lower than that of thiophene. By using dimer of EDOT as the monomer, the potential can be decreased further to 0.84 V versus SCE [42].

The polymerization potential of bithiophene is 1.20 V versus SCE, decreased by 0.45 V in comparison with that of thiophene. The lower polymerization potential makes the bithiophene easy to electropolymerize. Therefore, many researchers prepare PTh by the electropolymerization of bithiophene [43, 44]. The oxidative polymerization potential of terthiophene decreases further [45]. However, the molecular weight and conductivity of the PTh films prepared from bithiophene and terthiophene are lower than that of the PTh films obtained from thiophene, probably because of lower molecular weight of the conducting polythiophene due to the larger size of bithiophene and terthiophene.

In 1995, Shi et al. [46] electropolymerized thiophene in very strongly acidic boron trifluoride/ethylene ether (BFEE) solution, and obtained high quality PTh films with strong mechanical properties. The polymerization potential of thiophene dropped to ca. 1 V versus SCE in the BFEE solution. To improve further the ionic conductivity of the BFEE solution, Li et al. [47] used a mixed solution of BFEE and acetonitrile. By adding 10–20 % acetonitrile to the 0.1 mol/L Bu_4PF_6 BFEE electrolyte solution, the oxidative polymerization potential is decreased to 1.2–1.3 V versus SCE, and a PTh film with conductivity of ca. 120 S/cm was prepared by such electropolymerization.

The electropolymerization mechanism of thiophene should be very similar to that of the pyrrole electropolymerization [33].

2.8.2 Chemical Polymerization of Conducting Polymers

The chemical oxidation preparation of conducting polymers is performed in solution by using oxidants such as $FeCl_3$ and $(NH_4)_2S_2O_8$, etc., and it is easy to enlarge the production scale with the chemical polymerization. Chemical polymerization is the most important method for the preparation of PAn, and it can also be used to produce conducting PPy.

2.8.2.1 Chemical Preparation of PAn

Similar to the electropolymerization of aniline, the chemical polymerization of aniline also has to be performed in a strongly acidic aqueous solution. A typical chemical polymerization method for the preparation of PAn in laboratory was reported by MacDiarmid et al. in 1986 [48]. They performed the chemical polymerization in 1 mol/L HCl solution by using $(NH_4)_2S_2O_8$ as oxidant. The detailed preparation processes are as follows: 2 mL (0.022 mol) aniline was dissolved in 120 mL, 1 mol/L HCl solution, and the solution was cooled to 5 °C in an ice-water bath. At the same time, 0.025 mol $(NH_4)_2S_2O_8$ was added to 40 mL 1 mol/L HCl solution to prepare the $(NH_4)_2S_2O_8$ solution. Under vigorous stirring of the aniline HCl solution, the 40 mL $(NH_4)_2S_2O_8$ HCl solution was dropped into the aniline HCl solution drop by drop, then allowed to react for 8 h with vigorous stirring at 0 ° C. The precipitate was collected and dried to obtain the HCl doped conducting polyaniline (PAn-HCl) powder.

The PAn-HCl powder can be further treated by putting it into 0.1 mol/L NH_3 aqueous solution for 3 h under stirring to get dedoped emeraldine base PAn. The emeraldine base PAn can be dissolved in NMP, then the solution can be used to prepare PAn films. The emeraldine base PAn can also be used to prepare conducting PAn solution by appropriate proton-acid doping [9].

2.8.2.2 Chemical Preparation of PPy

PPy can also be prepared by chemical polymerization in aqueous solutions. However, the product is conducting PPy powder, different from the conducting PPy film prepared on electrodes by the electropolymerization mentioned above.

He et al. [49] prepared conducting PPy by the chemical polymerization of pyrrole with $FeCl_3$ as oxidant in an aqueous solution. By using 3 % poly(vinyl-alcohol) (PVA) additive in the solution of 0.18 mol/L dodecyl-benzenesulfonic acid (DBSA, used as dopant), 0.18 mol/L pyrrole, and 0.26 mol/L $FeCl_3$ (used as oxidant), conducting PPy powder with a conductivity of 43.18 S/cm (for the compressed pellets of the PPy powder) was obtained.

The chemical polymerization method can also be used to prepare conducting PPy nanotubes or nanofibrils by performing the chemical polymerization of pyrrole in polycarbonate template membranes [50].

2.9 Summary

Conjugated polymers possess a doped conducting state and a neutral semiconducting state, which lead to different applications. The doped conducting state could be found applications in the fields of electrode materials for batteries, electrochromics and super-capacitors, anti-static and anti-corrosion materials, electrolyte

Table 2.4 Properties and applications of some representative conjugated polymers

Conjugated polymers	Stable structure	Preparation method	Solubility	Application fields
Polypyrrole (PPy)	p-Doped conducting state	PPy films can be obtained by electrochemical oxidation polymerization from aqueous or organic solution. PPy powder can be prepared by chemical oxidation polymerization	Insoluble	Modified electrode, enzyme electrodes (biosensors), electrochromics, conducting polymer films
Polyaniline (PAn)	p-Doped conducting state (proton-acid doping)	Electrochemical or chemical oxidation polymerization from strong acidic aqueous solution	Insoluble for the PAn doped with common anions, but it can become soluble by counteranions doping induced solubility	Modified electrodes, enzyme electrodes (biosensors), electrochromics, electrode materials for batteries and solid capacitors, anti-corrosion, microwave absorption, electrode buffer layer for optoelectronic devices
Polythiophene (PTh)	Intrinsic semiconducting state	Electrochemical oxidation polymerization from organic solution, or chemical oxidation polymerization in organic solvent	Insoluble	Electrochromics, conducting polymer films
PEDOT:PSS	p-Type doped conducting state	Electrochemical oxidation polymerization from organic solution, chemical oxidation or chemical synthesis in organic solvent	Aqueous solution	Transparent conducting polymer films, anode buffer layer materials in organic/polymer light-emitting diodes and organic/polymer solar cells, anti-static-electricity coating layer materials, electrode materials in solid state capacitors, etc.
Poly(3-hexylthiophene) (P3HT)	Intrinsic semiconducting state	Chemical synthesis in organic solvents	Soluble in THF, chlorobenzene, dichlorobenzene, etc.	Donor material in polymer solar cells, semiconductors in field effect transistors
MEH-PPV and MDMO-PPV	Intrinsic semiconducting state	Chemical synthesis in organic solvents	Soluble in organic solvents	Orange electroluminescent material for polymer light-emitting diodes

capacitors, transparent electrodes, chemical and biosensors, etc. The neutral (intrinsic) semiconducting state makes these materials applicable in PLEDs, PSCs, polymer field effect transistors, etc. Table 2.4 summarizes the properties and applications of some representative conjugated polymers.

In comparison with inorganic semiconductors for optoelectronic applications, conjugated polymers possess the advantages of easy structural design, easy synthesis, good film-forming properties, and flexibility. However, there are some drawbacks to the conjugated polymers, such as poorer stability, lower charge carrier mobility, difficulty in forming ordered structure, etc. How to overcome the drawbacks by structural design and to realize large scale applications is the challenge for researchers in the field of conjugated polymers. At present, the hot research topics are the design and synthesis of highly stable conjugated polymers with high luminescent quantum efficiency, broad absorption, narrow bandgap, high charge carrier mobility, and suitable LUMO and HOMO energy levels for the applications in PSCs and PLEDs etc.

Since the discovery of conducting polyacetylene in 1977, the studies of conducting polymers have achieved great progress and have developed into a multidiscipline science. Conducting polymers including intrinsic semiconducting conjugated polymers are still a hot research fields in polymer science, materials science, and optoelectronic materials and devices. I am sure the studies on conducting polymers will make further great progress in the near future. Optoelectronic materials and devices will reach every corner of our lives.

References

1. Shirakawa H, Louis EL, MacDiarmid AG et al (1997) J Chem Soc Chem Commun 578
2. Chiang CK, Fincher CR Jr, Park YW et al (1977) Phys Rev Lett 39:1098
3. Burroughes JH, Bradley DDC, Brown AR et al (1990) Nature 347:539
4. Yu G, Gao J, Hummelen JC et al (1995) Science 270:1789
5. Li YF, Cao Y, Gao J et al (1999) Synth Met 99:243
6. Bredas JL, Street GB (1985) Acc Chem Res 18:309
7. Naarmann H, Theophilou N (1987) Synth Met 22:1
8. Epstein AJ (2001) In: Farchioni R, Grosso G (eds) Organic electronic materials—conjugated polymers and low molecular weight organic solids. Springer, Berlin, pp 3–37
9. Cao Y, Smith P, Heeger AJ (1992) Synth Met 48:91
10. Bredas JL, Heeger AJ (1994) Chem Phys Lett 217:507
11. Li YF, Qian RY (1993) J Electroanal Chem 362:267
12. Li YF, Qian RY et al (1994) Polym J 26:535
13. Li YF, Qian RY (2000) Electrochim Acta 45:1727
14. Li YF, Qian RY (1989) Synth Met 28:C127
15. Li YF, Qian RY (1988) Synth Met 26:139
16. Li YF, Liu ZF (1998) Synth Met 94:131
17. Li YF (1997) Electrochim Acta 42:203
18. Li YF, Yan BZ, Yang J, Cao Y, Qian RY (1988) Synth Met 25:79
19. Genies EM, Lapkowski M, Penneau JF (1988) J Electroanal Chem 249:97
20. Li YF, Qian RY (1993) Synth Met 53:149

21. Sun QJ, Wang HQ, Yang CH, Li YF (2003) J Mater Chem 13:800
22. Hou JH, Tan ZA, Yan Y et al (2006) J Am Chem Soc 128:4911
23. Diaz AF, Kanazawa KK, Gardini GP (1979) J Chem Soc Chem Commun 578
24. Qian RY, Qiu JJ (1987) Polym J 19:157
25. Warren LF, Anderson DP (1987) J Electrochem Soc 134:101
26. Ouyang JY, Li YF (1997) Polymer 38:1971
27. Wernet W, Monkenbusch M, Wegner G (1995) Mol Cryst Liq Cryst 118:193
28. Ouyang JY, Li YF (1997) Polymer 38:3997
29. Li YF, Ouyang JY (2000) Synth Met 113:23–28
30. Maddison DS, Unsworth J (1989) Synth Met 30:47
31. Genies EM, Bidan G, Diaz AF (1983) J Electroanal Chem 149:101
32. Qian RY, Pei QB, Huang ZT (1991) Makromol Chem 192:1263
33. Li YF (1997) J Electroanal Chem 433:181
34. Li YF, Fan YF (1996) Synth Met 79:225
35. Horanyi G, Inzelt G (1989) J Electroanal Chem 264:259
36. Desilvestro J, Scheifele W (1993) J Mater Chem 3:263
37. Waltman RJ, Bargon J, Diaz AF (1983) J Phys Chem 87:1459
38. Tourillon G, Garnier F (1984) J Electroanal Chem 161:51
39. Tourillon G, Garnier F (1982) J Electroanal Chem 135:173
40. Sato M, Tanaka S, Kaeriyama K, (1985) J Chem Soc Chem Commun 713
41. Fall M, Aaron JJ, Sakmeche N, Dieng MM, Jouini M, Aeiyach S, Lacroix JC, Lacaze PC (1998) Synth Met 93:175
42. Akoudad S, Roncali J (1998) Synth Met 93:111
43. Hu X, Wang GM, Wong TKS (1999) Synth Met 106:145
44. Bock A, Topeters A, Kryschi C (1995) Synth Met 75:133
45. Yumoto Y, Yoshimura S (1986) Synth Met 13:185
46. Shi GQ, Jin S, Xue G, Li C (1995) Science 276:994
47. Li XH, Li YF (2003) J Appl Polym Sci 90:940–946
48. Chiang JC, MacDiarmid AG (1986) Synth Met 13:193
49. He C, Yang CH, Li YF (2003) Synth Met 139:539–545
50. Martin CR (1996) Chem Mater 8:2382

Chapter 3
Organic Semiconductors for Field-Effect Transistors

Weifeng Zhang and Gui Yu

Abstract An important application of organic semiconductors is to fabricate organic field-effect transistors (OFETs) which are essential building blocks for the next generation of organic circuits. In terms of molecular size or molecular weight, organic semiconductors can be divided into small-molecule and polymer semiconductors, and thus their corresponding OFETs can also be categorized into organic small molecule OFETs and polymer field-effect transistors (PFETs). On the basis of the main charge carriers transporting in OFET channels, organic semiconductors can be further divided into p-type, n-type, and ambipolar semiconducting materials. According to the characteristic of the organic semiconductors, the OFETs can be classified into two types: organic thin film transistors (OTFTs) and organic single crystal transistors. In any kind of OFET devices, organic semiconductor materials are the core; their properties determine the performance of the electronic devices. Therefore, the design and synthesis of high performance organic semiconductor materials are the basis and premise of the wide application of OFET devices. In the past few decades, great progress has been made in developing organic semiconductors. Besides organic semiconducting materials, there are many other factors influencing the performance of OFETs including device configuration, processing technique, and other devices physical factors, etc. In the following, a brief review of the development of p-type, n-type, ambipolar organic semiconductors and their field-effect properties is given. The history, mechanism, configuration, and fabrication methods of OFET devices and main performance influencing factors of OFETs are also introduced.

Keywords Organic semiconductors · p-Type materials · n-Type materials · Ambipolar transport · Field-effect transistors · Mobilities

---

W. Zhang · G. Yu (✉)
Institute of Chemistry, Chinese Academy of Sciences, Beijing 100190, People's Republic of China
e-mail: yugui@iccas.ac.cn

Y. Li (ed.), *Organic Optoelectronic Materials*,
Lecture Notes in Chemistry 91, DOI 10.1007/978-3-319-16862-3_3

3.1 Introduction

3.1.1 Overview

Organic conjugated molecules are the basis of the design of new organic functional materials. These large π-conjugated systems provide chemists with plenty of inspiration for developing new functional materials. At the beginning of the twentieth century, McCoy and Moore predicted that "people are likely to produce organic metal conductors free of metal elements". In 1970, Wudl successfully synthesized the organic conjugated electron-donor tetrathiafulvalene (TTF). In 1973, Ferraris et al. found that TTF and organic conjugated electron-acceptor 7,7,8,8-tetracyanoquinodimethane (TCNQ) formed a charge-transfer complex with high conductive properties, then inaugurated a new era of organic conductors and superconductors. Organic semiconductors are a new and important extension of the organic conjugated molecules research field, and their excellent optical, electrical, and magnetic properties have received close attention in both academia and industry in recent years.

An important application of organic semiconductors is to fabricate organic field-effect transistors (OFETs) which are essential building blocks for the next generation of organic circuits. OFETs have wide potential applications, such as radio frequency identification tags, flexible displays, electronic paper, electronic skin and sensors, and so forth [1]. Compared with traditional silicon-based materials, organic semiconductors have attracted particular attention because of their unique designability of structure and tunability of properties, light weight, and ability to be flexible and transparent [4–6]. According to the characteristics of organic semiconductors, OFETs can be subdivided into two types: organic thin film transistors (OTFTs) fabricated by thin films of organic semiconductors and organic single crystal transistors fabricated by single crystal organic semiconductors. OTFTs can be fabricated over a large area at low cost. This advantage is the reason behind the rapid progress of OFETs based on thin film technology [7, 8]. However, because their characteristics are often strongly affected by imperfect thin film structure and insufficient purity of organic materials, thin film transistors commonly exhibit an exponential decrease of the mobility of field-effect charge carriers. Therefore, thin film transistors cannot reflect intrinsic electronic properties of organic semiconductors [9, 10]. On the other hand, because of their perfect molecular arrangements, free of grain boundaries and minimized charge traps, single crystal transistors tend to display higher performances than those of thin films counterparts, and are suitable to be used as a tool to study intrinsic electronic properties of organic semiconductors and the physical limitations in the performance of OFETs. Hence, well-ordered, continuous films, especially crystalline films, are being seriously investigated for high performance OFET applications.

In terms of molecular size or weight, organic semiconductors can be subdivided into small-molecule semiconductors and polymer semiconductors, and thus their corresponding OFETs can also be divided into organic small molecule OFETs and polymer field-effect transistors (PFETs). On the basis of the main charge carriers transporting in OFET channels, organic semiconductors can be further divided into p-type, n-type, and ambipolar semiconducting materials. The three kinds of organic semiconductors have made great progress in the past decade. However, in comparison to p-type organic semiconducting materials, n-type, and bipolar counterparts are still rare.

3.1.2 History and Work Principle of OFETs

3.1.2.1 Brief History

In 1930, Lilienfeld [11] first put forward the principle of the field-effect transistor (FET). In the patent he proposed that a field-effect transistor behaves as a capacitor with a conducting channel between a source and a drain electrode. When voltage was applied to the gate electrode, the amount of charge carriers flowing through the system could be controlled. In 1960, Kahng and Atalla designed and prepared the first field-effect transistor using a metal-oxide-semiconductor [12]. Afterwards, field-effect transistors based on inorganic semiconductors had been expensively investigated and introduced the world to the age of silicon-based semiconductors. However, with the rising costs of materials and manufacturing, as well as public interest in more environmentally friendly electronics materials, scientists also began the discovering work of organic semiconductor-based electronics in the following years. In 1982, Ebisawa et al. [13] prepared a capacitor using organic semiconductors . The device was fabricated by using polyacetylene as the semiconductor, polysiloxane as dielectric, aluminum as gate, and gold as source and drain electrodes. Though the device only showed a few percent current modulations when it worked, the potential of thin film transistors was recognized. In 1986, Tsumura et al. [14] developed the first OFETs utilizing an insoluble film of an organic macromolecule, polythiophene, as a semiconductor. In 1988, a soluble form of polythiophene was developed by Jen et al. and applied to fabricate OFETs by Assadi et al. [15]. The report ignited excitement about the possibility of printable semiconductor systems which could be made with the same economies of scale as printed paper media.

In recent years, lots of soluble small-molecule and polymer semiconductors have been developed. The studies related to electrodes and insulators also continue to be developed and refined. Moreover, better fabrication techniques have contributed to improved device performance and reduced deposition cost and time. In addition, continuing effort has led to an improved understanding of the relationship between material structures and their charge transport properties. Nowadays, OFETs stand at another new starting point. Remarkable progress has been made. For example, some polymer-based and small-molecule-based devices both exhibited mobilities over 10 $cm^2 V^{-1} s^{-1}$, which can be competitive with amorphous silicon semiconductors. However, the overall development of OFETs still lags behind the demand in the organic electronics field. The inadequacy mainly affects device stability and replicability as well as mobility.

3.1.2.2 Work Principle of OFETs

An OFET can be regarded as a plate capacitor consisting of a gate electrode and organic semiconducting layer. The device is operated as follows. When a gate voltage is pulsed, the carriers accumulate near the dielectric layer/organic layer interface and form a conductive channel. Carriers are then injected from the source electrode into the organic layer and transport takes place through the conductive channel to the drain electrode. The work principle of OFETs is illustrated in more detail in Fig. 3.1:

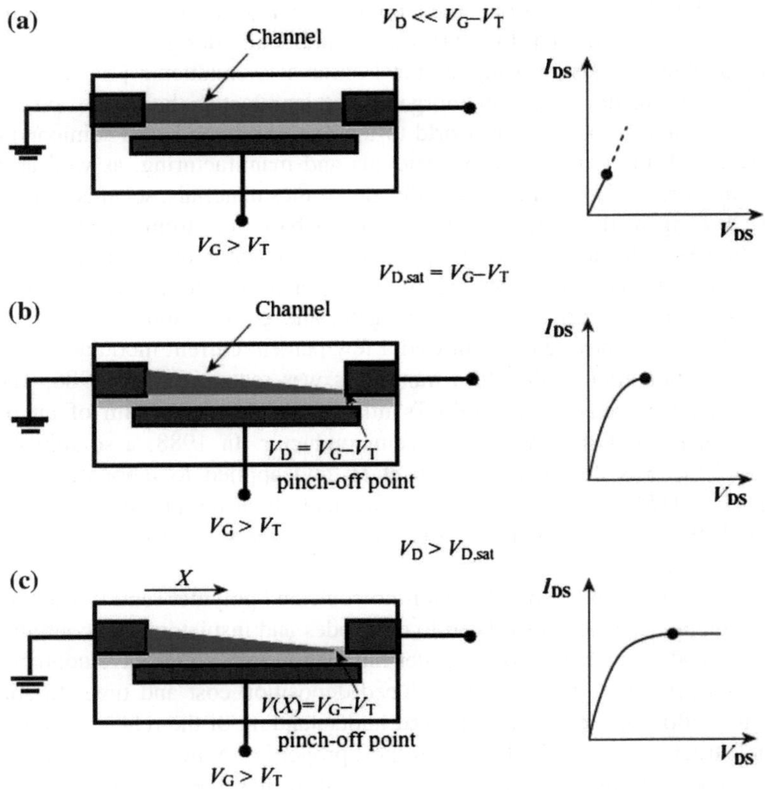

Fig. 3.1 Schematic work principles of OFETs: **a** linear regime; **b** start of saturation regime at pinch-off; **c** saturation regime and corresponding current–voltage characteristics. Reprinted with permission from [16]; © 2007, American Chemical Society

1. Linear regime. When no source–drain bias is applied, the charge carrier concentration in the transistor channel is uniform. When a small source–drain voltage ($V_D \ll V_G - V_T$) is applied, the charge concentration becomes distributed as a linear gradient in the channel, in which the current flowing through the channel is directly proportional to V_D (Fig. 3.1a).
2. Pinched off. As the source–drain voltage is further increased, a point $V_D = V_G - V_T$ is reached, at which the channel is "pinched off." There is no longer any potential difference between the gate and the drain electrode in the conductive channel near the drain electrode; this point is regarded as "pinched off" voltage. A depletion area is also formed near the drain electrode, in which there are no charge carriers left (Fig. 3.1b).
3. Saturation regime. Further increasing the source–drain voltage does not substantially increase the current but leads to an expansion of the depletion region and thus a slight shortening of the channel. Under this circumstance, the transistor operates in the saturation regime (Fig. 3.1c).

Different to inorganic semiconductors, organic semiconductors often show a low conductivity, and consequently OFETs usually operate in accumulation mode. In the case of *p*-channel OFETs, holes should be accumulated in the conductive channel by injecting holes from the source electrode into the semiconductor, which requires the semiconductor molecules to be stable as cations [17]. In contrast, in the case of *n*-channel OFETs, the electrons should be accumulated in the conducting channel, and the semiconductor molecules should be stable as anions. Hence, *p*-channel OFETs operate with negative gate voltages, whereas *n*-channel transistors operate with positive gate voltages (see Fig. 3.2). In fact, an organic semiconductor can transport holes and electrons in different degrees at the same time; however, it is difficult to measure both hole and electron transport properties most times. Moreover, charge transport may be heavily affected by other factors such as work function of the source–drain electrode, the interface between gate insulator and semiconducting layer, device configuration, etc., one organic semiconductor material affording different behaviors in different conditions. For example, pentene is a common p-type material; however, n-type [18] and even ambipolar characteristics [19] can be observed in FET devices with different configurations. Thus, deciding one semiconductor is p-type, n-type, or ambipolar should only depend on the property of the FET device.

3.1.2.3 Performance Evaluation System of OFETs

The electric properties of OFETs are generally gauged by two characteristics and three parameters. The two characteristics are (1) the output characteristics, $I_{DS} - V_D$ and (2) the transfer characteristics, $I_{DS} - V_G$ (see Fig. 3.3). The three parameters are:

1. Charge transport mobility, μ, which is the drift velocity of carriers under unit electric field. Depending on the nature of the semiconductor and electrodes

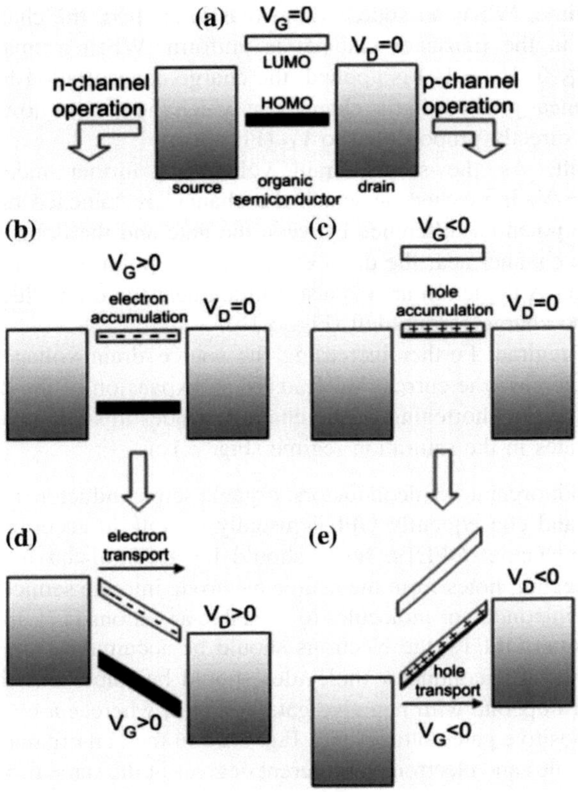

Fig. 3.2 Ideal energy level diagram of an OFET. **a** When no voltage is applied; **b–e** when operated in electron accumulation (**b**, **d**) and hole accumulation (**c**, **e**) mode. Reprinted with permission from [17]; American Chemical Society

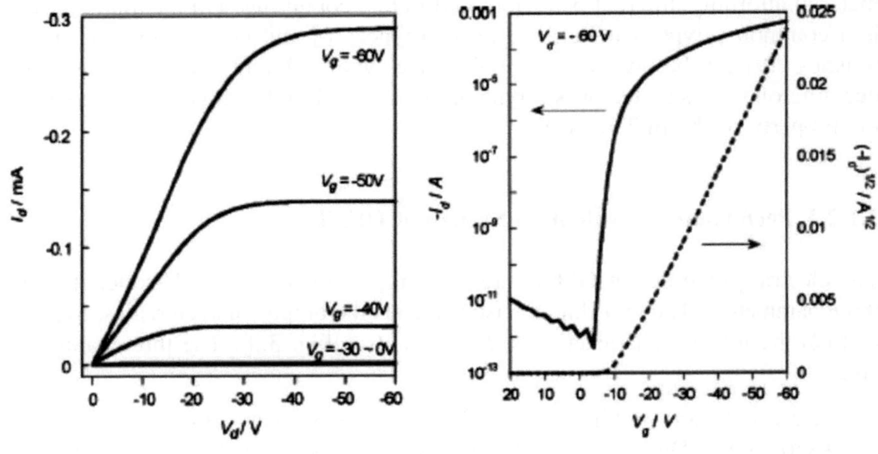

Fig. 3.3 Representative FET characteristics of p-type OFET: output characteristics (*left*) and transfer characteristics (*right*). Reprinted with permission from [20]; American Chemical Society

used, the channel formed can be p-type; μ_h is used when holes (electron deficient species) are the charge carriers, or n-type, and μ_e is used where electrons function as carriers. Charge transport mobility can be subdivided into μ of linear regime and μ of saturation regime which is very often used. The two kinds of charge transport mobilities are calculated by the following equations.
For linear regime:

$$I_{DS} = \frac{WC_i}{L}\mu(V_G - V_T)V_D$$

For saturation regime:

$$I_{DS} = \frac{WC_i}{2L}\mu(V_G - V_T)^2$$

where W is channel width, L is channel length, C_i is capacitance of the insulator, μ is field-effect mobility, V_G is gate voltage, and V_T is threshold voltage.

2. On/off current ratio, I_{on}/I_{off}, the ratio of the maximum on state current to the minimum off-state current.
3. Threshold voltage, V_T, the minimum gate voltage required to turn on the transistor.

A representative output and transfer characteristic of p-type OFET is shown in Fig. 3.3 [20]. The ultimate goal for organic semiconductors should be application in electronics, and thus the stability and replicability of OFETs are elements of the performance evaluation system. In a certain sense, these two elements are the most important.

3.1.3 Device Configuration and Processing Technique of OFETs

3.1.3.1 Device Configuration

The device configuration of OFETs can be subdivided into four different types as shown in Fig. 3.4 [21]: bottom gate/top contact (BGTC—Fig. 3.4a); bottom gate/bottom contact (BGBC—Fig. 3.4b); top gate/bottom contact (TGBC—Fig. 3.4c); and top gate/top contact (TGTC—Fig. 3.4d). The device configuration has an important influence on the performance of OFETs. For thin film transistors, bottom contact devices typically exhibit less than half the effective drive current of top contact devices, although bottom contact devices are more easily integrated into low-cost manufacturing processes, and smaller device feature sizes can be obtained through photolithographic techniques. The device configurations also have as great an influence on single crystal transistors as on thin film transistors. Moreover, besides affording minimal contact resistance, the device configuration adopted also needs to avoid the adverse effects brought about by subsequent deposition on single crystals because of their low thermal stability.

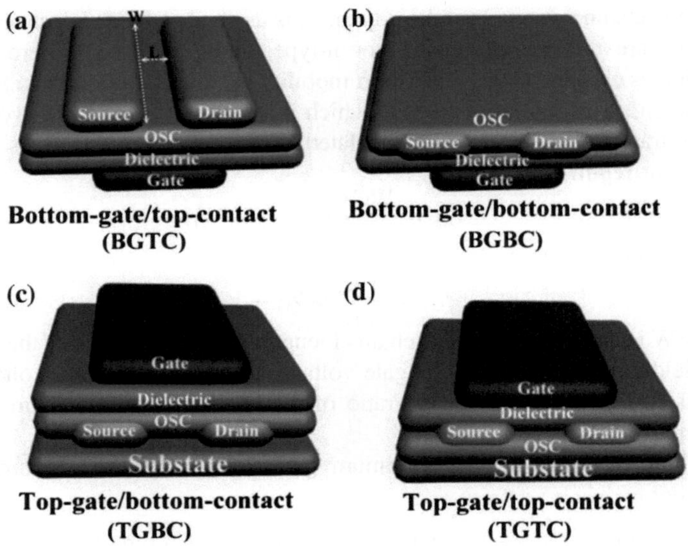

Fig. 3.4 Four types of OFETs device configuration (OSC: Organic Semiconductor). **a** Bottom gate/top contact (BGTC). **b** Bottom gate/bottom contact (BGBC). **c** Top gate/bottom contact (TGBC). **d** Top gate/top contact (TGTC).

3.1.3.2 Processing Technique for OFETs

Processing Technique for Thin Film OFETs

A wide diversity of methods has been adopted to prepare organic thin films for the fabrication of OFET devices since the first OFET was reported [14, 15]. The vacuum deposition technique may result in high-quality thin films, and has been widely adopted for the preparation of small-molecule semiconductor OFETs. Because of its freedom from solvents, the vacuum deposition method usually results in high-quality thin films. To avoid decomposition and degradation of small-molecule semiconductors during the sublimation process, a high-vacuum environment, typically lower than 10^{-4} Pa, should be adopted. In addition, deposition conditions, including sublimation rate, substrate temperature, etc., have an important influence on thin film morphology. Solution-processed techniques are widely used to deposit organic thin films mainly because of the low cost of such processes and the facility of obtaining large-area films by printing techniques. Moreover, when either decomposition or degradation happens during the sublimation process, it is necessary to adopt solution-processed techniques. Many elements such as solvent, concentration, evaporation temperature and rate, and substrate properties as well as the organic semiconductor itself can also affect the quality of the thin films. Solution-processed techniques for the preparation of organic thin films include spin-coating, dip-coating, drop-casting, zone-casting, the Langmuir–Blodgett (LB) technique, printing, and solution-shearing. Of these, spin-coating is the most commonly used solution-processed technique. The technique can provide uniform thin films of both the polymers and small molecules. The rotation speed adopted has a great influence on the thickness and morphology of thin films. Dip-coating, drop-casting, and zone-casting are three very convenient methods without the requirement of complex and/or expensive equipment. The techniques relate to self-assembled processes and can afford high orientation thin films. The LB technique is mainly used for amphiphilic molecules. Among the diverse methods used to fabricate organic thin films, the most exciting is the printing technique. Among the types of printing technique available, microcontact printing and ink-jet printing are considered to show the most promise. Solution-shearing is a new technique for organic semiconductors in which lattice strain is used to increase charge carrier mobilities by introducing greater electron orbital overlap between organic molecules. Using solution processing to modify molecular packing through lattice strain should aid the development of high-performance, low-cost organic semiconductor devices [22].

Processing Technique for Single Crystal OFETs

The growth methods of single crystals can be subdivided into the physical vapor transport technique (PVT) [23] and the solution-processed technique. The PVT can be used to provide high-quality organic single crystals. The equipment is very similar to the tube furnace, which requires a carrier gas to achieve an improved removal of impurities. Solution-processed techniques are also used to prepare organic single crystals. The solubility and intermolecular interactions are internal factors that determine the crystal growth, the solvent, the concentration, the deposition temperature, and the atmosphere being the major external factors. The solution-processed techniques could also be subdivided into the single solvent system and the solvent-exchange system. Of these, the solvent-exchange system involves two solvents, in one of which the organic semiconductor material is highly soluble and in the other almost insoluble. In the next step, the single crystals need to be deposited to obtain the corresponding transistors. There are also several kinds of techniques, which are different to those of their thin film counterparts. Of these, the electrostatic-bonding technique is where organic single crystals are placed onto pre-deposited electrodes to form a contact via an electrostatic force [24]. The technique could eliminate any potential damage to crystals during device fabrication, and is also convenient when one needs to change the position of the crystals in order to investigate the anisotropic properties. Nonetheless, electrostatic bonding always leads to a poor contact between the crystal and the drain/source electrodes. The drop-casting technique is where the single crystals grow directly onto pre-patterned drain/source electrode pairs [25]. Because the technique can improve contact between the semiconductor and the drain/source electrodes, it is widely used for the deposition of organic thin films and single crystals. In order to avoid the adverse effects brought about by subsequent deposition on single crystals, the BGBC configuration is usually adopted in FET devices. The deposition parylene dielectric technique uses parylene as the dielectric layer, avoiding either thermal- or solvent-based damage to the organic single crystals [26]. Upon this, the device configuration of top gate could be used for organic single crystals transistors. Shadow mask techniques using a copper grid, gold wire, fiber, or organic ribbon as a shadow mask are widely adopted to deposit drain/source electrodes of crystal FET devices with BGTC configuration [27, 28]. This technique could exhibit a better contact between the organic single crystals and the drain/source electrodes, and between the crystals and the dielectric layer. Moreover, the technique makes it possible to investigate the anisotropy on an individual single crystal because it permits significant scaling down of the channel length in order to adapt the crystal size and to form a good contact. The gold layer glue technique involves placing a gold film directly onto the single crystals as drain/source electrodes [27]. This can eliminate the thermal irradiation and maintain good contact of the top contact mode. When using this technique, asymmetric electrodes can be obtained with relative ease using only two different electrode films.

3.1.4 Factors Influencing the Performance of OFETs

3.1.4.1 Factors Related to Semiconducting π-Conjugated Systems

There are many factors influencing the FET properties of organic semiconductors; however, there is no doubt that the most important factor is the semiconductor itself. All factors influencing the charge carrier injection and transport related to organic semiconductors are discussed in the following content.

HOMO and LUMO Energy Level

The whole motion process of charge carriers in the semiconduting material involves two stages including charge injection and charge transport. The energy levels of the highest occupied molecular orbitals (HOMOs) and the lowest unoccupied molecular orbitals (LUMOs) have a large influence on charge carrier injection. In theory, all organic semiconductors should be able to conduct both holes and electrons, but the differences in internal reorganization energies or work functions of the electrodes relative to the HOMO and LUMO energy levels of the semiconducting material in the transistors can favor one type of charge transport [29]. At present, the most commonly used metal for source and drain electrodes in OFETs is gold, with a high work function of around 5.1 eV. Some low work function metals such as calcium, magnesium, or aluminum are also used to help electron injection. However, these metals are not environmentally stable. Thus, in order to obtain effective charge injection, p-type, n-type, and ambipolar semiconductors should have suitable HOMO, LUMO, and HOMO/LUMO energy levels to match with the high work functions of source and drain electrodes, respectively. This is because the mismatch between the HOMO or LUMO energy level and work function of the electrodes may cause the reduction of the measured mobility, though not affect the intrinsic mobility directly. The performance of OFET devices depends largely on the efficiency of the charge-transport processes; at the microscopic level, one of the major parameters governing the transport properties is the amplitude of the electronic transfer integrals between adjacent organic semiconducting molecules. Transfer integral means the splitting of the frontier molecular orbital between adjacent molecules (the HOMO and LUMO energy levels) and is extremely sensitive to the molecular packing (which is discussed in Section "Packing Mode in Solid State"). There is no doubt that increasing the transfer integral helps to increase the mobility. In other words, large splitting of the HOMO/LUMO energy levels tends to produce the higher hole/electron mobility [6, 30].

In addition, the HOMO and LUMO energy levels also have an important influence on the device stability. Nowadays, there are fewer accounts of n-type than p-type organic semiconductors, primarily because of the inherent instability of organic anions in the presence of air and water [8, 31], and problems with oxygen trapping within these materials [32, 33]. In most cases, the mobilities of the n-type OFETs can be one or even several orders of magnitude higher when taking precautionary measures to exclude atmospheric oxygen and water in vacuum or inert atmosphere. Thus, in order to fabricate OFETs with high performance, the organic semiconductors used should have suitable HOMO or LUMO energy levels. Specifically, when gold is the most commonly used metal for source and drain electrodes in OFETs, the HOMO energy level of p-type semiconductors should locate around -5.1 ± 0.3 eV, and the LUMO energy level of n-type semiconductors should locate close to -4.0 eV [6].

Packing Mode in Solid State

The packing mode of organic semiconductors in the solid state also have an important influence on their FET properties. Besides the transfer integral, reorganization energy can also affect intrinsic mobility of organic semiconductors [34]. The reorganization energy is the energy loss when a charge carrier passes through a molecule. The reorganization energy must be sufficiently low to facilitate charge transport. Reorganization energy is also extremely dependent on the the packing mode of the organic molecules in combination with the conjugation length, degree, and packing of the organic molecules [30]. A highly dense and ordered molecule packing motif is beneficial to get a small reorganization energy. A large π-conjugated system also exhibits low reorganization energy. The reasons for this are that a larger π-conjugated system can stabilize the +1 cation, and larger π-conjugated systems with electron-withdrawing groups can stabilize the −1 anion.

The packing mode of organic small-molecule semiconductors in the solid state can be divided into four types as shown in Fig. 3.5: (1) herringbone packing without π–π overlap between adjacent molecules; (2) slipped π-stacking between

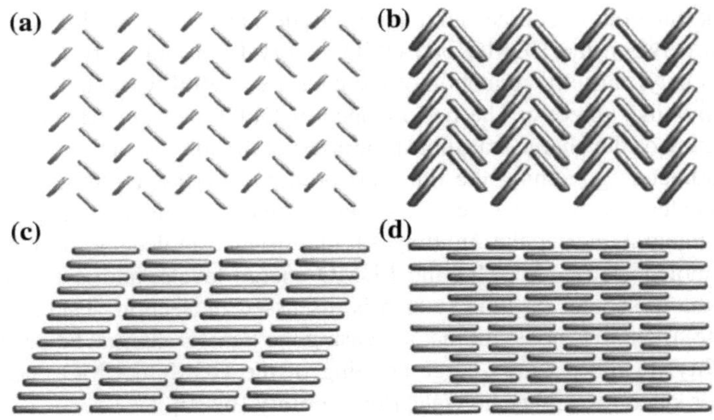

Fig. 3.5 Molecular packing motifs in crystals. **a** Herringbone packing without π–π overlap. **b** Slipped π-stacking between adjacent molecules. **c** One-dimensional lamellar packing. **d** Two-dimensional lamellar packing [38]. Reprinted with permission from [6]. Copyright 2012, The American Chemical Society

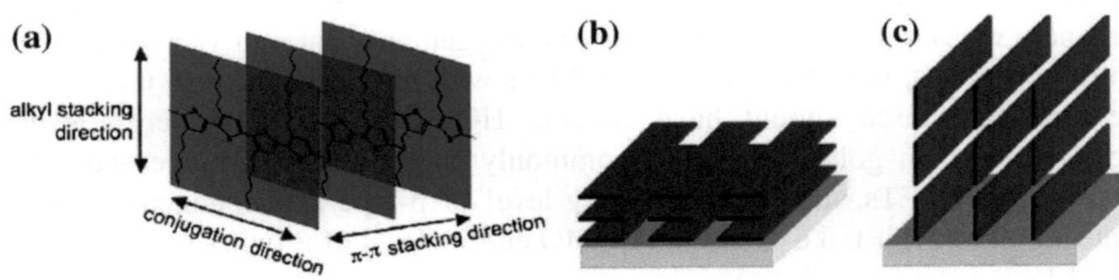

Fig. 3.6 **a** Possible charge transport mechanisms in crystalline polymer films (using P3HT for illustration): intrachain transport, along the π-conjugation direction, interchain transport, along the π-stacking direction and alkyl stacking direction. **b** Face-on and **c** edge-on orientation of the polymer molecules on the substrates [39]. Reprinted with permission from [6]. Copyright 2012, The American Chemical Society

adjacent molecules; (3) one-dimensional lamellar packing, and (4) two-dimensional lamellar packing. Of the four kinds of packing modes, one-dimensional and two-dimensional lamellar packing mean main one-dimensional and two-dimensional charge carrier hopping pathways, respectively. Of these, the two-dimensional charge carrier hopping pathway is believed to be the most efficient for charge transport because it can increase the transfer integrals to the maximum and transport the charge carriers through the shortest route [35–37]. For polymer semiconductors, the packing modes include face-on and edge-on orientation of the polymer molecules on the substrates (Fig. 3.6). The possible charge transport pathways of polymer films can be subdivided into intrachain transport, along the π-conjugation direction, interchain transport, along the π-stacking direction or alkyl stacking direction. Although high performance FET devices usually adopt an edge-on orientation, face-on orientation can also afford high performance.

Size/Molecular Weight Related to Polymers

More recently, the influence of molecular weight on the electronic properties of donor–acceptor conjugated copolymers has been examined, with field-effect mobilities improving significantly with increasing molecular weight [40]. For example, the thin film transistors based on (poly[(4,4-dihexadecylcyclo-penta-[2,1-*b*:3,4-*b*′]dithiophene)-2,6-diyl-*alt*-(2,1,3-benzothiadiazole)-4,7-diyl] with molecular weight of 11, 16, 25, and 35 kg mol^{-1} exhibited hole mobilities of 0.28, 0.59, 1.2, and 3.3 $cm^2\ V^{-1}\ s^{-1}$, respectively, displaying a nearly linear increase in transistor performance with increasing molecular weight [41]. The increased performances are accompanied by important changes in the thin film morphology: systems with high molecular weight often exhibit reduced crystallinity and more isotropic films, potentially leading to larger values for charge-carrier mobility [42]. Many operations, including optimizing organic and organometallic reaction conditions, have been used to achieve high molecular weights of polymers [43–45].

Impurities and Purification of Organic Semiconductors

Impurities here refer to compounds that have a different chemical structure than the compound nominally under investigation and which appear in small concentrations mainly as side products of the chemical synthesis. The presence of impurities can have a serious negative impact on the function of organic semiconductors because impurities can introduce charge carriers or traps within the material, leading to erroneous results [4]. Thus, from a synthetic perspective, preparative routes that minimize difficult-to-remove by-products are growing in importance [46]. In order to remove the impurities or side products effectively, a number of techniques have been used in the synthetic processing of organic semiconductors. The common techniques are as follows:

1. Recrystallization: widely used in purification of organic semiconductor samples. The solubility difference between the target compound and impurities in typical solvents is utilized to eliminate gradually any impurities. This procedure can be carried out using hot filtration or solvent evaporation at room temperature. However, for semiconductor samples showing a minimal association between temperature and solubility, a kind of "multi-solvent recrystallization" precipitation technique can be adopted, which purifies by slow diffusion of a poor solvent into a saturated solution of a target compound in a good solvent.
2. Column chromatography and gel permeation chromatography: for soluble organic small-molecule semiconductors, column chromatography techniques could be used, which make use of the different polarities of the target compound and the by-products. Silica gel- and alumina-based column chromatography provides an efficient means of purifying an individual compound from a complicated mixture. Different eluents and stationary phases are chosen to cooperate in the separation of different mixtures. With the advancing eluent, individual

components elute at different rates, and thus are separated from each other. For soluble polymer semiconductors, the gel permeation chromatography technique could be used, which makes use of the different molecular size or molecular weight of the target polymer and the by-products. The technique is necessary to separate polymers, both to analyze them and to purify the desired product.

3. Vacuum sublimation: for insoluble organic semiconductor samples, vacuum sublimation technique could be used, which makes use of the different sublimation points of individual compounds in vacuum. In most cases, temperature-gradient sublimation is adopted. Sometimes, an inert carrier gas is used to facilitate the procedure, as well as to protect the deposition from atmospheric impurities.
4. Soxhlet extraction: often used in purification of polymeric semiconductors to remove low-molecular-weight fractions of the material and residual catalytic metal. Normally, a polymer sample is placed inside a thimble made from thick filter paper, which is loaded into the main chamber of the Soxhlet extractor. A warm poor extraction solvent, such as hexane, acetone, methanol, chloroform, etc., is first used to remove the undesired impurities, then a hot good solvent, for example, chloroform, chlorobenzene, dichlorobenzene, etc., is used to extract the desired high molecular weight and soluble polymer. This procedure involves the repeated dissolution of a particular compound from a mixture, which may be repeated many times, over hours or days. After the solvent is removed, the desired polymers is obtained by precipitating the resulting solid residue in certain solvent(s) by filtering.

3.1.4.2 Factors Related to Device Physics

Device Configuration

The device configuration has an important influence on the performance of OFETs. Generally, the less the contact resistance, caused by intimate contact between the semiconductor and the electrodes, the higher the charge transport mobility tends to be [46]. The device configurations of BGTC and TGBC always give better performances than those of BGBC and TGTC. The better performances were attributed to the improved contact between the organic semiconducting layer and the electrodes [21]. Bottom contact devices typically exhibit less than half the effective driving current of top contact devices because of contact resistance and the difficulty in preparing highly ordered films on an irregular surface [8, 47, 48]. Because of shadowing effects, top contact has a limit as to how small the channel dimensions can be, and this process is not readily amenable to large-scale manufacturing; however, bottom contact devices are more easily integrated into low-cost manufacturing processes, and smaller device feature sizes can be obtained through photolithographic techniques. Bottom gate devices are a conventional device structure for material-testing purposes. Because the organic semiconducting layers are exposed to air and easily penetrated by oxygen and moisture, the performances of this kind of device often degrade during storage. In contrast, in top gate devices, the active layer is encapsulated by the gate dielectric layer and the gate electrode, and therefore the device stability is significantly improved.

Morphology

Morphology is a crucial factor in achieving high performance FETs. Specifically, morphology of high order, continuous thin film with lamellar packing is very important to obtain high mobility and stability. The reason is that the morphology leads to better molecular orbital overlap between neighboring molecules, and favors better carrier transfer; and the morphology can also inhibit the charge carrier trapping under ambient conditions by H_2O or O_2, and/or impurities of the atmosphere. The π-stacking distances of lamellar packing are usually in the range of 3.3–3.6 Å for small-molecule semiconductors and 3.6–4.0 Å for polymer semiconductors. There are many methods adopted to improve the morphology. For example, as mentioned above, sublimation rate and substrate temperature have an important influence on thin film morphology of small molecules. Specifically, higher substrate temperature and/or low sublimation favors the formation of large grains. Thus, in order to obtain high quality thin film, varying substrate temperature or sublimation rate are always used in a single deposition process. Sometimes a low sublimation rate is used first to form large grains near the interface and subsequently a high sublimation rate is used to fill the interspaces between grains, leading to high-quality thin films [49]. Occasionally, a higher substrate temperature is used first to obtain a large grain size close to the interface, followed by a lower substrate temperature used to fill the interspaces [50]. Annealing is another important tool in improving the thin film morphology of both small-molecule and polymer semiconductors. Annealing, naturally, is a self-assembly process of organic molecules in thin film. After annealing the semiconducting layer at selected temperatures, the performance of OFETs tends to be substantially improved. The reason is that the morphology/molecular packing order in thin films becomes more ordered. However, once beyond an optimal temperature, the mobility is found to decrease owing to discontinuities in the film created by an increase in intergranular spaces when the grain size becomes very large [51]. The contact between the organic semiconductor and the gate insulator are both improved in the process of annealing.

Interfaces and Their Modification

The two major processes, carrier injection and carrier transport, occur at the electrode/organic layer interface and the dielectric/organic layer interface, respectively. Therefore, the properties of these interfaces influence the device characteristics dramatically. Modification of the electrode/organic layer or dielectric/organic layer interfaces remains the most widely investigated approach to improving device performance [21].

The electrode/organic layer interface has a key influence on carrier injection. Many methods are used to modify the electrode/organic interface to improve the carrier injection. Introduction of a buffer layer between the source–drain electrode and organic layer is a common approach for modification of BGTC OFETs. The technique is frequently applied with the aim of both reducing the energy barrier and preventing metal atom penetration into organic layers. For example, when a very thin layer of MoO_3 is inserted as a buffer layer between an Al electrode and pentacene, the injection barrier of pentacene-based OFETs reduces dramatically, and thus the hole mobility increases from 2.8×10^{-3} to 0.4 $cm^2 V^{-1} s^{-1}$ [52]. In addition, the formation of a Cu_XO buffer layer enabled Cu to be used as an electrode in high performance OFETs. Chemical modification of the electrode is another effective way to improve OFET performance. For example, Cu and Ag source–drain electrodes were chemically modified with 7,7,8,8-tetracyanoquinodimethane (TCNQ), when the formation of Cu-TCNQ and Ag-TCNQ reduced the hole injection barrier and improved electrodes/organic layer contact, which reduced contact resistances. Taking pentacene-based OFETs with Ag-TCNQ modified electrodes as an example, the mobility increased from 0.02 to 0.18 $cm^2 V^{-1} s^{-1}$ [53]. Gundlach et al. [54] demonstrated induced crystallization of an organic layer by electrode interface modification. By using pentafluorobenzene thiol (PFBT)-modified gold electrodes, 5,11-bis(triethylsilylethynyl) anthradithiophene is induced to grow with large grain domains near the source–drain electrodes, indicating that optimization of the source–drain electrodes is an alternative way to improve carrier transport.

The dielectric/organic semiconductor interface implies an important influence on device stability in three ways [21]. First, trap density on the dielectric layer surface affects the device performance dramatically, especially for n-type OFETs. Second, the dielectric layer can influence the morphology of the organic semiconductor layers which in turn affect the device stability. The third way that the dielectric/organic semiconductor interface influences the stability is its influence on the aggregation of organic grains. For example, *N,N*-dioctyl-3,4,9,10-perylene tetracarboxylic diimide (PTCDI-C8) is generally believed to be unstable in air. Modification of the SiO_2 surface with hydroxyl-free polymer insulators, such as poly(methyl methacrylate) (PMMA), afforded significant improvements in device stability [55]. For another example, it has been observed that the stability of pentacene-based OFETs is also strongly related to the dielectric layer [56]. For pentacene-based OFETs with octadecyltrichlorosilane (OTS)-modified dielectric layers, the surface energy of the modified surface is much lower than that of the pentacene layer. As a result, the pentacene layer readily aggregates, leading to rapid degradation of device performance. In contrast, the high surface energy of polystyrene (PS) impedes pentacene aggregation and ensures excellent device stability for the devices with PS-modified dielectrics. Thus, optimization of the dielectric/organic interface is an effective way to realize high performance OFETs with excellent stability.

There are some other factors affecting the performance of OFETs, for example channel length/width ratio, and other environmental factors such as light, dust, ambient temperature, humidity, oxygen, and so forth. In certain cases, each of such factors could play a decisive role in the performance degradation of OFETs.

3.2 p-Type Semiconductors

As mentioned above, for p-type semiconductors, the charge carriers, holes, migrate through HOMOs of the molecules. Because p-type semiconductors are always comprised of electron-rich groups/π-conjugated systems, which are reactive to electrophilic substitution reactions, the performance of most p-type semiconductor-based devices exhibits small changes in different test circumstances. Great progress has been achieved in p-type semiconductors, including p-type small-molecule semiconductors and p-type polymer ones in past years. Nowadays, because new materials and forms of known materials are being quickly synthesized and developed as processing and synthetic techniques improve, it is, to be frank, difficult to catalog all p-type semiconductors. In the following, selected p-type small-molecules and polymer semiconductors are discussed.

3.2.1 Selected p-Type Small-Molecule Semiconductors

3.2.1.1 Polycyclic Aromatic Hydrocarbons and Derivatives

Polycyclic aromatic hydrocarbons (PAHs) are an important class of small-molecule semiconductors. These materials are composed of fused benzene rings in a linearly or nonlinearly constructing manner. The π-conjugation of these materials increases with elongation and expansion of the molecular structures. Linear PAHs always form a planar and rigid molecular structure, resulting in strong intermolecular interactions and tight packing mode in the solid state, in which charge carrier transport is favored, but solubility is poor in common solvents. For that reason, many substituted linear PAH derivatives were developed in order to improve the solubility, reactant yields, and chemical stability, while maintaining an efficient charge transport. Nonlinear PAHs with large π-conjugated systems had also received attention. Nonetheless, because nonplanar structures could probably be formed in these kinds of materials, some of them afford low mobilities. In terms of their molecular structures, PAHs can be further subdivided into acenes, pyrene, perylene, and other fused aromatic hydrocarbons.

Acenes and Derivatives

The acenes refer to PAHs that are composed of linearly fused benzene rings (see Chart 3.1 and Table 3.1). Anthracene, **1** is one of the simplest acenes with three fused benzene rings. Its FET properties were examined with single crystal devices exhibiting temperature dependence with the maximum mobility of 0.02 $cm^2\ V^{-1}\ s^{-1}$ at ~170–180 K [57]. With the expansion of the π-conjugated dimension to four or five fused benzene rings, the intermolecular overlap of molecule orbitals is increased, and this results in larger transfer integral and lower reorganization energy, and thus higher field-effect mobilities in the corresponding OFETs. Thermally evaporated thin films of naphthacene, **2** showed a mobility of 0.1 $cm^2\ V^{-1}\ s^{-1}$ with an on/off current ratio over 10^6 [58]. Studies showed a high density of submicron-sized grains with a surprisingly high degree of molecular order in a thin film of acene, **2**. Its single crystal transistors were also fabricated by Reese et al. with the use of a spin-coated poly(dimethylsiloxane) as gate dielectric and photolithographically defined source and drain electrodes, and it exhibited a mobility of 2.4 $cm^2\ V^{-1}\ s^{-1}$ [59]. Pentacene, **3** with five fused benzene rings has developed into a benchmark material because high-performance thin film transistor devices are easily obtained from vacuum deposited thin films. Thin film transistors based on **3** demonstrated mobilities of up to 1.5 $cm^2\ V^{-1}\ s^{-1}$ with an on/off current ratio over 10^8 [60] whereas its polycrystalline films afforded mobility as high as 5.0 $cm^2\ V^{-1}\ s^{-1}$ with an on/off current ratio over 10^6 [61]. With thermally evaporated pentacene on flexible substrates, a higher mobility of up to 23.2 $cm^2\ V^{-1}\ s^{-1}$ was observed [62]. Moreover, with 6,13-pentacenequinone films as gate insulators

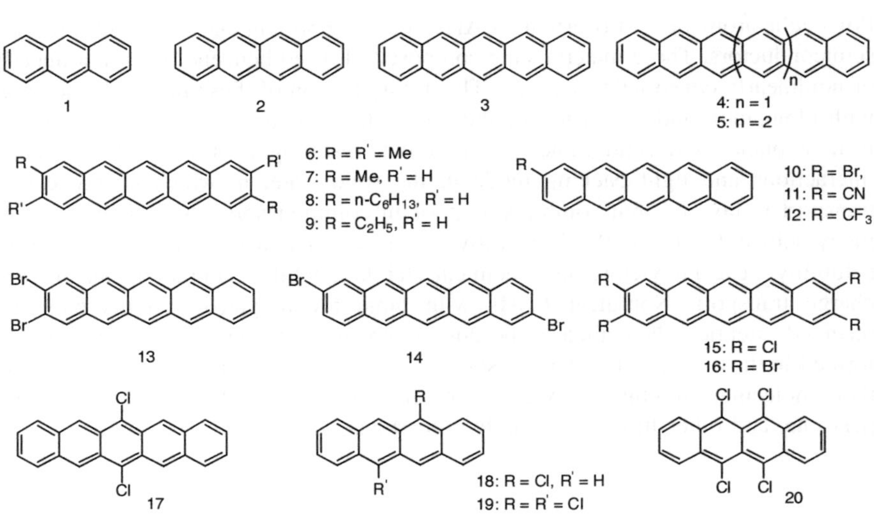

Chart 3.1 Linearly acene-based small-molecule semiconductors

Table 3.1 OFET device data for p-type small-molecule semiconductors

	HOMO (eV)	Deposition process	Max μ_h ($cm^2\ V^{-1}\ s^{-1}$) (m.d.)[a]	I_{on}/I_{off}	V_T (V)	Device structure[b]	Refs.
1		Single crystal	0.02 (vacuum)			BGBC; Au; SiO_2/Si; NBT; PDMS	[57]
2		Evaporation	0.12	10^7	−3	BGBC; Pd; SiO_2/Si; OTS-8	[58]
		Single crystal	2.4	$10^5 \sim 10^8$	$-40 \sim -10$	BGBC; Au/Ti; SiO_2/Si; PDMS	[59]
3		Evaporation	1.5	10^8	0	BGBC; Au; SiO_2/Si; ODTS	[60]
		Evaporation	23.2	3×10^4	−0.77	TGBC; Au; silk/Au	[62]
		Single crystal	40 (vacuum)	10^6		TGTC; Ag–epoxy; PQ/Ag–epoxy	[63]
6	4.41	Evaporation	0.30	6.3×10^3	12.5	BGBC; Au; SiO_2/Si; HMDS	[67]
7		Evaporation	2.5	2×10^6	−3.8	BGTC; Au; SiO_2/Al; PPHD	[68]
10	4.93	Evaporation	0.13	2.5×10^4	−11	BGBC; Au; SiO_2/Si; OTMS	[70]
11	5.05	Evaporation	0.032	5.5×10^3	−13		[70]
12	5.02	Evaporation	0.030	1.6×10^5	−27	BGBC; Au; SiO_2/Si; OTMS	[70]
13	−4.99	Evaporation	0.23	1.9×10^4	−8		[70]
17	−5.38	Evaporation	0.206 (air)	$10^4 \sim 10^5$	−1.15	BGTC; Au; SiO_2/Si; OTS-8	[72]
		Crystalline ribbons	9.0 (air)			BGTC; Au; SiO_2/Si	[73]
19		Single crystal	1.6	10^5		TGTC; graphite; parylene/graphite	[37]
20		Single crystal	1.7 (air)	$\sim 10^4$	−12	TGTC; graphite; parylene/graphite	[74]
22		Single crystal	15.4 (α-axe)	10^6		BGBC; Au/Ti; PDMS	[24]
			4.4 (β-axe)	10^6			
23		Evaporation	0.1			BGTC; Au;SiO_2/Si; OTS	[79]
26		Evaporation	0.4	10^6		BGBC; Au;SiO_2/Si; OTS	[81]
		Drop-casting	1.8	10^7		BGBC; Au; SiO_2/Si; PFBT; HMDS	[82]
		Single crystal	1.42	10^5	$-10 \sim -22$	BGTC; Au; SiO_2/Si	[83]
27	−5.19	Drop casting	2.5	10^7		BGTC; Au; SiO_2/Si; OTS	[84]
28	−5.39	Drop casting	4×10^{-4}			BGTC; Au;SiO_2/Si; OTS	[84]

(continued)

Table 3.1 (continued)

	HOMO (eV)	Deposition process	Max μ_h ($cm^2 V^{-1} s^{-1}$) (m.d.)[a]	I_{on}/I_{off}	V_T (V)	Device structure[b]	Refs.
29		Evaporation	0.033			BGTC; Au; Al_2O_3/Al	[85]
30		Single crystal	0.52	10^5		BGTC; Au; SiO_2/Si	[86]
31	−5.13	Evaporation	0.11	$>10^3$	−37	BGTC; Au; SiO_2/Si; OTS	[87]
35	−5.50	Evaporation	0.21	5.0×10^6	−16.9	BGTC; Au; SiO_2/Si; OTS	[93]
36	−5.26	Evaporation	0.11	3×10^6	−8.9	BGTC; Au; SiO_2/Si; OTS	[94]
37	5.05	Evaporation	2.1	7.6×10^6	−17.5	BGTC; Au; SiO_2/Si; OTS	[95]
38	−5.45	Evaporation	0.1	5×10^4	−0.6	BGTC; Au; SiO_2/Si	[96]
40		Evaporation	0.05	1.2×10^5	−6.3	BGTC; Au; SiO_2/Si; OTS	[100]
		Micrometer wires	0.8	1.7×10^3	−6.0		
41		Single crystal	2.66	2.3×10^5		BGTC; Au; SiO_2/Si OTS	[101]
42	−5.3	Single crystal	2.13	10^6	−14	BGTC; Au; SiO_2/Si; OTS	[102]
43	−5.40	Evaporation	0.12	3×10^5		BGTC; Au; SiO_2/Si; OTS	[103]
44	−4.08	Evaporation	0.42	10^6		BGTC; Au; SiO_2/Si; OTS	[104]
45		Evaporation	3.2 (O_2)			BGTC; Au; SiO_2/Si	[105]
46	−5.22	Spin-coating	1.6×10^{-4}	10^3	−15	BGTC; Au; SiO_2/Si; OTS-8	[106]
48		Spin-coating	0.02	10^6	−3	BGTC; Au; SiO_2/Si; OTS	[108]
49a		Spin-coating	1.8	10^7	−17	BGTC; Au; SiO_2/Si	[109]
		Inkjet printing	31.3	$10^5 \sim 10^7$	−10	TGTC; Au; parylene C/Au	[110]
		Spin-coating	43			BGTC; Ag; PVP/HDA; ITO	[111]
49b		Spin-coating	1.71	10^8	−20	BGTC; Au; SiO_2/Si	[109]
		Evaporation	3.9	$10^6 \sim 10^8$		BGTC; Au; SiO_2/Si; ODTS	[112]
49c		Spin-coating	2.75	10^7	−27	BGTC; Au; SiO_2/Si	[109]
50	−5.6	Evaporation	2.0	$>10^7$		BGTC; Au; SiO_2/Si; OTS	[20]
51		Evaporation	17.2	1.6×10^7	−2.7	BGTC; Al; SiO_2/Si; AlO_x/C_{14}-PA	[113]

(continued)

Table 3.1 (continued)

	HOMO (eV)	Deposition process	Max μ_h ($cm^2\ V^{-1}\ s^{-1}$) (m.d.)[a]	I_{on}/I_{off}	V_T (V)	Device structure[b]	Refs.
52	−5.30	Evaporation	0.15	10^6		BGTC; Au; SiO_2/Si; OTS	[114]
53		Evaporation	0.20	10^6	−19	BGTC; Au; SiO_2/Si; OTS	[115]
54		Evaporation	0.015	10^6	−20	BGTC; Au; SiO_2/Si; OTS	[115]
55	5.5	Evaporation	0.30	10^6	−19	BGTC; Au; SiO_2/Si; HMDS	[116]
57	−5.43	Evaporation	0.14	10^6	−19	BGTC; Au; SiO_2/Si OTS	[117]
58	−5.41	Evaporation	0.1	10^6	−15	BGTC; Au; SiO_2/Si; OTS	[117]
59	−5.39	Evaporation	0.09	10^6	−28	BGTC; Au; SiO_2/Si; OTS	[117]
60	−5.35	Evaporation	0.41 (air)	10^5		BGTC; Au; SiO_2/Si; OTS	[118]
61	−5.34	Microribbons	1.66 (air)	10^6	−10	BGTC; Au; SiO_2/Si; OTS	[119]
62	−5.68	Polycrystals	1.1	$10^4 \sim 10^5$	−20	BGTC; Au; SiO_2/Si; DTS	[120]
65	−5.6	Evaporation	0.51	4.5×10^6	−67	BGTC; Au; SiO_2/Si; OTS	[49]
		Single crystal	1.8	10^7		BGBC; Au; SiO_2/Si	[123]
66	−5.64	Single crystal	0.6	10^6		BGBC; Au; SiO_2/Si	[124]
67	−5.17	Evaporation	0.31	1.3×10^6	7	BGTC; Au; SiO_2/Si; OTS	[114]
72	−5.56	Spin-coating	1.7 (N_2)	10^7	−22 ± 9	BGTC; Au; SiO_2/Si	[126]
73	−5.33	Evaporation	0.045	$10^2 \sim 10^3$		BGTC; Au; SiO_2/Si; PTA	[127]
74	−5.62	Evaporation	0.15	5.9×10^6	9	BGTC; Au; SiO_2/Si; OTS	[128]
75	−5.70	Evaporation	0.047	1.9×10^6	−7	BGTC; Au; SiO_2/Si; OTS	[128]
76	−5.44	Evaporation	2.9	10^7	−11	BGTC; Au; SiO_2/Si; OTS	[129]
		Single crystal	8.3	$>10^8$		BGTC; Au/TTF-TCNQ; SiO_2/Si; CYTOP	[130]
77	5.38	Evaporation	1.9	5×10^6	−7.5	BGTC; Au; SiO_2/Si; OTS	[129]
78	5.38	Evaporation	8.0	$>10^8$		BGTC; Au; SiO_2/Si; ODTS	[131]
79		Evaporation	3.43	10^9	−9.4	BGTC; Au; SiO_2/Si; ODTS	[132]
81	−5.36	Single crystal	0.39 (air)	10^6	−16	BGTC; Au; SiO_2/Si; PMMA	[134]

(continued)

Table 3.1 (continued)

	HOMO (eV)	Deposition process	Max μ_h ($cm^2 V^{-1} s^{-1}$) (m.d.)[a]	I_{on}/I_{off}	V_T (V)	Device structure[b]	Refs.
82	−5.14	Single crystal	0.1 (air)	10^5	−3	BGTC; Au; SiO_2/Si; PMMA	[134]
83		Evaporation	0.11	10^5		BGTC; Au; SiO_2/Si	[136]
84	−5.09	Evaporation	0.26	4.5×10^5	9.8	BGTC; Au; SiO_2/Si; OTS	[137]
85	−5.13	Evaporation	0.1	1.3×10^6	14	BGTC; Au; SiO_2/Si; OTS	[137]
86	−5.1	Evaporation	0.46			BGTC; Au; PVP/Si	[138]
87		Evaporation	0.81			BGTC; Au; PVP/Si	[138]
88	−5.35	Evaporation	0.04	10^7	−1.4	BGTC; Au; SiO_2/Si	[139]
90	5.57 (film)	Evaporation	0.012	10^5		BGTC; Au; SiO_2/Si; OTS	[140]
90	−4.92	Evaporation	0.019	10^6	−30	BGBC; Au; SiO_2/Si	[141]
93	−4.83	Evaporation	0.074	10^8	−33	BGBC; Au; SiO_2/Si	[141]
94	−5.35 (film)	Evaporation	0.4	10^4	−29	BGTC; Au; SiO_2/Si; OTS	[142]
95	−5.35 (film)	Evaporation	0.4	10^4	−55	BGTC; Au; SiO_2/Si; OTS-8	[142]
96		Single crystal	2.1	2×10^5	−7	BGBC; Au; SiO_2/Si; PMMA	[143]
97	5.0	Single crystal	0.42	10^4	2.9	BGTC; Au; SiO_2/Si; OTS	[144]
99	−5.26	Evaporation	0.072	10^6		BGTC; Au; SiO_2/Si; OTS	[145]
101	−5.7	Evaporation	9×10^{-3} (N_2)	10^6	−45	BGBC; Au; SiO_2/Si	[147]
102	−5.2	Evaporation	1×10^{-3} (N_2)	10^6	−10	BGBC; Au; SiO_2/Si	[147]
103		Evaporation	0.42	5×10^6	−23.4	BGTC; Au; SiO_2/Si; OTS	[148]
104		evaporation	0.12	5×10^5	−20.2	BGTC; Au; SiO_2/Si; OTS	[148]
105	−5.5	Evaporation	0.54	2.2×10^6	16	BGTC; Au; SiO_2/Si	[149]
106	−5.5	Evaporation	0.12	4.2×10^3	3	BGTC; Au; SiO_2/Si	[149]
		Single crystal	1.1	6.5×10^4	−8	BGTC; Au; SiO_2/Si; OTS	[149]
107	−5.18	Evaporation	0.14	10^4	3.6	BGTC; Au; SiO_2/Si; OTS	[148]
108		Evaporation	0.05			BGTC; Au; SiO_2/Si	[150]

(continued)

Table 3.1 (continued)

	HOMO (eV)	Deposition process	Max μ_h ($cm^2 V^{-1} s^{-1}$) (m.d.)[a]	I_{on}/I_{off}	V_T (V)	Device structure[b]	Refs.
109		Evaporation	0.03	10^4		BGBC; Au; SiO_2/Si	[151]
100	5.33	Evaporation	0.08	10^3		BGBC; Au; SiO_2/Si; OTS	[152]
111	5.49	Evaporation	0.005			BGBC; Au; SiO_2/Si; OTS	[152]
112	−5.39	Evaporation	0.42	10^7	−14	BGBC; Au; SiO_2/Si; OTS	[152]
113	−5.41	Evaporation	2.0	10^8	−31	BGTC; Au; SiO_2/Si; OTS	[154]
114	−5.71	Evaporation	0.007	9.3×10^7	−20.5	BGTC; Au; SiO_2/Si; OTS	[153]
115	−5.53	Evaporation	0.50	4.8×10^6	−26	BGTC; Au; SiO_2/Si; OTS	[153]
116	−5.73	Evaporation	0.67	6.4×10^4	−40	BGTC; Au; SiO_2/Si; ODTS	[155]
117	−5.0	Evaporation	0.12	10^5	−0.4	BGTC; Au; SiO_2/Si; OTS-8	[156]
118	−5.42	Evaporation	0.1	10^6	−12∼−14	BGTC; Au; SiO_2/Si; HMDS	[157]
119	5.57	Evaporation	0.15	10^8		BGTC; Au; SiO_2/Si; OTS	[158]
122		Evaporation	0.28			BGTC; Au; SiO_2/Si	[160]
123		Evaporation	0.06	4×10^2	0	BGTC; Au; PMMA/Ta_2O_5/Si	[161]
124		Evaporation	0.08	10^4	−25	BGTC; Au; PMMA/Ta_2O_5/Si	[162]
		Evaporation	1.1	10^4		BGTC; Au; SiO_2/Si; OTS	[48]
126		Evaporation	0.09	10^6		BGTC; Au; SiO_2/Si	[164]
128		Evaporation	0.05	10^3		BGTC; Au; SiO_2/Si	[166]
130		Spin-coating	1.0	10^7		BGBC; Au; SiO_2/Si; PFBT	[167]
132			0.4	10^5		BGBC; Au; SiO_2/Si	[168]
135		Spin-coating	1.5			BGBC; Au; SiO_2/Si; PFBT	[169]
		Single crystal	6	10^8		BGBC; Au; SiO_2/Si; OTS	[170]
139	−5.18	Evaporation	7×10^{-3}	2×10^4	−10∼−20	BGTC; Au; SiO_2/Si; OTS	[172]
142	−5.21	Evaporation	1.05	10^5	0∼5	BGTC; Au; SiO_2/Si	[172]

(continued)

Table 3.1 (continued)

	HOMO (eV)	Deposition process	Max μ_h ($cm^2 V^{-1} s^{-1}$) (m.d.)[a]	I_{on}/I_{off}	V_T (V)	Device structure[b]	Refs.
143	−5.34	Drop casting	2×10^{-3}			BGBC; Au; SiO_2/Si; PFBT	[173]
144	−5.15	Drop casting	5×10^{-2}			BGBC; Au; SiO_2/Si PFBT	[173]
145	−5.03	Drop casting	1.8			BGBC; Au; SiO_2/Si; PFBT	[173]
147	5.62	Evaporation	0.012	10^6	−41	BGTC; Au; SiO_2/Si; OTS	[174]
148		Single crystal	1.2 (α-phase)			BGBC; Au; SiO_2/Si	[25]
		Single crystal	0.23 (β-phase)			BGBC; Au; SiO_2/Si	[25]
149		Zone-casting	0.08	10^4		BGTC; Au; SiO_2/Si_3O_4/Si	[182]
150		Single crystal (solution)	3.65				[183]
152		Evaporation	0.42	6×10^3	12	BGTC; Au; SiO_2/Si; OTS	[185]
153		Evaporation	0.2	10^6	36	BGTC; Au/Cr; Al_2O_3/Al	[185]
155	−5.70	Evaporation	0.20	4×10^5	−50	BGTC; Au; SiO_2/Si; HMDS	[186]
156	−5.75	Evaporation	0.64	3.3×10^5	−33	BGTC; Au; SiO_2/Si; HMDS	[186]
158		Single crystal (vapor)	11.2				[179]
		Single crystal (solution)	10.4				[179]
159	−5.42	Evaporation	0.40	$10^6 \sim 10^8$		BGTC; Au; SiO_2/Si; OTS	189]
161	−4.90	Evaporation	8.4×10^{-3}	10^5	−27	BGTC; Au; SiO_2/Si; HMDS	[192]
162	−4.91	Evaporation	0.27	10^6	−1.9	BGTC; Au; SiO_2/Si; HMDS	[192]
163	4.63	Evaporation	0.45			BGTC; Au; SiO_2/Si; OTS	[195]
164		Evaporation	1.4			BGTC; Au; pentene/NTS/Si	[196]
165		Evaporation	0.13			BGTC; Au; pentene/NTS/Si	[196]
166	5.19	Single crystal	1.0	10^4		BGBC; Au/Cr; SiO_2/Si	[197]
167	5.21	Evaporation	0.02	10^2	7.5	BGTC; Au; SiO_2/Si	[198]
168	5.14	Evaporation	0.01	10^5	−9.4	BGTC; Au; SiO_2/Si	[198]

(continued)

Table 3.1 (continued)

	HOMO (eV)	Deposition process	Max μ_h ($cm^2 V^{-1} s^{-1}$) (m.d.)[a]	I_{on}/I_{off}	V_T (V)	Device structure[b]	Refs.
169	5.44	Evaporation	0.34	$10^6 \sim 10^7$	−28	BGTC; Au; SiO_2/Si; OTS	[199]
		Single crystal	3.6	1.9×10^6	6.7	BGTC; Au; SiO_2/Si; OTS	[200]
170	5.12	Evaporation	0.12	10^7	−7	BGTC; Au; SiO_2/Si; OTS-8	[201]
171	5.26	Evaporation	0.14	10^7		BGTC; Au; SiO_2/Si; OTS-8	[202]
172	5.12	Single crystal	0.084		−8.7	BGTC; Au; SiO_2/Si; OTS	[203]
174	5.09	Evaporation	5.3×10^{-3}	10^6	−4.7	BGTC; Au; SiO_2/Si; OTS	[204]
175	5.22	Evaporation	0.015	10^7		BGTC; Au; SiO_2/Si; OTS	[206]
177		Evaporation	0.02	4×10^5		BGBC; Au; SiO_2/Si	[208]
		Single crystal	0.5			BGTC; Au; SiO_2/Si	[209]
178		Evaporation	3.31	1.2×10^7	−17.8	BGTC; Au; SiO_2/Si; OTS	[210]
182		Evaporation	0.32	10^3	−9	BGTC; Au; SiO_2/Si	[214]
184		Drop casting	0.13		7.5	BGTC; Au; SiO_2/Si	[216]
186		LB	0.68			BGTC; Au; SiO_2/Si	[218]

[a] *m.d.* measurement condition

[b] Device configuration; S/D electrodes; dielectric/gate; modification of substrate/electrode

in **3**-based FET devices fabricated with TGBC configuration, much higher mobilities of up to 40 $cm^2 V^{-1} s^{-1}$ were achieved [63].

However, because of the narrow energy gap and high-lying HOMO energy level, pentacene exhibited high sensitivity to light and high oxidation sensitivity to oxygen [64]. Moreover, the low solubility of pentacene in common organic solvents makes purification of pentacene highly challenging. For similar reasons, the higher-order π-conjugated systems based OFETs, such as hexacene and heptacene, have never been addressed [65, 66]. Encouraged by the high performance of acenes, many acene derivatives, especially pentacene derivatives, have been synthesized and examined as the active layer in FET devices with the hope of improving solubility and high stability while maintaining high performance. Studies on acene derivatives are usually focused on the improvement of solubility and stability. For example, alkyl derivatives 2,3,9,10-tetramethyl-pentacene **6**, 2,9-dimethylpentacene, **7**, and 2,9-dihexylpentacene **8** exhibited mobilities of 0.30, 2.5, and 0.251 $cm^2 V^{-1} s^{-1}$, respectively [67–69]. There are some other pentacene derivatives, for example, 2,9-diethylpentacene **9**, which showed even poorer field-effect performance than those of derivatives **6**, **7**, and **8** [69]. With the introduction of alkyl substituents, the solubility of the resulting materials has been significantly improved. However, because of the donating effect of alkyl substituents, the HOMO energy levels have also been driven up, and thus the stability of these derivatives becomes worse. For this reason, some electron-withdrawing groups including halogen, cyano, and trifluoromethyl, etc., had been introduced to the backbones of acenes as substituents in order to lower their HOMO energy levels and the reactivity of the resulting acene derivatives. For example, Bao et al. and Wudl et al. reported series of this kind of pentacene derivatives, **10–16** [70, 71]. Of these, derivative **13** afforded field-effect mobilities as high as 0.23 $cm^2 V^{-1} s^{-1}$ and improved device stability compared to pentacene. The **13**-based FET devices exhibited no significant decrease in mobility or on/off ratio when stored in air, with and without light exposure, even after 3 months. In addition, the thin film transistor based on 6,13-dichloropentacene **17** showed a mobility of 0.21 $cm^2 V^{-1} s^{-1}$ [72] and its crystalline ribbons-based FET devices afforded a high mobility of up to 9.0 $cm^2 V^{-1} s^{-1}$, which is one of the highest values reported for organic semiconducting materials [73]. This kind of naphthacene derivative had also been reported by several groups [37, 74]. For example, 5,11-dichlorotetracene, **19** has the slipped π-stacking structure in contrast to herringbone type tetracene between its neighboring molecules. The single crystal transistors based on derivative **19** showed high mobility as high as 1.6 $cm^2 V^{-1} s^{-1}$ [37]. 5,6,11,12-Tetrachloronaphthacene **20** also has slip π-stacks, and its single crystal transistors exhibited p-type behavior with a field-effect mobility of 1.7 $cm^2 V^{-1} s^{-1}$ [74]. It is notable that higher mobilities were afforded in peri-substituted acene derivatives than in their end-substituted counterparts in combination with improved stability. These changes in mobility could be attributed to the improved π–π packing in the solid state of resulting peri-substituted derivatives. Similar phenomena could be distinctly observed in acene derivatives with alkyl or aryl substituents in the peri-position, for example, 9,10-diphenylanthracene, **21** and rubrene, **22**. Time-of-flight results showed that the single crystals of anthracene **21** could afford high hole mobility of up to 3.7 $cm^2\ V^{-1}\ s^{-1}$ and electron mobility of up to 13 $cm^2\ V^{-1}\ s^{-1}$ [75]. Moreover, rubrene, **22** is another benchmark material for organic semiconductors. The single crystal of **22** exhibited higher mobility of 15.4 $cm^2\ V^{-1}\ s^{-1}$ along the *b*-axis and 4.4 $cm^2\ V^{-1}\ s^{-1}$ along the *a*-axis [24]. These results indicate that substitutions in the peri-positions of acenes are in fact favorable to strong π–π intermolecular interactions and a large π–π overlap. However, it is pity that thin films of **22** could show only low mobilities of 0.07 $cm^2\ V^{-1}\ s^{-1}$ when using pentacene as buffer layer [76], and 0.7 $cm^2\ V^{-1}\ s^{-1}$ when mixed with ultra-high-molecular-weight polymer [77]. The low mobility could be attributed to poor planarity structure of **22**. In fact, derivative **21** also has a nonplanar structure, in which the dihedral angle between the planes of substitution (benzene ring) and core (anthracene) is about 67° [75, 78].

Similar to the effect of phenyl substitution in **21** and **22**, 6,13-di(2′-thienyl) pentacene, **23** showed a strengthened π-stacking mode and exhibited a low mobility of 0.1 $cm^2\ V^{-1}\ s^{-1}$ [79]. To avoid the dilemma, carbon–carbon triple bonds were introduced into this kind of molecule affording nearly planar acene derivatives while maintaining strengthened π–π stacking. Several groups synthesized a lot of peri-alkynyl derivatives and examined their FET properties. Anthony et al. considered that the length ratio of the substituents and acene cores affected the packing of the materials [80, 81]. If the length of the substituents was approximately half the length of the acene core, the acene would form a lamellar π–π stacking mode. Otherwise, a slipped π-stacking or herringbone packing structure could be adopted. Indeed, laboratory findings are in good agreement with their assumption. For example, acene derivative **26** showed a two-dimensional brick layer structure, and its vacuum deposited technique films afforded high mobilities up to 0.4 $cm^2\ V^{-1}\ s^{-1}$ [82]. Its drop-casting films showed high mobility of up to 1.8 $cm^2\ V^{-1}\ s^{-1}$ with an on/off current ratio over 10^7, and self-assembled ribbons exhibited high mobilities of 1.42 $cm^2\ V^{-1}\ s^{-1}$ [83]. The tetramethyl-substituted acene derivative, **27**-based thin film transistors fabricated by a solution-processed technique also showed mobility as high as 2.5 $cm^2\ V^{-1}\ s^{-1}$ [84]. The results showed that the introduction of trialkylsilylethynyl groups on the peri-position of acene could afford nearly planar molecule structures, improved solubility and stability while maintaining strengthened π–π stacking as anticipated. Arylethynyl group as substituents were also introduced in acene-based π-conjugated systems affording new acene derivatives such as **29–31**. Nonetheless, these derivatives exhibited low mobilities of 0.033 $cm^2\ V^{-1}\ s^{-1}$ for vacuum deposited films of **29** [85], 0.52 $cm^2\ V^{-1}\ s^{-1}$ for single crystalline micro/nanoribbons of **30** [86], and 0.11 $cm^2\ V^{-1}\ s^{-1}$ for vacuum deposited films of **31** [87], although they all formed improved π–π stacking in solid states. The low mobilities could arise from either the mismatching the lengths of substituents and acenes, or large twisting molecule structure (Chart 3.2).

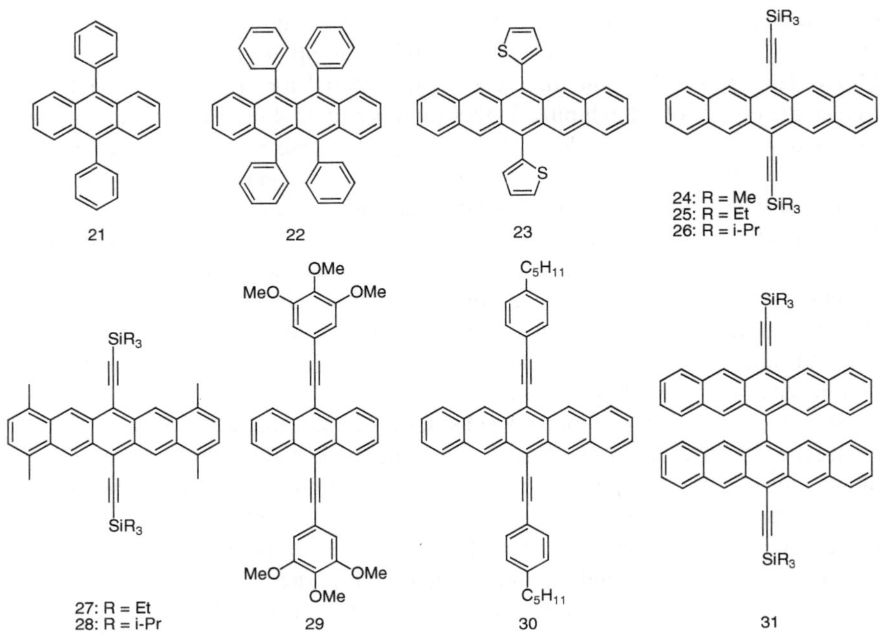

Chart 3.2 Peri-substituted linearly acene-based small-molecule semiconductors

Pyrene and Derivatives

Pyrene is a kind of nonlinear PAH with the same number of benzene rings as tetracene (see Chart 3.3). Pyrene exhibits π–π stacking mode in crystals. Highly pure pyrene single crystals showed low mobilities of $(3.8 \pm 0.1) \times 10^{-3}$ $cm^2\ V^{-1}\ s^{-1}$ for electrons and $(3.3 \pm 0.4) \times 10^{-3}$ $cm^2\ V^{-1}\ s^{-1}$ for holes at 345 K [88]. Nonetheless, high mobilities of 1.2 $cm^2\ V^{-1}\ s^{-1}$ for holes and 3 $cm^2\ V^{-1}\ s^{-1}$ for electrons were obtained by judging from the transient photocurrent of the pyrene crystal and calculating from the overlap integrals between two molecules at room temperature [89]. Because of the nonplanar structure, the tetrasubstituted-derivatives using pyrene as core including **33** and **34** tend to afford low mobilities [90–92]. However, the derivatives **35** and **36** using pyrene as an end-capped group and linear pyrene derivative **37** showed high mobilities [93–95]. For example, thin film transistors based on derivatives **35**, **36**, and **37** exhibited mobilities of 0.21, 0.11, and 2.1 $cm^2\ V^{-1}\ s^{-1}$, respectively. It is notable that an organic radical-based pyrene, namely 1-imino nitroxide pyrene, **38** was reported [96]. Vapor-deposited films of the **38**-based FET devices showed excellent p-type FET characteristics, with a mobility up to 0.1 $cm^2\ V^{-1}\ s^{-1}$ and an on/off current ratio at 5×10^4. The highest performance of OFETs based on **38** indicates the potential of organic radicals for the application in OFETs.

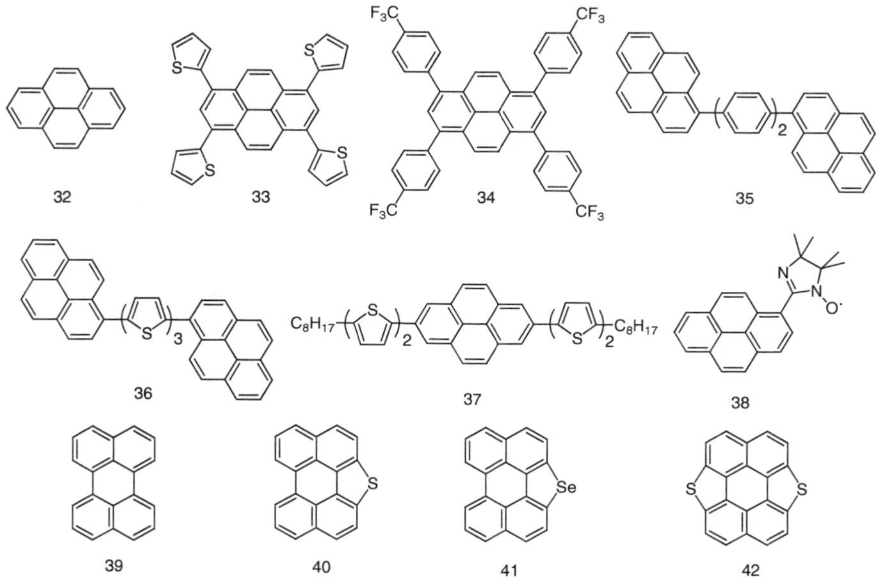

Chart 3.3 Pyrene- and perylene-based small-molecule semiconductors

Perylene and Derivatives

Perylene, **39** is an isomer of pentacene with five benzene rings (see Chart 3.3). The charge carrier transport property of **39** was investigated by both thin film- and single-crystal-based FET devices [97–99]. The single crystal of **39** showed a hole mobility of 0.12 $cm^2\ V^{-1}\ s^{-1}$, which is comparable with the mobilities of 0.15–0.3 $cm^2\ V^{-1}\ s^{-1}$ obtained from time-of-flight measurements. Compared with pyrene, there are relatively less derivatives of perylene synthesized and applied in OFETs. However, some derivatives exhibited high mobilities. For example, two chalcogen-heterocyclic perylene derivatives **40** [100] and **41** [101] showed strong chalcogen–chalcogen and π–π intermolecular interactions in single crystals. Single crystal transistors based on an individual assembled microwire of **40** showed a mobility as high as 0.8 $cm^2\ V^{-1}\ s^{-1}$. It is worth noting that individual microribbons of **41** exhibited a high mobility of 2.66 $cm^2\ V^{-1}\ s^{-1}$. Furthermore, perylene derivative, **42** afforded a compressed highly ordered packing mode, and the FET based on individual nanoribbon exhibited a high mobility of 2.13 $cm^2\ V^{-1}\ s^{-1}$ [102]. These results make perylene derivatives particularly attractive for electronic applications.

Other Polycyclic Aromatic Hydrocarbons

Besides pyrene, perylene, and their derivatives, other nonlinear PAHs were also developed and used in FET devices (see Chart 3.4). For example, two novel phenanthrene-based conjugated oligomers including derivative **43** were synthesized [103]. Field-effect transistors based on **43** showed a mobility as high as 0.12 $cm^2\ V^{-1}\ s^{-1}$ with an on/off current ratio at 10^5. These devices exhibited excellent stability during long-time ambient storage and under UV irradiation. It is interesting that α,ω-dihexyldithienyl-dihydrophenanthrene **44** also gave high mobility of 0.42 $cm^2\ V^{-1}\ s^{-1}$ [104]. Picene **45**, a nonlinear isomer of pentacene, showed *p*-channel characteristics with a mobility of 1.1 $cm^2\ V^{-1}\ s^{-1}$ and an on/off current ratio of 10^5 under atmospheric conditions [105]. Moreover, when the devices are located under 500 Torr of oxygen for oxygen doping, a higher mobility of 3.2 $cm^2\ V^{-1}\ s^{-1}$ was achieved. The excellent stability could be attributed to its high ionization potential of 5.5 eV and large energy bandgap of 3.3 eV. In theory, larger π-conjugated systems should afford higher mobilities because the increased transfer integrals, lowered reorganization energies, and strengthened π–π stacking would be afforded in these systems. However, triphenylenes, coronenes, and their derivatives with large π-conjugated systems showed low charge transport performance. For example, star triphenylene derivative **46**-based transistors exhibited hole mobility of 1.4×10^{-4} $cm^2\ V^{-1}\ s^{-1}$ [106]. Although hexabenzocoronene (HBC) derivative **47** formed highly oriented films under magnetic field, FET devices based on the oriented films showed charge carrier mobilities of 10^{-4} $cm^2\ V^{-1}\ s^{-1}$ [107]. Derivative **48**-based FETs fabricated by spin casting technique also afforded a low mobility of 10^{-2} $cm^2\ V^{-1}\ s^{-1}$ with an on/off current ratio of about 10^6 [108]. The low performance of the three PAHs was probably

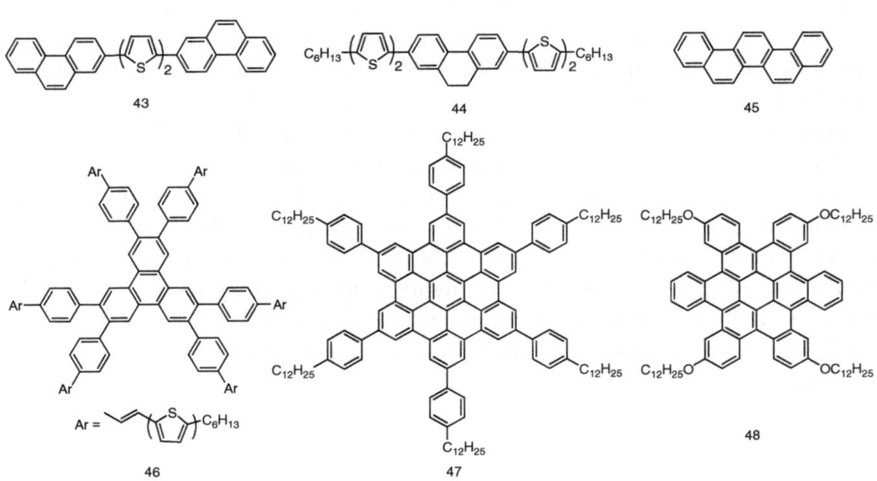

Chart 3.4 Other polycyclic aromatic hydrocarbon semiconductors

caused by the nonplanar structure of the compounds concerned. The results also indicated that there are many other factors heavily influencing the mobility of FETs besides the size of the π-conjugated systems, such as solubility, stability, planar property, and molecular packing, etc.

3.2.1.2 Chalcogen-Containing Heterocyclic Aromatic Hydrocarbons and Derivatives

Sulfur-containing heterocyclic aromatic hydrocarbons and derivatives play an important role in the development of OFETs. Plenty of these materials tend to exhibit excellent OFET performance and high stability. The high performance could be attributed to planar structure and ordering molecular arrangement of chalcogen-containing PAHs concerned in solid state, which results in an effectively intra-stack electronic coupling via strong intermolecular π–π, S–S, and CH–π interactions. The high stability can be explained by the absence of a Diels–Alder cyclization active center and low-lying HOMO energy level of these materials. It is believed that the large atomic radius of sulfur and the high electron densities of the sulfur atoms in the HOMO are two of the fundamental reasons for these improvements. As a result, sulfur-containing materials have become one of the most important semiconductors for OFET applications. Oxygen-, selenium-, and tellurium-containing derivatives have all attracted a great deal of attention too.

Thienoacenes

Thienoacenes consist of fused thiophene rings in a ladder-type molecular structure and have been intensively studied in OFETs (see Chart 3.5). For example, Takimiya et al. used a series of benzothieno[3,2-*b*]benzothiophene (BTBT) derivatives, **49** and **50** in OFETs [20, 109, 110]. Spin-coated films of derivatives **49** showed mobilities of over 0.1 $cm^2\ V^{-1}\ s^{-1}$.

Of these, spin-coated films of **49a** exhibited hole mobility of 1.8 $cm^2\ V^{-1}\ s^{-1}$ with an on/off current ratios of 10^7 [109]. Inkjet printed single crystal films of **49a** gave an average mobility as high as 16.4 $cm^2\ V^{-1}\ s^{-1}$ [110]. When its thin film was grown by an off-center spin-coating method, much higher mobilities up to 43 $cm^2\ V^{-1}\ s^{-1}$ were obtained [111]. Vacuum evaporated films of derivative **49b** also showed high mobility, for example, vacuum evaporated films of the dodecyl-substituted derivative afforded a mobility as high as 3.9 $cm^2\ V^{-1}\ s^{-1}$ [112]. Spin-coated films of **49c** gave high device performance with mobilities of 2.75 $cm^2\ V^{-1}\ s^{-1}$ with an on/off current ratios of 10^7 [109]. Phenyl-substituted derivative **50** also showed high OFETs performance. Vacuum evaporated films afforded high mobilities of up to 2.0 $cm^2\ V^{-1}\ s^{-1}$ with on/off current ratios of 10^8, and excellent stability [20]. In stability tests of the devices, there was almost no change in mobility after 250 days for the devices fabricated on the bare Si/SiO_2 substrates, whereas the devices on the OTS-treated substrates showed a slight

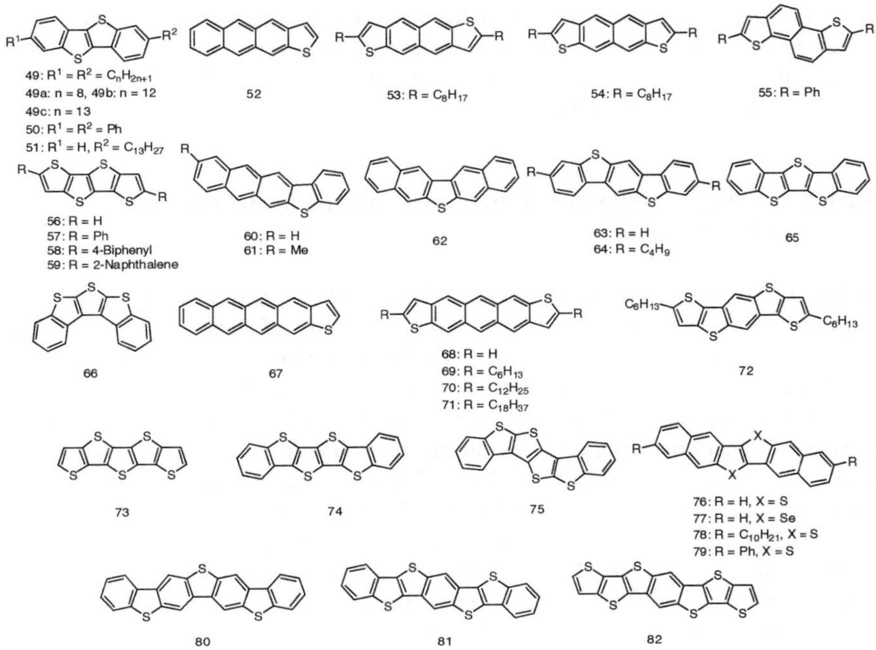

Chart 3.5 Selected thienoacene-based p-type small-molecule semiconductors

decrease in mobility. In 2012, Amin et al. reported a mono-substituted BTBT derivative, **51**. The material showed excellent p-type characteristics with large hole mobilities up to 17.2 cm^2 V^{-1} s^{-1} [113]. However, linear and angular analogues of tetracenes, **52–55** showed low mobilities of 0.15, 0.20, 0.015, and 0.30 cm^2 V^{-1} s^{-1}, respectively [114–116]. Thieno[3,2-*b*]thieno[2′,3′:4,5]thieno-[2,3-*d*]thiophene, **56** and derivatives, **57–59** showed mobilities larger than 0.1 cm^2 V^{-1} s^{-1} with ideal solubility and environmental stability [117]. The analogues of pentacene with one thiophene unit, **60–62** were synthesized and examined [118–120]. They show low HOMO energy levels of −5.34 to −5.68 eV, respectively. Thienoacenes **60**- and **62**-based transistors fabricated by vapor-deposited methods afforded hole mobilities of 0.41 and 1.1 cm^2 V^{-1} s^{-1}, respectively. A single crystal FET device of **62** gave a mobility up to 1.5 cm^2 V^{-1} s^{-1}. It is notable that phototransistors based on the individual **61** microribbons prepared by a solution-phase self-assembly process showed a high mobility of 1.66 cm^2 V^{-1} s^{-1}, a large photoresponsivity of 12,000 A W^{-1}, and a photocurrent/dark current ratio of 6000 even under low light power condition of 30 mW cm^{-2}.

Benzo[1,2-*b*:4,5-*b*′]bis[*b*]benzothiophene **63** and its butyl-substituted derivative **64** showed the HOMO energy level of −5.4 eV. Solution-processed devices based on **63** and **64** showed a hole mobility up to 0.01 cm^2 V^{-1} s^{-1} [121, 122]. Analogue of pentacene containing three thiophene rings, **65**, showed the HOMO energy level of −5.6 eV [49]. The thin film transistors based on **65** exhibited a hole mobility of 0.51 $cm^2 V^{-1} s^{-1}$ and high on/off ratio of 4.5 × 10^6. On the other hand, its single crystals demonstrated mobility as high as 1.8 $cm^2 V^{-1} s^{-1}$ and high on/off ratio greater than 10^7 [123]. However, isomer **66**-based single crystal ribbons FET devices afforded a mobility of 0.6 $cm^2 V^{-1} s^{-1}$ with on/off ratios over 10^6 [122]. Bao's group synthesized tetraceno[2,3-*b*]thiophene, **67**, which showed a mobility of 0.04–0.31 $cm^2 V^{-1} s^{-1}$ [114]. In 1998, Katz's group reported the synthesis and application of anthradithiophene **68** and alkylated derivatives **69–71** in OFETs. They all form highly ordered polycrystalline vacuum-evaporated films with mobilities as high as 0.15 $cm^2 V^{-1} s^{-1}$ [125]. Gao et al. [127] reported new five-ring-fused pentacene analogs, with four symmetrically fused thiophene-ring units. Employing the hexyl-substituted derivative **72** in solution-processed OFETs produced mobilities as high as 1.7 $cm^2 V^{-1} s^{-1}$, and an on/off current ratio of 10^7 for long-range ordered thin film [126]. Liu's group reported the application of pentathienoacene **73** in OFETs. The thienoacene has a HOMO energy level of −5.3 eV, which is lower than that of pentacene, and its energy gap of 3.2 eV, which is much larger than that of pentacene (1.8 eV). Highly extended thienoacenes were also synthesized and investigated. For example, thienoacenes, **74** and **75** with six fused rings were reported [128]. They exhibited hole mobilities of 0.15 and 0.047 $cm^2 V^{-1} s^{-1}$, respectively. Dianthra[2,3-*b*:2′,3′-*f*]thieno[3,2-*b*]thiophenes (DNTTs), **76–78** are another kind of high performance semiconductor developed by Takimiya et al. besides BTBT derivatives. For example, vacuum-sublimated films of thienoacene **76** exhibited a mobility up to 2.9 $cm^2 V^{-1} s^{-1}$ with an on/off current ratio of 10^7 [129]. FETs based on its single crystal showed a mobility as high as 8.3 $cm^2 V^{-1} s^{-1}$ with an on/off current ratio of up to 10^9 [130]. Similarly, derivative **77** showed a mobility as high as 1.9 $cm^2 V^{-1} s^{-1}$ [127]. Vacuum deposited films of **78** and **79** gave mobilities as high as 8.0 and 3.4 $cm^2 V^{-1} s^{-1}$, respectively [131, 132]. The fused seven ring compounds exhibited low charge transport behaviors. Thin film of compound **80** afforded hole mobility of 0.15 $cm^2 V^{-1} s^{-1}$ [133] and the single crystal microribbons of compounds **81** and **82** exhibited hole mobilities of 0.47 and 0.10 $cm^2 V^{-1} s^{-1}$, respectively [132].

Highly Fused Chalcogen-Containing Heterocyclic Aromatic Hydrocarbons

Many highly fused chalcogen-containing heterocyclic aromatic hydrocarbons had also been synthesized and applied as semiconducting layers in OFETs (see Chart 3.6). For example, fused heteroarene **83**, in which the core is isoelectronic with pyrene, was developed by Takimiya et al. [135, 136]. Compared to pyrene, the HOMO energy levels of the heteroarene greatly increased, whereas the LUMO energy levels were lowered. Thin films of **83** demonstrated a mobility as high as 0.11 $cm^2 V^{-1} s^{-1}$. Two dioxapyrene-based semiconductors **84** and **85** containing oxygen atoms formed well-ordered polycrystalline films, the devices affording a hole mobility as high as 0.25 $cm^2 V^{-1} s^{-1}$ for **84** and 0.1 $cm^2 V^{-1} s^{-1}$ for **85** [137]. Vacuum deposited thin films of 3,9-diphenyl-peri-xanthenoxanthene, **86** and

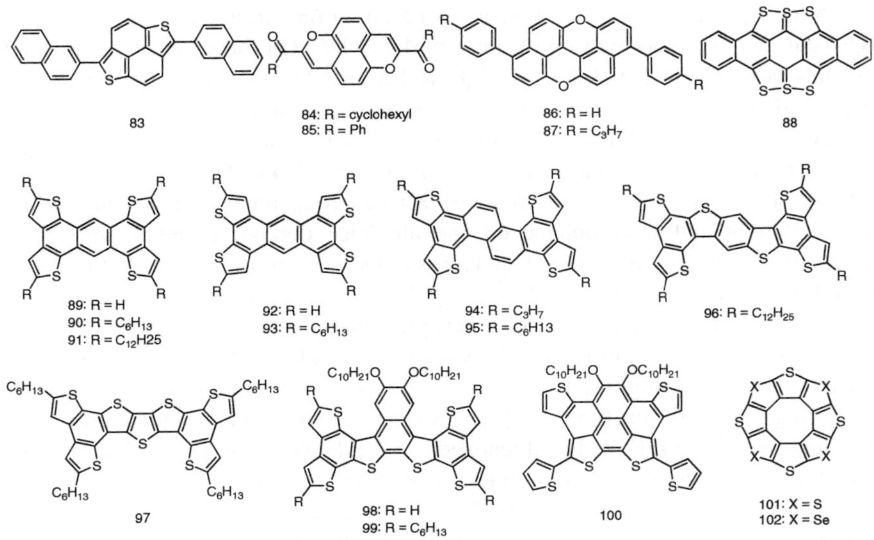

Chart 3.6 Highly fused chalcogen-containing heterocyclic aromatic hydrocarbons

3,9-bis(*p*-propylphenyl)-peri-xanthenoxanthene, **87** showed mobilities as high as 0.4 and 0.81 $cm^2\ V^{-1}\ s^{-1}$, respectively [138]. It is notable that these thin film FETs were also environmentally and thermally stable. Especially, for **86**-based devices, no significant degradation was observed after 5 months of storage in ambient conditions. The FET devices based on hexathiapentacene, **88** exhibited mobilities in a range of 0.005–0.04 $cm^2\ V^{-1}\ s^{-1}$ [139]. Single crystals of **88** showed strong S–S intermolecular interactions (3.37–3.41 Å).

Some "butterfly" shaped fused chalcogen-containing heterocyclic aromatic hydrocarbons, **89–100** were designed and synthesized with the aim of enlarging π-conjugated system, thus forming compressed π–π stacking, which is good for charge carrier transport. However, it is a pity that the thin films of these materials can only afford mobilities in a range of 10^{-3}–10^{-1} $cm^2\ V^{-1}\ s^{-1}$. The vacuum evaporated films of **90** showed a mobility of 0.012 $cm^2\ V^{-1}\ s^{-1}$ after thermal annealing at 230 °C for 20 min [140]. The isomeric hexylated compounds **93** exhibited a mobility of 0.074 $cm^2\ V^{-1}\ s^{-1}$ [141]. With the extension of π-conjugated system, alkyl functionalized chrysene derivatives **94** and **95** exhibited mobilities as high as 0.4 $cm^2\ V^{-1}\ s^{-1}$ based on their vacuum deposited films [142]. In addition, the benzo[1,2-*b*:4,5-*b*′]bis[*b*]benzothiophene derivative, **96** showed good self-assembly characteristics [143]. Transistors based on individual **96** microwires obtained via slow crystallization showed typical *p*-channel behavior with a mobility of 2.1 $cm^2\ V^{-1}\ s^{-1}$. Crystalline nanoribbons of **97** formed through solution self-assembly also exhibited excellent performance with a mobility of 0.42 $cm^2\ V^{-1}\ s^{-1}$ [144]. In 2012, a series of larger fused π-conjugated systems, **98–100** was synthesized upon highly regioselective bromination of naphthodithiophene in different solvents. The solution-processed OFETs based on **99** exhibited a hole mobility of 0.072 $cm^2\ V^{-1}\ s^{-1}$ with an on/off current ratio of 10^6 under ambient atmosphere [145]. In addition, two novel "sunflower" small molecules, **101** and **102** were developed [146, 147]. OFETs using **100** as a semiconducting layer exhibited mobilities of about $9 \times 10^{-3}\ cm^2\ V^{-1}\ s^{-1}$. On the other hand, the tetrathiotetraseleno [8] circulene, **102** provided a maximum mobility of $1 \times 10^{-3}\ cm^2\ V^{-1}\ s^{-1}$, about tenfold lower than that of **101**.

Thienoacene-Based Oligomers and Oligothiophenes

Oligomers of the smaller thienoacenes and oligothiophenes were widely investigated because of their high planar structure and property of easy-to-modify with alkyl substituents. Many oligothiophenes afforded promising FET performances. For example, oligomers **103–107** gave good mobilities [148, 149]. Thin film FET devices based on oligomers of **103**, **104**, **105**, and **107** demonstrated mobilities as high as 0.42, 0.12, 0.54, and 0.14 $cm^2\ V^{-1}\ s^{-1}$, respectively. The device based on **104** maintained a hole mobility of about 0.1 $cm^2\ V^{-1}\ s^{-1}$ even after 9 weeks [148]. The single crystal transistor based on **106** exhibited a mobility of 1.1 $cm^2\ V^{-1}\ s^{-1}$ with a current on/off ratio of 6.5×10^4 [149]. Although there are similar mobilities between α,α′-bis(dithieno[3,2-*b*:2′,3′-*d*]-thiophenes) **108**, with mobilities of $10^{-2}\ cm^2\ V^{-1}\ s^{-1}$, and the vinylene-bridged dimer, **110** with mobility of 0.08 $cm^2\ V^{-1}\ s^{-1}$ [150–152] a large difference in mobility exists between dimer **111** and vinylene-bridged dimer **112**, which afforded the mobilities of 5×10^{-3} and 0.42 $cm^2\ V^{-1}\ s^{-1}$, respectively. The similar phenomenon can also be found in thieno[3,2-*b*] [1] benzothiophene-based dimers **114** and **115** [153]. The double bond-linked derivatives **115** exhibited mobilities as high as 0.50 $cm^2\ V^{-1}\ s^{-1}$ with on/off current ratios of 4.8×10^6. Mobility was 10 times larger than the $7 \times 10^{-3}\ cm^2\ V^{-1}\ s^{-1}$ of the dimer **114**. The results showed that the vinylene bridge is a useful π-conjugated building block for organic semiconductors. Oligomer, **113** end-capped with phenyl groups, showed mobilities even as large as 2.0 $cm^2\ V^{-1}\ s^{-1}$, on/off current ratios up to 10^8, and high environmental stability [154]. No obvious degradation was observed after storing in air for 7 months. In combination with device stability of other oligomers, for example, **104**, a conclusion can be reached that FET devices based on the oligomers end-capped with phenyl tend to have high stability under ambient conditions. Some other oligomers, **116–119** based on thienothiophene were also investigated. OFETs with **116** as the active layer showed a mobility of 0.67 $cm^2\ V^{-1}\ s^{-1}$ in air [155]. OFETs based on **117–119** showed mobilities as high as 0.12, 0.10, and 0.15 $cm^2\ V^{-1}\ s^{-1}$, respectively [156–158]. Planar oligothiophenes **120–122** were also used as semiconducting materials in OFETs. Of these, oligothiophene **121** gave mobility of up to 0.1 $cm^2\ V^{-1}\ s^{-1}$ [159]. In addition, oligothiophene **122** showed a higher mobility of 0.28 $cm^2\ V^{-1}\ s^{-1}$ [48, 160]. However, with the increase of thiophene unit number, the resulting oligothiophenes demonstrated distinctly decreased solubility in common solvents, which is not good for the synthesis and purification of these materials and their application in solution-processed FET devices. Thus, many alkyl-substituted oligothiophenes were synthesized. Sometimes, oligothiophenes end-capped with alkyl chain showed highly increased mobilities. For example, dihexyl quaterthiophene, **123** afforded a mobility of 0.06 $cm^2\ V^{-1}\ s^{-1}$, which is tenfold higher than the 2.5×10^{-3} of quaterthiophene **120** [161]. Dihexyl sexithiophenederivative, **124** showed an increased mobility of 0.08 $cm^2\ V^{-1}\ s^{-1}$ [162], and even gave mobilities as high as 1.0 $cm^2\ V^{-1}\ s^{-1}$ [48]. Besides the linearly alkyl groups, cycloalkyl groups, the benzene ring, and the naphthalene ring have also been introduced into the molecular structures as end-capping groups [163–165]. The single crystal OFETs based on phenyl-capped quaterthiophene **125** and quinquethiophene **126** showed mobilities of 0.09 and 0.13 $cm^2\ V^{-1}\ s^{-1}$. Naphthalene-capped quaterthiophene **127** demonstrated a mobility of 0.4 $cm^2\ V^{-1}\ s^{-1}$. In addition, a novel cyclic oligothiophene of tetrathia-[22]annulene[2,1,2,1], **128** was synthesized and used as active material in OFETs [166]. It adopted a nearly planar structure and its evaporated films showed a mobility of 0.05 $cm^2\ V^{-1}\ s^{-1}$ (Chart 3.7).

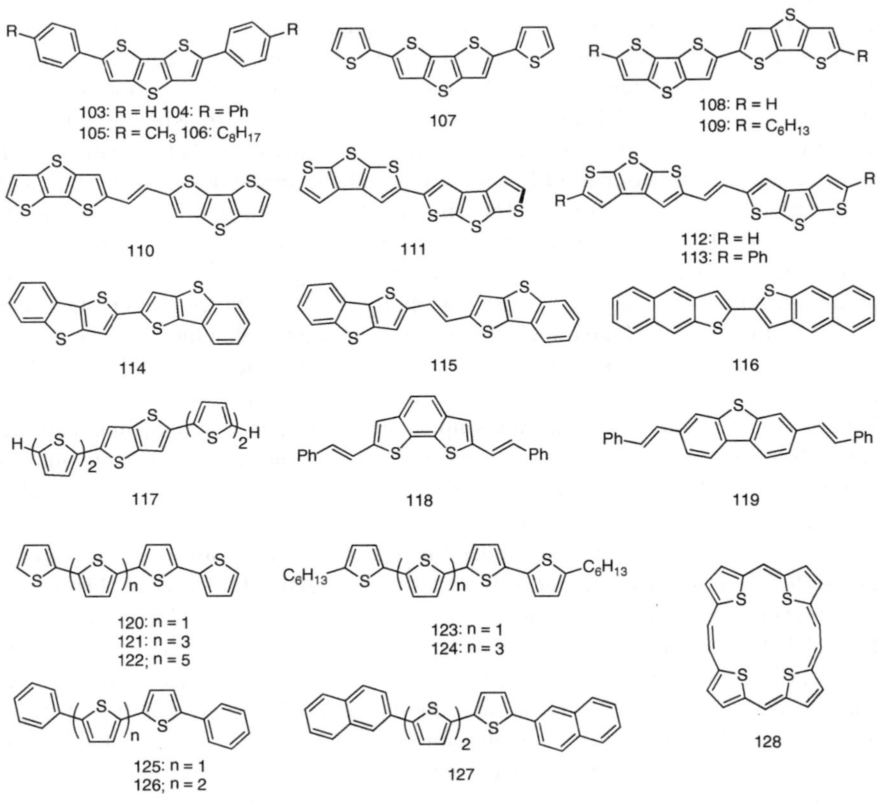

Chart 3.7 Thienoacene-based oligomers and oligothiophenes

Trialkylsilylethynyl Substituted Thienoacenes

As mentioned above, the introduction of trialkylsilylacetylene groups at the peri-position of acenes leads to improved π–π packing as well as increased solubility and stability. Thus the kind of thienoacene derivatives had also been widely studied by several groups. For instance, Anthony et al. [167] synthesized a series of anthradithiophene derivatives, **129–131** through similar synthetic procedures. Drop-casted films of **131**, a mixture of *syn*- and *trans*-isomer, showed high mobilities of up to 1.0 $cm^2\ V^{-1}\ s^{-1}$, which was attributed to its close two-dimensional (2D) π-stacking arrangement in the solid state. However, when small alkyl groups were introduced to the π-conjugated backbone of derivative **131**, the resulting compounds, **132–134** afforded lower mobilities [168]. Of these, compound **132** formed one-dimensional (1D) π-stacking arrangement and its solution-casting films showed hole mobilities in a range from 0.1 to 0.4 $cm^2\ V^{-1}\ s^{-1}$. Compounds **133** and **134** adopted 1D and 2D π-stacking arrangement, respectively. Nonetheless, the two derivatives showed low mobilities, probably because of their poor morphology and loose π-stacking mode. Partial fluorinated derivatives, **135** and **136** still behaved as p-type semiconductors but with dramatic increases in thermal and photostability compared to the non-fluorinated derivatives [169, 170]. Derivative **135** forms highly crystalline films even from spin-cast solutions, leading to devices with high mobility up to 1.5 $cm^2\ V^{-1}\ s^{-1}$. Its single crystal OFETs showed mobilities as high as 6 $cm^2\ V^{-1}\ s^{-1}$ with an on/off current ratios of 10^8. The high mobilities could arise from the non-covalent interactions of F–F and F–S which was observed in single crystals. Encouraged by the results above, two isomerically pure *syn*-anthradithiophene derivatives, **135a** and **136a** were synthesized and investigated [171]. The performances of the two *syn*-isomers **135a** and **136a** and *syn/anti* mixture, **135** and **136** in single crystal OFET devices are comparable, showing mobility values of 1.07 $cm^2\ V^{-1}\ s^{-1}$ for **135a** and 0.30 $cm^2\ V^{-1}\ s^{-1}$ for **136a** (1.01 $cm^2\ V^{-1}\ s^{-1}$ for **135** and 0.41 $cm^2\ V^{-1}\ s^{-1}$ for **136** under the same conditions). The lack of any difference between devices based on isomerically pure *syn*-isomer and syn/anti mixture was attributed to the fact that both materials suffer from similar positional disorder of the thiophene moieties in the solid state (Chart 3.8).

Bao et al. [172] reported a series of trialkylsilylethynyl-substituted thienoacene derivatives including anthra[2,3-*b*]thiophenes (ADT), **137–139** and tetraceno[2,3-*b*] thiophenes (TDT), **140–142**. Of these, derivative **142** showed the highest mobilities of up to 1.25 $cm^2\ V^{-1}\ s^{-1}$ on bare SiO_2 substrate. To investigate the impact of backbone size on solid-state order and electrical properties, 5-, 6-, and 7-ringed tri-*sec*-butylsilylethynyl-substituted ADT, TDT, and pentacenedithiophene (PDT) derivatives, **143–145** were synthesized by Jurchescu et al. [173]. Studies demonstrated that increasing the backbone size in fluorinated functionalized acenedithiophenes had a dramatic effect on the structural and electrical properties, leading to improved crystalline order, better π-stacking, and superior device performance. Typical field-effect mobilities for drop-cast films were distinctly different over a large range from $10^{-3}\ cm^2\ V^{-1}\ s^{-1}$ for the ADT derivative **143** to $10^{-2}\ cm^2\ V^{-1}\ s^{-1}$ for TDT derivative **144**, and 1.0 $cm^2\ V^{-1}\ s^{-1}$ for PDT derivative **145**. The field-effect

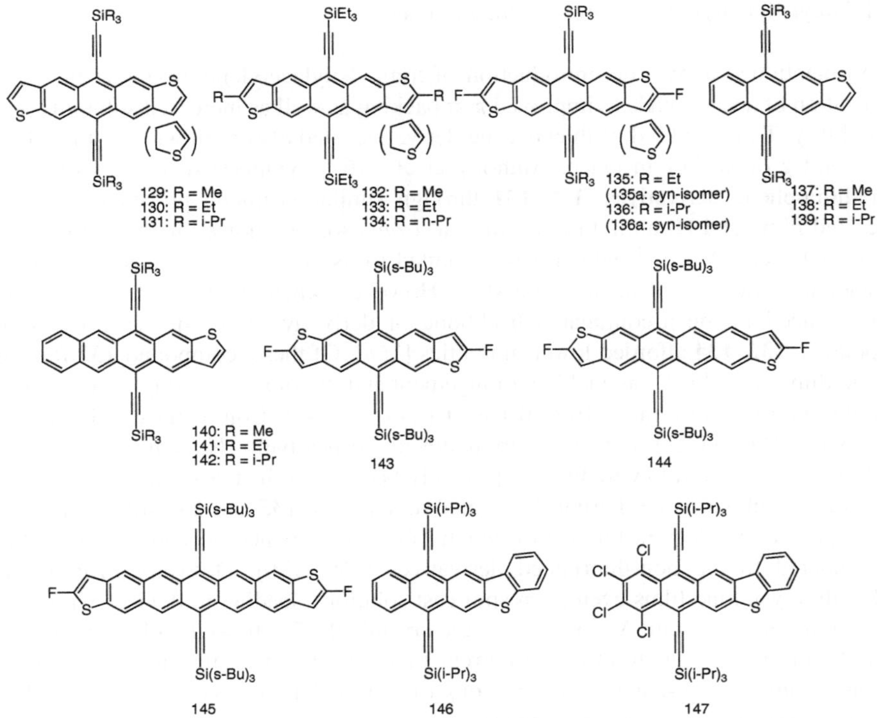

Chart 3.8 Trialkylsilylethynyl-substituted thienoacenes

mobility correlated well with the solid state order presented in the thin films, where a 2D π-stacking and superior long range order yielded good electronic performance, and a 1D stacking presented in the derivative **143**, as well as film disorder in derivative **144**, were reflected in lower mobilities. Recently, Yu group reported a series of trialkylsilylacetylene functionalized anthra[2,3-*b*]benzo[*d*]thiophene derivatives, **146** and **147** [174]. Evaporated films of **147** showed hole mobility of 0.012 $cm^2 V^{-1} s^{-1}$, which is significantly larger than that of the non-chlorinated counterpart **146**. The improvements in mobility verified that chlorine atoms could effectively tune intermolecular interaction. It is interesting that by utilizing the rotation of large isopropyl groups, two types of single crystals of **146** with different molecular conformation were successfully grown and separated. The related quantum-chemical calculations predicted that the two conformers have nearly 200 times difference in mobility. The results first demonstrated the important influence of molecular conformation on charge carrier transport.

Tetrathiafulvalene and Derivatives

Tetrathiafulvalene (TTF) and its derivatives are sulfur-rich small-molecules, and have been widely investigated as organic conductors and superconductors (see Chart 3.9) [175–177]. In 1993, TTF derivatives were first used in OFETs as semiconducting layers [178]. The FETs based on TTF and its derivatives could afford high mobilities up to 10 cm^2 V^{-1} s^{-1} [179]. Of these, the devices based on symmetrical derivatives tend to exhibit higher mobilities than those of unsymmetrical ones, which can be attributed to more orderly arrangement in solid states. Studies on crystals of TTF, **148** have found that the crystals which belonged to monoclinic α-phase of TTF demonstrated a mobility of 1.2 cm^2 V^{-1} s^{-1}, higher than the 0.23 cm^2 V^{-1} s^{-1} of those belonging to triclinic β-phase TTF [25]. This was attributed to the strong π-stacking along the b axis in the α-phase. Mas-Torrent et al. had developed a series of TTF derivatives [180, 181]. Because the long alkyl chains promoted intermolecular π–π overlapping because of their extremely closely packed nature, TTF-4SC18, **149** easily assembled into one-dimensional stacks and can be easily solution-processed by zone-casting. The FETs based on aligned films of **149** showed a mobility up to 0.08 cm^2 V^{-1} s^{-1} with an on/off current ratios of 10^4 [182]. The single crystals transistors of **150** exhibited the highest mobility of 3.6 cm^2 V^{-1} s^{-1} [183]. Single crystals of **151** also displayed high mobility up to 1.0 cm^2 V^{-1} s^{-1} [184]. As the extension of π-conjugation system, thin films of derivatives **152–156** showed hole mobilities of 0.42, 0.2, 0.11, 0.2, and 0.64 cm^2 V^{-1} s^{-1}, respectively. Mobilities of chlorinated derivatives were higher than the corresponding unsubstituted one which might be attributed to the stronger

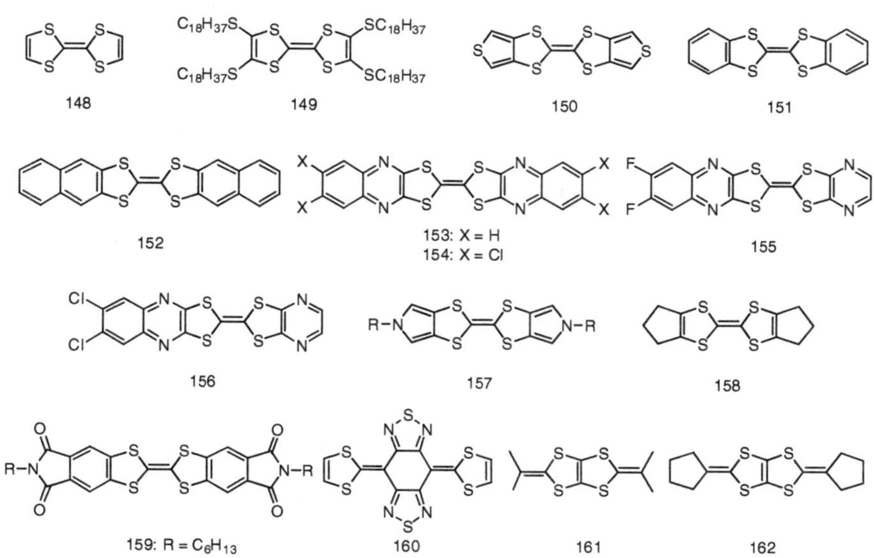

Chart 3.9 Tetrathiafulvalene and derivatives

intermolecular interactions promoted by chlorine atoms [185, 186]. However, the spin-coated films of *N*-alkylsubstituted bis(pyrrolo[3,4-*d*])tetrathiafulvalenes, **157** only showed a mobility of 0.013 $cm^2\ V^{-1}\ s^{-1}$ and an on/off current ratio of about 10^4 [187]. Takahashi and co-workers reported another TTF derivative, **158**, whose single transistor employing thermally evaporated TTF-TCNQ films as the source/drain electrodes exhibited mobilities over 10 $cm^2\ V^{-1}\ s^{-1}$ [179]. The high performance was attributed to the close two-dimensional lamellar arrangement and ideal contact between active semiconducting layer and the TTF-TCNQ electrodes. A hexyl group-substituted dibenzotetrathiafulvalene bisimide derivative, **159** showed a hole mobility of 0.4 $cm^2\ V^{-1}\ s^{-1}$ with an on/off current ratio of up to 10^6–10^8 [188]. The derivative, **160** with the TTF units separated by a cyclohexane unit exhibited a mobility as high as 0.2 $cm^2\ V^{-1}\ s^{-1}$ for thin films and 4 $cm^2\ V^{-1}\ s^{-1}$ for single crystals [189–191]. In addition, TTP derivatives, **161** and **162** were synthesized. The transistors fabricated by vacuum deposited film of **162** showed a mobility of 0.27 $cm^2\ V^{-1}\ s^{-1}$ [192].

3.2.1.3 Nitrogen-Containing Heterocyclic Aromatic Hydrocarbons and Derivatives

Azaacenes

The nitrogen atom is a versatile heteroatom, which can be used as an sp^2 hybridization form, such as pyrrole-nitrogen and pyridine-nitrogen, sp^3 and sp hybridization forms [193]. Because the lone pair of pyrrole-nitrogen is delocalized in aromatic systems, the related semiconductors have high HOMO and LUMO energy levels, and thus tend to exhibit p-type transport characteristics. However, because the lone pair of pyridine-nitrogen does not take a part in a π-electron system in combination with the stronger electronegativity of the nitrogen atom, the resulting semiconductors often have low HOMO and LUMO energy levels, and thus could exhibit n-type characteristics. Nuckolls et al. synthesized a series of nitrogen-containing derivatives including dihydrodiazapentacene **163** (see Chart 3.10) [194]. The vacuum deposited film FET devices of **163** exhibited low hole mobility of $5 \times 10^{-5}\ cm^2\ V^{-1}\ s^{-1}$. Later, Miao et al. found that derivative **163** could form three crystalline polymorphs according to (001) spacing and the field-effect mobility of **163** is extremely sensitive to the polymorphs with the "12.9 Å phase" yielding a mobility of 0.45 $cm^2\ V^{-1}\ s^{-1}$, which is over 5000 times higher than those of the other two phases [195]. Thin film transistors based on **164** and **165** afforded mobilities of 1.4 and 0.13 $cm^2\ V^{-1}\ s^{-1}$, respectively, when active films were deposited on n-nonyltrichlorosilane-modified SiO_2/Si substrates using a thin film of pentacene as the buffer layer [196]. Novel nitrogen-containing heptacyclic bisindoloquinoline **166** was synthesized via an intramolecular cyclization of anthrazoline derivatives. Its single crystal FETs had carrier mobility as high as 1.0 $cm^2\ V^{-1}\ s^{-1}$ with on/off current ratios greater than 10^4 [197]. Tetraazapentacene derivatives, **167** and **168** exhibited mobilities of 0.02 and 0.01 $cm^2\ V^{-1}\ s^{-1}$ in air, respectively [198].

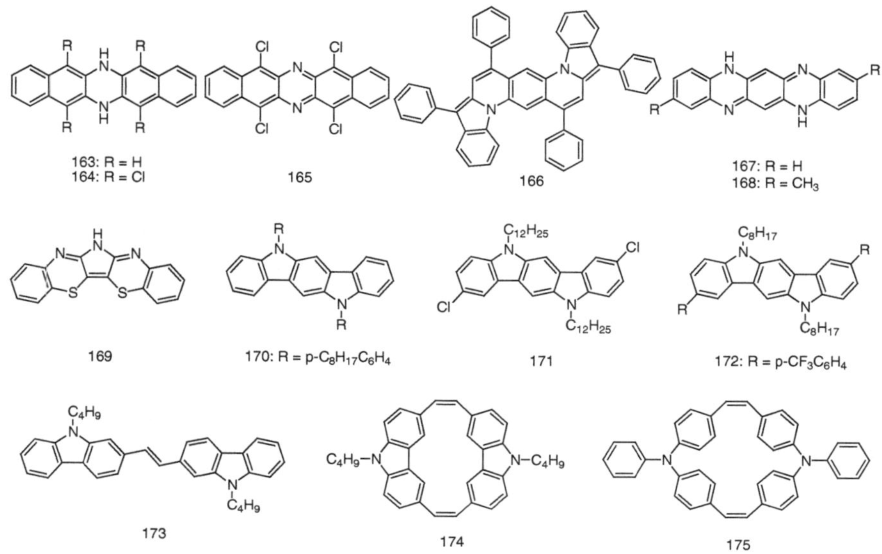

Chart 3.10 Selected azaacene semiconductors

A thiazine derivative, **169** which contains sulfur and nitrogen atoms was reported by Zhu and coworkers. A thin film of **169** exhibited a mobility of 0.34 $cm^2\ V^{-1}\ s^{-1}$ [199]. However, its single crystals mobilities afforded high mobilities up to 3.6 $cm^2\ V^{-1}\ s^{-1}$ [200]. 5,11-Bis(4-octylphenyl)indolo[3,2-*b*]carbazole, **170** readily self-organized into highly crystalline layered structures during vacuum deposition. FET devices based on the thin film of **170** exhibited mobility of 0.12 $cm^2\ V^{-1}\ s^{-1}$ [201]. The chlorinated derivative **171** also easily formed highly crystalline films with large interconnecting terrace-like layered domain structures, and showed a mobility of 0.14 $cm^2\ V^{-1}\ s^{-1}$ and an on/off current ratio of 10^7 under ambient conditions [202]. Yu and Tao also reported a similar ditrifluoromethylphenyl-substituted derivative **172**, which possesses a close π–π stacking with an intermolecular distance of 3.38 Å. The single crystal transistors afforded a hole mobility of 0.084 $cm^2\ V^{-1}\ s^{-1}$ [203]. Linear and cyclic carbazolenevinylene dimers **173** and **174** were also synthesized [204]. Multicrystalline thin films were observed from compounds **174**, but only amorphous thin films could be obtained for the linear compound **173**. Another important difference between the cyclic and linear compounds was the reduced reorganization energy for the cyclic compounds. These two facts have resulted in improved FET mobility for the cyclic compound compared with the linear counterpart. The FET devices based on **174** gave the highest mobility of $5.3 \times 10^{-3}\ cm^2\ V^{-1}\ s^{-1}$. The similar improvement in FET property could be found in triphenylamine derivatives. Because the nonplanar structure of triphenylamine derivatives results in the formation of amorphous films which are not beneficial for charge transport, these compounds were rarely applied in field-effect transistors and showed low mobilities [205]. However, cyclic triphenylamine dimer, **175** favored the formation of crystalline thin films and hence exhibited a mobility of 0.015 $cm^2\ V^{-1}\ s^{-1}$, which is much higher than $10^{-4}\ cm^2\ V^{-1}\ s^{-1}$ of its linear analogue [206].

Phthalocyanines, Porphyrins, and Organometallic Complexes

Phthalocyanines and their organometallic complexes have also been investigated as semiconducting layers in OFETs (see Chart 3.11). For example, OFETs based on phthalocyanine, **176** showed low mobilities of $10^{-3}\ cm^2\ V^{-1}\ s^{-1}$ [207]. However, its organometallic complexes such as copper phthalocyanine (CuPc), **177**, titanyl phthalocyanine (TiOPc), **178**, and vanadyl-phthalocyanine (VOPc), **179** exhibited higher mobilities. Highly-ordered CuPc thin films obtained by vacuum deposition showed a hole mobility of 0.02 $cm^2\ V^{-1}\ s^{-1}$ and an on/off current ratio of 10^5 [208]. The mobility of a single crystalline submicrometer-sized ribbon could even reach as

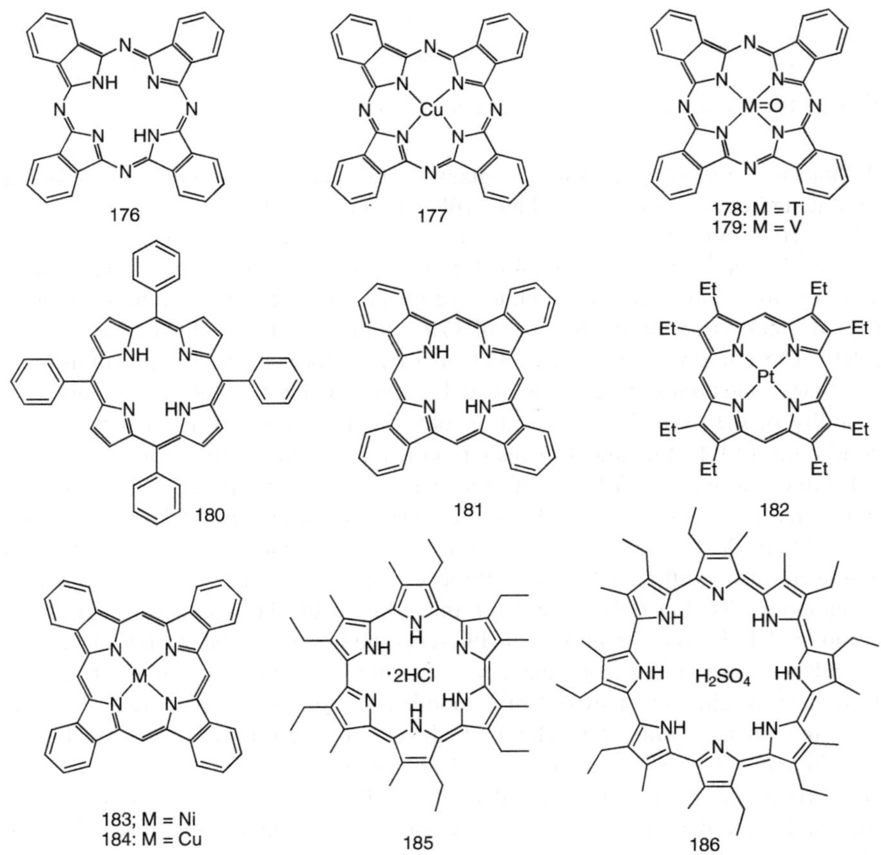

Chart 3.11 Phthalocyanines, porphyrins, and organometallic complexes

high as 0.5 $cm^2 V^{-1} s^{-1}$ [209]. Complexes **178** and vanadyl-phthalocyanine **179** are nonplanarly polar molecules. Thin film transistors based on **178** showed high mobilities of 3.3 $cm^2 V^{-1} s^{-1}$ [210]. High quality films of **179** showed a hole mobility of 1.5 $cm^2 V^{-1} s^{-1}$ [211]. Likewise, porphyrin and their complexes were also used in OFETs. Devices of the tetraphenyl porphyrin **180** and tetrabenzoporphyrin **181** exhibited improved field-effect performance with mobilities of 0.012 and 0.017 $cm^2 V^{-1} s^{-1}$, respectively [212, 213]. Annealing the as-deposited **182** film leads to the formation of a polycrystalline film which exhibited excellent overall charge transport properties with a charge mobility of up to 0.32 $cm^2 V^{-1} s^{-1}$ [214]. In addition, the spin-coated films of nickel and copper tetrabenzoporphyrins, **183** and **184** demonstrated mobilities of up to 0.2 and 0.1 $cm^2 V^{-1} s^{-1}$, respectively [215, 216]. Two porphyrin analogues, cyclo [6] pyrrole, **185** with a 22-π-electron system, and cyclo [8] pyrrole, **186** with a 30-π-electron system were developed and used as active layers of FETs. The thin films of **185** and **186**, which were prepared by the Langmuir–Blodgett (LB) method, exhibited mobilities of 0.014 and 0.68 $cm^2 V^{-1} s^{-1}$, respectively [217, 218]. In addition, there were some more complex phthalocyanine derivatives such as rare earth and triple-decker complexes examined [219].

3.2.2 Selected p-Type Polymer Semiconductors

3.2.2.1 Polythiophene-Based Semiconductors

Since the first OFET based on **187** was reported in the 1980s, 3-alkylated polythiophenes were most widely studied as active layers in PFETs because they were solution-processable (see Chart 3.12 and Table 3.2) [13–15]. Of these, P3HT, **188** is

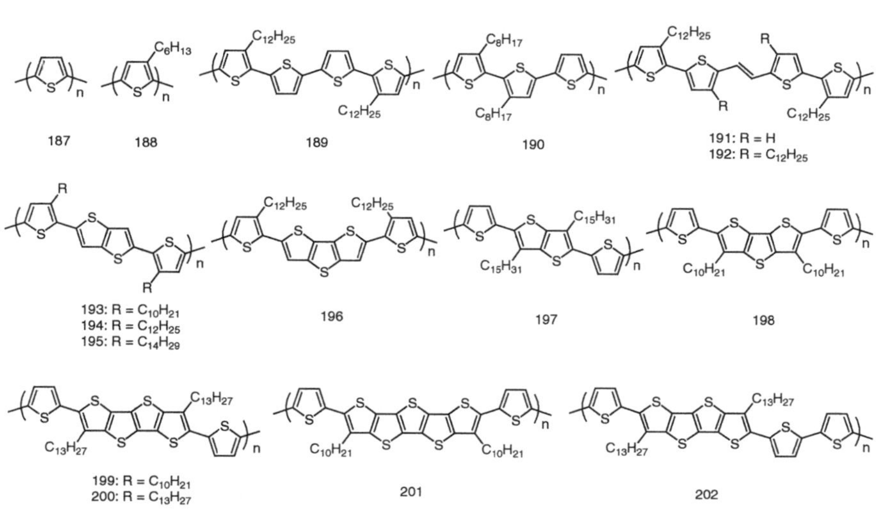

Chart 3.12 Polythiophene-based semiconductors

Table 3.2 OFET device data for p-type polymer semiconductors

	HOMO (eV)	Deposition process	Max μ_h ($cm^2 V^{-1} s^{-1}$) (m.d.)[a]	I_{on}/I_{off}	V_T (V)	Device structure[b]	Refs.
191	−5.2	Spin-coating	0.15	3.67×10^4	−8.5	TGBC; Au/Ni; PMMA/Al	[229]
192	−5.02	Spin-coating	1.05 (N_2)	5×10^4	−26.1	TGBC; Au/Ni; PMMA/Al	[230]
195		Spin-casting	1.0			TGBC; Au; SiO_2/Si; OTS	[234]
196		Drop-casting	0.32	10^6		BGBC; Au; SiO_2/Si; OTS-8	[238]
197	−5.23	Spin-casting	0.25	10^7		BGBC; Au; Si/SiO_2	[239]
198	−5.15	Spin-casting	0.0017	10^3	−11	BGTC; Au; SiO_2/Si; HMDS	[241]
199	−5.20	Spin-casting	0.087	10^6	−4	BGTC; Au; SiO_2/Si; HMDS	[241]
200		Spin-coating	0.33	10^5	−10	BGTC; Au; SiO_2/Si; HMDS	[240]
201	−5.21	Spin-casting	0.0023	$10^3 \sim 10^4$	−9	BGTC; Au; SiO_2/Si; HMDS	[241]
202		Spin-casting	0.027	$10^4 \sim 10^5$	5	BGTC; Au; SiO_2/Si; HMDS	[243]
203		Drop-casting	0.13	120	−21	BGBC; Au/Ti; SiO_2/Si; OTS	[244]
204	−5.22	Drop-casting	0.04	$10^5 \sim 10^6$	−20	BGBC; Au; SiO_2/Si; OTS	[240]
205	−5.22	Drop-casting	0.14	$10^5 \sim 10^6$	−21	BGBC; Au; SiO_2/Si; OTS	[240]
206	−5.0	Drop-casting	0.050	$10^5 \sim 10^7$	−17 ∼ −27	BGBC; Au; SiO_2/Si; OTS-8	[247]
207	−5.1	Drop-casting	0.23	$10^5 \sim 10^7$	−22 ∼ −31	BGBC; Au; SiO_2/Si; OTS-8	[247]
208	−5.1	Drop-casting	0.3	$10^5 \sim 10^7$	−23 ∼ −30	BGBC; Au; SiO_2/Si; OTS-8	[247]
209	−5.19	Spin-coating	0.33			BGTC; Au; SiO_2/Si; OTS	[248]
210	−5.1	Spin-coating	0.02	10^5	−4	BGTC; Au; SiO_2/Si; PV	[249]
212	−5.2	Spin-coating	0.26	10^7		BGTC; Au; SiO_2/Si; FDTS	[250]
213	−5.5	Spin-coating	0.04	10^6		TGBC; Au; CYTOP/Al	[251]

(continued)

Table 3.2 (continued)

	HOMO (eV)	Deposition process	Max μ_h ($cm^2\ V^{-1}\ s^{-1}$) (m.d.)[a]	I_{on}/I_{off}	V_T (V)	Device structure[b]	Refs.
214		Drop-casting	0.17	10^5		BGBC; Au; SiO_2/Si; HMDS	[255]
		Spin-coating	0.67	2×10^4		BGTC; Au; SiO_2/Si; PTES	[256]
		Dip-coating	1.4	$10^4 - 10^5$			
		Drop casting	3.3	$10^5 - 10^6$		BGBC; Au; SiO_2/Si; HMDS	[20]
215	−5.4	Spin casting	1.2	$10^3 - 10^4$	−25 ~ −34	TGBC; Au; CYTOP/Al	[257]
216	−5.16	Spin casting	0.6	2×10^4		BGTC; Ag; SiO_2/Si; OTS-8	[258]
		Directional solvent evaporation	6.7/23.7			BGBC; Au; SiO_2/Si; DTS	[259, 260]
217	−5.7	Spin-coating	0.79	~10^7		BGTC; Au; SiO_2/Si; OTS	[261]
218	−5.60	Spin-coating	0.40	>10^5	−10	BGTC; Au; SiO_2/Si; OTS	[263]
219	−5.52	Spin-coating	3.62	>10^6	−2	BGTC; Au; SiO_2/Si; OTS	[263]
220	−5.50	Spin-coating	1.76	>10^6	−5	BGTC; Au; SiO_2/Si; OTS	[263]
221	−5.20	Spin-coating	2.48	>10^6	3	BGTC; Au; SiO_2/Si; OTMS	[264]
222	−4.6	Spin-coating	2.5			BGBC; Au; SiO_2/Si; DTS	[265]
223	−5.16	Spin-coating	0.56	10^7	−10 ~ −2	BGTC; Au; SiO_2/Si; FDTS	[266]
224	−5.2	Spin-coating	10.5	>10^6	−6 ~ 4	BGBC; Au; SiO_2/Si; OTS; PMMA	[269]
225	−5.06	Spin-coating	1.95	~10^5	~0	TGBC; Au; PMMA/Au	[270]
226	−5.2	Spin-coating	0.97	~10^6	−2.7	BGTC; Au; SiO_2/Si; OTS	[271]
227	−5.12	Spin-coating	3.46	10^8	−10 ~ 5	BGTC; Au; SiO_2/Si; OTS	[272]
228	−5.30	Spin-coating	4.5	$10^5 - 10^7$	5 ~ 8	BGBC; Au; SiO_2/Si; OTS; PMMA	[273]

(continued)

Table 3.2 (continued)

	HOMO (eV)	Deposition process	Max μ_h ($cm^2\ V^{-1}\ s^{-1}$) (m.d.)[a]	I_{on}/I_{off}	V_T (V)	Device structure[b]	Refs.
229	−5.28	Spin-coating	8.2	$10^5 - 10^7$	5 ~ 8	BGBC; Au; SiO_2/Si; OTS; PMMA	[273]
230	−5.26	Spin-coating	4.97	10^7	2.92	BGTC; Au; Si/SiO_2; ODTS	[274]
231	−5.25	Spin-coating	10.54	>10^6		BGTC; Au; Si/SiO_2; CYTOP	[275]
232	−5.27	Spin-coating	12.04	>10^6		BGTC; Au; Si/SiO_2; CYTOP	[275]

[a]*m.d.* measurement condition
[b]Device configuration; S/D electrodes; dielectric/gate; modification of substrate/electrode

the most famous. Studies on polymer **188** showed that (1) the regioregular form of P3HT with head-to-tail (HT) coupling affords higher mobilities than regiorandom forms (HH and TT) in its three kinds of couplings: head-to-head (HH), tail-to-tail (TT), and head-to-tail (HT) coupling; (2) the regioregular form can adopt two kinds of arrangements on substrates: face-on and edge-on orientations; (3) edge-on orientation affords mobilities in the range 0.05–0.2 $cm^2\ V^{-1}\ s^{-1}$, usually about two orders of magnitude higher than those of the face-on orientation; and (4) the practical applications of polymer **188** were limited by its stability, mainly caused by its higher HOMO energy level. Most FETs based on the polymer were fabricated and examined in a glovebox with a N_2 atmosphere [47, 220–224]. Generally, there are two approaches used to improve the properties of FETs based on polythiophenes. One is regioregular introduction of alkyl chain on the π-conjugated backbone of polymer **187**. Alkyl groups can improve solubility, but also raise the HOMO energy levels of polythiophenes, making them sensitive to oxygen. Ong et al. developed a class of polythiophenes, **189**, in which only half of the thiophene rings were substituted by alkyl chains [225, 226]. The thin films afforded high mobilities up to 0.14 $cm^2\ V^{-1}\ s^{-1}$ and showed lower HOMO energy levels than that of polymer **188**, and thus better stability. Only slight degradation was observed after storing its devices in air for 1 month. It is pity that the mobility of related devices decreases under increasing humidity. However, alkyl groups were introduced with head-to-head (HH) coupling, the resulting polythiophenes tending to afford low mobilities. For example, polythiophene **190** exhibited the highest mobility of around 0.03 $cm^2\ V^{-1}\ s^{-1}$ though its HOMO energy level was found to be 0.20 eV lower than that of P3HT [227]. The decrease in mobility can be attributed to torsion of the polythiophene backbones caused by head-to-head (HH) coupling of side chains [228].

The other way is the introduction of highly coplanar π-conjugated groups on polythiophene backbones. The Kim group introduced vinylene groups to the main chain of polymer **189**, affording polythiophene **191** [229]. The increase of rotational freedom by incorporation of the vinylene group, in combination with the reduction of the alkyl group, made polymer **191** have lower HOMO energy level. In addition, the wider side-chain spacing induced a shorter interchain lamellar *d*-spacing than that of polythiophene **189**. The FETs based on polymer **191** showed a high mobility of 0.15 $cm^2\ V^{-1}\ s^{-1}$ with a relatively high oxidative stability. Nonetheless, because of highly packed molecular structure, polymer **191** exhibited relatively poor solubility in common organic solvents. Two alkyl-substituted conjugated polymers including **192** were developed by the same group [230]. Both polymers showed high solubility in common organic solvents as well as thin film crystallinity. Of these, thin films of the polymer **192** afforded a high mobility of over 1.0 $cm^2\ V^{-1}\ s^{-1}$. McCulloch and co-workers synthesized a series of novel polythiophenes, **193–195** by regiospecific incorporation of thieno[3,2-*b*]thiophene into the polymer backbone [231–236]. The HOMO energy levels of **193–195** were −5.1 eV, approximately 0.3 eV lower than that of P3HT. The low HOMO energy levels arise mainly from the reduced delocalization from the thienothiophene aromatic ring in comparison with thiophene, along with reduced electron donation from the fewer alkyl groups on the backbone than polymer **188**, as mentioned above. After annealing, the thin films of polymers **193–195** exhibited classical liquid-crystalline phases, of which crystalline domains are large—up to 200 nm. Transistors based on polymers **193–195** exhibited mobilities as high as 0.7 $cm^2 V^{-1} s^{-1}$, and 1.0 $cm^2 V^{-1} s^{-1}$ was found for the polymer **195**, even using Pt as the source and drain electrodes. Ong et al. introduced the dithieno[3,2-*b*:2′,3′-*d*] thiophene unit to the backbone, affording copolymer **196** [237, 238]. FET devices based on **196** afforded a decreased mobility of 0.3 $cm^2 V^{-1} s^{-1}$. The decrease in mobilities is probably because of the stronger rigid structure of dithieno[3,2-*b*:2′,3′-*d*]thiophene than that of thieno[3,2-*b*]thiophene. Copolymers based on alkylated thieno[2,3-*b*]thiophene cores usually showed lower HOMO energy levels than that of polymer **195**. Thin film transistors based on **197** showed a comparable performance to that of polymer **195** with a hole mobility of 0.25 $cm^2 V^{-1} s^{-1}$ [239, 240]. The Malliaras group developed a series of alkylated thienothiophene-based copolymers, **198–202** [241–243]. The trithienoacene- and pentathienoacene-based copolymers, **198** and **201** exhibited mobilities of 0.0017–0.0023 $cm^2 V^{-1} s^{-1}$, but the tetrathienoacene-based copolymers, **199** and **200** showed a higher mobility of 0.087 and 0.33 $cm^2 V^{-1} s^{-1}$, respectively. They ascribed this phenomenon to the influence of C2 symmetry on device performance. Indeed, when the backbone of the polymer **199** increased by one thiophene unit, affording the polymer **202**, the FETs based on polymer **202** showed a low mobility of 0.027 $cm^2 V^{-1} s^{-1}$. The results verified the importance of C2 symmetry on device performance.

3.2.2.2 Nitrogen-Containing Heteroaromatic Polymer Semiconductors

In addition to thiophene rings, other fused heteroaromatic building blocks were also explored and used in organic semiconductors. Of these, nitrogen-containing heteroaromatic building blocks including alkyldithieno[3,2-*b*:2′,3′-*d*]pyrrole, oxadiazole, thiazole, and benzobis(thiadiazole) were widely applied in OFETs. McCullough et al. reported a series of copolymers based on *N*-alkyldithieno[3,2-*b*:2′,3′-*d*]pyrroles (DTP) units [244]. The incorporation of soluble substituted thiophenes and planar DTP units resulted in low bandgap, highly conductive polymers. As-cast **203** thin films exhibited a poorly defined and randomly ordered lamellar structure which improved significantly after thermal annealing. FET devices based on polymer **203** showed typical *p*-channel transistor behavior. Interestingly, the mobilities of as-cast less ordered samples were much higher than those observed after annealing. The highest mobilities and average values were observed for the polymer **203** to be 0.21 and 0.13 $cm^2 V^{-1} s^{-1}$, respectively. However, the copolymer **203** has high HOMO energy level between −4.68 and −4.96 eV, so the related FET device showed low stability. In order to improve the stability, thiazole units were introduced, affording copolymers **204** and **205** [245]. The thiazole-based copolymers **204** and **205** were expected to possess high stability because of their donor–acceptor systems. Transistors based on polymers **204** and **205** exhibited mobilities of 0.04 and 0.14 $cm^2 V^{-1} s^{-1}$, respectively. In fact, the devices were very stable and no obvious degradation was observed even after the devices were stored in air for 60 days. A series of thiazolothiazole-based polymers, **206–208** showed hole mobilities of 0.05, 0.23, and 0.3 $cm^2 V^{-1} s^{-1}$, respectively [246, 247]. Liquid-crystalline semiconducting copolymer, poly(didodecylquaterthiophene-*alt*-didodecylbithiazole), **209**, which contains thiazole units, was developed by the Bao group [248]. Thin films of this polymer adopt preferential well-ordered intermolecular π–π stacking parallel to the substrate surface and showed mobilities as high as 0.33 $cm^2 V^{-1} s^{-1}$. Ong et al. synthesized homopolymers **210** and **211** [249]. Polymer **210** showed a mobility of 0.02 $cm^2 V^{-1} s^{-1}$; however, no FET performance was observed from **211**. Osaka et al. [250] reported benzobisthiazole-based copolymers **212**. FET devices showed excellent environmental stability in high-humidity air, which is an unusual performance for semiconducting polymers, along with hole mobilities as high as 0.26 $cm^2 V^{-1} s^{-1}$, even with disordered thin film structures. In addition, polytriarylamines constituted a class of polymer semiconductors with high stability. The devices based on polymer **213** showed excellent stability under ambient condition with a mobility of 0.04 $cm^2 V^{-1} s^{-1}$ [251] (Chart 3.13).

3.2.2.3 p-Type Donor–Acceptor Copolymer Semiconductors

In recent years, donor–acceptor (D–A) copolymers have been widely developed and applied in organic optoelectronics (see Chart 3.14) [252–254]. The properties of D–A copolymers, such as solubility, photophysics, electrochemical properties, crystallinity, packing mode, and device performance, can easily be modified by tuning the structures of donor, acceptor, and side chains. Moreover, the intra- and intermolecular "push-drag" effect between donor and acceptor units could effectively improve the packing mode in thin film of polymers affording close and orderly π–π arrangements of polymer, thus giving high device performance. In 2007, Müllen et al. reported a D–A copolymer, **213** based on benzothiadiazole (BT) and cyclopenta[2,1-*b*:3,4-*b*′]dithiophene (CDT) [255]. With a number-average molecular weight (M_n) of 10.3 kDa, thin films of **213** gave a mobility of 0.17 $cm^2 V^{-1} s^{-1}$ with an on/off current ratio of 10^5 after annealing at 200 °C for 2 h. After purifications of BT and CDT by passing the CDT monomer through a recycling high performance liquid chromatography/gel permeation chromatography (HPLC/GPC) system before polymerization, high molecular weight polymers with M_n of 50.0 kDa were obtained, the thin films of which showed a hole mobility of up to 0.67 $cm^2 V^{-1} s^{-1}$ for their spin-coated films. By using a dip-coating method, the macroscopic organization was further improved, resulting in hole mobilities of 1.0–1.4 $cm^2 V^{-1} s^{-1}$ along the dip-coating direction and 0.5–0.9 $cm^2 V^{-1} s^{-1}$ perpendicular to the dip-coated direction [256]. In 2011, Müllen et al. [41] further modified polymers by tuning the molecular weight, side alkyl chain, and morphology of thin film. Mobility was further improved to 3.3 $cm^2 V^{-1} s^{-1}$, which is the world record for p-type thin film PFETs at that time. D–A copolymer, **215** based on fused ladder indacenodithiophene (IDT) and benzothiadiazole (BT) units

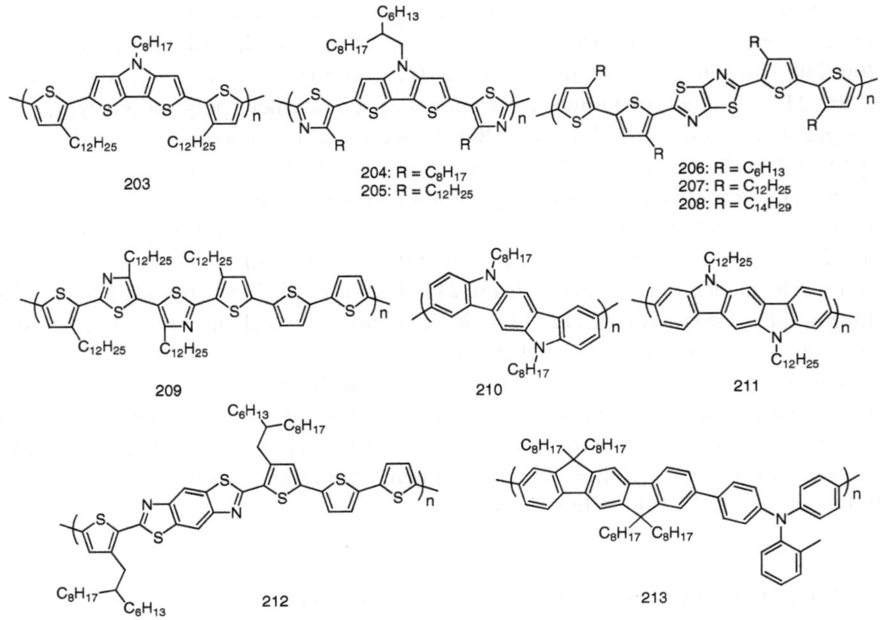

Chart 3.13 Nitrogen-containing heteroaromatic polymer semiconductors

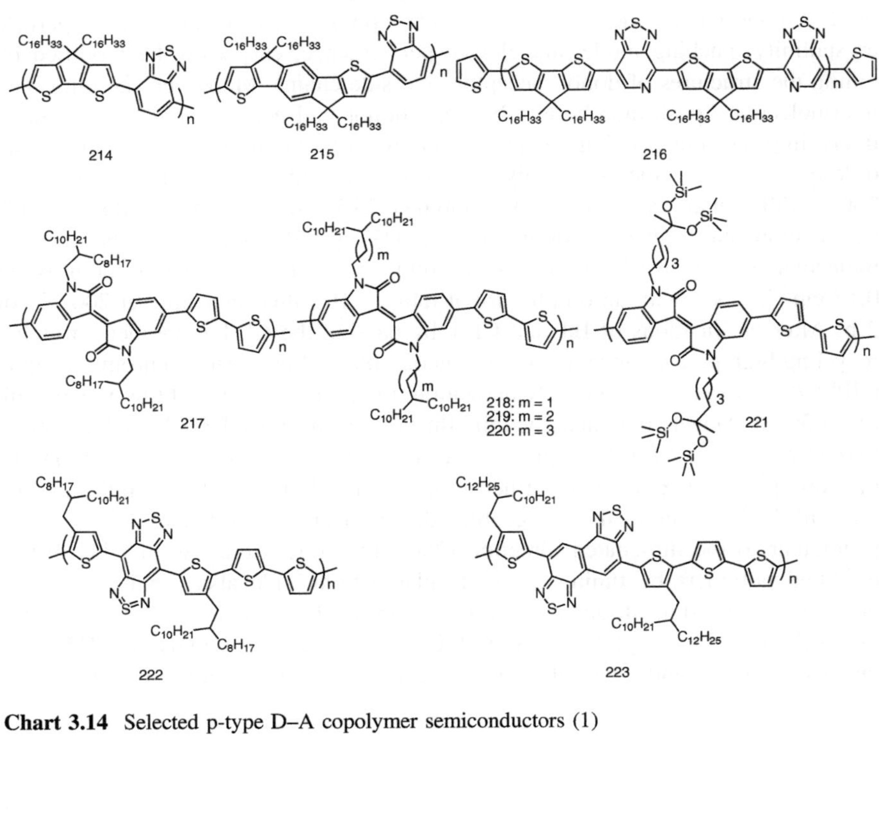

Chart 3.14 Selected p-type D–A copolymer semiconductors (1)

exhibited a lack of pronounced thin-film crystallinity, combined with the suboptimal orientation of the backbone and relatively large π-stacking distances, which would not be expected to be favorable for charge transport [257]. However, these thin film devices fabricated by spin-coating exhibited high mobilities in the range 0.8–1.2 $cm^2\ V^{-1}\ s^{-1}$ with an on/off current ratio between 10^3 and 10^4. The results demonstrated that the indacenodithiophene (IDT) unit has a huge potential for application in high performance FETs. Bazan group developed a series of regioregular pyridal [1–3] thiadiazole–cyclopenta[2,1-*b*:3,4-*b*′]dithiophene copolymers using the unsymmetrical pyridal [1–3] thiadiazole unit [258]. When the transistors were fabricated by the polymer **216** with an M_n of 34 kDa, a high mobility of 0.6 $cm^2\ V^{-1}\ s^{-1}$ was afforded. Similarly, when polymer molecular weight was increased, the corresponding hole mobilities also increased, from 0.8 $cm^2\ V^{-1}\ s^{-1}$ (M_n = 100 kDa) to 2.5 $cm^2\ V^{-1}\ s^{-1}$ (M_n = 300 kDa) [259]. The mobility of polymer **216** with high molecular weight (M_n = 300 kDa) can be further improved to 6.7 $cm^2\ V^{-1}\ s^{-1}$ using the macroscopic alignment method, by which long-range orientation of the polymer chains was obtained. The transport is anisotropic in the long-range alignment of polymer fibers, where the mobility for a direction parallel to the polymer backbone is six times higher than that for a direction perpendicular to the polymer backbone. Interestingly, the macroscopic alignment method is a combination of directional solvent evaporation through a tunnel-like configuration and slow drying, of the polymer on a nanostructured insulator, and operationally simple, not requiring sophisticated equipment or instrumentation. Moreover, a systematic study of OFETs based on polymer **216** with several weight average molecular weights (M_w) ranging from 30, 50, 80, 160, to 300 kDa demonstrated much higher mobilities of up to 23.7 $cm^2\ V^{-1}\ s^{-1}$ [260]. Studies showed that the high mobility values are relatively insensitive to M_w, which is consistent with long-range alignment of the semiconducting polymer chains such that the transport occurs dominantly along the conjugated backbone with occasional π–π hopping to neighboring chains.

Isoindigo is a kind of pigment with planar structure and strong electron-withdrawing ability. In recent years it had also been used as a building block to synthesize polymer semiconductors with high performance [261–264]. In 2011, Pei et al. reported an isoindigo-based D–A conjugated polymer, **217**. Thin film of the polymer had a low-lying HOMO energy level of −5.8 eV and showed crystalline fibrillar intercalating networks. Its FET devices exhibited a high mobility up to 0.79 $cm^2\ V^{-1}\ s^{-1}$, an on/off current ratio of 10^7, and good stability. They also investigated how the branching position of alkyl side chains affects FET performance of conjugated polymers **218–220**. High mobility of 3.62 $cm^2\ V^{-1}\ s^{-1}$ was achieved for polymer **219**, in comparison to the 1.76 $cm^2\ V^{-1}\ s^{-1}$ for polymer **220** and 0.40 $cm^2\ V^{-1}\ s^{-1}$ for polymer **218**. The developed branching position variation strategy revealed the significance of sophisticated side chain molecular engineering of conventional alkyl chains and its dependence on the backbone structures. In addition, the influence of symmetry and backbone curvature of polymers on interchain π–π stacking, lamellar packing, and crystallinity of isoindigo-based polymers was also investigated by introducing different donor units. Bao et al. introduced a novel siloxane-terminated solubilizing group to isoindigo-based conjugated polymer, affording **221**. The solution-processed thin film transistors showed an average mobility of 2.00 $cm^2\ V^{-1}\ s^{-1}$ (with a maximum mobility of 2.48 $cm^2\ V^{-1}\ s^{-1}$). The improvement in mobility was attributed to a combination of a small π-stacking distance, mixed crystallographic orientation (texture), and a large crystalline coherence length, all of which were induced by the introduction of siloxane-terminated side chains. D–A polymer **222** containing a co-benzobisthiadiazole-quaterthiophene unit showed typical *p*-channel OFET characteristics with a high carrier mobility up to 2.5 $cm^2\ V^{-1}\ s^{-1}$, which was attributed to the effectively long π-conjugation system, the very tight packing pattern, and the high-order orientation in the crystalline state [265]. Because of the highly π-extended structure and strong electron affinity of naphthobisthiadiazole (NTz), polymers **223** exhibited a deeper HOMO energy level of −5.54 eV. A thin film of **223** exhibited a hole mobility of 0.56 $cm^2\ V^{-1}\ s^{-1}$ with an on/off current ratio of ~10^7 [266] (Chart 3.14).

Diketopyrrolopyrrole (DPP)-based polymers are currently displaying some of the highest mobilities because of the remarkable aggregating properties of the DPP moieties (see Chart 3.15) [254]. In 2008, the Winnewisser group first applied DPP-based polymers as the active layer in OFETs. Solution-processed ambipolar field-effect transistors based on this material exhibited hole and electron mobilities of 0.1 and up to 0.09 $cm^2\ V^{-1}\ s^{-1}$, respectively [267]. DPP-based copolymers, **224** comprising thieno[3,2-*b*]thiophene moieties was developed in 2010 [268]. The strong intermolecular interactions led to the formation of interconnected polymer

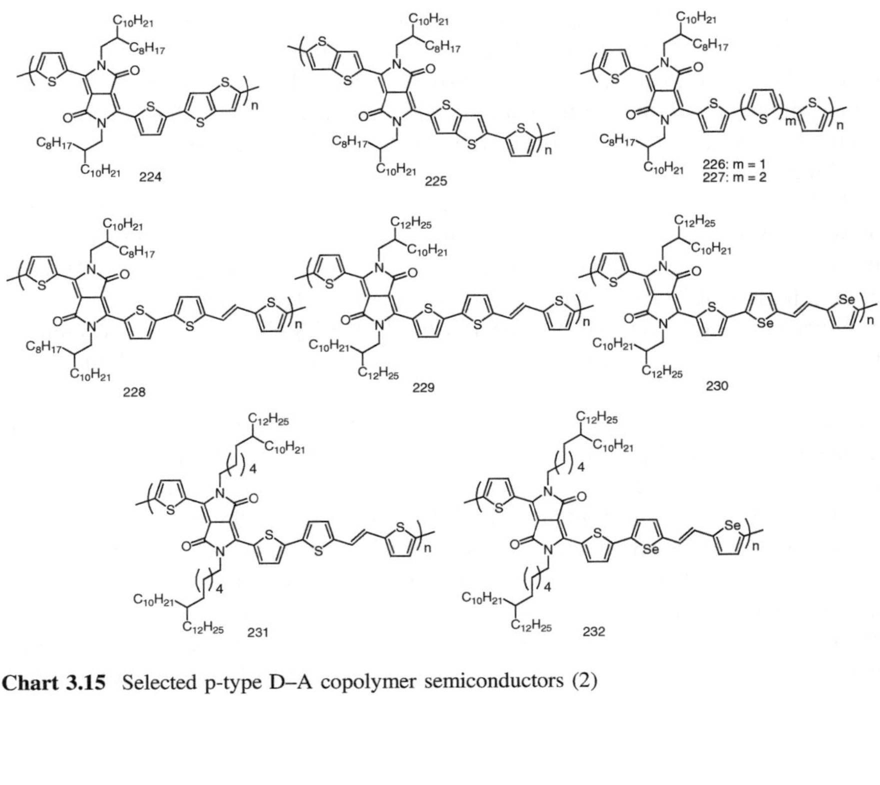

Chart 3.15 Selected p-type D–A copolymer semiconductors (2)

networks with an ordered lamellar structure and a small π–π stacking of 3.71 Å, which established highly efficient pathways for charge carrier transport. Polymer **224**-based FET devices showed a high hole mobility of up to 0.94 $cm^2 V^{-1} s^{-1}$. However, high molecular weight polymer **224**, obtained by an optimized synthetic procedure, afforded a tighter π–π stacking distance of 3.43 Å [269]. OTFT devices fabricated with this polymer semiconductor have exhibited very high mobilities of up to 10.5 $cm^2 V^{-1} s^{-1}$ with an on/off current ratio of 10^6. Copolymerization of a more rigid DPP-based building block starting from thieno[3,2-*b*]thiophene with the thiophene unit afforded a polymer, **225** with a maximum hole mobility of 1.95 $cm^2 V^{-1} s^{-1}$ [270]. The D–A polymer, **226** comprising DPP and a *β*-unsubstituted quaterthiophene unit formed ordered layer-by-layer lamellar packing with an edge-on orientation in thin films even without thermal annealing [271]. The well-interconnected crystalline grains formed efficient intergranular charge transport pathways. The polymer afforded a high hole mobility up to 0.97 $cm^2 V^{-1} s^{-1}$ in FETs with polymer thin films annealed at a mild temperature of 100 °C and similarly high mobility of 0.89 $cm^2 V^{-1} s^{-1}$ for polymer thin films, even without thermal annealing. When the donor unit was further extended from *β*-unsubstituted quaterthiophene to *β*-unsubstituted quintetthiophene, higher mobilities were achieved in the resulting copolymer **227** [272]. Its hole mobility of the thin film without thermal annealing reached 1.08 $cm^2 V^{-1} s^{-1}$, and a higher hole mobility of 3.46 $cm^2 V^{-1} s^{-1}$ was obtained for the film annealed at 200 °C directly in an air atmosphere. The improved performance was attributed to the longer *β*-unsubstituted quinquethiophene unit reducing the steric hindrance of the bulk side-chain groups, which is favorable to enhance the molecular ordering capability.

In 2012, Yu group developed two new DPP-based copolymers, **228** and **229** by copolymerization with (*E*)-2-(2-(thiophen-2-yl)vinyl)thiophene [273]. Polymers **228** exhibited high hole mobilities up to 4.5 $cm^2 V^{-1} s^{-1}$. For polymer **229**, higher mobilities up to 8.2 $cm^2 V^{-1} s^{-1}$ were achieved. The high performance of polymer **229** could be attributed to the strong intermolecular interactions in the solid state. Corn leaf-like interconnected networks and ordered lamellar structure with interlayer distances of 21.11 Å, and small π–π stacking distances of 3.66 Å were formed in thin films of polymer **229**. Additionally, polymer **229** took predominantly edge-on orientation, which was regarded as another major reason. A similar DPP-based copolymer containing (*E*)-2-(2-(selenophen-2-yl)vinyl)selenophene, polymer **230** was reported by Kang et al. [274]. FET devices based on the polymer exhibited hole mobilities of 4.97 $cm^2 V^{-1} s^{-1}$, higher than the value of 2.77 $cm^2 V^{-1} s^{-1}$ of the counterparts containing S atoms under the same conditions. The increased mobility was attributed to the stronger interaction between neighboring polymers chains, resulting from the more mobile lone pair of selenophene atoms compared to thiophene. In 2013, the π-conjugated backbones were further modified by a longer branched alkyl chain with six methylenes spacer between the branching point and the backbone, affording polymers **231** and **232** [275]. These showed record high mobilities of 10.54 and 12.04 $cm^2 V^{-1} s^{-1}$, respectively. The improved performances were attributed to the denser main chain packing and stronger intermolecular interaction in the two polymers. In thin films, smaller π–π stacking distances of 3.62 Å for polymer **231** and 3.58 Å for polymer **232**, were obtained.

3.3 n-Type Semiconductors

Compared with the breakthrough progress made in p-type semiconductors, the development of n-type counterparts is lagging behind, and has been regarded as one of the huge challenges of organic electronics. As mentioned above, electrons are charge carriers in channels in this kind of FET. Thus high electron affinities to facilitate charge injection and ambient stability are necessary for n-type semiconductors. To obtain ideal n-type semiconducting material, many electron-withdrawing groups or systems were synthesized and used to tune the LUMO energy level.

3.3.1 Selected n-Type Small-Molecule Semiconductors

3.3.1.1 Fullerenes

Fullerenes and their derivatives are spherical molecules with cage-like fused-ring structures. Because of electron affinities, these kinds of molecules can afford n-type characteristic in OFETs (see Chart 3.16 and Table 3.3). For example, the single crystal of C_{60}, **233** exhibited the highest electron mobility of 0.7 $cm^2\ V^{-1}\ s^{-1}$ and the highest hole mobility of 1.9 $cm^2\ V^{-1}\ s^{-1}$ measured by the time of flight (TOF) method [276]. Under ultrahigh vacuum conditions, the C_{60}-based FET devices showed electron mobilities of 0.08 $cm^2\ V^{-1}\ s^{-1}$ with on/off ratios of 10^6 [33]. It is notable that when amorphous alumina was used as insulating layer to shield C_{60} film from the air, the FET devices exhibited electron mobility of 0.1 $cm^2\ V^{-1}\ s^{-1}$ and improved stability [277]. No degradation had been detected even for more than 1 month. With the introduction of pentacene as buffer layer, the resulting FET devices based on C_{60} demonstrated high electron mobilities of 3.23 $cm^2\ V^{-1}\ s^{-1}$ [278]. Moreover, when using divinyltetramethyldisiloxane-bis(benzocyclobutene) (BCB) as the dielectric layer, the corresponding devices gave the highest mobility of 6 $cm^2\ V^{-1}\ s^{-1}$ [279]. In addition, a high mobility of 5.3 $cm^2\ V^{-1}\ s^{-1}$ was obtained when utilizing crystalline self-assembled monolayers of octadecyltrimethoxysilane-modified Si/SiO_2 substrates [280].

Methanofullerene[6,6]-phenyl-C_{61}-butyric acid methyl ester (PCBM), a soluble C_{60} derivative, **234** is usually used as an acceptor in organic photovoltaic cells. Using Ca as the drain and source electrodes, spin-coated films of PCBM showed a mobility of 4.5×10^{-3} $cm^2\ V^{-1}\ s^{-1}$ [281]. By optimizing dielectric materials, a high mobility of 0.2 $cm^2\ V^{-1}\ s^{-1}$ could be afforded [282]. Long-chain alkyl-substituted C_{60}, C_{60}-fused *N*-methylpyrrolidine-meta-C12 phenyl (C_{60}MC12), **235** exhibited an electron mobility of 0.067 $cm^2\ V^{-1}\ s^{-1}$, whereas **234** showed a mobility of

233

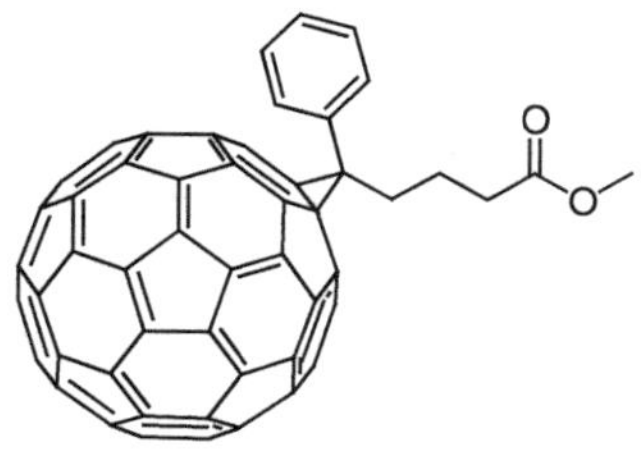

234

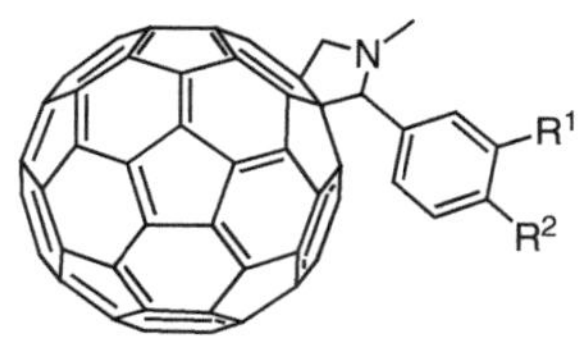

235: $R^1 = C_{12}H_{25}$, $R^2 = H$
236: $R^1 = H$, $R^2 = C_{12}F_{25}$

237

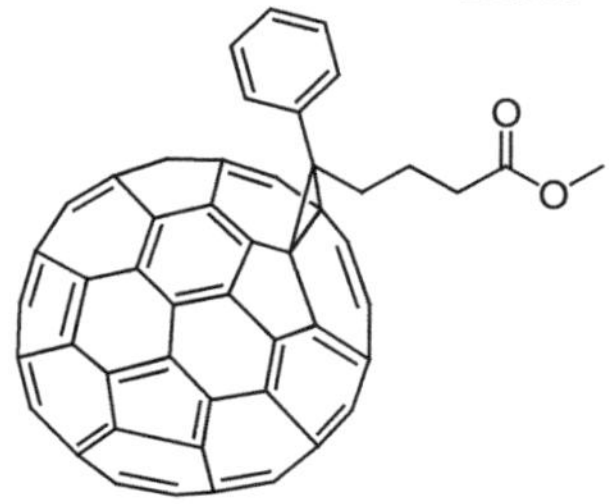

238

Chart 3.16 C_{60}, C_{70} and their derivatives

0.023 cm^2 V^{-1} s^{-1} under the same conditions [283]. Fluorinated C_{60} derivatives demonstrated increased mobilities and improved stabilities [284]. For example, derivative **236** displayed a high mobility of 0.25 cm^2 V^{-1} s^{-1} in a vacuum and 0.078 cm^2 V^{-1} s^{-1} in air. Similar to C_{60} and PCBM, C_{70}, **237** and its derivative PC$_{71}$BM, **238** were also used as semiconducting layers in FET devices [285, 286]. Soluble PC$_{71}$BM are also widely used as acceptor in organic solar cells. PC$_{71}$BM afforded electron mobility of 0.028 cm^2 V^{-1} s^{-1} whereas PCBM showed electron mobility of 0.022 cm^2 V^{-1} s^{-1} under the same conditions. The FET performance of other fullerene derivatives had also been examined, affording electron mobilities in the range 10^{-4}–10^{-3} cm^2 V^{-1} s^{-1} [287–290].

3.3.1.2 Diimides

Diimides are electron deficient because of the substitution of an aromatic core with two sets of strong electron-withdrawing carboxylic imide rings. Aromatic diimides are one of the classical n-type organic semiconductors with high electron affinities, high mobilities, and excellent stabilities. These diimides are subdivided into naphthalene diimide, perylenediimide, and other diimide derivatives, etc.

Table 3.3 OFET device data for n-type organic semiconductors

	LUMO (eV)	Deposition process	*Max* μ_e ($cm^2 V^{-1} s^{-1}$) (m.d.)[a]	I_{on}/I_{off}	V_T (V)	Device structure[b]	Refs.
233	−4.5	Evaporation	0.08 (vacuum)	10^6		BGTC; Au/Cr; SiO_2/Si	[33]
233		Evaporation	3.23 (N_2)	4×10^6	17.1	BGBC; Au; SiO_2/Si; HMDS	[277]
233		Evaporation	6.0			BGBC; LiF/Al; BCB/ITO	[278]
233		Evaporation	5.3	10^7	30	BGTC; Au; SiO_2/Si; OTMS	[280]
234		Spin-coating	0.2	10^3	−20	BGTC; LiF/Al; Mowiol 40-88/ITO	[282]
235		Spin coating	0.067	1.6×10^5	26.1	BGTC; Au; SiO_2/Si; HMDS	[284]
236	−3.63	Spin coating	0.25	4×10^5	28	BGTC; Au; SiO_2/Si; HMDS	[284]
			0.078	4×10^4	51		
237			0.002		27	BGBC; Au/Cr; SiO_2/Si	[285]
238		Spin casting	0.028			BGTC; Al; SiO_2/Si, TiO_x	[286]
240b		Evaporation	0.7 (Ar)	10^6	55	BGTC; Au; SiO_2/Si; OTS	[295]
240c		Evaporation	0.16 (vacuum)			BGTC; Au; SiO_2/Si	[294]
240d		Evaporation	0.01 (vacuum)			BGTC; Au; SiO_2/Si	[294]
240e		Evaporation	6.2 (Ar)	10^8	58	BGTC; Au; SiO_2/Si; OTS	[296]
240f	−3.77	Zone-casting	0.18 (air)	10^4		TGBC; Au; Parylene C/Ag	[297]
240g	−3.72	Evaporation	0.34 (N_2)	10^7	19	BGTC; Au; SiO_2/Si; OTS	[298]
			0.27 (air)				
	−4.02	Single crystal	0.7 (vacuum)	10^4	2.2		[299]
240h	−3.71	Evaporation	0.70 (air)	10^6	3–13	BGTC; Au; SiO_2/Si; OTS	[300]
240i		Evaporation	0.57 (air)	10^7		BGTC; Au; SiO_2/Si; OTMS	[302]
240j		Evaporation	0.025	10^5		BGBC; Au; SiO_2/Si; PαMS	[302]
		Evaporation	0.87	10^7		BGTC; Au; SiO_2/Si; PαMS	

(continued)

Table 3.3 (continued)

	LUMO (eV)	Deposition process	*Max* μ_e ($cm^2\ V^{-1}\ s^{-1}$) (m.d.)[a]	I_{on}/I_{off}	V_T (V)	Device structure[b]	Refs.
241a	−4.01	Evaporation	0.86 (N_2)	10^5	20	BGTC; Au; SiO_2/Si	[298]
			0.91 (air)	10^6	25		
		Solution shearing	4.26	5×10^6	18	BGTC; Au; SiO_2/Si; OTES	[303]
		Single crystal	8.6	7×10^7	9	BGTC; Au; SiO_2/Si; OTES	[304]
241b	−4.01	Evaporation	1.26 (N_2)	10^7	15	BGTC; Au; SiO_2/Si; OTS	[298]
			1.43 (air)	10^7	23		
242		Evaporation	4.7×10^{-3} (vacuum)	10^5		BGTC; Au; SiO_2/Si	[305]
243		Evaporation	0.15 (vacuum)	10^3		BGTC; Au; SiO_2/Si	[305]
244a	−4.3	Spin coating	1.2 (air)	10^8	−4.8 ~ 6.2	BGBC; Au; SiO_2/Si; OTS	[306]
244b		Spin coating	0.34 (air)	10^7	−5.1	BGBC; Au; SiO_2/Si; OTS	[307]
244c			3.5 (air)	10^8	−2.5		[307]
244d			0.25 (air)	10^7	7.07		[307]
245	−4.22	Spin coating	0.22 (air)	10^6	−1 ~ 5	BGBC; Au; SiO_2/Si; OTS	[311]
246	−4.35	Spin coating	0.17	10^6	−5 ~ 0	BGBC; Au; SiO_2/Si; OTS	[311]
247	−3.79	Evaporation	0.35 (vacuum)	10^6	28	BGTC; Au; SiO_2/Si; HMDS	[51]
			0.1 (air)	10^7	50		
248	−3.95	Evaporation	0.1 (vacuum)	10^3	0	BGTC; Ca; Si/SiO_2; β-PTS	[313]
249		Spin coating	0.15 (N_2)	10^4	4.9	TGBC; Au; CYTOP/Al	[313]
		Ink-jet printing	0.17 (N_2)	10^4	1.7	TGBC; Ag; CYTOP/Al; PCE	
250		Spin coating	1.2 (N_2)		13	TGBC; Au; CYTOP/Al	[314]

(continued)

Table 3.3 (continued)

	LUMO (eV)	Deposition process	*Max* μ_e ($cm^2 V^{-1} s^{-1}$) (m.d.)[a]	I_{on}/I_{off}	V_T (V)	Device structure[b]	Refs.
252b		Evaporation	1.7 (vacuum/H_2)	10^7	10 ~ 15	BGBC; Ag; SiO_2/Si; PαMS	[319]
252d	−3.4	Evaporation	0.58 (N_2)	10^7	44	BGTC; Cr; SiO_2/Si; OTS	[320]
252f	−3.85	Evaporation	1.44	10^6	39 ~ 47	BGTC; Au; SiO_2/Si; OTES	[324]
			1.24	10^6	47 ~ 57		
252g	−3.84	Evaporation	0.061			BGTC; Au; SiO_2/Si; OTS	[325]
252j	−4.27		0.066 (vacuum)			BGTC; Au; SiO_2/Si; OTS	[326]
			0.052 (air)				
252l	−4.1	Evaporation	0.11 (air)	10^5	29	BGTC; Au; SiO_2/Si; OTS	[327]
252m	−3.79	Evaporation	0.62 (vacuum)	10^6	21 ~ 29	BGTC; Au; SiO_2/Si; OTES	[324]
			0.37 (air)	10^7	28 ~ 39		
253a		Inkjet printing	0.056 (N_2)	10^6	−1.4	BGBC; Au; SiO_2/Si	[332]
253b	−4.33	Evaporation	0.86			BGTC; Au; SiO_2/Si; OTS	[333]
253c	−4.3	Evaporation	0.64 (air)	10^4	−20 ~ −30	BGTC; Au; SiO_2/Si; HMDS	[334]
		Solution processed	1.3 (air)	10^5	−8	BGTC; Au; SiO_2/Si; F-SAM	[334]
		Single crystal	6 (vacuum)			BGBC; Au; SiO_2/Si; PMMA	[335]
			3 (air)				
254	−3.9	Evaporation	0.11 (air)	10^6	36	BGTC; Au; SiO_2/Si; OTS	[327]
255	−4.11	Evaporation	0.38 (N_2)	10^7	19 ~ 23	BGTC; Au; SiO_2/Si; OTES	[324]
			0.27 (air)	10^7	17 ~ 23		
256	−4.23	Evaporation	0.91 (N_2)	2.1×10^7	28	BGTC; Au; SiO_2/Si; OTS-18	[337]

(continued)

Table 3.3 (continued)

	LUMO (eV)	Deposition process	*Max* μ_e ($cm^2 V^{-1} s^{-1}$) (m.d.)[a]	I_{on}/I_{off}	V_T (V)	Device structure[b]	Refs.
			0.82 (air)	1.5×10^8	28		
257a	−3.76	Evaporation	0.74 (N_2)	10^7	19 ~ 28	BGTC; Au; SiO_2/Si; OTES	[324]
257b	−3.88	Evaporation	0.66 (N_2)	10^6	9 ~ 15		[324]
257c	−3.94	Evaporation	0.85 (N_2)	10^7	8 ~ 14		[324]
258	−3.93	Evaporation	0.058 (vacuum)	10^6	1 ~ 16	BGTC; Au; SiO_2/Si; OTES	[324]
			0.056 (air)	10^6	4 ~ 22		
259	−3.88	Evaporation	0.15 (vacuum)	10^7	27	BGTC; Au, SiO_2/Si; HMDS	[51]
			0.08 (air)	6×10^7	42		
260a	−4.30	Spin coating	0.70 (air)	4×10^7	21	BGTC; Au; SiO_2/Si; OTMS	[338]
260b	−4.22	Single crystal	4.65 (air)			BGTC; Ag; SiO_2/Si; OTS	[339]
261	−4.26	Spin coating	0.25 (air)	10^7		BGTC; Au; SiO_2/Si; OTS	[341]
262a	−3.87	Evaporation	0.074 (air)	10^6	22.4	BGTC; Au; SiO_2/Si; OTS	[341]
262b		Evaporation	0.079 (air)	10^6	14.4	BGTC; Au; SiO_2/Si; OTS	[341]
262c	−3.97	Evaporation	0.030 (air)	10^5	13.7	BGTC; Au; SiO_2/Si; OTS	[341]
263		Evaporation	0.515 (N_2)	10^5	5 ~ 20	BGTC; Au; SiO_2/Si; OTS	[342]
265d	−4.1	Evaporation	0.03 (vacuum)	6×10^6	10	BGTC; Au; SiO_2/Si; HMDS	[345]
			0.02 (air)	2×10^7	15		
265e	−4.2	Evaporation	0.06 (vacuum)	10^5	10	BGTC; Au; SiO_2/Si; HMDS	[345]
			0.04 (air)	4×10^4	12		
268b	−4.15	Spin coating	0.08 (N_2)	$10^6 \sim 10^7$	−5 ~ 0	BGTC; Au; SiO_2/Si; OTS	[347]
271c	−3.90	Spin coating	1.0 (N_2)	$10^4 \sim 10^5$	−15	BGTC; Au; SiO_2/Si; OTMS	[351]
			0.51 (air)	$10^2 \sim 10^3$	4 ~ 40		
273	−3.70	Spin coating	0.12	$10^5 \sim 10^6$	20.9	BGBC; Au; SiO_2/Si; OTS-8	[353]

(continued)

Table 3.3 (continued)

	LUMO (eV)	Deposition process	*Max* μ_e ($cm^2 V^{-1} s^{-1}$) (m.d.)[a]	I_{on}/I_{off}	V_T (V)	Device structure[b]	Refs.
275		Spin coating	0.16 (air)	10^4	−1.2	BGTC; Au; SiO_2/Si; OTS-8	[354]
276	−4.2	Spin coating	0.015 (air)	10^5	5.1	BGTC; Au; SiO_2/Si; OTS-8	[355]
277	−4.37	Spin coating	0.014 (air)	10^4	8.3	BGTC; Au; SiO_2/Si; OTS	[356]
278	−4.3	Drop casting	0.9 (air)	10^5		BGTC; Au; SiO_2/Si; OTS-18	[357]
279		Evaporation	0.55 (air)	10^6	9.5	BGTC; Au; SiO_2/Si; OTS	[358]
280		Evaporation	0.02	10^6	9.4	BGTC; Ag; C_{16}-Al_2O_3; OTS	[359]
281	−4.45	Crystal	0.96	10^5	−25	BGTC; Au; SiO_2/Si	[360]
		Evaporation	0.64	$10^3 \sim 10^4$	−19	BGTC; Au; SiO_2/Si; ODTS	
282	−3.68	Evaporation	0.34 (vacuum)	6×10^3	35	BGTC; Au; SiO_2/Si	[361]
283	−3.88	Evaporation	0.044	10^6	8	BGTC; Au; SiO_2/Si; TPA	[362]
284	−4.04	Spin coating	0.16 (air)	$10^5 \sim 10^6$	23	TGBC, Au; CYTOP/Ag	[363]
285	−3.37	Evaporation	0.11 (vacuum)	10^5		BGTC; Au; SiO_2/Si; OTS	[364]
286	−3.2	Evaporation	0.016 (vacuum)	10^4		BGTC; Al/Ca; SiO_2/Si	[366]
287	−3.51	Evaporation	3.1 (crystal)			BGTC; Au; SiO_2/Si	[366]
			0.6 (film)				
288	−2.54	Evaporation	0.3 (vacuum)	10^7	56	BGTC; Au; SiO_2/Si; OTS	[367]
289		Evaporation	0.3 (vacuum)	10^5	25	BGTC; Au; SiO_2/Si; OTS	[368]
290	−3.94	Evaporation	2.14 (N_2)	10^7	34	BGTC; Au; SiO_2/Si; OTS	[369]
291	−3.76	Evaporation	0.61 (N_2)	10^6	32	BGTC; Au; SiO_2/Si; OTS	[369]
292	−3.8	Evaporation	0.16 (crystal)	10^4	21	BGTC; Au; SiO_2/Si	[370]
			0.03 (film)	10^4	11	BGTC; Au; SiO_2/Si; OTS	
295	−3.53	Evaporation	0.16 (vacuum)	10^5	9.2	BGTC; Au; Si/SiO_2; PS	[372]

(continued)

Table 3.3 (continued)

	LUMO (eV)	Deposition process	*Max* μ_e ($cm^2 V^{-1} s^{-1}$) (m.d.)[a]	I_{on}/I_{off}	V_T (V)	Device structure[b]	Refs.
296	−3.51	Evaporation	0.18 (air)			BGTC; Au; SiO_2/Si; OTMS	[372]
297	−3.96		4.6 (N_2)			BGTC; Au; SiO_2/Si; PαMS	[373, 374]
298		Evaporation	1.2 (vacuum)	10^7	63	BGTC; Au; SiO_2/Si; OTS	[376, 377]
299		Evaporation	0.64 (vacuum)		24	BGTC; Au; SiO_2/Si; OTS	[377]
300	−4.1	Evaporation	0.15 (vacuum)	10^4	17	BGTC; Au; SiO_2/Si; HMDS	[378]
			0.12 (air)	10^4	27		
301	−3.69	Evaporation	0.39 (vacuum)	10^6	23	BGTC; Au; SiO_2/Si; OTS	[379]
			0.14 (air)	10^7	23		
302	−3.67	Evaporation	0.07	10^6	60	BGTC; Au; SiO_2/Si; OTS-8	[380]
303	−3.67	Evaporation	0.03	10^6	58	BGTC; Au; SiO_2/Si; OTS-8	[380]
304	−3.79	Single crystal	3.39 (air)	10^4		TGTC; graphite; parylene/graphite; OTS	[381]
305	−4.01	Evaporation	3.3 (vacuum)			BGTC; Au; SiO_2/Si; OTMS	[382]
			0.5 (air)				[382]
306	−3.78	Evaporation	0.12 (air)			BGTC; Au; SiO_2/Si; OTMS	[365]
307	−4.2	Nanobelt	0.007 (air)			BGBC; Au; SiO_2/Si	[385]
308	−3.54	Spin coating	0.0026 (inert atmosphere)	10^4	5	BGTC; Au; SiO_2/Si; OTS	[386]
309	−4.0	Spin coating	0.85 (air)	10^7	5 ~ 10	TGBC; Au; D2200/Au	[387]
310	−3.76	Spin coating	0.076	10^5	7	BGTC; Au; SiO_2/Si; OTS	[388]
311	−4.0	Spin coating	0.5 (air)	10^5		TGBC; Au; PMMA/Au	[389]
312	−3.79	Spin coating	0.05 (N_2)	10^4	14	BGTC; Au; SiO_2/Si; OTS	[390]
313	−3.47	Spin coating	0.19 (vacuum)	10^5		TGBC; Au; D2200/Au	[391]

(continued)

Table 3.3 (continued)

	LUMO (eV)	Deposition process	*Max* μ_e ($cm^2 V^{-1} s^{-1}$) (m.d.)[a]	I_{on}/I_{off}	V_T (V)	Device structure[b]	Refs.
314		Spin coating	0.017	10^6	13	BGTC; Au; SiO_2/Si; OTS	[392]
315	−3.75	Spin coating	0.05 (N_2)	10^5	8	BGTC; Au; SiO_2/Si, OTS	[391]
316	−4.00	Spin coating	0.01 (air)	10^5	38	BGTC; Au; SiO_2/Si; OTS	[393]
317	−4.00	Spin coating	3.4 (air)	20	8.4	BGBC; Au; Si/SiO_2; HMDS	[394]
318	−3.66	Spin coating	3 (N_2)	10^4		TGBC; Al, CYTOP/Al	[395]
319	−3.64	Spin coating	2.36			BGTC; Au, Si/SiO_2, OTS-18	[396]
320	−4.24	Spin coating	1.1 (air)		5	TGBC; Au/Ti, CYTOP/Al	[397]
323	−4.15	Spin coating	1.74 (air)	$10^4 \sim 10^5$	44	TGBC; Au/Ti, CYTOP/Al	[399]

[a] *m.d.* measurement condition
[b] Device configuration; S/D electrodes; dielectric/gate; modification of substrate/electrode

Naphthalene Diimide and Derivatives

The precursor, naphthalene tetracarboxylic dianhydride, **239** of naphthalene diimide is one of the first successes of this generation of n-type materials (see Chart 3.17). Vacuum deposited films of **239** showed an electron mobility of 3×10^{-3} cm^2 V^{-1} s^{-1} under vacuum [291, 292]. In air, however, there was one to two orders of magnitude decrease in the electron mobilities. The naphthalene diimide (NDI), **240a** also showed *n*-channel properties with mobilities of 10^{-4} cm^2 V^{-1} s^{-1}. Although the electron mobility of NDI has obvious decreased, the nitrogen atoms of NDI could be substituted by alkyls or other groups affording us versatile ways to obtain solution-processable n-type semiconductors with different electronic properties. Indeed, a lot of alkyls or other groups substituting NDIs exhibited high electron mobility [293–296]. For example, the hexyl-substituted NDI, **240b** showed a high electron mobility up to 0.7 cm^2 V^{-1} s^{-1}. The octyl-substituted NDI, **240c** and dodecyl-substituted NDI, **240d** showed electron mobilities of 0.16 and 0.01 cm^2 V^{-1} s^{-1}, respectively. It is notable that the cyclohexylsubstituted derivative, **240e**, developed by Shukla et al., afforded high n-type performance with mobilities of 6.2 cm^2 V^{-1} s^{-1} with an on/off current ratio of 6×10^8. In addition, when the **240e**-based thin film devices were tested in argon atmosphere under low humidity conditions, their mobility further increased to 7.5 cm^2 V^{-1} s^{-1}, which is one of the highest values reported for n-type semiconductors so far. The high performance was attributed to the bulk phase crystalline packing with two-dimensional lamellar stacking and long axis of the molecule nearly vertically oriented out of the substrate plane, which results in efficient charge transport. Alkylphenyl-substituted NDI, **240f** was also investigated [297]. Its thin film FETs showed air stable electron transport properties with a mobility of 0.18 cm^2 V^{-1} s^{-1}. Fluorinated substituents including fluorinated alkyls and alkylphenyls were also introduced on the nitrogen-positions of NDIs. The resulting derivatives exhibited improved air stability because of the lowered LUMO energy levels. For example, fluoroalkyl-substituted NDI, **240g** showed the highest electron mobility of 0.38 cm^2 V^{-1} s^{-1} measured in nitrogen and 0.27 cm^2 V^{-1} s^{-1} in air, with an on/off ratio of over 10^7 [298]. The single crystal OFETs of **240g** exhibited electron mobility as high as 0.45 or 0.70 cm^2 V^{-1} s^{-1} along different crystal directions when tested in vacuum [299]. Derivative **240h** with a longer fluorinated alkyl chain afforded electron mobility approaching 0.7 cm^2 V^{-1} s^{-1} in air [300]. Its mobilities were stable in air for 1 week. After 100 days in air, the average mobility of three OTFTs decreased from 0.62 to 0.12 cm^2 V^{-1} s^{-1}, but stabilized thereafter. In addition, the NDI derivatives with fluorinated phenyl groups or phenyl group-substituted fluorinated alkyls chain, **240i** and **240j** afforded the highest electron mobilities of 0.57 and 0.87 cm^2 V^{-1} s^{-1}, respectively [301, 302]. The chlorine-substituted NDIs **241a** and **241b**, which were reported by Oh et al., showed excellent electron mobilities of 0.86 and 1.26 cm^2 V^{-1} s^{-1} in nitrogen, respectively [298]. When tested in ambient air, NDIs **241a** and **241b** showed some degree of increased electron mobilities of 0.91 and 1.43 cm^2 V^{-1} s^{-1}, respectively. The improvement in mobilities was attributed to the trapping of the charge carriers by

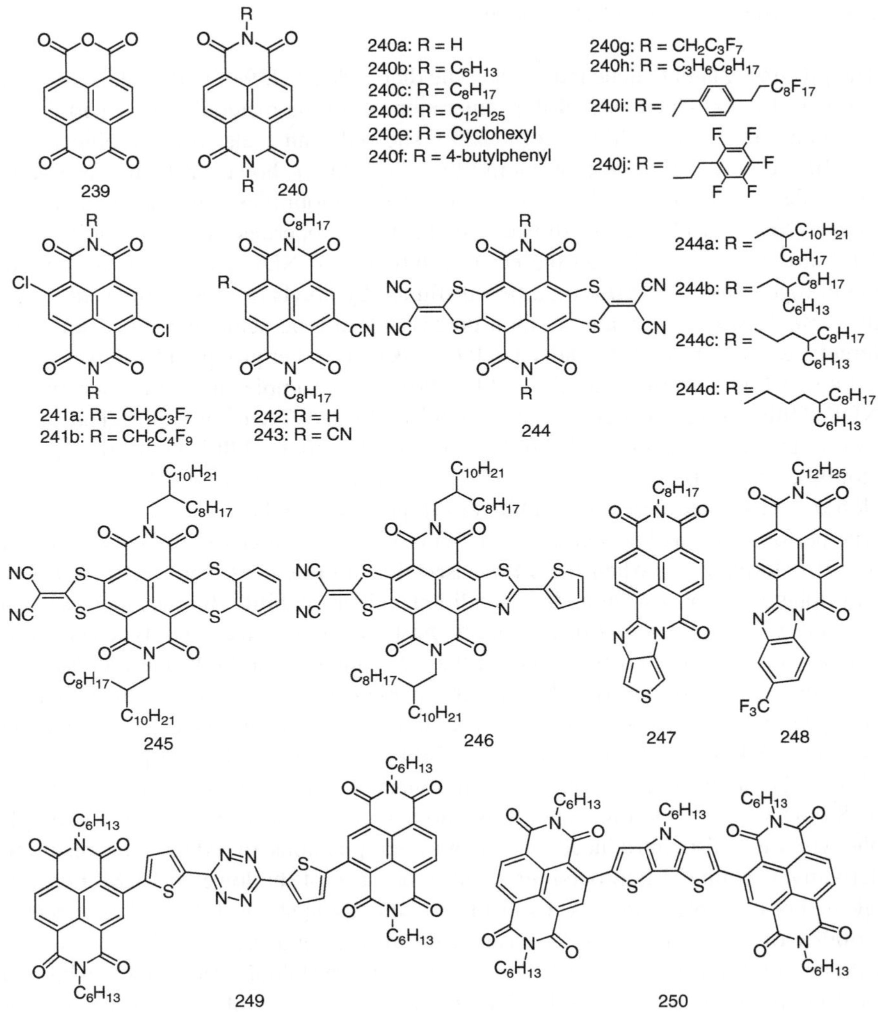

Chart 3.17 Naphthalenediimide and derivatives

ambient oxidants. Furthermore, **241a**-based OFETs fabricated by solution shearing technique, afforded electron mobility increased up to 4.26 $cm^2 V^{-1} s^{-1}$ under bias stress after 1000 cycles [303], and the single crystal FET devices of **241a** showed higher mobilities of up to 8.6 $cm^2 V^{-1} s^{-1}$ [304]. The cyano-substituted derivative, **242** afforded a low mobility of about 4.7×10^{-3} $cm^2 V^{-1} s^{-1}$, whereas dicyano substituted derivative **243** exhibited an electron mobility of 0.15 $cm^2 V^{-1} s^{-1}$, which is comparable to that of **240c** unsubstituted by cyano groups [305]. However, because of the low-lying LUMO energy level, derivative **243**-based devices exhibited improved air stability with electron mobilities of 0.11 $cm^2\ V^{-1}\ s^{-1}$ in ambient air.

Core-expanded NDI derivatives were also synthesized and examined, for example, symmetrical derivatives **244** and unsymmetrical derivatives **245** and **246**. The **244c**-based thin film FETs fabricated by the spin-coating technique exhibited n-type performance with high mobilities up to 1.2 $cm^2\ V^{-1}\ s^{-1}$ and an on/off current ratio of 10^8 [306]. Moreover, these devices showed excellent stability with the annealing process taking place in ambient air. The low LUMO energy level of −4.3 eV and close π-packing in combination with long alkyl chains, which can inhibit the ingress of oxygen, was thought to be the reasons behind the excellent performance and stability. In 2013, three similar compounds, **244a**, **244b**, and **244d** were also synthesized [307]. Of these, **244d**-based FETs afforded a electron mobility of 3.5 $cm^2\ V^{-1}\ s^{-1}$, which is one of the highest mobilities for solution-processed air stable n-type OFETs. The kind of core-expanded NDIs with different functional groups on the nitrogen of imide was also reported and showed similar performance [308]. More recently, Zhang's group reported another symmetrical thiazole-fused NDI derivatives, which gave a mobility as high as 0.15 $cm^2\ V^{-1}\ s^{-1}$ with a high on/off current ratio in air [309]. Unsymmetrical core-expanded NDIs, **245** and **246** were also developed by the same group [310, 311]. Derivatives **245**- and **246**-based devices showed electron mobilities of 0.22 and 0.17 $cm^2\ V^{-1}\ s^{-1}$ with a high on/off current ratio under ambient conditions after annealing. A new family of NDIs, such as **247** and **248**, also showed typical n-type OFET behavior. For example, NDI derivative **247** afforded electron mobilities as high as 0.35 $cm^2\ V^{-1}\ s^{-1}$ [51], whereas derivative **248** exhibited electron mobility as high as 0.10 $cm^2\ V^{-1}\ s^{-1}$ with good bias stress stability [312]. In 2012, two kinds of NDI-based oligomers, such as **249** and **250**, were synthesized by Kippelen et al. [313, 314]. Optimized spin-coated OFETs based on **249** showed electron mobility of up to 0.15 $cm^2\ V^{-1}\ s^{-1}$. Inkjet-printed OFETs have also been fabricated in ambient atmosphere on flexible plastic substrates, and exhibited mobility of up to 0.17 $cm^2\ V^{-1}\ s^{-1}$. The flexible printed-OFETs exhibited excellent operational stability. OFETs with a solution-processed derivative **250** showed ambipolar transistor properties with an average electron mobility value of 1.2 $cm^2\ V^{-1}\ s^{-1}$ and an average hole mobility value of 0.01 $cm^2\ V^{-1}\ s^{-1}$.

Perylene Diimide and Derivatives

Similar to naphthalene tetracarboxylic dianhydride **239**, perylene tetracarboxylic dianhydride, **251** also showed n-type charge transport properties (see Chart 3.18). The thin film-based and single crystal-based FETs afforded electron mobilities of 10^{-4} and 10^{-3} $cm^2\ V^{-1}\ s^{-1}$, respectively [315, 316]. *N*-Alkylated perylene diimides (PDIs) exhibited good electron transport performance and high device stability. For example, *N*-pentyl derivative **252a** showed mobilities up to 0.1 $cm^2\ V^{-1}\ s^{-1}$ [317]. Polycrystalline films of octyl-substituted derivative **252b** afforded a high mobility of 1.7 $cm^2\ V^{-1}\ s^{-1}$ [318, 319]. The dodecyl- and tridecyl-substituted derivatives

252c and **252d** showed mobilities up to 0.52 and 2.1 $cm^2 V^{-1} s^{-1}$, respectively [320–322]. The high electron mobility of **252d** arises from its thin film morphology of crystallinity and flat and large tile-like grains. A series of *N*-fluorinated alkyl substituted PDIs, **252e**, **252f**, and **252g**, were developed by Bao et al. Of these, **252e** and **252g** showed low mobilities of 0.049 and 0.11 $cm^2 V^{-1} s^{-1}$, respectively [323, 324]. However, derivative **252f** exhibited high mobilities up to 1.44 and 1.24 $cm^2 V^{-1} s^{-1}$ respectively when tested in vacuum and in air. It is notable that solution-processed FET devices based on **252f** exhibited a mobility as high as 1.42 $cm^2 V^{-1} s^{-1}$, which is much higher than the value of 0.72 $cm^2 V^{-1} s^{-1}$ of the vapor-processed technique. In addition, **252g**-based OFETs were not air stable in ambient air until the thickness of semiconducting layer increased up to ten

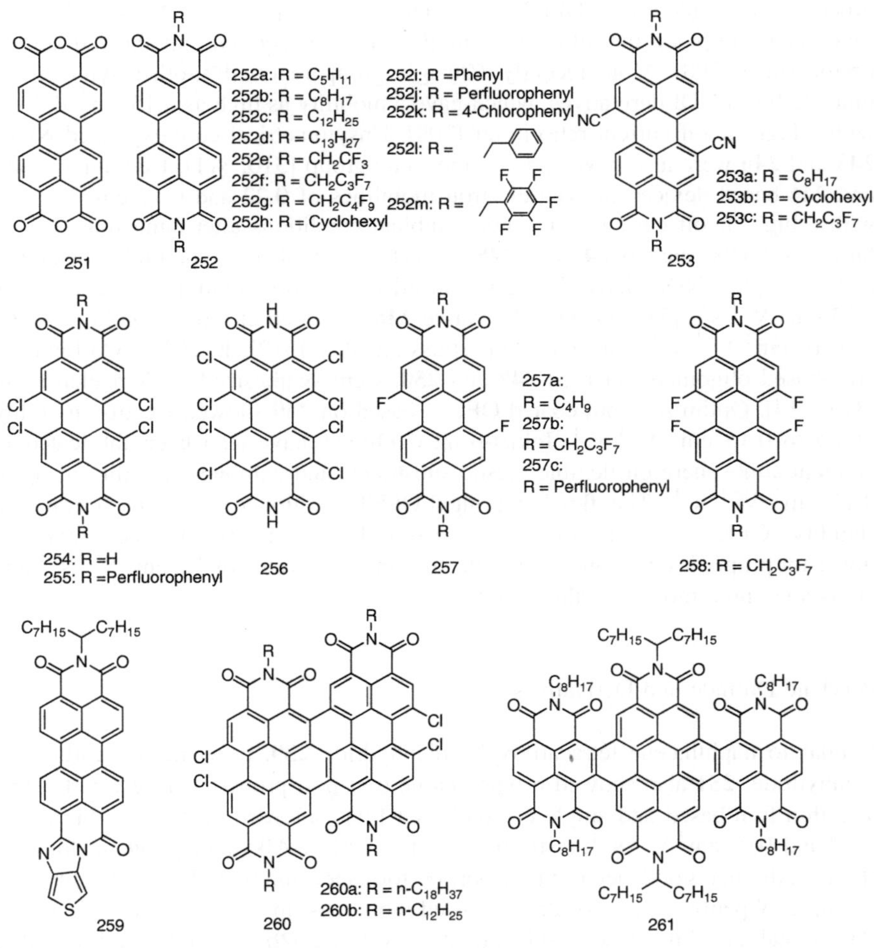

Chart 3.18 Perylene diimide and derivatives

monolayers. The result demonstrated the importance of the thickness of semiconducting layer in OFETs [325]. Different to *N*-cyclohexyl substituted **240e**, the *N*-cyclohexyl substituted PDI **252h**, exhibited lower mobility of only about 1.9×10^{-4} cm^2 V^{-1} s^{-1} [163]. The phenyl and perfluorophenyl substituted PDIs, **252i** and **252j** also afforded low mobilities of 0.017 and 0.068 cm^2 V^{-1} s^{-1}, respectively [326]. Nonetheless, the two derivatives have improved stability. For example, device performance of **252j** exhibited no significant change, even after storage in air for 72 days. The phenyl-ethyl substituted compounds, **252l** exhibited n-type charge transport properties with mobility of about 0.11 cm^2 V^{-1} s^{-1} and good air-stability [327]. The nanowire FET devices of **252l** showed the highest mobility of 1.4 cm^2 V^{-1} s^{-1} [328]. Perfluorophenyl-ethylsubstituted derivative, **252m** exhibited electron mobility of about 0.62 cm^2 V^{-1} s^{-1} tested in vacuum and 0.37 cm^2 V^{-1} s^{-1} in air [324]. Dicyano substituted PDIs were also synthesized and applied in OFETs [329–331]. For example, the vacuum deposited thin film of **253a** showed mobilities of 0.16 cm^2 V^{-1} s^{-1}, and its OFETs fabricated by inkjet-printing technique exhibited a mobility of 0.056 cm^2 V^{-1} s^{-1} [332]. The evaporated films of **253b** and **253c** showed mobilities of 0.16 and 0.64 cm^2 V^{-1} s^{-1}, respectively [333]. The crystalline films of **253c** afforded a mobility of 1.3 cm^2 V^{-1} s^{-1} in ambient air [334]. Its single crystal transistors exhibited the highest mobilities of 6 cm^2 V^{-1} s^{-1} under a vacuum and 3.0 cm^2 V^{-1} s^{-1} in ambient air [335]. The excellent performance in air could be attributed to the low-lying LUMO energy level of −4.3 eV in combination with the dense packed cores and fluoroalkyl chains. Furthermore, its vacuum-gap single crystal could afford high mobility up to 10.8 cm^2 V^{-1} s^{-1} [336]. The core-chlorinated and fluorinated PDIs also showed high stability. For example, tetrachloro-substituted PDI **254** had the low LUMO energy level of −3.9 eV [327]. The **254**-based FETs afforded electron mobility as high as 0.18 cm^2 V^{-1} s^{-1}. After storage in air for 80 days, the mobility slightly decreased to 10^{-2} cm^2 V^{-1} s^{-1}. *N*-Perfluorophenyl tetrachloro-substituted PDI **255** showed electron mobility of 0.38 cm^2 V^{-1} s^{-1} with an on/off ratio of 10^7 [324]. Octachlorosubstituted PDI **256** showed a much lower LUMO energy level of −4.23 eV [337]. The **256**-based FETs exhibited electron mobilities of 0.91 cm^2 V^{-1} s^{-1} when measured in vacuum and 0.82 cm^2 V^{-1} s^{-1} in ambient air. It is notable that electrical parameters were almost unchanged even when keeping the devices in air for about 20 months. The high stability was also attributed to the low-lying LUMO energy level and high packing density. Difluorinated PDIs usually showed higher electron mobilities than those of PDIs only substituted at the nitrogen atoms. For example, difluorinated PDI derivatives **257a**, **257b**, and **257c** showed electron mobilities of 0.74, 0.66, and 0.85 cm^2 V^{-1} s^{-1}, respectively [324]. However, the tetrafluorinated PDI derivative **258** showed a decreased electron mobility of 0.056 cm^2 V^{-1} s^{-1} and an increased stability. The decreased mobility is probably because of the nonplanar structures caused by too many substituents in the bay positions, and the increased stability because of the lower LUMO energy level. In addition, the LUMO energy level of PDI derivative **259** is located at −3.88 eV [51]. The derivative showed an electron mobility of 0.15 cm^2 V^{-1} s^{-1} in vacuum and 0.08 cm^2 V^{-1} s^{-1} in air.

Wang et al. reported series of large π-conjugated systems, such as dimer and trimer of NDI and PDI derivatives. The extension of π-conjugated systems not only leads to a dramatically broadened absorption spectrum but also increases the electron affinities to facilitate electron injection and transport with ambient stability. Dimers **260** exhibited good solubility in common organic solvents. Of these, **260a** demonstrated excellent thin film electron performance in air with the mobility as high as 0.70 $cm^2 V^{-1} s^{-1}$ and an on/off current ratio of 4×10^7 [338]. Moreover, single crystal ribbons of **260b** were easily grown by a solvent vapor diffusion strategy 339. The FET devices based on these individual ribbons exhibited electron mobilities of over 1.0 $cm^2 V^{-1} s^{-1}$ with the highest mobility of 4.65 $cm^2 V^{-1} s^{-1}$. Moreover, the devices exhibited excellent air stability. No obvious degradation was observed, even after the devices had been stored in air for more than 6 weeks. The molecular structure of highly fused derivative **261** is composed of two NDI units and one PDI unit [340]. The spin-coating thin films of derivative **261** showed a moderate electron mobility of 0.02 $cm^2 V^{-1} s^{-1}$ under ambient conditions. Thermal annealing of the thin films led to an improved performance with the highest electron mobility up to 0.25 $cm^2 V^{-1} s^{-1}$ and an on/off current ratio of 10^7.

Other Diimides and Derivatives

Besides NDIs and PDIs, other diimides were also developed and used as semiconducting layers in the n-type OFETs (see Chart 3.19). For example, pyromellitic diimides have a benzene ring in the center, and the tetracarboxylic diimides on both sides of the benzene ring. In 2008, Kate et al. synthesized a series of *N*-substituted pyromellitic diimides [341]. The thin films of pyromellitic diimide **262a** and **262b** exhibited electron mobilities of 0.074 and 0.079 $cm^2 V^{-1} s^{-1}$ when tested in vacuum, respectively. In addition, the on/off current ratios of *n*-channel devices are as high as 10^6. The electrical parameters of the devices had a slight decrease when measured in air. For example, the mobility and on/off current ratio for **262b** dropped to 0.054 $cm^2 V^{-1} s^{-1}$ and 10^4, respectively, because of the presence of water and oxygen. The LUMO energy levels of **262a** and **262c** were estimated to be about −3.9 eV. Angular-shaped naphthalene tetracarboxylic diimide, **263** afforded the highest mobility of 0.515 $cm^2 V^{-1} s^{-1}$ [342]. The high electron mobility of the OFETs was attributed to the improved crystallinity and enlarged grain sizes at the high substrate temperature. Anthracene diimides have nearly planar structure with potential application in organic electronics [343]. For example, the single crystal of anthracene diimide **264** exhibited dense packing with a small π–π distance of 3.45 Å. Linear anthracene diimides **265** were reported in 2007 [344, 345]. Vapor-deposited films of **265a**, **265b**, and **265c** showed n-type charge transport performance with mobilities in the range 0.01–0.02 $cm^2 V^{-1} s^{-1}$ with an on/off current ratio of 10^6–10^7 in vacuum. However, core-cyanated diimide **265d** showed a high electron mobility of 0.03 $cm^2 V^{-1} s^{-1}$ in vacuum and 0.02 $cm^2 V^{-1} s^{-1}$ in air with an on/off current ratio of 10^7. The device performance of **265d**-based devices stored in air with exclusion of light remained stable for at least 4 months after fabrication.

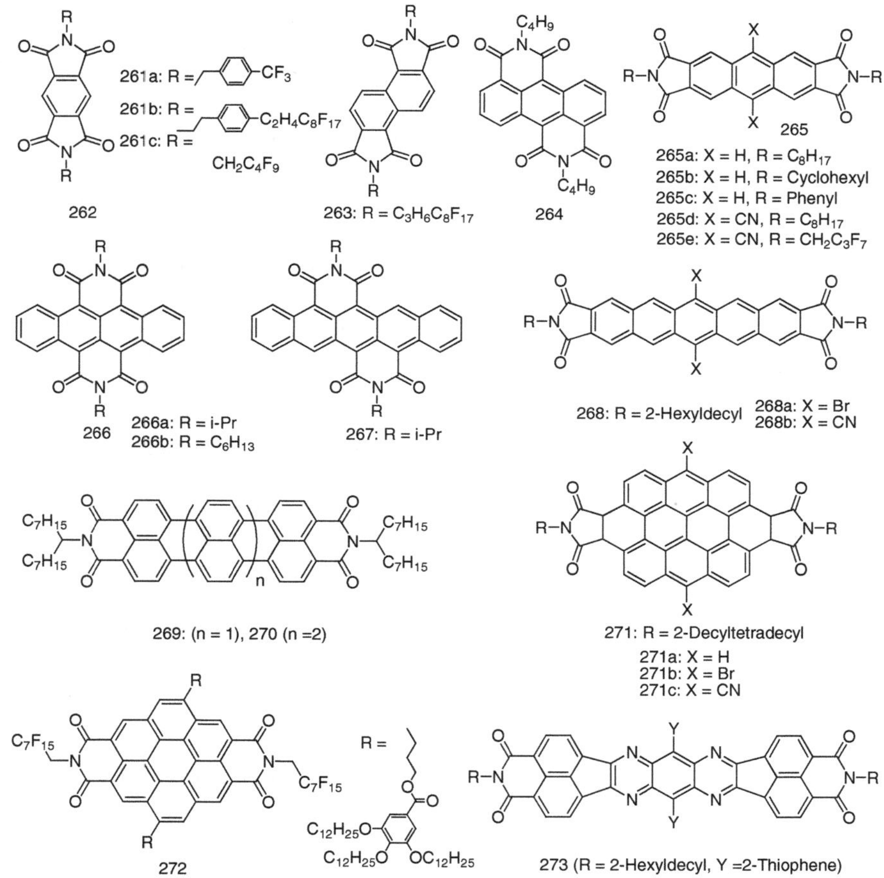

Chart 3.19 Other diimides and derivatives

Core-cyanated diimide **265d** also showed improved solubility. The improved performance of **265d** was attributed to the low-lying LUMO energy level induced by the nitrile groups. Angular-shaped tetracene- and pentacene-bisimides, **266** and **267** were synthesized via a similar double-cyclization reaction [346]. Vapor deposited films of **266b** showed low electron mobility of 3.3×10^{-3} $cm^2\ V^{-1}\ s^{-1}$. Linear pentacene-diimides **268** was also synthesized [347]. The core-cyanated diimide **268b** showed a low-lying LUMO energy level of −4.15 eV. The thin film transistors based on **268b** showed an electron mobility of 0.08 $cm^2\ V^{-1}\ s^{-1}$ and a high on/off current ratio of 10^6–10^7. The devices exhibited good air stability when exposed to air for several months. Terylene tetracarboxdiimides (TDI) and quaterylene tetracarboxdiimides (QDI), homologues of PDI, were investigated by several groups [348–350]. For example, **269** and **270** exhibited electron mobilities of 10^{-2}–10^{-3} $cm^2\ V^{-1}\ s^{-1}$. It is interesting to note that ambipolar behavior could be observed in some FET devices based on these materials. In 2012, Wu et al. developed a series of large disc-like ovalene diimides of **271** [351]. Because of the attachment of electron-withdrawing imide and cyano groups, diimide **271c** showed a low-lying LUMO energy level of −3.9 eV and a lamellar packing with a small π–π distance of 3.42 nm existing in thin films. The **271c**-based FET devices fabricated by solution processing exhibited high electron mobilities up to 0.51 $cm^2 V^{-1} s^{-1}$ in air and 1.0 $cm^2 V^{-1} s^{-1}$ in nitrogen atmosphere. Other diimides had also been reported. For example, the coronene-based diimide **272** exhibited mobilities as high as 6.7 $cm^2 V^{-1} s^{-1}$ as measured using the space-charge limited current method [352]. The tetraazabenzodifluoranthene diimide **273** showed an electron mobility of 0.12 $cm^2 V^{-1} s^{-1}$ in solution-processed thin film transistor devices [353].

3.3.1.3 Cyano-, Halogen-Containing and Other n-Type Semiconductors

Besides diimides, other small molecule semiconductors with electron-withdrawing groups and/or electron-deficient π-conjugated systems also exhibit good electron-transporting properties. The electron-withdrawing groups include cyano, halogen, halogenated alkyl, carbonyl, etc. The electron-deficient π-conjugated systems involve pyrazine, oxadiazole, thiazole, and benzobisthiadiazole derivatives, etc.

Cyano-Substituted n-Type Small-Molecule Semiconductors

The cyano group has a strong electron-withdrawing ability and is widely used in constructing n-type semiconductors (see Chart 3.20). The most famous is 7,7,8,8-tetracyanoquinodimethane (TCNQ), **274**, which itself is a very strong electron acceptor. Later, the tetracyanothienoquinodimethane systems were found to have a great potential as optoelectronics materials and their versatile functions are endowed with the extended thienoquinoidal conjugated structures. A great drawback of these compounds is, however, that they become insoluble in common solvents with quinoid extension. Developing soluble terthienoquinoid derivatives became important in the field. Dicyanomethylene-substituted terthienoquinoid compound **275** was highly soluble in common organic solvents [354]. Its highly crystalline thin film showed strong intermolecular interaction and exhibited electron mobilities up to 0.16 $cm^2 V^{-1} s^{-1}$. (Alkyloxy)carbonyl)cyanomethylene-substituted thienoquinoidal derivatives, for example **276**, showed high solubility and low-lying LUMO energy levels in a range of −4.0 ~ −4.2 eV, which is beneficial for injection and transport of electrons and for getting n-type OFETs with high stability [355]. The solution-processed thin films of **276** afforded the highest electron mobility of 0.015 $cm^2 V^{-1} s^{-1}$ in air. Zhu et al. also developed a series of pyrrole-containing quinoids with dicyanomethylene groups [356]. The incorporating pyrrole unit to quinoidal core structure presented two main advantages: first, *N*-alkyl substituents of pyrrole can serve as solubilizing groups for the realization of solution-

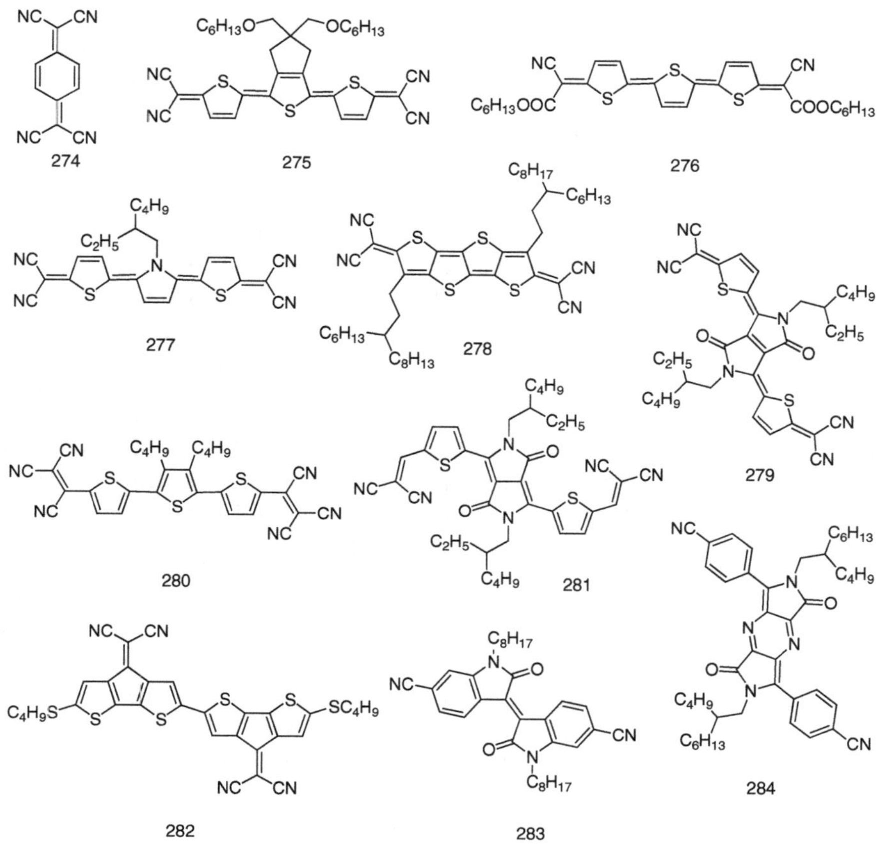

Chart 3.20 Cyano-substituted n-type small-molecule semiconductors

processability; second, the properties of dicyanomethylene derivatives could easily be modified by variation of *N*-alkyl substituents. Meanwhile, the retained quinoidal core structure and the electron-withdrawing end-capped group ensures a sufficiently low LUMO energy level, meeting the requirement for air stable *n*-channel organic semiconductors. For example, the compound **277** showed a low-lying LUMO energy level of –4.37 eV and its solution-processed thin films exhibited typical *n*-channel FET characteristics with the electron mobility of 0.014 $cm^2\ V^{-1}\ s^{-1}$ under ambient conditions. Dicyanomethylene-substituted fused tetrathienoquinoid **278** also showed a low LUMO energy level of −4.3 eV [357]. It means that the compound **278** could form high crystallinity film without thermal or solvent annealing, and adopts a nearly perpendicular orientation on the substrate in the thin film. The corresponding thin film transistors demonstrated high stability and high performance with an average electron mobility of 0.43 $cm^2\ V^{-1}\ s^{-1}$ and the highest value up to 0.9 $cm^2\ V^{-1}\ s^{-1}$ in ambient conditions. In addition, a novel series of diketopyrrolopyrrole (DPP)-containing quinoidal small molecules were developed [358]. Of these, compound **279** showed a very low LUMO energy level of −4.51 eV, well within the requirement for air stable n-type OFET materials. Under ambient conditions, the vapor processed films of **279** afforded the maximum electron mobility of 0.55 $cm^2\ V^{-1}\ s^{-1}$ with an on/off current ratio of 10^6, whereas the solution-processed films exhibited the highest electron mobility of 0.35 $cm^2\ V^{-1}\ s^{-1}$ with an on/off current ratio of 10^5–10^6. Besides dicyanomethylene, the strong electron-withdrawing tricyanovinyl groups were also introduced into the π-conjugated systems affording n-type semiconductors [359]. For example, derivative **280** showed the highest performance with an electron mobility of 0.02 $cm^2\ V^{-1}\ s^{-1}$ on hydrophobic C_{16}-alkane chain-terminated Al_2O_3 substrates. Cyano-groups substituted into the backbone of π-conjugated systems had also been reported. For example, Park et al. reported dicyanovinyl-substituted DPP based *n*-channel organic semiconductor **281** [360]. The strong electron-withdrawing dicyanovinyl units in **281** lowered the LUMO energy level to −4.45 eV. The solubility and crystallinity are slightly higher than those of dicyanomethylene-substituted quinoidal molecule **279**. The crystals of **281** displayed a dense and well-defined lamellar packing with a uniform terrace step height corresponding to a molecular monolayer. As a result of outstanding self-assembly characteristics derived from the conformational planarity, the solution-processed single crystal FET devices of **281** exhibited electron mobility up to 0.96 $cm^2\ V^{-1}\ s^{-1}$, whereas polycrystalline OFETs afforded electron mobility of 0.64 $cm^2\ V^{-1}\ s^{-1}$. Cross-conjugated aromatic quaterthiophene **282** with strongly electron-accepting tetracyanomethylene substituents has the HOMO/LUMO energy levels of −5.60/−3.68 eV [361]. The vacuum deposited films of **282** afforded an electron mobility of 0.34 $cm^2\ V^{-1}\ s^{-1}$. Core-cyanated isoindigo **283** showed a decreased LUMO energy level of −3.88 eV [362]. The thin film FET devices of **283** afforded ambient stable electron mobility up to 0.044 $cm^2\ V^{-1}\ s^{-1}$. It is interesting that the isoindigo derivative exhibited ambipolar charge transport behavior with electron and hole mobilities of 0.11 and 0.045 $cm^2\ V^{-1}\ s^{-1}$ on FOPA-modified substrates. In 2013, a series of cyano-disubstituted dipyrrolopyrazinedione (CNPzDP) small-molecules was developed [363]. By incorporating cyano substituents, the HOMO and LUMO energy levels of the electron-deficient PzDP **284** can be sufficiently decreased to −5.88 and −4.04 eV, which facilitate the electron injection and transport. In OTFT devices, **284** showed n-type semiconductor behavior with a high electron mobility of 0.16 $cm^2\ V^{-1}\ s^{-1}$ and an on/off current ratio of 10^6.

Halogen-Substituted and Other n-Type Small-Molecule Semiconductors

The introduction of halogen atoms into organic semiconductor molecules is another important strategy for the design of n-type semiconductors (see Chart 3.21). Among halogen atoms, the fluorine atom is the most widely used in the process of realizing n-type behaviors. The main reason is that the fluorine atom has stronger electronegativity compared to other halogen atoms and smaller atomic radius than that of the hydrogen atom; therefore, it can be used to tune the molecular properties without significantly changing the molecular geometry. For example, perfluoropentacene, **285** showed typical n-type charge transport properties in FET devices [364]. Perfluoropentacene possesses a planar structure as observed for pentacene. The reduction and oxidation peaks of perfluoropentacene shift positively relative to pentacene. A vacuum deposited film of **283** exhibited an electron mobility of 0.11 $cm^2\ V^{-1}\ s^{-1}$. Under the same conditions, pentacene showed *p*-channel transport behavior with a hole mobility of 0.45 $cm^2\ V^{-1}\ s^{-1}$. The similar change could also be seen in hexa-peri-hexabenzocoronene (HBC) and its fluorine-substituted derivative, **286** [365]. The introduction of fluorine atoms made the

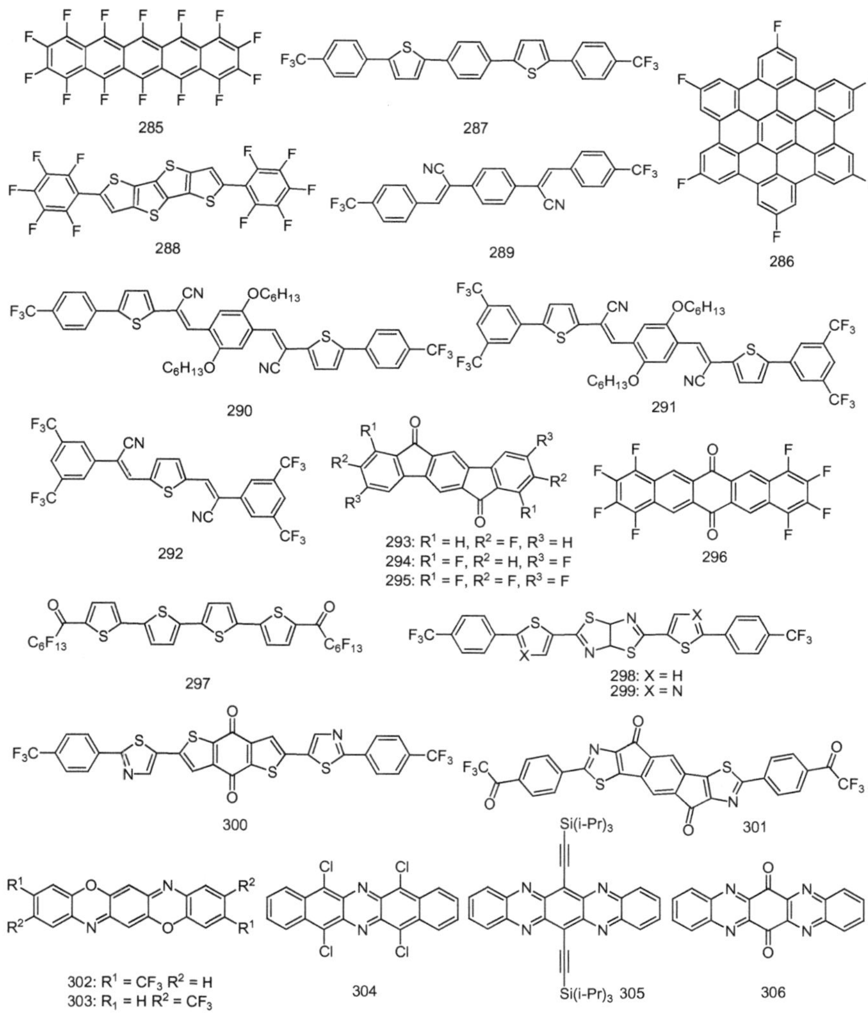

Chart 3.21 Halogen-substituted and other n-type small-molecule semiconductors

LUMO energy level of derivative **286** lower than that of HBC by about 0.5 eV. A vacuum deposited thin film of **286** showed an electron mobility of 0.016 $cm^2 V^{-1} s^{-1}$ with an on/off current ratio of 10^4. Fluorine-substituted and trifluoromethyl-substituted oligomers with *n*-channel transport behavior were also developed. Thin film transistors based on trifluoromethyl-substituted oligomer **287** afforded an electron mobility of 0.55 ± 0.05 $cm^2 V^{-1} s^{-1}$, whereas the single crystal transistors exhibited high electron mobilities up to 3.1 $cm^2 V^{-1} s^{-1}$ in vacuum on bare SiO_2 dielectric [366]. In 2012, a series of tetrathienoacene-based oligomers were developed by Youn et al. Of these, perfluorophenyl end functionalized derivative **288** showed an electron mobility as high as 0.30 $cm^2 V^{-1} s^{-1}$ with an on/off current ratio of 1.8×10^7 in vacuum [367]. For the phenyl-substituted analogue, *p*-channel transport was observed with a hole mobility as high as 0.21 $cm^2 V^{-1} s^{-1}$ under the same conditions. Although halogen atoms could effectively tune the energy levels and intermolecular interaction of semiconducting materials, and even change their charge transport type from p-type to n-type, most acenes containing halogen atoms have relatively high-lying LUMO energy levels, which lead to their FET devices having low stability in air. In order to obtain high air stability and high-performance n-type organic semiconductors, other electron-withdrawing groups and/or electron-deficient π-conjugated systems are also used in combination with halogen substituents. Shoji et al. [368] synthesized a series of oligomers containing trifluoromethyl and cyano groups. Derivative **289**-based FET devices fabricated with top contact configuration afforded the highest mobility of 0.30 $cm^2 V^{-1} s^{-1}$, when deposited on OTS-treated SiO_2 dielectric. Similar compounds such as **290–292** were also developed and used in OFETs [369, 370]. Compound **292** showed an electron mobility of 0.03 $cm^2 V^{-1} s^{-1}$ for OFET with evaporated film and 0.16 $cm^2 V^{-1} s^{-1}$ for single crystal OFET. However, compounds **290** and **291** showed high transport performances. The thin film FET devices based on **290** and **291** exhibited electron mobilities as high as 0.61 and 2.14 $cm^2 V^{-1} s^{-1}$ with on/off current ratios higher than 10^6. The high performances of **290** and **291** were attributed to tight molecular stacking and optimized energy levels, caused by incorporating these electron-withdrawing groups. Besides cyano group, carbonyl groups were also used in combination with halogen atoms affording n-type semiconductors. For example, indenofluorenediones, **293–295** with different numbers of fluorine substituents were synthesized and investigated as active layers of n-type OFETs [371]. Of them, the **295**-based FET devices fabricated on polystyrene (PS) substrates showed n-type performance with a field-effect mobility of 0.16 $cm^2 V^{-1} s^{-1}$ and an on/off current ratio of 10^6. The **293**-based FETs exhibited a dramatic decrease in the electron mobilities from 0.14 to 0.003 $cm^2 V^{-1} s^{-1}$ after 40 h of storage in air. However, the electron mobilities of **295**-based FETs decreased from 0.15 to 0.07 after 40 min of storage in air, but showed negligible changes even in the following 3 months. The high environmental stability of **295**-based FETs was attributed to the lower LUMO energy level, relative to that of **293**. Fluorinated pentacenequinone **296** was reported by Miao et al. [372]. X-ray crystallographic analysis revealed the molecular packing of **296** features weak C–H···O/F hydrogen bonds and π–π stacking with a small distance of

3.32 Å. Vacuum deposited film showed electron mobility up to 0.18 $cm^2\ V^{-1}\ s^{-1}$ when measured in vacuum. In addition, oligomer **297** containing both carbonyl and fluorohexyl groups showed the highest mobilities up to 2 $cm^2\ V^{-1}\ s^{-1}$ on the PS modified Si/SiO_2 substrate in nitrogen [373]. Using Au as top contacts, oligomer **297** afforded higher electron mobility up to 4.6 $cm^2\ V^{-1}\ s^{-1}$ [374]. Electron-deficient π-conjugated systems, such as oxadiazole, thiazole, etc., were also used in combination with halogen atoms for getting n-type semiconductors. For example, a series of thiazolothiazole derivatives were developed by Ando et al. [375]. Of these, thiazolothiazole derivative **298** has a nearly planar geometry, in which two thiophene rings take all-*trans* conformation. The thin film of **298**-based FET devices fabricated with top contact configuration exhibited an electron mobility of 0.30 $cm^2\ V^{-1}\ s^{-1}$ with an on/off current ratio of 10^6. The high mobility was related to the π-stacking structure with a distance of 3.53 Å and intermolecular short S⋯S contact of 3.25 Å. By employing a self-assembled monolayer on a SiO_2 gate insulator, the highest electron mobility of 1.2 $cm^2\ V^{-1}\ s^{-1}$ and on/off current ratio of 10^7 were achieved in **298**-based FET devices [376]. Thiazole-thiazolothiazole conjugated molecule **299** also showed a nearly planar geometry [377]. However, the HOMO-LUMO energy gap of **299** obtained from its absorption edge was 2.57 eV, larger than the 2.48 eV of **298**. The thin film of **299**-based FET devices fabricated with top contact configuration exhibited high electron mobility of 0.64 $cm^2\ V^{-1}\ s^{-1}$. Benzo[1,2-*b*:4,5-*b*′]-dithiophene-4,8-dione derivative **300** has a deeper LUMO energy level of −4.1 eV, leading to efficient charge-carrier injection and air stability [378]. A columnar structure with efficient intermolecular π-π and horizontal direction interactions, leading to high electron mobilities, is formed in its thin film. The **300**-based OFET devices showed n-type characteristics, where the electron mobility was 0.15 $cm^2\ V^{-1}\ s^{-1}$ under vacuum conditions and above 0.1 $cm^2\ V^{-1}\ s^{-1}$ in air. Oligomer **301** containing an electronegative unit, 4,9-dihydro-s-indaceno[1,2-*b*:5,6-*b*′]dithiazole-4,9-dione, formed an almost planar geometry with inter-ring dihedral angles less than 2° [379]. The C2-symmetric oligomer **301** showed a cross-oriented layer-by-layer packing and dense π-stacking with a distance of 3.38 Å in each layer. Moreover, the **301** film showed crystal-shaped submicrometer-sized grains with no observable grain boundaries, which are advantageous to electron transport. Its thin film showed electron mobility as high as 0.39 $cm^2\ V^{-1}\ s^{-1}$ and good stability. Under air-exposed conditions there was only a slight change in the mobility even after 3 months. In addition, Yu and coworkers developed two air stable n-type trifluoromethyl-substituted triphenodioxazines, **302** and **303** [380]. The vacuum deposited thin films of compounds **302** and **303** exhibited electron mobilities of 0.07 and 0.03 $cm^2\ V^{-1}\ s^{-1}$, respectively. Although the two compounds had relatively high-lying LUMO energy levels of −3.67 eV, the two compound-based FET devices exhibited outstanding environmental stability. For example, the mobility of compound **302** varied from 0.05 to 0.047 $cm^2\ V^{-1}\ s^{-1}$ even after its devices were stored for 30 days in air. The close molecular packing resulting from trifluoromethyl groups could be responsible for the high stability of these devices. Some pyrazine derivatives also afforded n-type characteristics. For example, pyrazine derivative **304**, which was synthesized by Islam et al., showed high n-type performance [381]. Single crystal FETs using **304** as the active semiconducting material and graphite as source/drain electrodes exhibited a very high electron mobility of 3.39 $cm^2 V^{-1} s^{-1}$ in ambient conditions. A vacuum deposited film of silylethynylated tetraazapentacene **305** exhibited an electron mobility of 3.3 $cm^2 V^{-1} s^{-1}$, which is one of the highest values for n-type organic semiconductors [382]. The high electron mobility of **305** was attributed to its low-lying LUMO energy level of −4.01 eV and highly ordered and denser two-dimensional brickwork arrangement with a small π-stack distance of 3.28 Å. Another pyrazine derivative **306** showed an electron mobility of 0.12 $cm^2 V^{-1} s^{-1}$ [365]. X-ray diffraction patterns revealed that derivative **306** formed polycrystalline film.

3.3.2 Selected n-Type Polymer Semiconductors

3.3.2.1 Selected Diimide-Based n-Type Polymer Semiconductors

In the same way that the development of n-type semiconductors is lagging behind that of p-type counterparts, the development of n-type polymer semiconductors has lagged far behind that of p-type counterparts. Most n-type semiconducting polymers exhibit low electron mobilities below 0.1 $cm^2 V^{-1} s^{-1}$ under ambient conditions. Despite this, many new n-type polymer semiconductors have also been reported recently and some of them exhibit high n-type charge transport properties (see Chart 3.22). Jenekhe et al. reported a ladder-shaped polymer, poly(benzobisimidazobenzophenanthroline) (BBL), **307**. The polymer has highly rigid structure and strong intermolecular interaction, with extremely high glass transition temperature up to 500 °C. The thin films of **307**, which were prepared by spin-coating from methanesulfonic acid solution, showed an electron mobility up to 0.1 $cm^2 V^{-1} s^{-1}$ under ambient air conditions and excellent air stability. The electrical parameters of **305**-based transistors were found to be constant over 4 years. In addition, self-assembled nanobelts with tight π-stacking of 3.36 Å of polymer **307** showed electron mobilities of 7×10^{-3} $cm^2 V^{-1} s^{-1}$. The high mobility and stability can be attributed to the low-lying LUMO energy level of −4.2 eV, highly crystalline morphology, and tight π-stacking. However, polymer **307** is insoluble in common solvents, which limits its wide application [383–385]. A solution-processable ladderized novel n-type NDI-based copolymer, **308** showed average electron mobilities of 0.0026 $cm^2 V^{-1} s^{-1}$, and on/off current ratios on the order of 10^4 [386]. In 2009, Facchetti et al. reported a solution-processable polymer **309** containing NDI and 2,2′-dithiophene unit. The N3-based transistor device with top gate configuration showed high n-type performance with an electron mobility up to 0.85 $cm^2 V^{-1} s^{-1}$ and remarkable stability [387]. Because of high performance, afterward, the polymer **309** became one of the most widely investigated n-type polymer semiconductors. Upon the polymer, the relationship between dielectric layers, morphology (molecule packing), device geometry, and electron transport was widely studied. In 2010, Luscombe et al. further synthesized another

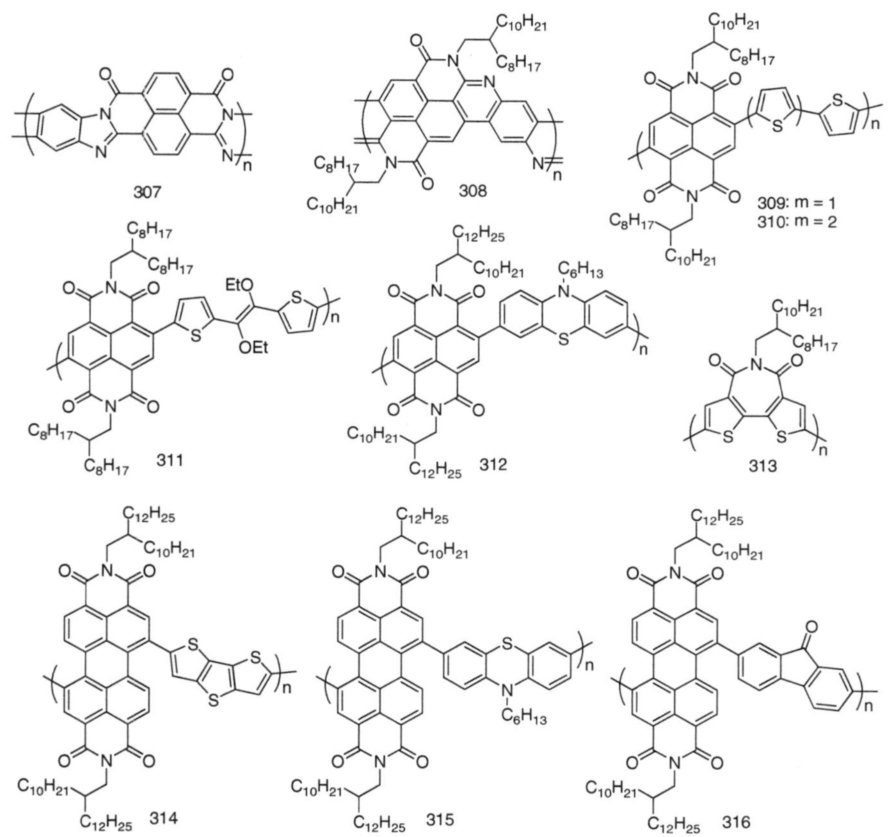

Chart 3.22 Selected diimide-based n-type polymer semiconductors

three NDI-based polymers differing only in the number of thiophene units, being zero, one, and three [388]. Of these, polymer **310**-based FETs with bottom gate top contact configuration afforded the highest electron mobility of 0.076 $cm^2\ V^{-1}\ s^{-1}$, whereas **309**-based FETs showed a mobility of 0.039 $cm^2\ V^{-1}\ s^{-1}$ with the same device geometry.

The NDI-based polymer, **311** containing electron-neutral (*E*)-1,2-diethoxy-1,2-di(thiophen-2-yl)ethene as donor unit, showed a low-lying LUMO energy level of −4.0 eV, which favors the injection of electrons and portends transistor device stability [389]. The **309**-based FET devices fabricated with bottom gate top contact configuration without annealing exhibited electron mobility of $2.3 \times 10^{-3}\ cm^2\ V^{-1}\ s^{-1}$, consistent with the low crystallinity. However, annealing at 200 °C increased the mobility to 0.2 $cm^2\ V^{-1}\ s^{-1}$. When **311**-based FET devices were fabricated with top gate bottom contact configuration, an increased electron mobility up to 0.5 $cm^2\ V^{-1}\ s^{-1}$ in ambient was achieved. The copolymer **312** comprised of NDI and phenothiazine (PTZ) units showed a slightly high LUMO energy level of −3.8 eV [390]. The copolymer **312**-based FETs with bottom gate top contact geometry exhibited electron mobility of 0.05 $cm^2\ V^{-1}\ s^{-1}$ and on/off current ratios of 10^5 in nitrogen. In 2008, Letizia et al. developed a series of novel *N*-alkyl-2,2′-bithiophene-3,3′-dicarboximide-based π-conjugated homopolymers [391]. Of these, homopolymer **313** exhibited extremely high crystallinity and an electron mobility of over 0.01 $cm^2\ V^{-1}\ s^{-1}$ with on/off current ratios of 10^7, which is remarkably independent of film-deposition conditions. It is notable that thin films of **313** also exhibited terracing in AFM images with a step height matching the X-ray diffraction *d*-spacing, a rare phenomenon for polymeric organic semiconductors. Similar to NDI-based polymers, PDI-based polymers can also afford n-type performance. For example, the first soluble PDI-based copolymer **314** was reported by Zhan et al. [392]. Polymer **314** showed high thermal stability and solution-processable properties. Its LUMO energy level locates at −3.9 eV. The FET devices with a bottom gate configuration afforded electron mobility of 0.013 $cm^2\ V^{-1}\ s^{-1}$, whereas top gate devices showed high electron mobility up to 0.06 $cm^2\ V^{-1}\ s^{-1}$. These top gate devices exhibited high stability, affording electron mobility of 0.005 $cm^2\ V^{-1}\ s^{-1}$ after being kept in air for 3 months. PDI-based copolymer **315** gave an electron mobility of 0.05 $cm^2\ V^{-1}\ s^{-1}$ with an on/off current ratio of 10^5 in nitrogen [390]. Another PDI-based copolymer **316**-based PDI incorporating planar electron-deficient fluorenone exhibited an air stable n-type performance with an electron mobility of 0.01 $cm^2\ V^{-1}\ s^{-1}$ [393].

3.3.2.2 Other n-Type Polymer Semiconductors

Besides NDI- and PDI-based n-type polymers, other high performance polymers have been developed (see Chart 3.23). For example, poly(pyridiniumphenylene)s are water-soluble and display high degrees of electroactivity [394]. When *n*-doped, these materials displayed in situ conductivities as high as 160 S/cm. The high conductivity was attributed to the planar structure, which was enforced by the cyclic structures of the polymer. Of these, poly(pyridiniumphenylene) **317** showed high electron mobility up to 3.4 $cm^2\ V^{-1}\ s^{-1}$. In recent years, DPP-based conjugated copolymers **318** and **319** were designed and synthesized for *n*-channel OFETs. For DPP–DPP copolymer **318** based on the DPP units functionalized with triethylene glycol side chains, spontaneous chain crystallization was induced, providing maximum solubility and allowing the synthesis of high molecular weight DPP–DPP copolymers [395]. Polymer **318** showed extended absorption characteristics up to 1100 nm. The thin films of **318**-based FETs with a top gate configuration and using CYTOP as dielectric layer showed high electron mobilities exceeding 3 $cm^2\ V^{-1}\ s^{-1}$. DPP-based polymer **319** was reported by the Jo group [396]. The polymer has a low-lying LUMO energy level of −4.18 eV, which favors the injection of electrons and means high air stability. The **319**-based transistor devices exhibited high electron mobility of 2.36 $cm^2\ V^{-1}\ s^{-1}$. Both the face-on and edge-on packing orientation on the substrate was observed in thin films. The electrical

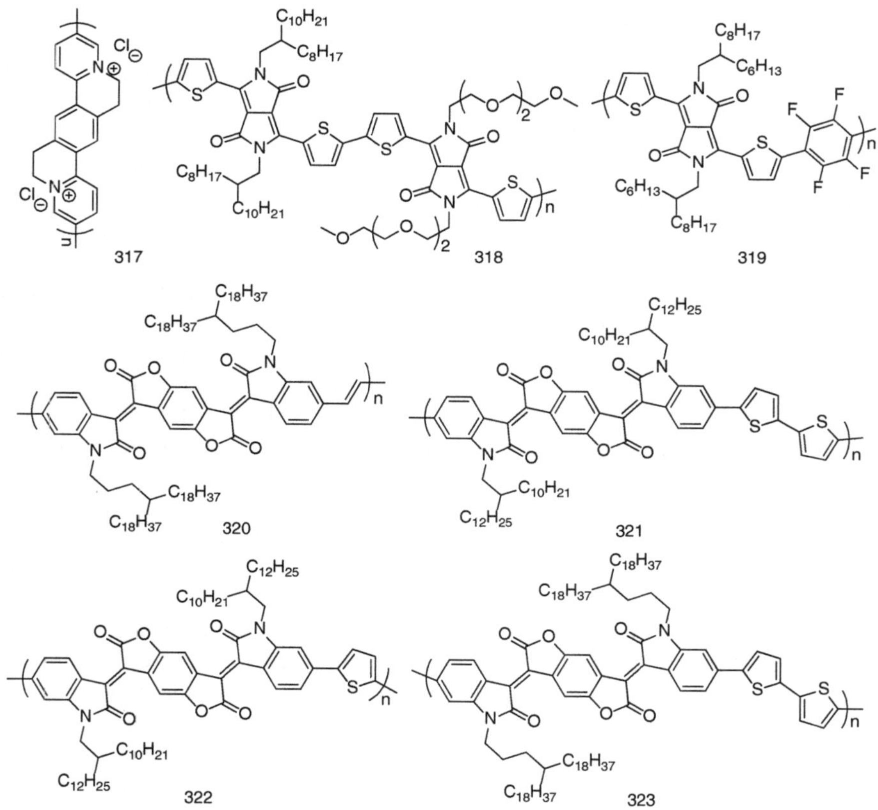

Chart 3.23 Other n-type polymer semiconductors

characteristics of **319**-based OFETs did not change significantly after 7 months storage in ambient atmosphere.

In 2013, a novel electron deficient unit, (3*E*,7*E*)-3,7-bis(2-oxoindolin-3-ylidene)-benzo[1,2-*b*:4,5-*b*′]difuran-2,6(3*H*,7*H*)-dione was synthesized by Lei and by Li et al. by different synthetic procedures [397–399]. The two groups had highly electron-deficient large fused aromatic backbones with electron-withdrawing carbonyl units, and contributed to intermolecular interaction and the overlaps of intermolecular frontier orbitals, thus facilitating interchain transport. Cyclic voltammetry (CV) curves displaying the electron deficient unit with 4-octadecyldocosyl groups has low HOMO/LUMO energy levels of −6.21/−4.24 eV. Moreover, Lei et al. thought that the aromatic backbone was an almost planar backbone with small dihedral angles of ~7.6°, which was attributed to the carbonyl groups forming intramolecular hydrogen bonds to prevent the conformational transformation of the double bonds, affording a giant "locked" aromatic plane through careful analysis of the optimized structure and ^{1}H NMR spectra of the new aromatic unit and its precursors. Based on the new aromatic unit, a series of copolymers including **320–323** were synthesized and used in OFETs. For example, copolymer **320** with 4-octadecyldocosyl groups, synthesized by Lei et al. [397], exhibited significantly lowered HOMO/LUMO levels of −6.12/−4.10 eV. Its FET performance was investigated with a top gate/bottom contact (TG/BC) device configuration. Copolymer **320** showed electron mobilities up to 1.1 $cm^2 V^{-1} s^{-1}$ with an average mobility of 0.84 $cm^2 V^{-1} s^{-1}$ under ambient conditions. The devices had high stability, affording an electron mobility of 0.31 $cm^2 V^{-1} s^{-1}$ after being stored for 30 days under room light and ambient conditions. Li et al. synthesized the copolymers **321** and **322** [398]. The HOMO and LUMO energy levels of **322** were estimated to be −5.79 and −4.11 eV, which favors the injection of electrons. The **322**-based FET device encapsulated by PMMA with a bottom gate, bottom contact configuration showed n-type characteristics with an electron mobility of 5.4×10^{-3} $cm^2 V^{-1} s^{-1}$ even though the polymer films are rather disordered. It is interesting that the non-encapsulated devices exhibited ambipolar charge transport behavior with balanced electron/hole mobilities of up to $8.2 \times 10^{-3}/1.0 \times 10^{-2}$ $cm^2 V^{-1} s^{-1}$. However, the polymer **321** is essentially insoluble in any solvent, although bearing very large branched 2-decyltetradecyl side chains. Lei et al. [399] synthesized polymer **323** with the same π-conjugated backbone of **321** and longer alkyl chains of 4-octadecyldocosyl groups. The cyclic voltammetry (CV) measurement of **323** gave HOMO/LUMO energy levels of −5.72/−4.15 eV. Polymer **323** had better solubility, for example, 3 mg/mL in 1,2-dichlorobenzene. The **323**-based FET devices fabricated in a glovebox with top gate/bottom contact configuration showed high electron mobilities up to 1.74 $cm^2 V^{-1} s^{-1}$ and an average mobility of 1.42 $cm^2 V^{-1} s^{-1}$ under ambient conditions. As with polymer **322**, ambipolar charge transport behavior could also be observed for polymer **323**. For devices fabricated under ambient conditions, the hole mobilities of **323** significantly increased. The highest hole mobility of 0.47 $cm^2 V^{-1} s^{-1}$ and an average mobility of 0.20 $cm^2 V^{-1} s^{-1}$ were obtained, whereas the highest electron mobility of **323** only slightly decreased to 1.45 $cm^2 V^{-1} s^{-1}$ (average: 1.20 $cm^2 V^{-1} s^{-1}$). From the results above, two points could be concluded. First, the new electron-deficient unit is an effective building block in tuning the energy level of polymers. Second, alkyl side chains have a huge influence on FET performance of the related polymers.

3.4 Ambipolar Semiconductors

Ambipolar semiconductors can provide both *n*- and *p*-channel performance in a single device. Upon these, large-area manufacturing of complementary integrated circuits can be obtained without requiring micro-patterning of the individual p- and n-type semiconductors. In addition, light emission can be achieved by recombination of holes and electrons within the transistor channel. To be efficient in proceeding with injection and transport of electron and hole, this kind of semiconductor should both satisfy the requirement of p-type semiconductors for the

HOMO energy level, and satisfy the requirements of n-type semiconductors for LOMO energy level. Specifically, the semiconductors are required to have a HOMO energy level below −5.0 eV and the LUMO level needs to be below or at least close to −4.0 eV.

3.4.1 Selected Ambipolar Small-Molecule Semiconductors

In 2008, Bao et al. introduced fluorine atoms to asymmetric silylethynylated tetraceno[2,3-*b*]thiophene affording thienoacene **324** (see Chart 3.24 and Table 3.4). The **324**-based thin film showed a balanced ambipolar performance with a hole mobility of 0.12 $cm^2\ V^{-1}\ s^{-1}$ in air and an electron mobility of 0.37 $cm^2\ V^{-1}\ s^{-1}$ inside the nitrogen glovebox [400]. The introduction of fluorine atoms at the terminal rings exhibited a reduced π–π stacking distance of 3.32 Å in a 2D brick layer structure of thienoacene **324**. Similarly, silylethynylated *N*-heteropentacene **325**,

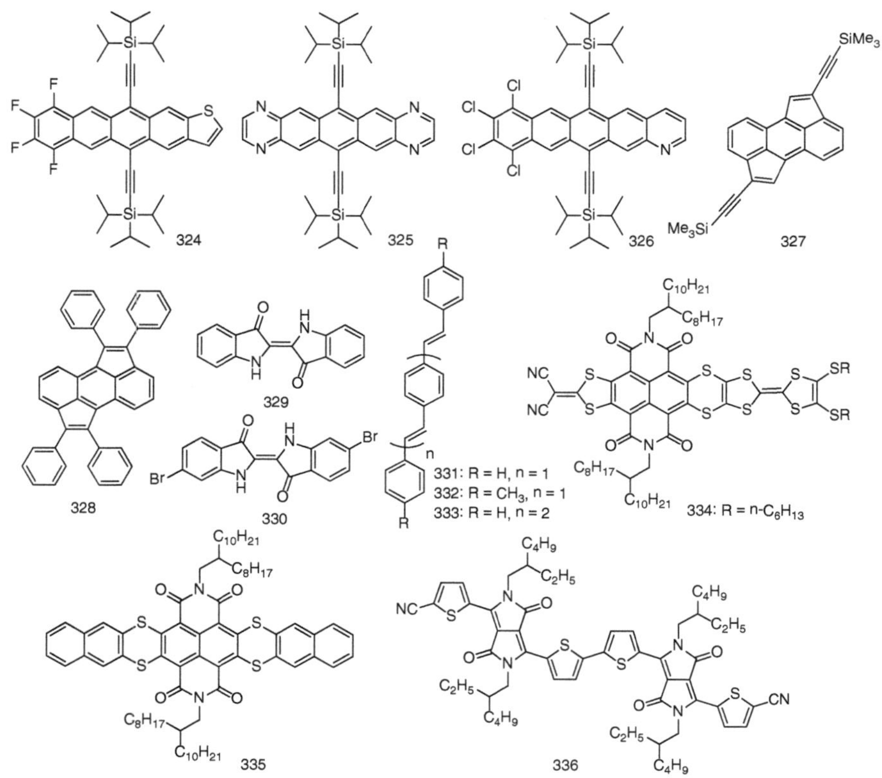

Chart 3.24 Ambipolar small-molecule semiconductors

Table 3.4 OFET device data for ambipolar transporting organic semiconductors

	Deposition Process	HOMO (eV)	LUMO (eV)	Max μ_h ($cm^2 V^{-1} s^{-1}$) (m.d.)[a]	Max μ_e ($cm^2 V^{-1} s^{-1}$)	Device structure[b]	Refs.
324	Evaporation	−5.39	−3.35	0.12 (air)	0.37 (N_2)	BGTC; Au; SiO_2/Si; OTS	[400]
325	Evaporation	−5.49	−3.68	0.22 (air)	1.1 (vacuum)	BGTC; Au; SiO_2/Si; OTMS	[401]
326	Evaporation	−5.5	−3.53	0.12 (N_2)	0.14 (N_2)	BGTC; Au, SiO_2/Si; OTS	[402]
327	Evaporation	−5.32	−3.67	0.10	0.10 (vacuum)	BGTC; Au, SiO_2/Si; OTMS	[403]
328		−5.42	−3.55	0.21	0.01 (vacuum)		[404]
329	Evaporation	− 5.5	−3.8	0.01 (vacuum) 0.01 (air)	0.01 (vacuum)	BGTC; Au; TTC/AlOx/Al	[404]
330	Evaporation	−5.8	−4.0	0.22 (vacuum) 0.20 (air)	0.03 (vacuum) 0.015 (air)	TC; Au; TTC/AlOx/Al	[405]
331	Single crystal	−5.8	−2.8	0.12	0.013	BGTC; Au/Ca	[406]
332		−5.6	−2.7	0.17	0.09		[406]
333		−5.7	−3.0	0.12	0.11		[406]
334	Spin-coating	−5.1	−4.3	0.3 (air)	0.003 (air)	BGBC; Au; SiO_2/Si; OTS	[408]
335	Spin-coating	−5.73	−3.91	0.047 (N_2)	0.016 (N_2)	BGBC; Au; SiO_2/Si; OTS	[311]
336	Spin-coating	−5.45	−3.74	0.066 (air)	0.033 (air)	BGTC; Au; SiO_2/Si; ODTS	[408]
337	Spin-casting	−5.5	−4.0	0.1 (N_2)	0.09 (N_2)	BGTC; Ba/Al; SiO_2/Si; OTS	[267]
338	Spin-coating	−5.2	−4.0	0.35 (N_2)	0.40 (N_2)	BGTC; Au; SiO_2/Si; OTS-8	[409]
339	Spin coating	−5.37	−3.74	0.23 (N_2)	0.56 (N_2)	BGTC; Au; SiO_2/Si; OTS-8	[411]
340	Spin coating	−5.16	−3.84	0.46 (N_2)	0.84 (N_2)	TGBC; Au; PMMA/Au	[412]
341	Spin coating	−4.65	−3.5	0.29 (N_2)	0.25 (N_2)	BGBC; Au; SiO_2/Si; DTS	[413]
342	Spin coating	−4.55	−3.9	0.83 (N_2)	1.36 (N_2)	BGBC; Au; SiO_2/Si; DTS	[414]
343		−4.55	−3.9	1.17 (N_2)	1.32 (N_2)		[414]
344	Spin Coating	−5.40	−3.68	0.024 (N_2)	0.056 (N_2)	BGTC; Au; SiO_2/Si; OTS-8	[415]

(continued)

Table 3.4 (continued)

	Deposition Process	HOMO (eV)	LUMO (eV)	Max μ_h ($cm^2\ V^{-1}\ s^{-1}$) (m.d.)[a]	Max μ_e ($cm^2\ V^{-1}\ s^{-1}$)	Device structure[b]	Refs.
		−5.40	−3.68	0.013 (N_2)	0.010 (N_2)		[415]
346	Spin coating	−5.67	−4.24	0.36	0.41	TGBC; Au; CYTOP/Al	[416]
347	Spin coating	−5.36	−3.56	0.37	0.24	BGTC; Au; SiO_2/Si; ODTS	[396]
348		−5.45	−3.57	0.30	0.26		[396]
224	Spin coating	−5.33	−4.07	1.36	1.58	TGBC; Au; PMMA/Au	[418]
349	Spin coating	−5.3	−3.4	0.11 (vacuum)	0.081 (vacuum)	BGTC; Au; SiO_2/Si; OTS	[419]
350		−5.3	−3.4	1.3 (vacuum)	0.1 (vacuum)		[419]
351	Spin coating	−5.09	−3.46	1.62 (N_2)	0.14 (N_2)	BGTC; Au; SiO_2/Si; OTS	[420]
352	Solution shearing	−5.07	−3.82	2.53 (N_2)	0.43 (N_2)	BGTC; Au; SiO_2/Si; OTS	[421]
353		−5.17	−3.56	6.16 (N_2)	3.07 (N_2)		[422]
354		−5.10	−3.49	8.84 (N_2)	4.34 (N_2)	BGTC; Au; SiO_2/Si; OTS	[422]
355		−5.09	−3.41	3.97 (N_2)	2.20 (N_2)		[421]
356	Spin coating	−5.46	−3.96	1.85 (N_2)	0.43 (N_2)	TGBC; Au; CYTOP/Al	[423]
				1.25 (air)	0.51 (air)		
357	Spin coating	−5.57	−3.84	1.05 (air)	0.72 (air)	TGBC; Au; CYTOP/Al	[424]
358	Spin coating	−4.8		0.16 (N_2)	0.14 (N_2)	TGBC; Au; PMMA/Au	[425]
359	Spin coating	−5.27	−3.79	0.022	0.0038	BGBC; Au/Cr; SiO_2/Si; OTS-8	[426]
360	Spin coating	−5.61	−3.93	0.23 (air)	1.13 (air)	TGBC; Au/Ti; PMMA/Al	[428]
361	Spin coating	−5.62	−3.90	0.30 (air)	1.57 (air)	TGBC; Au/Ti; PMMA/Al	[428]
362	Spin coating	−5.56	−3.7	0.04 (air)	0.3 (air)	TGBC; Au; PMMA/Au	[429]
363	Spin coating	−5.01	−3.7	0.003 (air)	0.03 (air)	TGBC; Au; PMMA/Au	[429]

(continued)

Table 3.4 (continued)

	Deposition Process	HOMO (eV)	LUMO (eV)	Max μ_h ($cm^2 V^{-1} s^{-1}$) (m.d.)[a]	Max μ_e ($cm^2 V^{-1} s^{-1}$)	Device structure[b]	Refs.
364	Spin coating	−4.36	−3.8	1.0 (N_2)	0.7 (N_2)	BGBC; Au; SiO_2/Si; DTS	[430]
365	Spin coating	−5.49	−4.17	0.053 (air)	0.021 (air)	BGBC; Au; SiO_2/Si; DTS; PMMA	[432]
366	Spin coating	−5.27	−4.24	0.21	0.10	TGBC; Au; CYTOP/Al	[436]

[a] *m.d.* measurement condition
[b] Device configuration; S/D electrodes; dielectric/gate; modification of substrate/electrode

with N atoms on the terminal rings of the pentacene backbone, exhibited a hole mobility up to 0.22 $cm^2 V^{-1} s^{-1}$ as measured in ambient air and an electron mobility up to 1.1 $cm^2 V^{-1} s^{-1}$ as measured under vacuum. It is worth noting that molecules of **325** form bilayer π-stacks with C–H–N hydrogen bonds, which enabled the herringbone packing mode to retain a possible pathway for charge transport [401]. When measured in ambient air, the electron mobility of **325** decreased to the region of 10^{-3} $cm^2 V^{-1} s^{-1}$, suggesting that most of the mobile electrons were trapped by oxygen or water. In 2011, two azapentacene derivatives were synthesized [402]. Of these, derivative **326** exhibited high and balanced ambipolar transport properties in the glovebox, with the hole and electron mobilities reaching up to 0.12 and 0.14 $cm^2 V^{-1} s^{-1}$ using Au as source/drain electrodes, respectively, and average electron and hole mobilities of 0.10 and 0.11 $cm^2 V^{-1} s^{-1}$ were also obtained by using Ag as source/drain electrodes, respectively. The high performance of **326** was attributed to the unoccupied 3d orbital in the chlorine atom which can delocalize electrons from the conjugated core and thus the chlorine atoms act as an additional electron pathway. More recently, Miao et al. reported a series of cyclopenta-fused anthracenes **327** and **328** [403]. The **327**-based devices showed ambipolar transport properties with balanced hole and electron mobilities of 0.1 $cm^2 V^{-1} s^{-1}$, and **328**-based devices afforded a hole mobility of 0.21 $cm^2 V^{-1} s^{-1}$, but lower electron mobility in the region of 0.01 $cm^2 V^{-1} s^{-1}$. Natural dye isoindigo, **327** and Tyrian purple, **330** were also used as semiconducting layer in OFETs [404, 405]. Sariciftci and coworkers fabricated **329**-based devices on AlOx passivated with tetratetracontane. Upon encapsulation, the devices showed mobilities of around 1×10^{-2} $cm^2 V^{-1} s^{-1}$ for electrons and 5×10^{-3} to 1×10^{-2} $cm^2 V^{-1} s^{-1}$ for holes, and good operational stability in air. The **330**-based FET devices with similar composites demonstrated high hole and electron mobilities of 0.22 and 0.03 $cm^2 V^{-1} s^{-1}$ and air stable operation. The good operational stability in air of **329** and **330** could be attributed to their low-lying LUMO energy levels. The ambipolar behavior of linearly oligo(*p*-phenylenevinylene) derivatives, **331–333** were reported by Nakanotani et al. [406]. Of these, the single crystal OFETs of **333** exhibited balanced hole and electron mobilities of higher than 0.1 $cm^2 V^{-1} s^{-1}$. Recently, a series of naphthalene-based semiconductors with ambipolar transport properties were developed by Zhang et al. [311, 407]. For example, compound **334**-based devices exhibited relatively high hole and electron mobilities in air, reaching 0.03 and 0.003 $cm^2 V^{-1} s^{-1}$, respectively. Compound **335**-based devices also showed relatively balanced hole and electron mobilities of 0.047 and 0.016 $cm^2 V^{-1} s^{-1}$, respectively. More recently, Wang et al. reported a cyano-terminated dithienyldiketopyrrolopyrrole dimer **336**. The thin film transistor devices based on **336** fabricated by solution processing showed ambipolar performance with hole and electron mobilities of 0.066 and 0.033 $cm^2 V^{-1} s^{-1}$, respectively, under ambient conditions [408].

3.4.2 Selected Ambipolar Polymer Semiconductors

3.4.2.1 DPP-Based Ambipolar Polymer Semiconductors

DPP-based polymers showed high p-type or n-type charge transport properties, and high ambipolar properties were observed in some DPP-based polymers (see Chart 3.25). The first ambipolar DPP-based polymer, **337** was reported by the Winnewisser group [267]. The FETs based on **337** exhibited hole and electron mobilities of 0.1 and 0.09 $cm^2 V^{-1} s^{-1}$, respectively. Ambipolarity in this material was not limited to one particular transistor architecture, but had been observed in five different configurations including transistors with solution-processed gate dielectrics in bottom gate as well as top gate structures. Copolymer **338**, containing DPP and electron-withdrawing benzothiadiazole units has ideal HOMO (5.2 eV) and LUMO

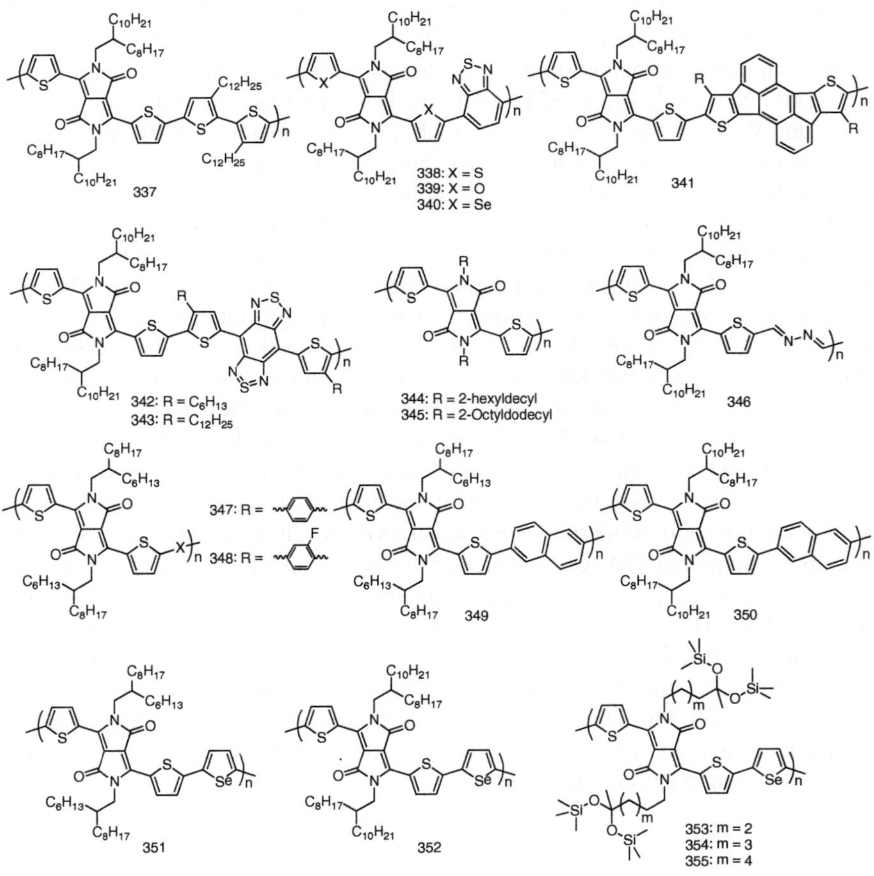

Chart 3.25 Ambipolar DPP-based polymer semiconductors

(4.0 eV) energy levels for the formation of electron and hole accumulation layers [409]. **338**-based transistor devices exhibited high and balanced ambipolar performance with a hole mobility of 0.35 $cm^2\ V^{-1}\ s^{-1}$ and electron mobility of 0.4 $cm^2\ V^{-1}\ s^{-1}$. Adopting a top gate configuration, **338**-based OFETs afforded even higher mobilities with both holes and electrons of over 0.5 $cm^2\ V^{-1}\ s^{-1}$ [410]. Beside the suitable energy levels, semicrystallinity and highly ordered lamellar packing also favor the high and balanced ambipolar performance in **338**-based OFETs. The counterparts containing furan and selenophene were also synthesized and examined [411, 412]. For example, with replacing the thiophene unit with the furan unit, the resulting polymer **339** has HOMO and LUMO energy levels of −5.37 and −3.74 eV, respectively. **339**-based transistor devices with bottom gate top contact configuration demonstrated a hole mobility of 0.20 $cm^2\ V^{-1}\ s^{-1}$ and electron mobility as high as 0.56 $cm^2\ V^{-1}\ s^{-1}$. The selenophene-substituted polymer, **340** afforded increased mobilities, with hole and electron mobilities of 0.46 and 0.84 $cm^2\ V^{-1}\ s^{-1}$, respectively. The improvement in mobilities of **340**-based FETs was attributed to the small π–π stacking distance of 3.64 Å.

Mohebbi et al. [413] developed a novel DPP-based copolymer **341** containing the emeraldicene (EMD) unit. As anticipated, enhanced intermolecular interaction forces were formed because of the π–π stacking of the fused rings and the D–A interaction between EMD and DPP units. The strong intermolecular interaction is better for getting high charge mobility. The thin films of polymer **341** showed equivalent hole and electron mobilities of 0.29 and 0.25 $cm^2\ V^{-1}\ s^{-1}$, respectively. Copolymers **342** and **343** comprised of DPP and benzobisthiadiazole (BBT) units showed high and balanced ambipolar performance [414]. Because of the strong electron-withdrawing ability, it is the BBT unit, not DPP, which plays 'acceptors' in the D–A copolymers, which have narrow bandgaps of 0.65 eV. Thin films of polymers **342** and **343** showed equivalent hole and electron mobilities above 0.5 $cm^2\ V^{-1}\ s^{-1}$. In particular, **343** demonstrated both hole and electron mobilities exceeding 1 $cm^2\ V^{-1}\ s^{-1}$.

DPP-bithiophene polymers **344** and **345**, reported by Li and coworkers, have HOMO and LUMO energy levels of −5.40 and −4.20 eV, respectively, which are suitable for injection and transport of both holes and electrons [415]. **345**-based devices also showed ambipolar characteristics with balanced hole mobility of 0.024 $cm^2\ V^{-1}\ s^{-1}$ and electron mobility of 0.056 $cm^2\ V^{-1}\ s^{-1}$. A new DPP-based polyazine **346** was developed by the same group [416]. The azine linkage was found to be a strong electron-withdrawing moiety, which is useful for lowering the LUMO energy level to achieve good electron transport characteristics. The FET devices based on **346** showed equivalent ambipolar performance with an electron mobility up to 0.41 $cm^2\ V^{-1}\ s^{-1}$ and a hole mobility up to 0.36 $cm^2\ V^{-1}\ s^{-1}$. The Jo group synthesized four alternating copolymers composed of DPP and a fluorinated phenyl unit, where the number of fluorine substitutions on phenylene varies from zero, one, two to four for n-type OFET application as mentioned before. When the number of fluorine substitutions is zero and one, the resulting polymers, **347** and **348**, both showed high ambipolar performance with hole and electron mobilities of about 0.30–0.40 $cm^2\ V^{-1}\ s^{-1}$ [396].

Copolymer **224** was previous reported by Li et al. and showed only hole transport performance [417]. In 2012, Chen et al. fabricated the **224**-based FETs using top gate and solvent-cleaned gold contact instead of previously reported bottom gate and O_2-plasma-cleaned gold contact [418]. The devices exhibited balanced ambipolar performance with a hole mobility of 1.36 $cm^2\ V^{-1}\ s^{-1}$ and electron mobility of 1.56 $cm^2\ V^{-1}\ s^{-1}$ under nitrogen atmosphere. The main reason for this is that the work function of solvent-cleaned gold (4.7–4.9 eV) is more suitable for electron injection compared with O_2-plasma-cleaned gold (5.0–5.5 eV). Zhao and coworkers also fabricated the FET devices based on **224** with high molecular weight. Ambipolar characteristics were observed in these devices under vacuum. The highest hole and electron mobilities at 373 K were respectively 13.5 and 1.58 $cm^2\ V^{-1}\ s^{-1}$ [269]. Similar copolymers **349** and **350** containing DPP and another fused ring, naphthalene, have optical bandgaps of ~1.4 eV. The highest hole and electron mobilities achieved in polymer **350** were as high as 1.3 and 0.1 $cm^2\ V^{-1}\ s^{-1}$, respectively [419].

A series of selenophene-DPP copolymers were developed by Chen et al. and Oh et al. The copolymer **351**-based FET fabricated by spin-coating process from 1,2,4-trichlorobenzene exhibited a dense nanofiber morphology with lamellar chain packing, leading to the relatively high hole and electron mobility up to 1.62 and 0.14 $cm^2\ V^{-1}\ s^{-1}$, respectively [420]. The copolymer **352** substituted by longer branched alky chains, 2-octyldedecyl groups, afforded similar ambipolar performance with those of **351** [421]. However, hybrid siloxane-substituted polymers, **353–355** showed higher ambipolar performance. For example, **355**-based FET devices fabricated by a solution shearing process showed balanced hole mobility as high as 3.97 $cm^2\ V^{-1}\ s^{-1}$ and electron mobility as high as 2.20 $cm^2\ V^{-1}\ s^{-1}$. Furthermore, Oh et al. systematically studied the -conjugated system with different alkyl spacer length of hybrid side chains [422]. The annealed films of **354** with pentyl spacer exhibited extraordinary hole and electron mobilities of up to 8.84 and 4.34 $cm^2\ V^{-1}\ s^{-1}$, respectively. These hole and electron mobilities are the highest reported ambipolar mobilities measured in organic or polymer-based semiconductors to date. The high performance of these polymers with hybrid siloxane side chains was attributed to the formation of efficient π–π stacking and three-dimensional conduction channels in thin films.

3.4.2.2 Other Ambipolar Polymer Semiconductors

As well as DPP-based ambipolar polymers, some other polymers with ambipolar performance were discovered (see Chart 3.26). For example, Lei et al. [423] synthesized a fluorinated isoindigo-based polymer **356**. The introduction of two fluorine atoms to the -conjugated backbone effectively lowered the LUMO energy level to −3.96 eV, and led to the formation of highly ordered thin film with lamellar structure and low π-stacking distance of 3.53 Å. The **356**-based devices showed a significant increase of electron mobility to 0.43 from 0.07 $cm^2\ V^{-1}\ s^{-1}$ obtained by unfluorinated polymer, while maintaining a high hole mobility of up to

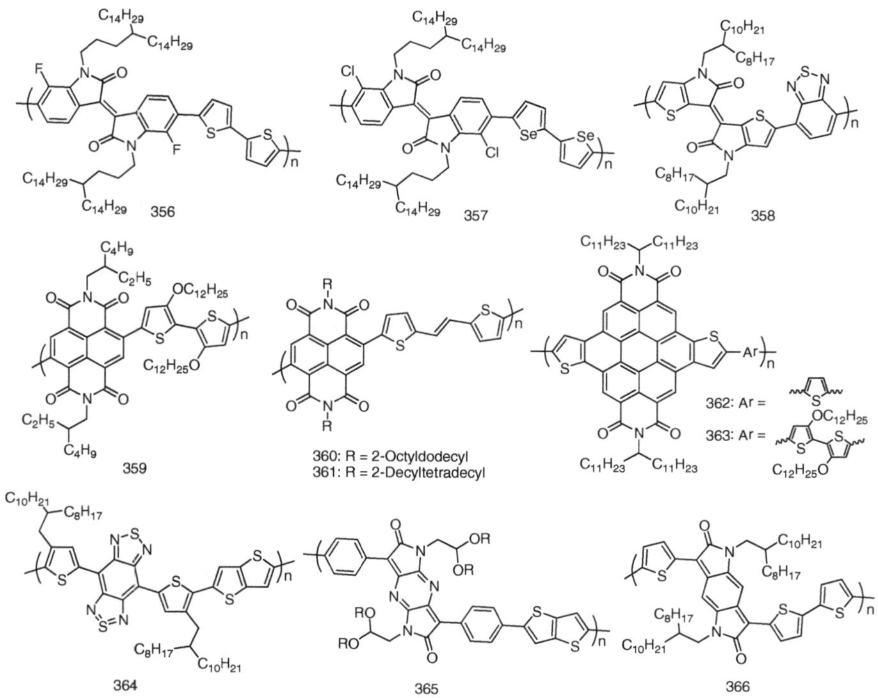

Chart 3.26 Other ambipolar polymer semiconductors

1.85 cm^2 V^{-1} s^{-1}. Then the group developed a chlorinated isoindigo-based copolymer **357**, in which 2,2′-diselenophene was used as donor unit [424]. The polymer **357**-based FETs demonstrated a rather balanced hole mobility of 1.05 cm^2 V^{-1} s^{-1} and electron mobility of 0.72 cm^2 V^{-1} s^{-1}. The results verified that chlorination and fluorination are useful methods for tuning the properties of organic semiconductors. Ashraf et al. [425] developed a novel copolymer, **358**-based on a new thienopyrrolone building block, in which the outer phenyl rings of isoindigo were replaced with thiophene rings. The thiophene–thiophene links along the backbone optimize planarity, thus maximizing π-conjugation and further enhancing close intermolecular contacts, which also promotes the polar functionality. The strong D–A character created a highly hybridized frontier molecular orbital system leading to low-lying LUMO and high-lying HOMO orbitals, optimal for ambipolarity. This polymer **358** exhibited high ambipolar charge transport with hole and electron mobilities of over 0.1 cm^2 V^{-1} s^{-1}.

Some naphthalene diimide (NDI)-based copolymers also showed ambipolar performance. For example, copolymer **359** is comprised of 2,2′-dithiophene as donor unit and naphthalene diimide (NDI) as acceptor unit [426]. The alkoxy groups successfully raised the HOMO energy level to −5.28 eV and kept the LUMO energy level at about −3.9 eV. Thin film FET devices based on the polymer

359 afforded ambipolar performance with hole and electron mobilities of 0.04 and 0.003 $cm^2\ V^{-1}\ s^{-1}$, respectively. Other NDI-based copolymers containing different thiophene moieties with varied electron-donating strength and conformations showed LUMO energy levels of −3.9 to −3.8 eV [427]. Ambipolar OFETs with electron mobilities of 0.006–0.02 $cm^2\ V^{-1}\ s^{-1}$ and hole mobilities of greater than $10^{-3}\ cm^2\ V^{-1}\ s^{-1}$ were observed in the NDI copolymers with high-lying HOMO energy levels higher than −5.4 eV. More recently, the Yu group reported two NDI-based copolymers, **360** and **361** [428]. The incorporation of vinyl linkages into polymer backbones maintains the LUMO energy levels at −3.90 eV, and the HOMO energy levels between −5.82 and −5.61 eV. The energy levels ensured the efficient injection of holes and electrons. The **360**-based devices exhibited good ambipolar characteristics in ambient conditions (20 ~ 40 % air humidity) with hole mobility up to 0.30 $cm^2\ V^{-1}\ s^{-1}$ and electron mobility up to 1.57 $cm^2\ V^{-1}\ s^{-1}$. These mobilities are among the highest values observed to date for NDI-based polymers. Based on polymer **361**, ambipolar inverters had been realized in ambient conditions, exhibiting a high gain of 155. The results provided important progress in solution-processed ambipolar NDI-based copolymeric FETs and complementary-like inverters. In addition, Usta and coworkers synthesized a class of new highly extended π-electron deficient diimide building blocks, coronenediimide (**DTCDI**) core [429]. The HOMO and LUMO energy levels of the **DTCDI** core are ~0.3–0.4 eV higher than those of well-studied PDI cores. DTCDI-based copolymers **362** and **363** demonstrated charge carrier mobilities up to 0.30 and 0.04 $cm^2\ V^{-1}\ s^{-1}$ for electrons and holes in ambient conditions, respectively. The efficient charge transport of the DTCDI-based polymer system was attributed to a highly planar polymeric backbone and enhanced regioregularity, which may facilitate intrachain charge delocalization, interchain charge transport, and thin-film microstructure.

A new BBT-based polymer **364** with a bi(thiophen-2-yl)-thieno[3,2-*b*]thiophene unit located between each pair of BBT units along the chain has the LUMO energy level of −3.8 eV, which is advantageous to the injection and transport of electrons [430]. The **364**-based FETs showed balanced ambipolar transport performance with hole mobility of 0.7 $cm^2\ V^{-1}\ s^{-1}$ and electron mobility of 1.0 $cm^2\ V^{-1}\ s^{-1}$. The highly ordered thin films with lamellar structure and low -stacking distance of 3.53 Å were responsible for the high charge transport performance. Other ambipolar copolymers containing the strongly accepting BBT moiety were also synthesized [431]. Hong et al. [432] developed two new low bandgap conjugated polymers, for example **365**, containing dipyrrolo[2,3-*b*:2′,3′-*e*]pyrazine-2,6(1*H*,5*H*)-dione (PzDP). The stronger electron-withdrawing inherence of the PzDP unit was expected to bring about strong intermolecular interactions and a short π–π stacking distance. The resulting polymers have HOMO energy levels of around −5.50 eV, and LUMO energy levels in the range −4.17 ~ −4.31 eV, which is favorable to the injection of electrons and shows the potential for ambipolar semiconductors. The **365**-based devices showed ambipolar transport with hole mobility of 0.053 $cm^2\ V^{-1}\ s^{-1}$ and electron mobility of 0.021 $cm^2\ V^{-1}\ s^{-1}$, and high on/off current ratios up to 10^6. In recent years, a series of phenyl-flanked benzodipyrrolidone (BPP)-based copolymers were synthesized by several groups

[433–435]. However, because the phenyl moieties of BPPs are often responsible for introducing steric twisting along the backbone, leading to large dihedral angles between linked units, the relating polymers often afforded low charge mobilities. In 2013, Rumer et al. [436] developed a thiophene-flanked benzodipyrrolidone (BPT) building block and synthesized a series of BPT-based copolymers, which all showed HOMO/LUMO energy levels of −5.4 eV and below −3.8 eV, respectively. The electrochemical data indicated that BPT is an effective unit for constructing low-bandgap polymers and the injection of electrons and holes is favorable in these polymers. **366**-based FETs with top gate configuration showed balanced ambipolar transport behavior with hole mobility of 0.2 $cm^2\ V^{-1}\ s^{-1}$ and electron mobility of 0.1 $cm^2\ V^{-1}\ s^{-1}$, respectively.

3.5 Outlook

During the past 20 years, especially the past 10 years, interdisciplinary research focused on the design and synthesis of organic semiconductors, analysis of various elemental mechanisms involved in charge carrier transport processes, and fabrication technology of organic field-effect transistors has made remarkable progress: plenty of semiconducting materials with mobility higher than that of amorphous silicon, and demonstration systems and products based on OFETs becoming available on a commercial basis. Nonetheless, the electrical parameters and stability of the OFETs still lag far behind that of their inorganic counterparts.

For semiconducting materials, new π-conjugated molecules with high stability and high performance still need to be developed. More clear relationships between molecular structure, aggregation structure, and properties still need to be set up. In addition, a general charge transport mechanism need to be put forward, although various types of transport mechanism have been proposed. For processing techniques, low cost fabricating techniques, mainly printing technique, for use in large-area production of OFETs, need to be constantly improved. The technique related to flexible OFETs devices also need to be studied further.

In brief, exploiting the properties of OFETs and translating them into killer applications are common goals of the OFET research community. Great opportunities exist in the field of OFETs; however, there are also a lot of challenges that we have to face on the road to success.

References

1. Brown AR, Pomp A, Hart CM, Deleeuw DM (1995) Logic gates made from polymer transistors and their use in ring oscillators. Science 270:972–974
2. Crone B, Dodabalapur A, Lin YY, Filas RW, Bao Z, LaDuca A, Sarpeshkar R, Katz HE, Li W (2000) Large-scale complementary integrated circuits based on organic transistors. Nature 403:521–523

3. Tee B, Wang C, Allen R, Bao Z (2012) An electrically and mechanically self-healing composite with pressure- and flexion-sensitive properties for electronic skin applications. Nat Nanotech 7:825–832
4. Murphy AR, Fréchet JMJ (2007) Organic semiconducting oligomers for use in thin film transistors. Chem Rev 107:1066–1096
5. Anthony JE (2006) Functionalized acenes and heteroacenes for organic electronics. Chem Rev 106:5028–5048
6. Wang C, Dong H, Hu W, Liu Y, Zhu D (2012) Semiconducting π-conjugated systems in field-effect transistors: a material odyssey of organic electronics. Chem Rev 112:2208–2267
7. Gelinck GH, Geuns TCT, de Leeuw DM (2000) High-performance all-polymer integrated circuits. Appl Phys Lett 77:1487–1498
8. Dimitrakopoulos D, Malenfant PRL (2002) Organic thin film transistors for large area electronics. Adv Mater 14:99–117
9. Horowitz G (2003) Tunneling current in polycrystalline organic thin-film transistors. Adv Funct Mater 13:53–60
10. Campbell IH, Smith L (2001) Physics of organic electronic devices. Solid State Phys 55:1–334
11. Lilienfeld JE (1930) Method and apparatus for controlling electric currents. US 1745175
12. Atalla MM, Tannenbaum E, Scheibner EJ (1959) Stabilization of silicon surfaces by thermally grown oxides. Bell Syst Tech J 38:749–783
13. Ebisawa F, Kurokawa T, Nara S (1983) Electrical properties of polyacetylene/polysiloxane interface. J Appl Phys 54:3255–3259
14. Tsumura A, Koezuka H, Ando T (1986) Macromolecular electronic device: field-effect transistor with a polythiophene thin film. Appl Phys Lett 49:1210–1212
15. Assadi A, Svensson C, Willander M, Inganas O (1988) Field-effect mobility of poly(3-hexylthiophene). Appl Phys Lett 53:195–197
16. Zaumseil J, Sirringhaus H (2007) Electron and ambipolar transport in organic field-effect transistors. Chem Rev 107:1296–1323
17. Newman CR, Frisbie CD, da Silva Filho DA, Bredas JL, Ewbank PC, Mann KR (2004) Introduction to organic thin film transistors and design of n-channel organic semiconductors. Chem Mater 16:4436–4451
18. Ahles M, Schmechel R, Seggern HV (2004) n-Type organic field-effect transistor based on interface-doped pentacene. Appl Phys Lett 85:4499–4501
19. Singh TB, Senkarabacak P, Sariciftci NS, Tanda A, Lackner C, Hagelauer R, Horowitz G (2006) Organic inverter circuits employing ambipolar pentacene field-effect transistors. Appl Phys Lett 89:033512
20. Takimiya K, Ebata H, Sakamoto K, Izawa T, Otsubo T, Kunugi Y (2006) 2,7-Diphenyl[1]benzothieno[3,2-*b*]benzothiophene. A new organic semiconductor for air-stable organic field-effect transistors with mobilities up to 2.0 $cm^2\ V^{-1}\ s^{-1}$. J Am Chem Soc 128:12604–12605
21. Di CA, Liu Y, Yu G, Zhu D (2009) Interface engineering: an effective approach toward high-performance organic field-effect transistors. Acc Chem Res 42:1573–1583
22. Giri G, Verploegen E, Mannsfeld SCB, Atahan-Evrenk S, Kim DH, Lee SY, Becerril HA, Aspuru-Guzik A, Toney MF, Bao Z (2011) Tuning charge transport in solution-sheared organic semiconductors using lattice strain. Nature 480:504–508
23. Briseno AL, Mannsfeld SCB, Ling MM, Liu S, Tseng RJ, Reese C, Roberts ME, Yang Y, Wudl F, Bao Z (2006) Patterning organic single-crystal transistor arrays. Nature 444:913–917
24. Sundar VC, Zaumseil J, Podzorov V, Menard E, Willett RL, Someya T, Gershenson ME, Rogers JA (2004) Elastomeric transistor stamps: reversible probing of charge transport in organic crystals. Science 303:1644–1646
25. Jiang H, Yang X, Cui Z, Liu Y, Li H, Hu W, Liu Y, Zhu D (2007) Phase dependence of single crystalline transistors of tetrathiafulvalene. Appl Phys Lett 91:123505

26. Minemawari H, Yamada T, Matsui H, Tsutsumi JY, Haas S, Chiba R, Kumai R, Hasegawa T (2011) Inkjet printing of single-crystal films. Nature 475:364–367
27. Tang Q, Li H, Liu Y, Hu W (2006) High-performance air-stable n-type transistors with an asymmetrical device configuration based on organic single-crystalline submicrometer/nanometer ribbons. J Am Chem Soc 128:14634–14639
28. Zhou Y, Wang L, Wang J, Pei J, Cao Y (2008) Highly sensitive, air-stable photodetectors based on single organic sub-micrometer ribbons self-assembled through solution processing. Adv Mater 20:3745–3749
29. Hutchison GR, Ratner MA, Marks TJ (2005) Intermolecular charge transfer between heterocyclic oligomers. Effects of heteroatom and molecular packing on hopping transport in organic semiconductors. J Am Chem Soc 127:16866–16881
30. Bredas JL, Calbert JP, da Silva DA, Cornil J (2002) Organic semiconductors: a theoretical characterization of the basic parameters governing charge transport. Proc Natl Acad Sci USA 99:5804–5809
31. De Leeuw DM, Simenon MMJ, Brown AR, Einerhand REF (1997) Stability of n-type doped conducting polymers and consequences for polymeric microelectronic devices. Synth Met 87:53–59
32. Brown AR, De Leeuw DM, Lous EJ, Havinga EE (1994) Organic n-type field-effect transistor. Synth Met 66:257–261
33. Haddon RC, Perel AS, Morris RC, Palstra TTM, Hebard AF, Fleming RM (1995) C_{60} thin film transistors. Appl Phys Lett 67:121–123
34. Bromley ST, Mas-Torrent M, Hadley P, Rovira C (2004) Importance of intermolecular interactions in assessing hopping mobilities in organic field effect transistors: pentacene versus dithiophene–tetrathiafulvalene. J Am Chem Soc 126:6544–6545
35. Wurthner F, Schmidt R (2006) Electronic and crystal engineering of acenes for solution-processible self-assembling organic semiconductors. Chem Phys Chem 7:793–797
36. Dong H, Wang C, Hu W (2010) High performance organic semiconductors for field-effect transistors. Chem Commun 44:5211–5222
37. Moon H, Zeis R, Borkent EJ, Besnard C, Lovinger AJ, Siegrist T, Kloc C, Bao ZN (2004) Synthesis, crystal structure, and transistor performance of tetracene derivatives. J Am Chem Soc 126:15322–15323
38. Allard S, Forster M, Souharce B, Thiem H, Scherf U (2008) Organic semiconductors for solution-processable field-effect transistors (OFETs). Angew Chem Int Ed 47:4070–4098
39. Salleo A (2007) Charge transport in polymeric transistors. Mater Today 10:38–45
40. Henson ZB, Müllen K, Bazan GC (2012) Design strategies for organic semiconductors beyond the molecular formula. Nat Chem 4:699–704
41. Tsao HN, Cho DM, Park I, Hansen MR, Mavrinskiy A, Yoon DY, Graf R, Pisula W, Spiess HW, Müllen K (2011) Ultrahigh mobility in polymer field-effect transistors by design. J Am Chem Soc 133:2605–2612
42. Kline RJ, McGehee MD, Kadnikova EN, Liu J, Fréchet JMJ (2003) Controlling the field-effect mobility of regioregular polythiophene by changing the molecular weigh. Adv Mater 15:1519–1522
43. Okamoto K, Luscombe CK (2011) Controlled polymerizations for the synthesis of semiconducting conjugated polymers. Polym Chem 2:2424–2434
44. Carsten B, He F, Son HJ, Xu T, Yu L (2011) Stille polycondensation for synthesis of functional materials. Chem Rev 111:1493–1528
45. Osaka I, McCullough RD (2008) Advances in molecular design and synthesis of regioregular polythiophenes. Acc Chem Res 41:1202–1214
46. Berrouard P, Najari A, Pron A, Gendron D, Morin P, Pouliot J, Veilleux J, Leclerc M (2012) Synthesis of 5-Alkyl[3,4-*c*]thienopyrrole-4,6-dione-based polymers by direct heteroarylation. Angew Chem Int Ed 51:2068–2071
47. Sirringhaus H, Tessler N, Friend RH (1998) Integrated optoelectronic devices based on conjugated polymers. Science 280:1741–1744

48. Halik M, Klauk H, Zschieschang U, Schmid G, Ponomarenko S, Kirchmeyer S, Weber W (2003) Relationship between molecular structure and electrical performance of oligothiophene organic thin film transistors. Adv Mater 15:917–922
49. Gao J, Li R, Li L, Meng Q, Jiang H, Li H, Hu W (2007) High-performance field-effect transistor based on dibenzo[*d,d'*]thieno[3,2-*b*;4,5-*b'*]dithiophene, an easily synthesized semiconductor with high ionization potential. Adv Mater 19:3008–3011
50. Meng Q, Jiang L, Wei Z, Wang C, Zhao H, Li H, Xu W, Hu W (2010) Development of organic field-effect properties by introducing aryl-acetylene into benzodithiophene. J Mater Chem 20:10931–10935
51. Ortiz RP, Herrera H, Blanco RL, Huang H, Facchetti A, Marks TJ, Zheng Y, Segura JL (2010) Organic n-channel field-effect transistors based on arylenediimide-thiophene derivatives. J Am Chem Soc 132:8440–8452
52. Chu CW, Li SH, Chen CW, Shrotriya V, Yang Y (2005) High-performance organic thin-film transistors with metal oxide/metal bilayer electrode. Appl Phys Lett 87:193508
53. Di CA, Yu G, Liu YQ, Guo YL, Wang Y, Wu WP, Zhu DB (2008) High performance organic field-effect transistors with low-cost copper electrodes. Adv Mater 20:1286–1290
54. Gundlach DJ, Royer JE, Park SK, Subramanian S, Jurchescu OD, Hamadani BH, Moad AJ, Kline RJ, Teague LC, Kirillov O, Richter CA, Kushmerick JG, Richter LJ, Parkin SR, Jackson TN, Anthony JE (2008) Contact-induced crystallinity for high-performance soluble acene-based transistors and circuits. Nat Mater 7:216–221
55. Chen FC, Liao CH (2008) Improved air stability of n-channel organic thin-film transistors with surface modification on gate dielectrics. Appl Phys Lett 93:103310
56. Kumaki D, Yahiro M, Inoue Y, Tokito S (2007) Air stable, high performance pentacene thin-film transistor fabricated on SiO_2 gate insulator treated with β–phenethyltrichlorosilane. Appl Phys Lett 90:133511
57. Aleshin AN, Lee JY, Chu SW, Kim JS, Park YW (2004) Mobility studies of field-effect transistor structures based on anthracene single crystals. Appl Phys Lett 84:5383–5385
58. Gundlach DJ, Nichols JA, Zhou L, Jackson TN (2002) Thin-film transistors based on well-ordered thermally evaporated naphthacene films. Appl Phys Lett 80:2925
59. Reese C, Chung WJ, Ling MM, Roberts M, Bao Z (2006) High-performance microscale single-crystal transistors by lithography on an elastomer dielectric. Appl Phys Lett 89:202108
60. Lin YY, Gundlach DJ, Nelson SF, Jackson TN (1997) Stacked pentacene layer organic thin-film transistors with improved characteristics. IEEE Electron Device Lett 18:606–608
61. Kelley TW, Muyres DV, Baude PF, Smith TP, Jones TD (2003) High performance organic thin film transistors. Mater Res Soc Symp Proc 771:169–179
62. Wang CH, Hsieh CY, Hwang JC (2011) Flexible organic thin-film transistors with silk fibroin as the gate dielectric. Adv Mater 23:1630–1634
63. Jurchescu OD, Popinciuc M, van Wees BJ, Palstra TTM (2007) Interface-controlled, high-mobility organic transistors. Adv Mater 19:688–692
64. Klauk H, Zschieschang U, Weitz RT, Meng H, Sun F, Nunes G, Keys DE, Fincher CR, Xiang Z (2007) Organic transistors based on di(phenylvinyl)anthracene: performance and stability. Adv Mater 19:3882–3887
65. Mondal R, Adhikari RM, Shah BK, Neckers DC (2007) Revisiting the stability of hexacenes. Org Lett 9:2505–2508
66. Mondal R, Shah BK, Neckers DC (2006) Photogeneration of heptacene in a polymer matrix. J Am Chem Soc 128:9612–9613
67. Meng H, Bendikov M, Mitchell G, Helgeson R, Wudl F, Bao Z, Siegrist T, Kloc C, Chen CH (2003) Tetramethylpentacene: remarkable absence of steric effect on field effect mobility. Adv Mater 15:1090–1093
68. Kelley TW, Boardman LD, Dunbar TD, Muyres DV, Pellerite MJ, Smith TYP (2003) High-performance OTFTs using surface-modified alumina dielectrics. J Phys Chem B 107:5877–5881
69. Kunugi Y, Ikari M, Okamoto K, Ogino K (2008) Organic field-effect transistors based on vapor deposited 2,9-dialkylpentacene films. J Photopolym Sci Technol 21:197–208

70. Okamoto T, Senatore ML, Ling MM, Mallik AB, Tang ML, Bao ZN (2007) Synthesis, characterization, and field-effect transistor performance of pentacene derivatives. Adv Mater 19:3381–3384
71. Perepichka DF, Bendikov M, Meng H, Wudl F (2003) A one-step synthesis of a poly (iptycene) through an unusual diels-alder cyclization/dechlorination of tetrachloropentacene. J Am Chem Soc 125:10190–10191
72. Li J, Wang M, Ren S, Gao X, Hong W, Li H, Zhu D (2012) High performance organic thin film transistor based on pentacene derivative: 6,13-dichloropentacene. J Mater Chem 22:10496–10500
73. Wang M, Li J, Zhao G, Wu Q, Huang Y, Hu W, Gao X, Li H, Zhu D (2013) High-performance organic field-effect transistors based on single and large-area aligned crystalline microribbons of 6,13-dichloropentacene. Adv Mater 25:2229–2233
74. Chi XL, Li DW, Zhang HQ, Chen YS, Garcia V, Garcia C, Siegrist T (2008) 5,6,11,12-tetrachlorotetracene, a tetracene derivative with π-stacking structure: the synthesis, crystal structure and transistor properties. Org Electron 9:234–240
75. Tripathi AK, Heinrich M, Siegrist T, Pflaum J (2007) Growth and electronic transport in 9,10-diphenylanthracene single crystals-an organic semiconductor of high electron and hole mobility. Adv Mater 19:2097–2101
76. Seo JH, Park DS, Cho SW, Kim CY, Jang WC, Whang CN, Yoo KH, Chang GS, Pedersen T, Moewes A, Chae KH, Cho SJ (2006) Buffer layer effect on the structural and electrical properties of rubrene-based organic thin-film transistors. Appl Phys Lett 89:163505
77. Stingelin-Stutzmann N, Smits E, Wondergem H, Tanase C, Blom P, Smith P, De Leeuw D (2005) Organic thin-film electronics from vitreous solution-processed rubrene hypereutectics. Nat Mater 4:601–606
78. Adams JM, Ramdas S (1979) The crystal structure of solution-grown 9,10-diphenylanthracene. a combined computational and X-ray study. Acta Crystallogr B 35:679–683
79. Miao Q, Chi X, Xiao S, Zeis R, Lefenfeld M, Siegrist T, Steigerwald ML, Nuckolls C (2006) Organization of acenes with a cruciform assembly motif. J Am Chem Soc 128:1340–1345
80. Anthony JE, Brooks JS, Eaton DL, Parkin SR (2001) Functionalized pentacene: improved electronic properties from control of solid-state order. J Am Chem Soc 123:9482–9483
81. Sheraw CD, Jackson TN, Eaton DL, Anthony JE (2003) Functionalized pentacene active layer organic thin-film transistors. Adv Mater 15:2009–2011
82. Park SK, Jackson TN, Anthony JE, Mourey DA (2007) High mobility solution processed 6,13-bis(triisopropyl-silylethynyl) pentacene organic thin film transistors. Appl Phys Lett 91:063514
83. Kim DH, Lee DY, Lee HS, Lee WH, Kim YH, Han JI, Cho K (2007) High-mobility organic transistors based on single-crystalline microribbons of triisopropylsilylethynyl pentacene via solution-phase self-assembly. Adv Mater 19:678–682
84. Llorente GR, Dufourg-Madec MB, Crouch DJ, Pritchard RG, Ogier S, Yeates SG (2009) High performance, acene-based organic thin film transistors. Chem Commun 43:3059–3061
85. Schmidt R, Gottling S, Leusser D, Stalke D, Krause AM, Wurthner F (2006) Highly soluble acenes as semiconductors for thin film transistors. J Mater Chem 16:3708–3714
86. Li Y, Wu Y, Liu P, Prostran Z, Gardner S, Ong BS (2007) Stable solution-processed high-mobility substituted pentacene semiconductors. Chem Mater 19:418–423
87. Zhang X, Jiang X, Luo J, Chi C, Chen H, Wu J (2010) A cruciform 6,6′-dipentacenyl: synthesis, solid-state packing and applications in thin-film transistors. Chem Eur J 16:464–468
88. Ohki K, Inokuchi H, Maruyama Y (1963) Charge mobility in pyrene crystals. Bull Chem Soc Jpn 36:1512–1515
89. Suzuki A, Inokuchi H, Maruyama Y (1976) Charge-carrier drift mobility in pyrene single crystals. Bull Chem Soc Jpn 49:3347–3351

90. Zhang H, Wang Y, Shao K, Liu Y, Chen S, Qiu W, Sun X, Qi T, Ma Y, Yu G, Su Z, Zhu D (2006) Novel butterfly pyrene-based organic semiconductors for field effect transistors. Chem Commun 40:755–757
91. Ashizawa M, Yamada K, Fukaya A, Kato R, Hara K, Takeya J (2008) Effect of molecular packing on field-effect performance of single crystals of thienyl-substituted pyrenes. Chem Mater 20:4883–4890
92. Anant P, Lucas NT, Ball JM, Anthopoulos TD, Jacob J (2010) Synthesis and characterization of pyrene-centered oligothiophenes. Synth Met 160:1987–1993
93. Kwon J, Hong JP, Lee S, Hong JI (2013) 4,4′-Di(pyren-1-yl)-1,10-biphenyl as an efficient material for organic light-emitting diodes and thin-film transistors. New J Chem 37:2881–2887
94. Kwon J, Hong JP, Noh S, Kim TM, Kim JJ, Lee C, Lee S, Hong JI (2012) Pyrene end-capped oligothiophene derivatives for organic thin-film transistors and organic solar cells. New J Chem 36:1813–1818
95. Cho H, Lee S, Cho NS, Jabbour GE, Kwak J, Hwang DH, Lee C (2013) High-mobility pyrene-based semiconductor for organic thin-film transistors. ACS Appl Mater Interfaces 5:3855–3860
96. Wang Y, Wang H, Liu Y, Di CA, Sun Y, Wu W, Yu G, Zhang D, Zhu D (2006) 1-Imino nitroxide pyrene for high performance organic field-effect transistors with low operating voltage. J Am Chem Soc 128:13058–13059
97. Choi TY, Kang HS, Park DH, Koo JM, Lee JK, Ahn SD, Joo J (2003) Trap distribution and field effect transistor (FET) of perylene by organic molecular beam deposition (OMBD). Synth Met 137:929–930
98. Ohta T, Nagano T, Ochi K, Kubozono Y, Fujiwara A (2006) Field-effect transistors with thin films of perylene on SiO_2 and polyimide gate insulators. Appl Phys Lett 88:103506
99. Kotani M, Kakinuma K, Yoshimura M, Ishii K, Yamazaki S, Kobori T, Okuyama H, Kobayashi H, Tada H (2006) Charge carrier transport in high purity perylene single crystal studied by time-of-flight measurements and through field effect transistor characteristics. Chem Phys 325:160–169
100. Sun YM, Tan L, Jiang SD, Qian HL, Wang ZH, Yan DW, Di CA, Wang Y, Wu WP, Yu G, Yan SK, Wang CR, Hu WP, Liu YQ, Zhu DB (2007) High-performance transistor based on individual single-crystalline micrometer wire of perylo[1,12-*b,c,d*]thiophene. J Am Chem Soc 129:1882–1883
101. Tan L, Jiang W, Jiang L, Jiang S, Wang Z, Yan S, Hu W (2009) Single crystalline microribbons of perylo[1,12-*b,c,d*]selenophene for high performance transistors. Appl Phys Lett 94:153306
102. Jiang W, Zhou Y, Geng H, Jiang S, Yan S, Hu W, Wang Z, Shuai Z, Pei J (2011) Solution-processed, high-performance nanoribbon transistors based on dithioperylene. J Am Chem Soc 133:1–3
103. Tian HK, Shi JW, Dong SQ, Yan DH, Wang LX, Geng YH, Wang FS (2006) Novel highly stable semiconductors based on phenanthrene for organic field-effect transistors. Chem Commun 40:3498–3500
104. Cho NS, Cho S, Elbing M, Lee JK, Yang R, Seo JH, Lee K, Bazan GC, Heeger AJ (2008) Organic thin-film transistors based on r,ω-Dihexyldithienyl-dihydrophenanthrene. Chem Mater 20:6289–6291
105. Kawasaki N, Kubozono Y, Okamoto H, Fujiwara A, Yamaji M (2009) Trap states and transport characteristics in picene thin film field-effect transistor. Appl Phys Lett 94:043310
106. Hoang MH, Cho MJ, Kim KH, Cho MY, Joo JS, Choi DH (2009) New semiconducting multi-branched conjugated molecules based on π-Extended triphenylene and its application to organic field-effect transistor. Thin Solid Films 518:501–506
107. Shklyarevskiy IO, Jonkheijm P, Stutzmann N, Wasserberg D, Wondergem HJ, Christianen PCM, Schenning A, de Leeuw DM, Tomovic Z, Wu JS, Müllen K, Maan JC (2005) High anisotropy of the field-effect transistor mobility in magnetically aligned discotic liquid-crystalline semiconductors. J Am Chem Soc 127:16233–16237

108. Xiao SX, Myers M, Miao Q, Sanaur S, Pang KL, Steigerwald ML, Nuckolls C (2005) Molecular wires from contorted aromatic compounds. Angew Chem Int Ed 44:7390–7394
109. Ebata H, Izawa T, Miyazaki E, Takimiya K, Ikeda M, Kuwabara H, Yui T (2007) Highly soluble [1]benzothieno[3,2-*b*]benzothiophene (BTBT) derivatives for high-performance, solution-processed organic field-effect transistors. J Am Chem Soc 129:15732–15733
110. Minemawari H, Yamada T, Matsui H, Tsutsumi J, Haas S, Chiba R, Kumai R, Hasegawa T (2011) Inkjet printing of single-crystal films. Nature 275:364–367
111. Yuan Y, Giri G, Ayzner AL, Zoombelt AP, Mannsfeld SCB, Chen J, Nordlund D, Toney MF, Huang J, Bao Z (2014) Ultra-high mobility transparent organic thin film transistors grown by an off-centre spin-coating method. Nat Comm 5:3005
112. Izawa T, Miyazaki E, Takimiya K (2008) Molecular ordering of high-performance soluble molecular semiconductors and re-evaluation of their field-effect transistor characteristics. Adv Mater 20:3388–3392
113. Amin AY, Khassanov A, Reuter K, Meyer-Friedrichsen T, Halik M (2012) Low-voltage organic field effect transistors with a 2–tridecyl[1]benzothieno[3,2-*b*][1]benzothiophene semiconductor layer. J Am Chem Soc 134:16548–16550
114. Tang ML, Okamoto T, Bao ZN (2006) High-performance organic semiconductors: asymmetric linear acenes containing sulphur. J Am Chem Soc 128:16002–16003
115. Shinamura S, Osaka I, Miyazaki E, Nakao A, Yamagishi M, Takeya J, Takimiya K (2011) Linear- and angular-shaped naphthodithiophenes: selective synthesis, properties, and application to organic field-effect transistors. J Am Chem Soc 133:5024–5035
116. Shinamura S, Miyazaki E, Takimiya K (2010) Synthesis, properties, crystal structures, and semiconductor characteristics of naphtho[1,2-*b*:5,6-*b*′]dithiophene and –diselenophene derivatives. J Org Chem 75:1228–1234
117. Liu Y, Wang Y, Wu WP, Liu YQ, Xi HX, Wang LM, Qiu WF, Lu K, Du CY, Yu G (2009) Synthesis, characterization, and field-effect transistor performance of thieno[3,2-b]thieno[2′,3′:4,5]thieno[2,3-*d*]thiophene derivatives. Adv Funct Mater 19:772–778
118. Du CY, Guo YL, Liu YQ, Qiu WF, Zhang HJ, Gao XK, Liu Y, Qi T, Lu K, Yu G (2008) Anthra[2,3-*b*]benzo[*d*]thiophene: an air-stable asymmetric organic semiconductor with high mobility at room temperature. Chem Mater 20:4188–4190
119. Guo YL, Du CY, Yu G, Di CA, Jiang SD, Xi HX, Zheng J, Yan SK, Yu CL, Hu WP, Liu YQ (2010) High-performance phototransistors based on organic microribbons prepared by a solution self-assembly process. Adv Funct Mater 20:1019–1024
120. Okamoto T, Mitsui C, Yamagishi M, Nakahara K, Soeda J, Hirose Y, Miwa K, Sato H, Yamano A, Matsushita T, Uemura T, Takeya J (2013) V-shaped organic semiconductors with solution processability, high mobility, and high thermal durability. Adv Mater 25:6392–6397
121. Ebata H, Miyazaki E, Yamamoto T, Takimiya K (2007) Synthesis, properties, and structures of benzo[1,2-*b*:4,5-*b*′]bis[*b*]benzothiophene and Benzo[1,2-*b*:4,5-*b*′]bis[*b*]benzoselenophene. Org Lett 9:4499–4502
122. Gao P, Beckmann D, Tsao HN, Feng XL, Enkelmann V, Pisula W, Müllen K (2008) Benzo[1,2-*b*:4,5-*b*′]bis[*b*]benzothiophene as solution processible organic semiconductor for field-effect transistors. Chem Commun 42:1548–1550
123. Li R, Jiang L, Meng Q, Gao J, Li H, Tang Q, He M, Hu W, Liu Y, Zhu D (2009) Micrometer-sized organic single crystals, anisotropic transport, and field-effect transistors of a fused-ring thienoacene. Adv Mater 21:4492–4495
124. Li R, Dong H, Zhan X, He Y, Li H, Hu W (2010) Single crystal ribbons and transistors of a solution processed sickle-like fused-ring thienoacene. J Mater Chem 20:6014–6018
125. Laquindanum JG, Katz HE, Lovinger AJ (1998) Synthesis, morphology, and field-effect mobility of anthradithiophenes. J Am Chem Soc 120:664–672
126. Gao P, Beckmann D, Tsao HN, Feng X, Enkelmann V, Baumgarten M, Pisula W, Müllen K (2009) Dithieno[2,3-*d*;2′,3′-*d*′]benzo[1,2-*b*;4,5-*b*′]dithiophene (DTBDT) as semiconductor for high-performance, solution-processed organic field-effect transistors. Adv Mater 21:213–216

127. Xiao K, Liu YQ, Qi T, Zhang W, Wang F, Gao JH, Qiu WF, Ma YQ, Cui GL, Chen SY, Zhan XW, Yu G, Qin JG, Hu WP, Zhu DB (2005) A highly π-stacked organic semiconductor for field-effect transistors based on linearly condensed pentathienoacene. J Am Chem Soc 127:13281–13286
128. Huang JY, Luo H, Wang LP, Guo YL, Zhang WF, Chen HJ, Zhu ML, Liu YQ, Yu G (2012) Dibenzoannelated tetrathienoacene: synthesis, characterization, and applications in organic field-effect transistors. Org Let 14:3300–3303
129. Yamamoto T, Takimiya K (2007) Facile synthesis of highly π-extended heteroarenes, dinaphtho[2,3-*b*:2′,3′-*f*]chalcogenopheno[3,2-*b*]chalco-genophenes, and their application to field-effect transistors. J Am Chem Soc 129:2224–2225
130. Haas S, Takahashi Y, Takimiya K, Hasegawa T (2009) High-performance dinaphtho-thieno-thiophene single crystal field-effect transistors. Appl Phys Lett 95:022111
131. Kang MJ, Doi I, Mori H, Miyazaki E, Takimiya K, Ikeda M, Kuwabara H (2011) Alkylated dinaphtho[2,3-*b*:2′,3′-*f*]thieno[3,2-b]thiophenes (Cn-DNTTs): organic semiconductors for high performance thin-film transistors. Adv Mater 23:1222–1225
132. Kang MJ, Miyazaki E, Osaka I, Takimiya K, Nakao A (2013) Diphenyl derivatives of dinaphtho[2,3-*b*:2′,3′-*f*]thieno[3,2-*b*]thiophene: organic semiconductors for thermally stable thin-film transistors. ACS Appl Mater Interfaces 5:2331–2336
133. Sirringhaus H, Friend RH, Wang C, Leuninger J, Müllen K (1999) Dibenzothienobisbenzothiophene-a novel fused-ring oligomer with high field-effect mobility. J Mater Chem 9:2095–2101
134. Yang YS, Yasuda T, Adachi C (2012) Organic single-crystal transistors based on π-extended heteroheptacene microribbons. Bull Chem Soc Jpn 85:1186–1191
135. Kunugi Y, Takimiya K, Yamashita K, Aso Y, Otsubo T (2002) Organic field-effect transistors using di(2-thienyl)naphthodithiophenes as active layers. Chem Lett 10:958–959
136. Takimiya K, Kunugi Y, Toyoshima Y, Otsubo T (2005) 2,6-diarylnaphtho[1,8-*bc*:5,4-*b′c′*] dithiophenes as new high-performance semiconductors for organic field-effect transistors. J Am Chem Soc 127:3605–3612
137. Shukla D, Welter TR, Robello DR, Giesen DJ, Lenhard JR, Ahearn WG, Meyer DM, Rajeswaran M (2009) Dioxapyrene-based organic semiconductors for organic field effect transistors. J Phys Chem C 113:14482–14486
138. Kobayashi N, Sasaki M, Nomoto K (2009) Stable peri-xanthenoxanthene thin-film transistors with efficient carrier injection. Chem Mater 21:552–556
139. Briseno AL, Miao Q, Ling MM, Reese C, Meng H, Bao ZN, Wudl F (2006) Hexathiapentacene: structure, molecular packing, and thin-film transistors. J Am Chem Soc 128:15576–15577
140. Liu WJ, Zhou Y, Ma Y, Cao Y, Wang J, Pei J (2007) Thin film organic transistors from air-stable heteroarenes: anthra-[1,2-*b*:4,3-*b′*:5,6-*b″′*:8,7-*b″′*]tetrathiophene derivatives. Org Lett 9:4187–4190
141. Brusso JL, Hirst OD, Dadvand A, Ganesan S, Cicoira F, Robertson CM, Oakley RT, Rosei F, Perepichka DF (2008) Two-dimensional structural motif in thienoacene semiconductors: synthesis, structure, and properties of tetrathienoanthracene isomers. Chem Mater 20:2484–2494
142. Wang JY, Zhou Y, Yan J, Ding L, Ma Y, Cao Y, Wang J, Pei J (2009) New fused heteroarenes for high-performance field-effect transistors. Chem Mater 21:2595–2597
143. Zhou Y, Lei T, Wang L, Pei J, Cao Y, Wang J (2010) High-performance organic field-effect transistors from organic single-crystal microribbons formed by a solution process. Adv Mater 22:1484–1487
144. Zhang S, Guo Y, Zhang Y, Liu R, Li Q, Zhan X, Liu Y, Hu W (2010) Synthesis, self-assembly, and solution-processed nanoribbon field-effect transistor of a fused-nine-ring thienoacene. Chem Commun 46:2841–2843
145. Zhang W, Sun X, Xia P, Yu G, Wong M, Liu Y, Zhu D (2011) novel butterfly-shaped fused heteroacenes: synthesis, properties, and device performance of solution-processed field-effect transistors. Org Let 14:4382–4385

146. Chernichenko KY, Sumerin VV, Shpanchenko RV, Balenkova ES, Nenajdenko VG (2006) "Sunflower": a new form of carbon sulfide. Angew Chem Int Ed 45:7367–7370
147. Dadvand A, Cicoira F, Chernichenko KY, Balenkova ES, Osuna RM, Rosei F, Nenajdenko VG, Perepichka DF (2008) Heterocirculenes as a new class of organic semiconductors. Chem Commun 42:5354–5356
148. Sun YM, Ma YW, Liu YQ, Lin YY, Wang ZY, Wang Y, Di CG, Xiao K, Chen XM, Qiu WF, Zhang B, Yu G, Hu WP, Zhu DB (2006) High performance and stable organic thin-film transistors based on fused thiophenes. Adv Funct Mater 16:426–432
149. Zhu M, Luo H, Wang L, Guo Y, Zhang W, Liu Y, Yu G (2013) The synthesis of 2,6-dialkylphenyldithieno[3,2-*b*:2′,3′-*d*]thiophene derivatives and their applications in organic field-effect transistors. Dyes Pigm 98:17–24
150. Li XC, Sirringhaus H, Garnier F, Holmes AB, Moratti SC, Feeder N, Clegg W, Teat SJ, Friend RH (1998) a highly π-stacked organic semiconductor for thin film transistors based on fused thiophenes. J Am Chem Soc 120:2206–2207
151. Hunziker C, Zhan X, Losio PA, Figi H, Kwon OP, Barlow S, Guenter P, Marder SR (2007) Highly ordered thin films of a bis(dithienothiophene) derivative. J Mater Chem 17:4972–4979
152. Zhang L, Tan L, Hu WP, Wang ZH (2009) Synthesis, packing arrangement and transistor performance of dimers of dithienothiophenes. J Mater Chem 19:8216–8222
153. Chen H, Yu QC, Yu G, Guo Y, Huang J, Zhu M, Guo X, Liu Y (2011) Synthesis and characterization of novel semiconductors based on thieno[3,2-*b*][1]benzothiophene cores and their applications in the organic thin-film transistors. J Phys Chem C 115:23984–23991
154. Zhang L, Tan L, Wang Z, Hu W, Zhu D (2009) High-performance, stable organic field-effect transistors based on trans-1,2-(Dithieno[2,3-*b*:3′,2′-*d*]thiophene)ethene. Chem Mater 21:1993–1999
155. Mamada M, Nishida JI, Kumaki D, Tokito S, Yamashita Y (2008) High performance organic field-effect transistors based on [2,2′]bi[naphtho[2,3-*b*]thiophenyl] with a simple structure. J Mater Chem 18:3442–3447
156. Zhang Y, Ichikawa M, Hattori J, Kato T, Sazaki A, Kanazawa S, Kato S, Zhang C, Taniguchi Y (2009) Fused thiophene-split oligothiophenes with high ionization potentials for OTFTs. Synth Met 159:1890–1895
157. Didane Y, Mehl GH, Kumagai A, Yoshimoto N, Videlot-Ackermann C, Brisset H (2008) A "kite" shaped styryl end-capped benzo[2,1-*b*:3,4-*b*′]dithiophene with high electrical performances in organic thin film transistors. J Am Chem Soc 130:17681–17683
158. Wang C, Wei Z, Meng Q, Zhao H, Xu W, Li H, Hu W (2010) Dibenzo[*b*,*d*]thiophene based oligomers with carbon–carbon unsaturated bonds for high performance field-effect transistors. Org Electron 11:544–551
159. Garnier F, Horowitz G, Peng XZ, Fichou D (1991) Structural basis for high carrier mobility in conjugated oligomers. Synth Met 45:163–171
160. Horowitz G, Hajlaoui ME (2000) Mobility in polycrystalline oligothiophene field-effect transistors dependent on grain size. Adv Mater 12:1046–1050
161. Deman AL, Tardy J, Nicolas Y, Blanchard P, Roncali J (2004) Structural effects on the characteristics of organic field effect transistors based on new oligothiophene derivatives. Synth Met 146:365–371
162. Garnier F, Hajlaoui R, El Kassmi A, Horowitz G, Laigre L, Porzio W, Armanini M, Provasoli F (1998) Dihexylquaterthiophene, a two-dimensional liquid crystal-like organic semiconductor with high transport properties. Chem Mater 10:3334–3339
163. Locklin J, Li DW, Mannsfeld SCB, Borkent EJ, Meng H, Advincula R, Bao ZN (2005) Organic thin film transistors based on cyclohexyl-substituted organic semiconductors. Chem Mater 17:3366–3374
164. Yoon MH, Facchetti A, Stern CE, Marks TJ (2006) Fluorocarbon-modified organic semiconductors: molecular architecture, electronic, and crystal structure tuning of arene- versus fluoroarene-thiophene oligomer thin-film properties. J Am Chem Soc 128:5792–5801

165. Yamao T, Juri K, Kamoi A, Hotta S (2009) Field-effect transistors based on organic single crystals grown by an improved vapor phase method. Org Electron 10:1241–1247
166. Zhao TY, Wei ZM, Song YB, Xu W, Hu WP, Zhu DB (2007) Tetrathia[22]annulene [1,1,2,2]: physical properties, crystal structure and application in organic field-effect transistors. J Mater Chem 17:4377–4381
167. Payne MM, Parkin SR, Anthony JE, Kuo CC, Jackson TN (2005) Organic field-effect transistors from solution-deposited functionalized acenes with mobilities as high as 1 cm^2 V^{-1} s^{-1}. J Am Chem Soc 127:4986–4987
168. Anthony JE, Subramanian S, Parkin SR, Park SK, Jackson TN (2009) Thin-film morphology and transistor performance of alkyl-substituted triethylsilylethynyl anthradithiophenes. J Mater Chem 19:7984–7989
169. Subramanian S, Park SK, Parkin SR, Podzorov V, Jackson TN, Anthony JE (2008) Chromophore fluorination enhances crystallization and stability of soluble anthradithiophene semiconductors. J Am Chem Soc 130:2706–2707
170. Jurchescu OD, Subramanian S, Kline RJ, Hudson SD, Anthony JE, Jackson TN, Gundlach DJ (2008) Organic single crystal field-effect transistors of a soluble anthradithiophene. Chem Mater 20:6733–6737
171. Lehnherr D, Waterloo AR, Goetz KP, Payne MM, Hampel F, Anthony JE, Jurchescu OD, Tykwinski RR (2012) Isomerically pure syn-anthradithiophenes: synthesis, properties, and fet performance. Org Lett 14:3660–3663
172. Tang ML, Reichardt AD, Siegrist T, Mannsfeld SCB, Bao ZN (2008) Trialkylsilylethynyl-functionalized tetraceno[2,3-*b*]thiophene and anthra[2,3-b]thiophene organic transistors. Chem Mater 20:4669–4676
173. Goetz KP, Li Z, Ward JW, Bougher C, Rivnay J, Smith J, Conrad BR, Parkin SR, Anthopoulos TD, Salleo A, Anthony JE, Jurchescu OD (2011) Effect of acene length on electronic properties in 5-, 6-, and 7-ringed heteroacenes. Adv Mater 23:3698–3703
174. Zhang W, Zhang J, Chen X, Mao Z, Xie X, Wang L, Liao Y, Yu G, Liu Y, Zhu D (2013) Bitrialkylsilylethynyl thienoacenes: synthesis, molecular conformation and crystal packing, and their field-effect properties. J Mater Chem C 1:6403–6410
175. Bendikov M, Wudl F, Perepichka DF (2004) Tetrathiafulvalenes, oligoacenenes, and their buckminsterfullerene derivatives: the brick and mortar of organic electronics. Chem Rev 104:4891–4946
176. Jérome D (2004) Organic conductors: from charge density wave TTF–TCNQ to superconducting $(TMTSF)_2PF_6$. Chem Rev 104:5565–5591
177. Kobayashi H, Kobayashi A, Sasaki Y, Saito G, Inokuchi H (1986) The crystal and molecular structures of bis(ethylenedithio)-tetrathiafulvalene. Bull Chem Soc Jpn 59:301–302
178. Bourgoin JP, Vandevyver M, Barraud A, Tremblay G, Hesto P (1993) Field-effect transistor based on conducting langmuir-blodgett films of EDTTTF derivatives. Mol Eng 2:309–314
179. Takahashi Y, Hasegawa T, Horiuchi S, Kumai R, Tokura Y, Saito G (2007) High mobility organic field-effect transistor based on hexamethylenetetrathiafulvalene with organic metal electrodes. Chem Mater 19:6382–6384
180. Mas-Torrent M, Durkut M, Hadley P, Ribas X, Rovira C (2004) High mobility of dithiophene-tetrathiafulvalene single-crystal organic field effect transistors. J Am Chem Soc 126:984–985
181. Mas-Torrent M, Hadley P, Bromley ST, Ribas X, Tarres J, Mas M, Molins E, Veciana J, Rovira C (2004) Correlation between crystal structure and mobility in organic field-effect transistors based on single crystals of tetrathiafulvalene derivatives. J Am Chem Soc 126:8546–8553
182. Miskiewicz P, Mas-Torrent M, Jung J, Kotarba S, Glowacki I, Gomar-Nadal E, Amabilino DB, Veciana J, Krause B, Carbone D, Rovira C, Ulanski J (2006) Efficient high area OFETs by solution based processing of a π-electron rich donor. Chem Mater 18:4724–4729
183. Leufgen M, Rost O, Gould C, Schmidt G, Geurts J, Molenkamp LW, Oxtoby NS, Mas-Torrent M, Crivillers N, Veciana J, Rovira C (2008) High-mobility tetrathiafulvalene organic field-effect transistors from solution processing. Org Electron 9:1101–1106

184. Mas-Torrent M, Hadley P, Bromley ST, Crivillers N, Veciana J, Rovira C (2005) Single-crystal organic field-effect transistors based on dibenzo-tetrathiafulvalene. Appl Phys Lett 86:012110
185. Nishida JI, Ando S, Yamaguchi J, Itaka K, Koinuma H, Tada H, Tokito S, Yamashita Y (2005) High-performance organic field-effect transistors based on π-extended tetrathiafulvalene derivatives. J Am Chem Soc 127:10142–10143
186. Nishida JI, Kumaki D, Tokito S, Yamashita Y (2006) High performance *n*- and p-type field-effect transistors based on tetrathiafulvalene derivatives. J Am Chem Soc 128:9598–9599
187. Doi I, Miyazaki E, Takimiya K, Kunugi Y (2007) Development of N-Alkyl-substituted bis (pyrrolo[3,4-*d*])tetrathiafulvalenes as organic semiconductors for solution-processible field-effect transistors, development of n-alkyl-substituted bis(pyrrolo[3,4-*d*])tetrathiafulvalenes as organic semiconductors for solution-processible field-effect transistors. Chem Mater 19:5230–5233
188. Gao X, Wang Y, Yang X, Liu Y, Qiu W, Wu W, Zhang H, Qi T, Liu Y, Lu K, Du C, Shuai Z, Yu G, Zhu D (2007) Dibenzotetrathiafulvalene bisimides: new building blocks for organic electronic materials. Adv Mater 19:3037–3042
189. Xue JG, Forrest SR (2001) Organic thin-film transistors based on bis(1,2,5-thiadiazolo)-p-quinobis(1,3-dithiole). Appl Phys Lett 79:3714
190. Takada M, Graaf H, Yamashita Y, Tada H (2002) BTQBT (bis-(1,2,5-thiadiazolo)-*p*-quinobis(1,3-dithiole)) thin films; a promising candidate for high mobility organic transistors. Jpn J Appl Phys Part 2(41):L4
191. Imaeda K, Yamashita Y, Li YF, Mori T, Inokuchi H, Sano M (1992) Hall-effect observation in the new organic semiconductor bis(1,2,5-thiadiazolo)-*p*-quinobis(1,3-dithiole)(BTQBT). J Mater Chem 2:115–118
192. Bando Y, Shirahata T, Shibata K, Wada H, Mori T, Imakubo T (2008) Organic field-effect transistors based on alkyl-terminated tetrathiapentalene (TTP) derivatives. Chem Mater 20:5119–5121
193. Zhang W, Yu G, Liu YQ (2014) Heteroatom substituted organic/polymeric semiconductors and their applications in field-effect transistors. Adv Mater Adma 201305297R1
194. Miao Q, Nguyen TQ, Someya T, Blanchet GB, Nuckolls C (2003) Synthesis, assembly, and thin film transistors of dihydrodiazapentacene: an isostructural motif for pentacene. J Am Chem Soc 125:10284–10287
195. Tang Q, Zhang D, Wang S, Ke N, Xu J, Yu JC, Miao Q (2009) A meaningful analogue of pentacene: charge transport, polymorphs, and electronic structures of dihydrodiazapentacene. Chem Mater 21:1400–1405
196. Weng SZ, Shukla P, Kuo MY, Chang YC, Sheu HS, Chao I, Tao YT (2009) Diazapentacene derivatives as thin-film transistor materials: morphology control in realizing high-field-effect mobility. ACS Appl Mater Interfaces 1:2071–2079
197. Ahmed E, Briseno AL, Xia Y, Jenekhe SA (2008) High mobility single-crystal field-effect transistors from bisindoloquinoline semiconductors. J Am Chem Soc 130:1118–1119
198. Ma YQ, Sun YM, Liu YQ, Gao JH, Chen SY, Sun XB, Qiu WF, Yu G, Cui GL, Hu WP, Zhu DB (2005) Organic thin film transistors based on stable amorphous ladder tetraazapentacenes semiconductors. J Mater Chem 15:4894–4898
199. Hong W, Wei Z, Xi HX, Xu W, Hu WP, Wang QR, Zhu DB (2008) 6H-Pyrrolo[3,2-*b*:4,5-*b′*] bis[1, 4]benzothiazines: facilely synthesized semiconductors for organic field-effect transistors. J Mater Chem 18:4814–4820
200. Wei Z, Hong W, Geng H, Wang C, Liu Y, Li R, Xu W, Shuai Z, Hu WP, Wang Q, Zhu DB (2010) Organic single crystal field-effect transistors based on 6H-pyrrolo[3,2–*b*:4,5–*b′*]bis[1, 4]benzothiazine and its derivatives. Adv Mater 22:2458–2462
201. Wu YL, Li YN, Gardner S, Ong BS (2005) Indolo[3,2-*b*]carbazole-based thin-film transistors with high mobility and stability. J Am Chem Soc 127:614–618
202. Li YN, Wu YL, Gardner S, Ong BS (2005) Novel peripherally substituted indolo[3,2-*b*] carbazoles for high-mobility organic thin-film transistors. Adv Mater 17:849–853

203. Guo YL, Zhao HP, Yu G, Di CA, Liu W, Jiang SD, Yan SK, Wang CR, Zhang HL, Sun XN, Tao XT, Liu YQ (2008) Single-crystal microribbons of an indolo[3,2-*b*]carbazole derivative by solution-phase self-assembly with novel mechanical, electrical, and optical properties. Adv Mater 20:4835–4839
204. Song YB, Di CA, Wei ZM, Zhao TY, Xu W, Liu YQ, Zhang DQ, Zhu DB (2008) Synthesis, characterization, and field-effect transistor properties of carbazolenevinylene oligomers: from linear to cyclic architectures. Chem-Eur J 14:4731–4740
205. Shirota Y, Kageyama H (2007) Charge carrier transporting molecular materials and their applications in devices. Chem Rev 107:953–1010
206. Song YB, Di CA, Yang X, Li S, Xu W, Liu YQ, Yang L, Shuai ZG, Zhang DQ, Zhu DB (2006) A cyclic triphenylamine dimer for organic field-effect transistors with high performance. J Am Chem Soc 128:15940–15941
207. Bao ZN, Lovinger AJ, Dodabalapur A (1997) Highly ordered vacuum-deposited thin films of metallophthalocyanines and their applications in field-effect transistors. Adv Mater 9:42–44
208. Bao ZN, Lovinger AJ, Dodabalapur A (1996) Organic field-effect transistors with high mobility based on copper phthalocyanine. Appl Phys Lett 69:3066
209. Tang QX, Li HX, Song Y, Xu W, Hu W, Jiang L, Liu Y, Wang X, Zhu D (2006) In situ patterning of organic single-crystalline nanoribbons on a SiO_2 surface for the fabrication of various architectures and high-quality transistors. Adv Mater 18:3010–3014
210. Li LQ, Tang QX, Li HX, Yang XD, Hu WP, Song YB, Shuai ZG, Xu W, Liu YQ, Zhu DB (2007) An ultra closely π-stacked organic semiconductor for high performance field-effect transistors. Adv Mater 19:2613–2617
211. Wang H, Song D, Yang J, Yu B, Geng Y, Yan D (2007) High mobility vanadyl-phthalocyanine polycrystalline films for organic field-effect transistors. Appl Phys Lett 90:253510
212. Checcoli P, Conte G, Salvatori S, Paolesse R, Bolognesi A, Berliocchi A, Brunetti F, D'Amico A, Di Carlo A, Lugli P (2003) Tetra-phenyl porphyrin based thin film transistors. Synth Met 138:261–266
213. Shea PB, Kanicki J, Ono N (2005) Field-effect mobility of polycrystalline tetrabenzoporphyrin thin-film transistors. J Appl Phys 98:014503
214. Che CM, Xiang HF, Chui SSY, Xu ZX, Roy VAL, Yan JJ, Fu WF, Lai PT, Williams ID (2008) A high-performance organic field-effect transistor based on platinum(ii) porphyrin: peripheral substituents on porphyrin ligand significantly affect film structure and charge mobility. Chem Asian J 3:1092–1103
215. Shea PB, Kanicki J, Pattison LR, Petroff P, Kawano M, Yamada H, Ono N (2006) Solution-processed nickel tetrabenzoporphyrin thin-film transistors. J Appl Phys 100:034502
216. Shea PB, Pattison LR, Kawano M, Chen C, Chen J, Petroff P, Martin DC, Yamada H, Ono N, Kanicki J (2007) Solution-processed polycrystalline copper tetrabenzoporphyrin thin-film transistors. Synth Met 157:190–197
217. Xu H, Wang Y, Yu G, Xu W, Song Y, Zhang DQ, Liu YQ, Zhu DB (2005) Organic field-effect transistors based on langmuir-blodgett films of an extended porphyrin analogue–cyclo [6]pyrrole. Chem Phys Lett 414:369–373
218. Xu H, Yu G, Xu W, Xu Y, Cui GL, Zhang DQ, Liu YQ, Zhu DB (2005) High-performance field-effect transistors based on langmuir–blodgett films of cyclo[8]pyrrole. Langmuir 21:5391–5395
219. Chen Y, Su W, Bai M, Jiang J, Li X, Liu Y, Wang L, Wang S (2005) High performance organic field-effect transistors based on amphiphilic tris(phthalocyaninato) rare earth triple-decker complexes. J Am Chem Soc 127:15700–15701
220. Sirringhaus H, Brown PJ, Friend RH, Nielsen MM, Bechgaard K, Langeveld-Voss BMW, Spiering AJH, Janssen RAJ, Meijer EW, Herwig P, de Leeuw DM (1999) Two-dimensional charge transport in self-organized, high-mobility conjugated polymers. Nature 401:685–688
221. Hao XT, Hosokai T, Mitsuo N, Kera S, Okudaira KK, Mase K, Ueno N (2007) Control of the interchain π–π interaction and electron density distribution at the surface of conjugated poly (3-hexylthiophene) thin films. J Phys Chem B 111:10365–10372

222. Bao ZN, Lovinger AJ (1999) Soluble regioregular polythiophene derivatives as semiconducting materials for field-effect transistors. Chem Mater 11:2607–2612
223. Wang GM, Swensen J, Moses D, Heeger AJ (2003) Increased mobility from regioregular poly(3-hexylthiophene) field-effect transistors. J Appl Phys 93:6137–6141
224. Chang JF, Sun BQ, Breiby DW, Nielsen MM, Solling TI, Giles M, McCulloch I, Sirringhaus H (2004) Enhanced mobility of poly(3-hexylthiophene) transistors by spin-coating from high-boiling-point solvents. Chem Mater 16:4772–4776
225. Ong BS, Wu Y, Liu P, Gardner S (2004) High-performance semiconducting polythiophenes for organic thin-film transistors. J Am Chem Soc 126:3378–3379
226. Chabinyc ML, Endicott F, Vogt BD, DeLongchamp DM, Lin EK, Wu Y, Liu P, Ong BS (2006) Effects of humidity on unencapsulated poly(thiophene) thin-film transistors. Appl Phys Lett 88:113514
227. McCulloch I, Bailey C, Giles M, Heeney M, Love I, Shkunov M, Sparrowe D, Tierney S (2005) Influence of molecular design on the field-effect transistor characteristics of terthiophene polymers. Chem Mater 17:1381–1385
228. Kong H, Jung YK, Cho NS, Kang IN, Park JH, Cho S, Shim HK (2009) New semiconducting polymers containing 3,6-dimethyl(thieno[3,2-b]-thiophene or selenopheno [3,2-*b*]selenophene) for organic thin-film transistors. Chem Mater 21:2650–2660
229. Lim B, Baeg KJ, Jeong HG, Jo J, Kim H, Park JW, Noh YY, Vak D, Park JH, Park JW, Kim DY (2009) A new poly(thienylenevinylene) derivative with high mobility and oxidative stability for organic thin-film transistors and solar cells. Adv Mater 21:2808–2814
230. Kim J, Lim B, Baeg KJ, Noh YY, Khim D, Jeong HG, Yun JM, Kim DY (2011) Highly soluble poly(thienylenevinylene) derivatives with charge-carrier mobility exceeding 1 cm^2 V^{-1} s^{-1}. Chem Mater 23:4663–4665
231. Heeney M, Bailey C, Genevicius K, Shkunov M, Sparrowe D, Tierney S, McCulloch I (2007) Stable polythiophene semiconductors incorporating thieno[2,3-*b*]-thiophene. J Am Chem Soc 127:1078–1079
232. McCulloch I, Heeney M, Bailey C, Genevicius K, MacDonald I, Shkunov M, Sparrowe D, Tierney S, Wagner R, Zhang W, Chabinyc ML, Kline RJ, McGehee MD, Toney MF (2006) Liquid-crystalline semiconducting polymers with high charge-carrier mobility. Nat Mater 5:328–333
233. DeLongchamp DM, Kline RJ, Lin EK, Fischer DA, Richter LJ, Lucas LA, Heeney M, McCulloch I, Northrup JE (2007) High carrier mobility polythiophene thin films: structure determination by experiment and theory. Adv Mater 19:833–837
234. Hamadani BH, Gundlach DJ, McCulloch I, Heeney M (2007) Undoped polythiophene field-effect transistors with mobility of 1 cm^2 V^{-1} s^{-1}. Appl Phys Lett 91:243512
235. Chabinyc ML, Toney MF, Kline RJ, McCulloch I, Heeney M (2007) X-ray scattering study of thin films of poly(2,5-bis(3-alkylthiophen-2-yl)thieno[3,2-b]thiophene). J Am Chem Soc 129:3226–3237
236. Kline RJ, DeLongchamp DM, Fischer DA, Lin EK, Heeney M, McCulloch I, Toney MF (2007) Significant dependence of morphology and charge carrier mobility on substrate surface chemistry in high performance polythiophene semiconductor films. Appl Phys Lett 90:062117
237. Li J, Qin F, Li CM, Bao QL, Chan-Park MB, Zhang W, Qin JG, Ong BS (2008) High-performance thin-film transistors from solution-processed dithienothiophene polymer semiconductor nanoparticles. Chem Mater 20:2057–2059
238. Li J, Bao QL, Zhang W, Gong C, Chan-Park MB, Qin JG, Ong BS (2010) Organic thin-film transistors processed from relatively nontoxic, environmentally friendlier solvents. Chem Mater 22:5747–5753
239. Li YN, Wu YL, Liu P, Birau M, Pan HL, Ong BS (2006) Poly(2,5-bis(2-thienyl)-3,6-dialkylthieno[3,2-*b*]thiophene)s–high-mobility semiconductors for thin-film transistors. Adv Mater 18:3029–3032
240. McCulloch I, Heeney M, Chabinyc ML, DeLongchamp D, Kline RJ, Coelle M, Duffy W, Fischer D, Gundlach D, Hamadani B, Hamilton R, Richter L, Salleo A, Shkunov M,

Sporrowe D, Tierney S, Zhong W (2009) Semiconducting thienothiophene copolymers: design, synthesis, morphology, and performance in thin-film organic transistors. Adv Mater 21:1091–1109
241. Fong HH, Pozdin VA, Amassian A, Malliaras GG, Smilgies DM, He MQ, Gasper S, Zhang FX, Sorensen M (2008) Tetrathienoacene copolymers as high mobility, soluble organic semiconductors. J Am Chem Soc 130:13202–13203
242. He MQ, Li JF, Sorensen ML, Zhang FX, Hancock RR, Fong HH, Pozdin VA, Malliaras GG (2009) Alkylsubstituted thienothiophene semiconducting materials: structure-property relationships. J Am Chem Soc 131:11930–11938
243. He MQ, Li JF, Tandia A, Sorensen ML, Zhang FX, Fong HH, Pozdin VA, Malliaras GG (2010) Importance of *C2* symmetry for the device performance of a newly synthesized family of fused-ring thiophenes. Chem Mater 22:2770–2779
244. Liu J, Zhang R, Sauve G, Kowalewski T, McCullough RD (2008) Highly disordered polymer field effect transistors: *n*-alkyl dithieno[3,2-*b*:2′,3′-*d*]pyrrole-based copolymers with surprisingly high charge carrier mobilities. J Am Chem Soc 130:13167–13176
245. Liu JY, Zhang R, Osaka I, Mishra S, Javier AE, Smilgies DM, Kowalewski T, McCullough RD (2009) Transistor paint: environmentally stable *n*-alkyldithienopyrrole and bithiazole-based copolymer thin-film transistors show reproducible high mobilities without annealing. Adv Funct Mater 19:3427–3434
246. Osaka I, Sauve G, Zhang R, Kowalewski T, McCullough RD (2007) Novel thiophene-thiazolothiazole copolymers for organic field-effect transistors. Adv Mater 19:4160–4165
247. Osaka I, Zhang R, Sauve G, Smilgies DM, Kowalewski T, McCullough RD (2009) High-lamellar ordering and amorphous-like π-network in short-chain thiazolothiazole-thiophene copolymers lead to high mobilities. J Am Chem Soc 131:2521–2529
248. Kim DH, Lee BL, Moon H, Kang HM, Jeong EJ, Park J, Han KM, Lee S, Yoo BW, Koo BW, Kim JY, Lee WH, Cho K, Becerril HA, Bao ZN (2009) Liquid-crystalline semiconducting copolymers with intramolecular donor-acceptor building blocks for high-stability polymer transistors. J Am Chem Soc 131:6124–6132
249. Li Y, Wu Y, Ong BS (2006) Polyindolo[3,2-*b*]carbazoles: a new class of *p*-channel semiconductor polymers for organic thin-film transistors. Macromolecules 39:6521–6527
250. Osaka I, Takimiya K, McCullough RD (2010) Benzobisthiazole-based semiconducting copolymers showing excellent environmental stability in high-humidity air. Adv Mater 22:4993–4997
251. Zhang W, Smith J, Hamilton R, Heeney M, Kirkpatrick J, Song K, Watkins SE, Anthopoulos T, McCulloch I (2009) Systematic improvement in charge carrier mobility of air stable triarylamine copolymers. J Am Chem Soc 131:10814–14815
252. Beaujuge PM, Fréchet JMJ (2011) Molecular design and ordering effects in π-functional materials for transistor and solar cell applications. J Am Chem Soc 133:20009–20029
253. Facchetti A (2011) π-conjugated polymers for organic electronics and photovoltaic cell applications. Chem Mater 23:733–758
254. Nielsen CB, Turbiez M, McCulloch I (2013) Recent advances in the development of semiconducting DPP-containing polymers for transistor applications. Adv Mater 25:1859–1880
255. Zhang M, Tsao HT, Pisula W, Yang CD, Mishra AK, Müllen K (2007) Field-effect transistors based on a benzothiadiazole-cyclopentadithiophene copolymer. J Am Chem Soc 129:3472–3473
256. Tsao HT, Cho D, Andreasen JW, Rouhanipour A, Breiby DW, Pisula W, Müllen K (2009) The influence of morphology on high-performance polymer field-effect transistors. Adv Mater 21:209–212
257. Zhang WM, Smith J, Watkins SE, Gysel R, McGehee M, Salleo A, Kirkpatrick J, Ashraf S, Anthopoulos T, Heeney M, McCulloch I (2010) Indacenodithiophene semiconducting polymers for high-performance, air-stable transistors. J Am Chem Soc 132:11437–11439

258. Ying L, Hsu BB, Zhan HM, Welch GC, Zalar P, Perez LA, Kramer EJ, Nguyen TQ, Heeger AJ, Wong WY, Bazan GC (2011) Regioregular pyridal[1–3]thiadiazole π-conjugated copolymers. J Am Chem Soc 133:18538–18541
259. Tseng HR, Ying L, Hsu BB, Perez LA, Takacs CJ, Bazan GC, Heeger AJ (2012) High mobility field effect transistors based on macroscopically oriented regioregular copolymers. Nano Lett 12:6353–6357
260. Tseng HR, Phan H, Luo C, Wang M, Perez LA, Patel SN, Ying L, Kramer EJ, Nguyen TQ, Bazan GC, Heeger AJ (2014) High-mobility field-effect transistors fabricated with macroscopic aligned semiconducting polymers. Adv Mater doi: 10.1002/adma.201305084
261. Lei T, Cao Y, Fan YL, Liu CJ, Yuan SC, Pei J (2011) High-performance air-stable organic field-effect transistors: isoindigo-based conjugated polymers. J Am Chem Soc 133:6099–6101
262. Lei T, Cao Y, Zhou X, Peng Y, Bian J, Pei J (2012) Systematic investigation of isoindigo-based polymeric field-effect transistors: design strategy and impact of polymer symmetry and backbone curvature. Chem Mater 24:1762–1770
263. Lei T, Dou JH, Pei J (2012) Influence of alkyl chain branching positions on the hole mobilities of polymer thin-film transistors. Adv Mater 24:6457–6461
264. Mei JG, Kim DH, Ayzner AL, Toney MF, Bao ZN (2011) Siloxane-terminated solubilizing side chains: bringing conjugated polymer backbones closer and boosting hole mobilities in thin-film transistors. J Am Chem Soc 133:20130–20133
265. Fan J, Yuen JD, Cui WB, Seifter J, Mohebbi AR, Wang MF, Zhou HQ, Heeger A, Wudl F (2012) High-hole-mobility field-effect transistors based on co-benzobisthiadiazole-quaterthiophene. Adv Mater 24:6164–6168
266. Osaka I, Shimawaki M, Mori H, Doi I, Miyazaki E, Koganezawa K, Takimiya K (2012) Synthesis, characterization, and transistor and colar cell applications of a naphthobisthiadiazole-based semiconducting polymer. J Am Chem Soc 134:3498–3507
267. Bürgi L, Turbiez M, Pfeiffer R, Bienewald F, Kirner HJ, Winnewisser C (2008) High-mobility ambipolar near-infrared light-emitting polymer field-effect transistors. Adv Mater 20:2217–2224
268. Li YN, Singh SP, Sonar PA (2010) A high mobility p-type dpp-thieno[3,2-*b*]thiophene copolymer for organic thin-film transistors. Adv Mater 22:4862–4866
269. Li J, Zhao Y, Tan HS, Guo YL, Di CA, Yu G, Liu YQ, Lin M, Lim SH, Zhou YH, Su HB, Ong BS (2012) A stable solution-processed polymer semiconductor with record high-mobility for printed transistors. Sci Rep 2:754
270. Bronstein H, Chen ZY, Ashraf RS, Zhang WM, Du JP, Durrant JR, Tuladhar PS, Song K, Watkins SE, Geerts Y, Wienk MM, Janssen RAJ, Anthopoulos T, Sirringhaus H, Heeney M, McCulloch I (2011) Thieno[3,2-*b*]thiophene-diketopyrrolopyrrole-containing polymers for high-performance organic field-effect transistors and organic photovoltaic devices. J Am Chem Soc 133:3272–3275
271. Li YN, Sonar P, Singh SP, Soh MS, Meurs MV, Tan J (2011) Annealing-free high-mobility diketopyrrolopyrrole-quaterthiophene copolymer for solution-processed organic thin film transistors. J Am Chem Soc 133:2198–2204
272. Yi ZY, Sun XN, Zhao Y, Guo YL, Chen XG, Qin JG, Yu G, Liu YQ (2012) Diketopyrrolopyrrole-based π-conjugated copolymer containing β-unsubstituted quintetthiophene unit: a promising material exhibiting high hole-mobility for organic thin-film transistors. Chem Mater 24:4350–4356
273. Chen HJ, Guo YL, Yu G, Zhao Y, Zhang J, Gao D, Liu HT, Liu YQ (2012) Highly π-extended copolymers with diketopyrrolopyrrole moieties for high-performance field-effect transistors. Adv Mater 24:4618–4622
274. Kang I, An TK, Hong J, Yun HJ, Kim R, Chung DS, Park CE, Kim YH, Kwon SK (2013) Effect of selenophene in a DPP copolymer incorporating a vinyl group for high-performance organic field-effect transistors. Adv Mater 25:524–528
275. Kang I, Yun HJ, Chung DS, Kwon SK, Kim YH (2013) Record high hole mobility in polymer semiconductors via side-chain engineering. J Am Chem Soc 135:14896–14899

276. Frankevich E, Maruyama Y, Ogata H (1993) Mobility of charge carriers in vapor-phase grown C_{60} single crystal. Chem Phys Lett 214:39–44
277. Horiuchi K, Nakada K, Uchino S, Hashii S, Hashimoto A, Aoki N, Ochiai Y, Shimizu M (2002) Passivation effects of alumina insulating layer on C_{60} thin-film field-effect transistors. Appl Phys Lett 81:1911–1912
278. Kitamura M, Aomori S, Na JH, Arakawa Y (2008) Bottom-contact fullerene C_{60} thin-film transistors with high field-effect mobilities. Appl Phys Lett 93:033313
279. Anthopoulos TD, Singh B, Marjanovic N, Sariciftci NS, Ramil AM, Sitter H, Colle M, de Leeuw DM (2006) High performance *n*-channel organic field-effect transistors and ring oscillators based on C_{60} fullerene films. Appl Phys Lett 89:213504
280. Ito Y, Virkar AA, Mannsfeld S, Oh JH, Toney M, Locklin J, Bao ZN (2009) Crystalline ultrasmooth self-assembled monolayers of alkylsilanes for organic field-effect transistors. J Am Chem Soc 131:9396–9404
281. Waldauf C, Schilinsky P, Perisutti M, Hauch J, Brabec CJ (2003) Solution-processed *n*-type organic thin-film transistors. Adv Mater 15:2084–2088
282. Singh TB, Marjanovic N, Stadler P, Auinger M, Matt GJ, Gunes S, Sariciftci NS, Schwodiauer R, Bauer S (2005) Fabrication and characterization of solution-processed methanofullerene-based organic field-effect transistors. J Appl Phys 97:083714
283. Chikamatsu M, Nagamatsu S, Yoshida Y, Saito K, Yase K, Kikuchi K (2005) Solution-processed *n*-type organic thin-film transistors with high field-effect mobility. Appl Phys Lett 87:203504
284. Chikamatsu M, Itakura A, Yoshida Y, Azumi R, Yase K (2008) High-performance *n*-type organic thin-film transistors based on solution-processable perfluoroalkyl-substituted C_{60} derivatives. Chem Mater 20:7365–7367
285. Haddon RC (1996) C_{70} thin film transistors. J Am Chem Soc 118:3041–3042
286. Cho S, Seo JH, Lee K, Heeger AJ (2009) Enhanced performance of fullerene n-channel field-effect transistors with titanium sub-oxide injection layer. Adv Funct Mater 19:1459–1464
287. Suglyama H, Nagano T, Nouchi R, Kawasaki N, Ohta Y, Imai K, Tsutsui M, Kubozono Y, Fujiwara A (2007) Transport properties of field-effect transistors with thin films of C_{76} and its electronic structure. Chem Phys Lett 449:160–164
288. Kubozono Y, Rikiishi Y, Shibata K, Hosokawa T, Fujiki S, Kitagawa H (2004) Structure and transport properties of isomer-separated C_{82}. Phys Rev B 69:165412
289. Shibata K, Kubozono Y, Kanbara T, Hosokawa T, Fujiwara A, Ito Y, Shinohara H (2004) Fabrication and characteristics of C_{84} fullerene field-effect transistors. Appl Phys Lett 84:2572–2574
290. Nagano T, Sugiyama H, Kuwahara E, Watanabe R, Kusai H, Kashino Y, Kubozono Y (2005) Fabrication of field-effect transistor device with higher fullerene, C88. Appl Phys Lett 87:023501
291. Laquindanum JG, Katz HE, Dodabalapur A, Lovinger AJ (1996) N-channel organic transistor materials based on naphthalene frameworks. J Am Chem Soc 118:11331–11332
292. Tanida S, Noda K, Kawabata H, Matsushige K (2009) N-channel thin-film transistors based on 1,4,5,8-naphthalene tetracarboxylic dianhydride with ultrathin polymer gate buffer layer. Thin Solid Films 518:571–574
293. Katz HE, Lovinger AJ, Johnson J, Kloc C, Siegrist T, Li W, Lin YY, Dodabalapur A (2000) A soluble and air-stable organic semiconductor with high electron mobility. Nature 404:478–481
294. Katz HE, Johnson J, Lovinger AJ, Li WJ (2000) Naphthalene tetracarboxylic diimide-based *n*-channel transistor semiconductors: structural variation and thiol-enhanced gold contacts. J Am Chem Soc 122:7787–7792
295. Shukla D, Nelson SF, Freeman DC, Rajeswaran M, Ahearn WG, Meyer DM, Carey JT (2008) Thin-film morphology control in naphthalene-diimide-based semiconductors: high mobility *n*-type semiconductor for organic, thin-film transistors. Chem Mater 20:7486–7491
296. Gawrys P, Boudinet D, Zagorska M, Djurado D, Verilhac JM, Horowitz G, Pecaud J, Pouget S, Pron A (2009) Solution processible naphthalene and perylene bisimides: synthesis,

electrochemical characterization and application to organic field effect transistors (OFETs) fabrication. Synth Met 159:1478–1485
297. Tszydel I, Kucinska M, Marszalek T, Rybakiewicz R, Nosal A, Jung J, Gazicki-Lipman M, Pitsalidis C, Gravalidi C, Logothetidis S, Zagorska M, Ulansk J (2012) High-mobility and low turn-on voltage *n*-channel otfts based on a solution-processable derivative of naphthalene bisimide. Adv Funct Mater 22:3840–3844
298. Oh JH, Suraru SL, Lee WY, Konemann M, Hoffken HW, Roger C, Schmidt R, Chung Y, Chen WC, Wurthner F, Bao ZN (2010) High-performance air-stable *n*-type organic transistors based on core-chlorinated naphthalene tetracarboxylic diimides. Adv Funct Mater 20:2148–2156
299. Lv A, Li Y, Yue W, Jiang L, Dong H, Zhao G, Meng Q, Jiang W, He Y, Li Z, Wang Z, Hu W (2012) High performance *n*-type single crystalline transistors of naphthalene bis (dicarboximide) and their anisotropic transport in crystals. Chem Commun 48:5154–5156
300. Jung BJ, Lee K, Sun J, Andreou AG, Katz HE (2010) Air-operable, high-mobility organic transistors with semifluorinated side chains and unsubstituted naphthalenetetracarboxylic diimide cores: high mobility and environmental and bias stress stability from the perfluorooctylpropyl side chain. Adv Funct Mater 20:2930–2944
301. See KC, Landis C, Sarjeant A, Katz HE (2008) Easily synthesized naphthalene tetracarboxylic diimide semiconductors with high electron mobility in air. Chem Mater 20:3609–3616
302. Sun J, Devine R, Dhar BM, Jung BJ, See KC, Katz HE (2009) Improved morphology and performance from surface treatments of naphthalenetetracarboxylic diimide bottom contact field-effect transistors. ACS Appl Mater Interfaces 1:1763–1769
303. Stolte M, Gsanger M, Hofmockel R, Suraru SL, Wurthner F (2012) Improved ambient operation of *n*-channel organic transistors of solution-sheared naphthalene diimide under bias stress. Phys Chem Chem Phys 14:14181–14185
304. He T, Stolte M, Würthner F (2013) Air-stable *n*-channel organic single crystal field-effect transistors based on microribbons of core-chlorinated naphthalene diimide. Adv Mater 25:6951–6955
305. Jones BA, Facchetti A, Marks TJ, Wasielewski MR (2007) Cyanonaphthalene diimide semiconductors for air-stable, flexible, and optically transparent *n*-channel field-effect transistors. Chem Mater 19:2703–2705
306. Zhao Y, Di CA, Gao X, Hu Y, Guo Y, Zhang L, Liu Y, Wang J, Hu W, Zhu D (2011) All-solution-processed, high-performance *n*-channel organic transistors and circuits: toward low-cost ambient electronics. Adv Mater 23:2448–2453
307. Zhang F, Hu Y, Schuettfort T, Di CA, Gao X, McNeill CR, Thomsen L, Mannsfeld SC, Yuan W, Sirringhaus H, Zhu D (2013) Critical role of alkyl chain branching of organic semiconductors in enabling solution-processed n-channel organic thin-film transistors with mobility of up to 3.50 $cm^2\ V^{-1}\ s^{-1}$. J Am Chem Soc 135:2338–2349
308. Hu Y, Qin Y, Gao X, Zhang F, Di CA, Zhao Z, Li H, Zhu D (2012) One-pot synthesis of core-expanded naphthalene diimides: enabling *n*-substituent modulation for diverse *n*-type organic materials. Org Lett 14:292–295
309. Chen X, Guo Y, Tan L, Yang G, Li Y, Zhang G, Liu Z, Xu W, Zhang D (2013) Dithiazole-fused naphthalene diimides toward new *n*-type semiconductors. J Mater Chem C 1:1087–1092
310. Luo H, Cai Z, Tan L, Guo Y, Yang G, Liu Z, Zhang G, Zhang D, Xu W, Liu Y (2013) Solution-processed core-extended naphthalene diimides toward organic *n*-type and ambipolar semiconductors. J Mater Chem C 1:2688–2695
311. Chen X, Wang J, Zhang G, Liu Z, Xu W, Zhang D (2013) New core-expanded naphthalene diimides with different functional groups for air-stable solution-processed organic n-type semiconductors. New J Chem 37:1720–1727
312. Deng P, Yan Y, Wang SD, Zhang Q (2012) Naphthoylene(trifluoromethyl-benzimidazole)-dicarboxylic acid imides for high-performance *n*-type organic field-effect transistors. Chem Commun 48:2591–2593

313. Hwang DK, Dasari RR, Fenoll M, Alain-Rizzo V, Dindar A, Shim JW, Deb N, Fuentes-Hernandez C, Barlow S, Bucknall DG, Audebert P, Marder SR, Kippelen B (2012) Stable solution-processed molecular *n*-channel organic field-effect transistors. Adv Mater 24:4445–4450
314. Tiwari SP, Kim J, Knauer KA, Hwang DK, Polander LE, Barlow S, Marder SR, Kippelen B (2012) Complementary-like inverters based on an ambipolar solution-processed molecular bis(naphthalene diimide)-dithienopyrrole derivative. Org Electron 13:1166–1170
315. Ostrick JR, Dodabalapur A, Torsi L, Lovinger AJ, Kwock EW, Miller TM, Galvin M, Berggren M, Katz HE (1997) Conductivity-type anisotropy in molecular solids. J Appl Phys 81:6804–6808
316. Yamada K, Takeya J, Takenobu T, Iwasa Y (2008) Effects of gate dielectrics and metal electrodes on air-stable *n*-channel perylene tetracarboxylic dianhydride single-crystal field-effect transistor. Appl Phys Lett 92:253311
317. Chesterfield RJ, McKeen JC, Newman CR, Frisbie CD, Ewbank PC, Mann KR, Miller LL (2004) Variable temperature film and contact resistance measurements on operating *n*-channel organic thin film transistors. J Appl Phys 95:6396–6405
318. Malenfant PRL, Dimitrakopoulos CD, Gelorme JD, Kosbar LL, Graham TO, Curioni A, Andreoni W (2002) *N*-type organic thin-film transistor with high field-effect mobility based on a *N*,*N*′-dialkyl-3,4,9,10-perylene tetracarboxylic diimide derivative. Appl Phys Lett 80:2517–2519
319. Chesterfield RJ, McKeen JC, Newman CR, Ewbank PC, da Silva Filho DA, Bredas JL, Miller LL, Mann KR, Frisbie CD (2004) Organic thin film transistors based on *N*-alkyl perylene diimides: charge transport kinetics as a function of gate voltage and temperature. J Phys Chem B 108:19281–19292
320. Gundlach DJ, Pernstich KP, Wilckens G, Gruter M, Haas S, Batlogg B (2005) High mobility *n*-channel organic thin-film transistors and complementary inverters. J Appl Phys 98:064502
321. Rost C, Gundlach DJ, Karg S, Riess W (2004) Ambipolar organic field-effect transistor based on an organic heterostructure. J Appl Phys 95:5782–5787
322. Tatemichi S, Ichikawa M, Koyama T, Taniguchi Y (2006) High mobility N-type thin-film transistors based on *N*,*N*-ditridecyl perylene diimide with thermal treatments. Appl Phys Lett 89:112108
323. Oh JH, Liu S, Bao Z, Schmidt R, Wurthner F (2007) Air-stable n-channel organic thin-film transistors with high field-effect mobility based on N,N′-bis(heptafluorobutyl)-3,4:9,10-perylene diimide. Appl Phys Lett 91:212107
324. Schmidt R, Oh JH, Sun YS, Deppisch M, Krause AM, Radacki K, Braunschweig H, Konemann M, Erk P, Bao ZN, Wurthner F (2009) High-performance air-stable *n*-channel organic thin film transistors based on halogenated perylene bisimide semiconductors. J Am Chem Soc 131:6215–6228
325. Oh JH, Sun YS, Schmidt R, Toney MF, Nordlund D, Konemann M, Wurthner F, Bao ZN (2009) Interplay between energetic and kinetic factors on the ambient stability of *n*-channel organic transistors based on perylene diimide derivatives. Chem Mater 21:5508–5518
326. Chen HZ, Ling MM, Mo X, Shi MM, Wang M, Bao Z (2007) Air stable n-channel organic semiconductors for thin film transistors based on fluorinated derivatives of perylene diimides. Chem Mater 19:816–824
327. Ling MM, Erk P, Gomez M, Koenemann M, Locklin J, Bao ZN (2007) Stable *n*-channel organic semiconductors based on perylene diimide derivatives without strong electron withdrawing groups. Adv Mater 19:1123–1127
328. Oh JH, Lee HW, Mannsfeld S, Stoltenberg RM, Jung E, Jin YW, Kim JM, Yoo JB, Bao ZN (2009) Solution-processed, high-performance *n*-channel organic microwire transistors. Proc Natl Acad Sci USA 106:6065–6070
329. Yoo B, Jung T, Basu D, Dodabalapur A, Jones BA, Facchetti A, Wasielewski MR, Marks TJ (2006) High-mobility bottom-contact *n*-channel organic transistors and their use in complementary ring oscillators. Appl Phys Lett 88:082104

330. Weitz RT, Amsharov K, Zschieschang U, Villas EB, Goswami DK, Burghard M, Dosch H, Jansen M, Kern K, Klauk H (2008) Organic *n*-channel transistors based on core-cyanated perylene carboxylic diimide derivatives. J Am Chem Soc 130:4637–4645
331. Rivnay J, Jimison LH, Northrup JE, Toney MF, Noriega R, Lu S, Marks TJ, Facchetti A, Salleo A (2009) Large modulation of carrier transport by grain-boundary molecular packing and microstructure in organic thin films. Nat Mater 8:952–958
332. Baeg KJ, Khim D, Kim JH, Kang M, You IK, Kim DY, Noh YY (2011) Improved performance uniformity of inkjet printed *n*-channel organic field-effect transistors and complementary inverters. Org Electron 12:634–640
333. Jones BA, Ahrens MJ, Yoon MH, Facchetti A, Marks TJ, Wasielewski MR (2004) High-mobility air-stable *n*-type semiconductors with processing versatility: dicyanoperylene-3,4:9,10-bis(dicarboximides). Angew Chem Int Ed 43:6363–6366
334. Soeda J, Uemura T, Mizuno Y, Nakao A, Nakazawa Y, Facchetti A, Takeya J (2011) High electron mobility in air for *N,N*'-1H,1H-perfluorobutyldicyanoperylene carboxydi-imide solution-crystallized thin-film transistors on hydrophobic surfaces. Adv Mater 23:3681–3865
335. Molinari AS, Alves H, Chen Z, Facchetti A, Morpurgo AF (2009) High electron mobility in vacuum and ambient for PDIF-CN2 single-crystal transistors. J Am Chem Soc 131:2462–2463
336. Minder NA, Ono S, Chen Z, Facchetti A, Morpurgo AF (2012) Band-like electron transport in organic transistors and implication of the molecular structure for performance optimization. Adv Mater 24:503–508
337. Gsänger M, Oh JH, Könemann M, Höffken HW, Krause A-M, Bao Z, Würthner F (2010) A crystal-engineered hydrogen-bonded octachloroperylene diimide with a twisted core: an *n*-channel organic semiconductor. Angew Chem Int Ed 49:740–743
338. Zhang J, Tan L, Jiang W, Hua W, Wang Z (2013) N-alkyl substituted di(perylene bisimides) as air-stable electron transport materials for solution-processible thin-film transistors with enhanced performance. J Mater Chem C 1:3200–3206
339. Lv A, Puniredd SR, Zhang J, Li Z, Zhu H, Jiang W, Dong H, He Y, Jiang L, Li Y, Pisula W, Meng Q, Hu W, Wang Z (2012) High mobility, air stable, organic single crystal transistors of an *n*-type diperylene bisimide. Adv Mater 24:2626–2630
340. Yue W, Lv A, Gao J, Jiang W, Hao L, Li C, Li Y, Polander LE, Barlow S, Hu W, Motta SD, Negri F, Marder SR, Wang Z (2012) Hybrid rylene arrays via combination of stille coupling and C–H transformation as high-performance electron transport materials. J Am Chem Soc 134:5770–5773
341. Zheng Q, Huang J, Sarjeant A, Katz HE (2008) Pyromellitic diimides: minimal cores for high mobility *n*-channel transistor semiconductors. J Am Chem Soc 130:14410–14411
342. Chen SC, Ganeshan D, Cai D, Zheng Q, Yin Z, Wang F (2013) High performance *n*-channel thin-film field-effect transistors based on angular-shaped naphthalene tetracarboxylic diimides. Org Electron 14:2859–2865
343. Mohebbi AR, Munoz C, Wudl F (2011) Synthesis and characterization of 2,8-diazaperylene-1,3,7,9-tetraone, a new anthracene diimide containing six-membered imide rings. Org Lett 13:2560–2563
344. Wang Z, Kim C, Facchetti A, Marks TJ (2007) Anthracenedicarboximides as air-stable *n*-channel semiconductors for thin-film transistors with remarkable current on-off ratios. J Am Chem Soc 129:13362–13363
345. Usta H, Kim C, Wang Z, Lu S, Huang H, Facchetti A, Marks TJ (2012) Anthracenedicarboximide-based semiconductors for air-stable, *n*-channel organic thin-film transistors: materials design, synthesis, and structural characterization. J Mater Chem 22:4459–4472
346. Katsuta S, Tanaka K, Maruya Y, Mori S, Masuo S, Okujima T, Uno H, Nakayama K, Yamada H (2011) Synthesis of pentacene-, tetracene- and anthracene disimides using double-cyclization reaction mediated by bismuth(III) triflate. Chem Commun 47:10112–10114

347. Chang J, Qu H, OOI Z-E, Zhang J, Chen Z, Wu J, Chi C (2013) 6,13-dicyano pentacene-2,3:9,10-bis(dicarboximide) for solution-processed air-stable *n*-channel field effect transistors and complementary circuit. J Mater Chem C 1: 456–462
348. Petit M, Hayakawa R, Shirai Y, Wakayama Y, Hill JP, Ariga K, Chikyow T (2008) Growth and electrical properties of *N,N*'-bis(n-pentyl)terrylene-3,4:11,12-tetracarboximide thin films. Appl Phys Lett 92:163301
349. Liu C, Liu Z, Lemke HT, Tsao HN, Naber RCG, Li Y, Banger K, Müllen K, Nielsen MM, Sirringhaus H (2010) High-performance solution-deposited ambipolar organic transistors based on terrylene diimides. Chem Mater 22:2120–2124
350. Tsao HN, Pisula W, Liu ZH, Osikowicz W, Salaneck WR, Müllen K (2008) From ambi- to unipolar behavior in discotic dye field-effect transistors. Adv Mater 20:2715–2719
351. Li JL, Chang JJ, Tan HS, Jiang H, Chen XD, Chen ZK, Zhang J, Wu JS (2012) Disc-like 7,14-dicyano-ovalene-3,4:10,11-bis(dicarboximide) as a solution-processible *n*-type semiconductor for air stable field-effect transistors. Chem Sci 3:846–850
352. An ZZ, Yu JS, Domercq B, Jones SC, Barlow S, Kippelen B, Marder SR (2009) Room-temperature discotic liquid-crystalline coronene diimides exhibiting high charge-carrier mobility in air. J Mater Chem 19:6688–6698
353. Li H, Kim FS, Ren G, Hollenbeck EC, Subramaniyan S, Jenekhe SA (2013) Tetraazabenzodifluoranthene diimides: building blocks for solution-processable *n*-type organic semiconductors. Angew Chem Int Ed 52:5513–5517
354. Handa S, Miyazaki E, Takimiya K, Kunugi Y (2007) Solution-processible n-channel organic field-effect transistors based on dicyanomethylene-substituted terthienoquinoid derivative. J Am Chem Soc 129:11684–16685
355. Suzuki Y, Miyazaki E, Takimiya K (2010) (Alkyloxy)carbonyl)cyanomethylene-substituted thienoquinoidal compounds: a new class of soluble *n*-channel organic semiconductors for air-stable organic field-effect transistors. J Am Chem Soc 132:10453–10466
356. Qiao YL, Zhang J, Xu W, Zhu DB (2012) Incorporation of pyrrole to oligothiophene-based quinoids endcapped with dicyanomethylene: a new class of solution processable *n*-channel organic semiconductors for air-stable organic field-effect transistors. J Mater Chem 22:5706–5714
357. Wu QH, Li RJ, Hong W, Li HX, Gao XK, Zhu DB (2011) Dicyanomethylene-substituted fused tetrathienoquinoid for high-performance, ambient-stable, solution-processable *n*-channel organic thin-film transistors. Chem Mater 23:3138–3140
358. Qiao Y, Guo Y, Yu C, Zhang F, Xu W, Liu Y, Zhu D (2012) Diketopyrrolopyrrole-containing quinoidal small molecules for high-performance, air-stable, and solution-processable *n*-channel organic field-effect transistors. J Am Chem Soc 134:4084–4087
359. Cai X, Burand MW, Newman CR, da Silva Filho DA, Pappenfus TM, Bader MM, Bredas JL, Mann KR, Frisbie CD (2006) *N*- and *p*-channel transport behavior in thin film transistors based on tricyanovinyl-capped oligothiophenes. J Phys Chem B 110:14590–14597
360. Yoon WS, Park SK, Cho I, Oh JA, Kim JH, Park SY (2013) High-mobility *n*-type organic transistors based on a crystallized diketopyrrolopyrrole derivative. Adv Funct Mater 23:3519–3524
361. Ortiz RP, Facchetti A, Marks TJ, Casado J, Zgierski MZ, Kozaki M, Hernandez V, Navarrete JTL (2009) Ambipolar organic field-effect transistors from cross-conjugated aromatic quaterthiophenes; comparisons with quinoidal parent materials. Adv Funct Mater 19:386–394
362. Yue W, He T, Stolte M, Gsängera M, Würthner F (2014) Cyanated isoindigos for *n*-type and ambipolar organic thin film transistors. Chem Commun 50:545–547
363. Hong W, Guo C, Sun B, Yan Z, Huang C, Hu Y, Zheng Y, Facchetti A, Li Y (2013) Cyano-disubstituted dipyrrolopyrazinedione (CNPzDP) dmall molecules for solution processed *n*-channel organic thin-film transistors. J Mater Chem C 1:5624–5627
364. Sakamoto Y, Suzuki T, Kobayashi M, Gao Y, Fukai Y, Inoue Y, Sato F, Tokito S (2004) Perfluoropentacene: high-performance p–n junctions and complementary circuits with pentacene. J Am Chem Soc 126:8138–8140

365. Kikuzawa Y, Mori T, Takeuchi H (2007) Synthesis of 2,5,8,11,14,17-Hexafluoro-hexa-perihexabenzocoronene for *n*-type organic field-effect transistors. Org Lett 9:4817–4820
366. Ichikawa M, Kato T, Uchino T, Tsuzuki T, Inoue M, Jeon HG, Koyama T, Taniguchi Y (2010) Thin-film and single-crystal transistors based on a triflkuoromethyl-substituted alternating (thiophene/phenylene)-co-oligomer. Org Electron 11:1549–1554
367. Youn J, Huang PY, Huang YW, Chen MC, Lin YJ, Huang H, Ortiz RP, Stern C, Chung MC, Feng CY, Chen LH, Facchetti A, Marks TJ (2012) Versatile α, ω-disubstituted tetrathienoacene semiconductors for high performance organic thin-film transistors. Adv Funct Mater 22:48–60
368. Shoji K, Nishida J, Kumaki D, Tokito S, Yamashita Y (2010) Synthesis and FET characteristics of phenylene-vinylene and anthracene-vinylene compounds containing cyano groups. J Mater Chem 20:6472–6478
369. Yun SW, Kim JH, Shin S, Yang H, An BK, Yang L, Park SY (2012) High-performance *n*-type organic semiconductors: incorporating specific electron-withdrawing motifs to achieve tight molecular stacking and optimized energy levels. Adv Mater 24:911–915
370. Kim JH, Chung JW, Jung Y, Yoon SJ, An BK, Huh HS, Lee SW, Park SY (2010) High performance *n*-type organic transistors based on a distyrylthiophene derivative. J Mater Chem 20:10103–10106
371. Park YI, Lee JS, Kim BJ, Kim B, Lee J, Kim DH, Oh SY, Cho JH, Park JW (2011) High-performance stable *n*-type indenofluorenedione field-effect transistors. Chem Mater 23:4038–4044
372. Liang ZX, Tang Q, Liu J, Li JH, Yan F, Miao Q (2010) *N*-type organic semiconductors based on π-deficient pentacenequinones: synthesis, electronic structures, molecular packing, and thin film transistors. Chem Mater 22:6438–6443
373. Yoon MH, Kim C, Facchetti A, Marks TJ (2006) Gate dielectric chemical structure-organic field-effect transistor performance correlations for electron, hole, and ambipolar organic semiconductors. J Am Chem Soc 128:12851–12869
374. Schols S, Van Willigenburg L, Müller R, Bode D, Debucquoy M, De Jonge S, Genoe J, Heremans P, Lu S, Facchetti A (2008) Influence of the contact metal on the performance of n-type carbonyl-functionalized quaterthiophene organic thin-film transistors. Appl Phys Lett 93:263303
375. Ando S, Nishida J, Tada H, Inoue Y, Tokito S, Yamashita Y (2005) High performance *n*-type organic field-effect transistors based on π-electronic systems with trifluoromethylphenyl groups. J Am Chem Soc 127:5336–5337
376. Kumaki D, Ando S, Shimono S, Yamashita Y, Umeda T, Tokito S (2007) Significant improvement of electron mobility in organic thin-film transistors based on thiazolothiazole derivative by employing self-assembled monolayer. Appl Phys Lett 90:053506
377. Mamada M, Nishida JI, Kumaki D, Tokito S, Yamashita Y (2007) n-Type organic field-effect transistors with high electron mobilities based on thiazole-thiazolothiazole conjugated molecules. Chem Mater 19:5404–5409
378. Mamada M, Kumaki D, Nishida J, Tokito S, Yamashita Y (2010) Novel semiconducting quinone for air-stable *n*-type organic field-effect transistors. ACS Appl Mater Interfaces 2:1303–1307
379. Ie Y, Ueta M, Nitani M, Tohnai N, Miyata M, Tada H, Aso Y (2012) Air-stable *n*-type organic field-effect transistors based on 4,9-dihydro-s-indaceno[1,2-*b*:5,6-*b*′]dithiazole-4,9-dione unit. Chem Mater 24:3285–3293
380. Di CA, Li J, Yu G, Xiao Y, Guo YL, Liu YQ, Qian XH, Zhu DB (2008) Trifluoromethyltriphenodioxazine: air-stable and high-performance *n*-type semiconductor. Org Lett 10:3025–3028
381. Islam MM, Pola S, Tao YT (2011) High mobility *n*-channel single-crystal field-effect transistors based on 5,7,12,14-Tetrachloro-6,13-diazapentacene. Chem Commun 47:6356–6358
382. Liang Z, Tang Q, Xu J, Miao Q (2011) Soluble and stable *N*-heteropentacenes with high field-effect mobility. Adv Mater 23:1535–1539

383. Babel A, Jenekhe SA (2003) High electron mobility in ladder polymer field-effect transistors. J Am Chem Soc 125:13656–13657
384. Briseno AL, Mannsfeld SCB, Shamberger PJ, Ohuchi FS, Bao ZN, Jenekhe SA, Xia YN (2008) Self-assembly, molecular packing, and electron transport in *n*-type polymer semiconductor nanobelts. Chem Mater 20:4712–4719
385. Briseno AL, Kim FS, Babel A, Xia YN, Jenekhe SA (2011) N-channel polymer thin film transistors with long-term air-stability and durability and their use in complementary inverters. J Mater Chem 21:16461–16466
386. Durban MM, Kazarinoff PD, Segawa Y, Luscombe CK (2011) Synthesis and characterization of solution-processable ladderized *n*-type naphthalene bisimide copolymers for OFET applications. Macromolecules 44:4721–4728
387. Yan H, Chen ZH, Zheng Y, Newman C, Quinn JR, Dötz F, Kastler M, Facchetti A (2009) A high-mobility electron-transporting polymer for printed transistors. Nature 457:679–686
388. Durban MM, Kazarinoff PD, Luscombe CK (2010) Synthesis and characterization of thiophene-containing naphthalene diimide *n*-type copolymers for OFET applications. Macromolecules 43:6348–6352
389. Huang H, Chen Z, Ortiz RP, Newman C, Usta H, Lou S, Youn J, Noh YY, Baeg KJ, Chen LX, Facchetti A, Marks TJ (2012) Combining electron-neutral building blocks with intramolecular "conformational locks" affords stable, high-mobility *p*- and *n*-channel polymer semiconductors. J Am Chem Soc 134:10966–10973
390. Zhou W, Wen Y, Ma L, Liu Y, Zhan X (2012) Conjugated polymers of rylene diimide and phenothiazine for *n*-channel organic field-effect transistors. Macromolecules 45:4115–4121
391. Letizia JA, Salata MR, Tribout CM, Facchetti A, Ratner MA, Marks TJ (2008) *N*-channel polymers by design: optimizing the interplay of solubilizing substituents, crystal packing, and field-effect transistor characteristics in polymeric bithiophene-imide semiconductors. J Am Chem Soc 130:9679–9694
392. Zhan X, Tan Z, Domercq B, An Z, Zhang X, Barlow S, Li Y, Zhu D, Kippelen B, Marder SR (2007) A high-mobility electron-transport polymer with broad absorption and its use in field-effect transistors and all-polymer solar cells. J Am Chem Soc 129:7246–7247
393. Zhao X, Wen Y, Ren L, Ma L, Liu Y, Zhan X (2012) An acceptor–acceptor conjugated copolymer based on perylene diimide for high mobility n-channel transistor in air. J Polym Sci, Part A: Polym Chem 50:4266–4271
394. Izuhara D, Swager TM (2009) Poly(pyridinium phenylene)s: water-soluble *n*-type polymers. J Am Chem Soc 131:17724–17725
395. Kanimozhi C, Yaacobi-Gross N, Chou KW, Amassian A, Anthopoulos TD, Patil S (2012) Diketopyrrolopyrrole-diketopyrrolopyrrole-based conjugated copolymer for high-mobility organic field-effect transistors. J Am Chem Soc 134:16532–16535
396. Park JH, Jung EH, Jung JW, Jo WH (2013) A fluorinated phenylene unit as a building block for high-performance *n*-type semiconducting polymer. Adv Mater 25:2583–2588
397. Lei T, Dou JH, Cao XY, Wang JY, Pei J (2013) Electron-deficient poly(p-phenylene vinylene) provides electron mobility over 1 $cm^2\ V^{-1}\ s^{-1}$ under ambient conditions. J Am Chem Soc 135:12168–12171
398. Yan Z, Sun B, Li Y (2013) Novel stable (3e,7e)-3,7-bis(2-oxoindolin-3-ylidene)-benzo[1,2-b:4,5-b′]difuran-2,6(3h,7h)-dione based donor-acceptor polymer semiconductors for n-type organic thin film transistors. Chem Commun 49:3790–3792
399. Lei T, Dou J, Cao X, Wang J, Pei J (2013) A BDOPV-based donor-acceptor polymer for high-performance *n*-type and oxygen-doped ambipolar field-effect transistors. Adv Mater 25:6589–6593
400. Tang ML, Reichardt AD, Miyaki N, Stoltenberg RM, Bao Z (2008) Ambipolar, high performance, acene-based organic thin film transistors. J Am Chem Soc 130:6064–6065
401. Liang Z, Tang Q, Mao R, Liu D, Xu J, Miao Q (2011) The position of nitrogen in *N*-heteropentacenes matters. Adv Mater 23:5514–5518

402. Song CL, Ma CB, Yang F, Zeng WJ, Zhang HL, Gong X (2011) Synthesis of tetrachloroazapentacene as an ambipolar organic semiconductor with high and balanced carrier mobilities. Org Lett 13:2880–2883
403. Xia H, Liu DQ, Xu XM, Miao Q (2013) Ambipolar organic semiconductors from electron-accepting cyclopenta-fused anthracene. Chem Commun 49:4301–4303
404. Irimia-Vladu M, Glowacki ED, Troshin PA, Schwabegger G, Leonat L, Susarova DK, Krystal O, Ullah M, Kanbur Y, Bodea MA, Razumov VF, Sitter H, Bauer S, Sariciftci NS (2011) Indigo-a natural pigment for high performance ambipolar organic field effect transistors and circuits. Adv Mater 24:375–380
405. Glowacki ED, Leonat L, Voss G, Bodea MA, Bozkurt Z, Ramil AM, Irimia-Vladu M Bauer S, Sariciftci NS (2011) Ambipolar organic field effect transistors and inverters with the natural material tyrian purple. AIP Adv 1: 042132
406. Nakanotani H, Saito M, Nakamura H, Adachi C (2009) Blue-light-emitting ambipolar field-effect transistors using an organic single crystal of 1,4-Bis(4-methylstyryl)benzene. Appl Phys Lett 95:033308
407. Tan LX, Guo YL, Yang Y, Zhang GX, Zhang DQ, Yu G, Xu W, Liu YQ (2012) New tetrathiafulvalene fused-naphthalene diimides for solution-processible and air-stable p-type and ambipolar organic semiconductors. Chem Sci 3:2530–2541
408. Wang L, Zhang X, Tian H, Lu Y, Geng Y, Wang F (2013) A cyano-terminated dithienyldiketopyrrolopyrrole dimer as a solution processable ambipolar semiconductor under ambient conditions. Chem Commun 49:11272–11274
409. Sonar P, Singh SP, Li Y, Soh MS, Dodabalapur A (2010) A low-bandgap diketopyrrolopyrrole-benzothiadiazole based copolymer for high-mobility ambipolar organic thin-film transistors. Adv Mater 22:5409–5413
410. Ha TJ, Sonar P, Cobb B, Dodabalapur A (2012) Charge transport and density of trap states in balanced high mobility ambipolar organic thin-film transistors. Org Electron 13:136–141
411. Sonar P, Foong TR, Singh SP, Li Y (2012) A furan-containing conjugated polymer for high mobility ambipolar organic thin film transistors. Chem Commun 48:8383–8385
412. Kronemeijer AJ, Gili E, Shahid M, Rivnay J, Salleo A, Heeney M, Sirringhaus HA (2012) A selenophene-based low-bandgap donor-acceptor polymer leading to fast ambipolar logic. Adv Mater 24:1558–1565
413. Mohebbi AR, Yuen JD, Fan J, Munoz C, Wang M, Shirazi RS, Seifter J, Wudl F (2011) Emeraldicene as an acceptor moiety: balanced-mobility, ambipolar, organic thin-film transistors. Adv Mater 23:4644–4688
414. Yuen JD, Fan J, Seifter J, Lim B, Hufschmid R, Heeger AJ, Wudl F (2011) High performance weak donor-acceptor polymers in thin film transistors: effect of the acceptor on electronic properties, ambipolar conductivity, mobility, and thermal stability. J Am Chem Soc 13:20799–20807
415. Li Y, Sun B, Sonar P, Singh SP (2012) Solution processable Poly(2,5-dialkyl-2,5-dihydro-3,6-di-2-thienylpyrrolo[3,4-*c*]pyrrole-1,4-dione) for ambipolar organic thin film transistors. Org Electron 13:1606–1613
416. Hong W, Sun B, Aziz H, Park WT, Noh YY, Li YA (2012) A conjugated polyazine containing diketopyrrolopyrrole for ambipolar organic thin film transistors. Chem Commun 48:8413–8415
417. Li Y, Singh SP, Sonar PA (2010) A high mobility p-Type DPP-thieno[3,2-*b*]thiophene copolymer for organic thin-film transistors. Adv Mater 22:4862–4866
418. Chen Z, Lee MJ, Ashraf RS, Gu Y, Albert-Seifried S, Nielsen MM, Schroeder B, Anthopoulos TD, Heeney M, McCulloch I, Sirringhaus H (2012) High-performance ambipolar diketopyrrolopyrrole-thieno[3,2-*b*]thiophene copolymer field-effect transistors with balanced hole and electron mobilities. Adv Mater 24:647–652
419. Lee HS, Lee JS, Cho S, Kim H, Kwak KW, Yoon Y, Son SK, Kim H, Ko MJ, Lee DK, Kim JY, Park S, Choi DH, Oh SY, Cho JH, Kim B (2012) Crystallinity-controlled naphthalene-*alt*-diketopyrrolopyrrole copolymers for high-performance ambipolar field effect transistors. J Phys Chem C 116:26204–26213

420. Lin HW, Lee WY, Chen WC (2012) Selenophene-DPP donor-acceptor conjugated polymer for high performance ambipolar field-effect transistor and nonvolatile memory applications. J Mater Chem 22:2120–2128
421. Lee J, Han AR, Kim J, Kim Y, Oh JH, Yang C (2012) Solution-processable ambipolar diketopyrrolopyrrole—selenophene polymer with unprecedentedly high hole and electron mobilities. J Am Chem Soc 134:20713–20721
422. Lee J, Han AR, Yu H, Shin T, Yang C, Oh JH (2013) Boosting the ambipolar performance of solution-processable polymer semiconductors via hybrid side-chain engineering. J Am Chem Soc 135:9540–9547
423. Lei T, Dou JH, Ma ZJ, Yao CH, Liu CJ, Wang JY, Pei J (2012) Ambipolar polymer field-effect transistors based on fluorinated isoindigo: high performance and improved ambient stability. J Am Chem Soc 134:20025–20028
424. Lei T, Dou JH, Ma ZJ, Liu CJ, Wang JY, Pei J (2013) Chlorination as a useful method to modulate conjugated polymers: balanced and ambient-stable ambipolar high-performance field-effect transistors and inverters based on chlorinated isoindigo polymers. Chem Sci 4:2447–2452
425. Ashraf RS, Kronemeijer AJ, James DI, Sirringhaus H, McCulloch I (2012) A new thiophene substituted isoindigo based copolymer for high performance ambipolar transistors. Chem Commun 48:3939–3941
426. Kim FS, Guo XG, Watson MD, Jenekhe SA (2010) High-mobility ambipolar transistors and high-gain inverters from a donor-acceptor copolymer semiconductor. Adv Mater 22:478–482
427. Guo XG, Kim FS, Seger MJ, Jenekhe SA, Watson MD (2012) *n*-type and ambipolar polymer semiconductors based on naphthalene diimide: synthesis, structure-property correlations, and field-effect transistors. Chem Mater 24:1434–1442
428. Chen HJ, Guo YL, Mao ZP, Yu G, Huang JY, Zhao Y, Liu YQ (2013) Naphthalenediimide-based copolymers incorporating vinyl-linkages for high-performance ambipolar field-effect transistors and complementary-like inverters under air. Chem Mater 25:3589–3596
429. Usta H, Newman C, Chen Z, Facchetti A (2012) Dithienocoronenediimide-based copolymers as novel ambipolar semiconductors for organic thin-film transistors. Adv Mater 24:3678–3684
430. Fan J, Yuen JD, Wang MF, Seifter J, Seo JH, Mohebbi AR, Zakhidov D, Heeger AJ, Wudl F (2012) High-performance ambipolar transistors and inverters from an ultralow bandgap polymer. Adv Mater 24:2186–2190
431. Yuen JD, Kumar R, Zakhidov D, Seifter J, Lim B, Heeger AJ, Wudl F (2011) Ambipolarity in benzobisthiadiazole-based donor-acceptor conjugated polymers. Adv Mater 23:3780–3785
432. Hong W, Sun B, Guo C, Yuen J, Li Y, Lu S, Huang C, Facchetti A (2013) Dipyrrolo[2,3-*b*:2′,3′-*e*]pyrazine-2,6(*1H,5H*)-dione based conjugated polymers for ambipolar organic thin-film transistors. Chem Commun 49:484–486
433. Weibin C, Yuen J, Wudl F (2011) Benzodipyrrolidones and their polymers. Macromolecules 44:7869–7873
434. Deng P, Liu L, Ren S, Li H, Zhang Q (2012) *N*-acylation: an effective method for reducing the LUMO energy levels of conjugated polymers containing five-membered lactam units. Chem Commun 48:6960–6962
435. Hong W, Guo C, Li Y, Zheng Y, Huang C, Lu S, Facchetti A (2012) Synthesis and thin-film transistor performance of benzodipyrrolinone and bithiophene donor-acceptor copolymers. J Mater Chem 22:22282–22289
436. Rumer JW, Levick M, Dai SY, Rossbauer S, Huang ZG, Biniek L, Anthopoulos TD, Durrant JR, Procter DJ, McCulloch I (2013) BPTs: thiophene-flanked benzodipyrrolidone conjugated polymers for ambipolar organic transistors. Chem Commun 49:4465–4467

Chapter 4
Organic Semiconductor Photovoltaic Materials

Zhi-Guo Zhang

Abstract Organic solar cells (OSCs) are an emerging alternative photovoltaic technology and thus they have recently gained much attention. In this chapter, recent developments in organic photovoltaic materials, involving small molecule donors and non-fullerene acceptors for vacuum and solution-processed photovoltaic cells, are summarized. A general overview of the structure–property relationships of these organic photovoltaic materials and the design rules for such materials is presented. Critical factors which determined their photophysical properties such as energy levels, absorption, and carrier mobilities are also highlighted.

Keywords Small molecule donors · Non-fullerene acceptors · Donor–acceptor systems · Organic photovoltaic properties

4.1 Introduction

In the past few decades, intensive research effort has been devoted to the development of organic solar cells (OSCs) based on solution-processed small-molecule (SM) donors or acceptors [1–5]. These materials are competitive alternative to widely used conjugated polymer-based donors or acceptors because of their potential advantages over conjugated polymer systems. The advantages of organic photovoltaic materials are easier purification and synthesis, defined structures without end group contaminants, and better batch-to-batch reproducibility [2–4]. With these advantages, a substantial amount of research in both academic and industrial circles has been directed towards OSCs in an effort to improve their processability, power conversion efficiency, and stability. Power conversion efficiency (PCE) is the most important parameter used to evaluate solar cell performance. PCE is determined by open-circuit voltage (V_{OC}), short-circuit current density (J_{SC}), and fill factors (FF), as shown by the

Z.-G. Zhang (✉)
CAS Key Laboratory of Organic Solids, Institute of Chemistry,
Chinese Academy of Sciences, Beijing 100190, China
e-mail: zgzhangwhu@iccas.ac.cn

Y. Li (ed.), *Organic Optoelectronic Materials*,
Lecture Notes in Chemistry 91, DOI 10.1007/978-3-319-16862-3_4

formula PCE = (J_{SC} × V_{OC} × FF)/P_{in}, where P_{in} is the input light power [6]. Recently, PCE of OSCs approached 8 % by synthesizing planar Acceptor-Donor-Acceptor-based conjugated molecule donor materials [7, 8], which makes solution-processed OSCs comparable to polymer solar cells (PSCs).

For device fabrication, two main approaches have been explored: (1) vacuum deposition for the planar-heterojunction (PHJ) and bulk-heterojunction (BHJ) solar cells and (2) solution processing (spin-coating, doctor blade, and dip-coating) for BHJ solar cells. The so-called BHJ represents the ideal case as bicontinuous donor-acceptor composites for large surface area.

In this chapter, the latest developments in SM donors are examined, namely, donor–acceptor SMs for vacuum, solution-processed OSCs and non-fullerene acceptors. By summarizing the structure–property relationships of representative photovoltaic materials, the design rules for such materials are highlighted. Chemical strategies for tuning their photophysical properties such as the optical bandgaps, energy levels, and charge mobilities are also discussed. For the donor materials, representative classes of materials include dyes (squaraine, boron dipyrromethene, porphyrin, and diketopyrrolopyrroles), triphenylamine derivatives, oligothiophenes, and push-pull type oligomers; for the non fullerene acceptor, those constructed from D-A structure and naphthalenediimide (NDI) and perylene diimide (PDI) derivatives are summarized.

4.2 Organic Solar Cells by Vacuum Deposition

Compared with solution processing, a distinct advantage of vacuum deposition is the ability to prepare multi-layer thin films. With this virtue, the interfaces for carrier collection can be carefully tuned, and multi-junction structures are easier to realize. For example, Heliatek GmbH reported certified PCE of 12.0 % for a tandem device fabricated by the connection of two BHJ devices, rendering OSCs more competitive [9].

Acenes are formed by fused benzene rings. Because of their intriguing photophysical and electronic properties, these materials have the potential to be used as organic electronics [10]. However, the numbers of higher order polyacenes reported so far are still limited because of their poor stability and solubility. Based on a pentacene (**1a,** Fig. 4.1)/C_{60} heterojunction, Kippelen and coworkers have fabricated an efficient OSC [11]. The efficiency is 2.7 % with a J_{SC} of 15 mA cm^{-2} and an FF of 0.50, but a low V_{OC} of 0.36 V. Brütting et al. reported vacuum deposited OSCs based on diindenoperylene (DIP, **1b**) as a new donor material and the fullerene C_{60} as an electron acceptor. The high HOMO of DIP and the favorable energy level offset with C_{60} produced large V_{OC} values close to 1 V and PCEs of about 4 % in the planar heterojunction and planar-mixed heterojunction cell architectures [12]. Adachi and coworkers fabricated vacuum deposited OSCs with tetraphenyldibenzoperiflanthene (**1C**) as donor and C_{60} as acceptors. To prevent exciton quenching, a layer of tris[4-(5-phenyl thiophen-2-yl)phenyl]amine was

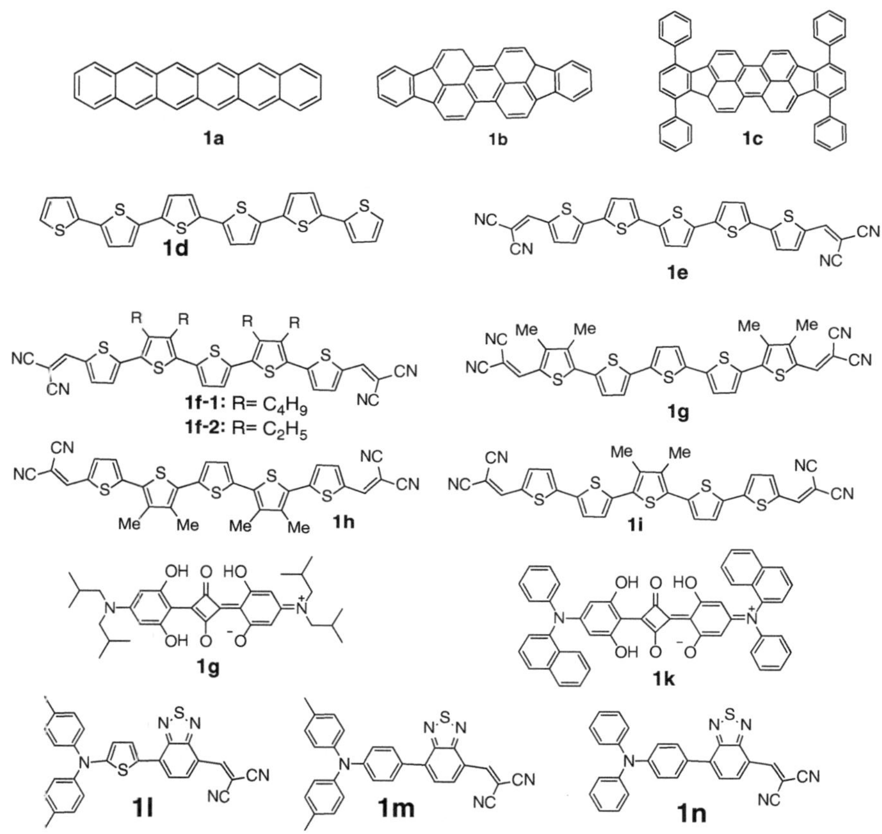

Fig. 4.1 Chemical structures of small molecules for vacuum deposited solar cells

deposited between the donor layer and the PEDOT:PSS layer, and a high PCE of 5.24 % was reached together with a V_{OC} of 0.94 V, J_{SC} of 7.25 mA cm^{-2}, and a high FF of 0.77 [13]. To make the less soluble organic materials compatible with solution processable techniques, Nakayama et al. reported the photoprecursor method. In addition, this approach enables the formation of multi-layer device structures using solution processes.

Structurally well-defined oligothiophenes have attracted much attention in terms of high mobility, environmental/thermal stability, and, most importantly, synthetic versatility. Without any chemical modifications or substitutions, α-sexithiophene (**1d**) was easy to crystallize by poor miscibility with C_{70}. BHJ OSCs using a blend of α-sexithiophene (**1d**) [14] and C_{70} were fabricated by the vacuum co-evaporation method. At a **1d**:C_{70} ratio of 1:5 (w/w), the OSC devices showed a PCE of 2.38 %. Notably, the wide bandgap and relatively high HOMO level of **1d** are the two factors limiting its efficiency. Bäuerle prepared a family of terminally acceptor

Table 4.1 Properties and device characteristics of molecules in Fig. 4.1

Molecules	HOMO/LUMO (eV/eV)	E_g (eV)	V_{oc} (V)	J_{sc} (mA cm^{-2})	FF (%)	PCE (%)	Reference
1a	−3.0/−4.9	1.77	0.36	15.0	0.50	2.70	[11]
1b	N.A.	N.A.	0.91	8.40	0.52	4.10	[12]
1c	N.A.	1.90	0.94	7.25	0.77	5.24	[13]
1d	−3.1/−5.3	N.A.	0.58	9.20	0.45	2.38	[14]
1e	−3.7/-5.7	1.69	0.97	11.1	0.49	5.20	[15]
1f-1	N.A.	N.A.	1.00	8.90	0.51	3.40	[16]
1f-2	N.A.	N.A.	1.00	8.40	0.40	2.50	[16]
1g	−3.73/−5.64	2.01	0.61	9.60	0.63	4.80	[17]
1h	−3.71/−5.61	2.03	0.95	9.40	0.62	4.80	[17]
1i	−3.73/−5.62	1.99	0.95	11.5	0.63	6.90	[17]
1j	N.A./ −5.10	1.77	0.76	7.01	0.56	3.10	[18]
1k	−3.70/−5.30	1.75	0.90	10.0	0.64	5.70	[19]
1l	−3.44/−5.30	1.86	0.79	15.08	0.48	5.70	[20]
1m	−3.36/−5.50	2.14	0.93	13.48	0.53	6.60	[20]
1n	−3.35/−5.43	2.08	1.00	11.28	0.45	5.00	[20]

dicyanovinyl-substituted oligothiophenes (**1e**) without solubilizing side chains, and implemented them in vacuum-deposited planar heterojunction solar cells. Optimization of BHJ devices based on a 40-nm active layer of **1e** and C_{60} in 2:1 ratio resulted in a PCE of 5.2 % for a 5.06 mm^2-sized and masked device [15]. Alkyl side chains of butyl groups and ethyl groups were appended on the backbone **1e** and tested in PHJ solar cells. Compared to the ethyl-substituted analogue **1f-2** (2.5 %), devices with butyl-substituted SMs (**1f-1**) showed a higher PCE (3.4 %). The differences in the molecular packing and the hole mobilities of the two oligothiophenes [16] was acceptable as the hole injection between the hole-transport layer and the oligothiophene are the reasons for the differences in their device performances (Table 4.1).

To improve the efficiency further, Bäuerle synthesized a series of methyl-substituted **1e** oligothiophenes. In these materials, the positions of methyl substituents were systematically varied [17]. These oligomers exhibited PCEs of 4.8–6.1 % in vacuum-deposited p-i-n-type BHJ solar cells. The PCE of oligomer **1i**-based device was improved to 6.9 % by device optimizations. The authors point out that the morphology of D-A blends and consequently the device performance can be effectively tuned by the methyl substitution in oligothiophenes.

Squaraine (SQ) dyes have a unique aromatic four-membered ring with resonance-stabilized zwitterionic structures, and are characterized for their high absorption coefficients and broad absorptions which can extend from the green to the NIR region [18, 19, 21–24]. Forrest, Thompson, and coworkers reported a squaraine-based molecule (**1j**), for structure see Fig. 4.1) as a photovoltaic donor (C_{60} as acceptor) in vacuum-deposited BHJ OSCs [18]. Under AM 1.5G simulated solar irradiation, devices exhibited a PCE of 3.1 % with a V_{OC} of 0.76 ± 0.01 V, a J_{SC} of 7.01 ± 0.05 mA cm^{-2}, and an FF of 0.56 ± 0.05. The attaching of *N,N*-dialkyl groups can produce soluble squaraines, although the carrier transport can be hindered. To improve the molecular stacking and hence charge transport, Forrest, Thompson, and coworkers synthesized **1k** by substitution of isobutylamines in the common "parent Squaraine" with arylamines. The strong electron-withdrawing arylamine group results in a lower HOMO of −5.3 eV, compared to −5.1 eV for the parent SQ (**1j**), thereby leading to an increased V_{OC}. Heterojunction **1k**/C_{60}/bathocuproine solar cells shows an improved efficiency of 5.7 % with a V_{OC} of 0.90 V, FF of 0.64, and J_{SC} of 1.1 mA cm^{-2} [19].

Organic dyes with a D-A-A (donor−acceptor−acceptor) structure have both a smaller bandgap and lower-lying highest occupied molecular orbital (HOMO) level as compared with their analogs. Wong, Lin, and coworkers recently reported a new D-A-A donor materials **1l-1n**. In these materials, an electron-donating ditolylaminothienyl group and an electron-deficient dicyanovinylene group are bridged by another electron acceptor of benzothiadiazole unit [20].

Using **1l** as donor and C_{70}, a vacuum-deposited solar cell as acceptor gives a PCE of 5.81 %. This respectable PCE value is attributed to its broad solar response range extending to the near-IR region and the ultra compact absorption dipole stacking of **1l** in a thin film. This delicate molecular structure tuning along with device engineering allows manipulation of the trade-off between the V_{OC} and J_{SC}. Planar heterojunction cells using **1m** as the donor and C_{70} as the acceptor demonstrated the best performance with a PCE of 6.6 ± 0.2 % (the highest PCE of 6.8 %), along with a V_{OC} of 0.93 ± 0.02 V, and a J_{SC} of 13.48 ± 0.27 mA cm^{-2}.

4.3 Organic Solar Cells by Solution Processing

4.3.1 Dyes

Boron dipyrromethene (BODIPY) dyes are characterized by unique chemical and photochemical stabilities, redox activities, and optical features. Their photophysical properties can easily be tuned by chemical modification, allowing them to act as prominent donors in BHJ solar cells. In 2009, Roncali and coworkers described the first examples of solar cells using boron dipyrromethene (**2c-1** and **2c-2** Fig. 4.2) donors involving one (**2c-1**) or two styryl units (**2c-2**) in the structure [25]. The more extensively conjugated compound **2c-2** has a lower oxidation potential and red-shifted absorption. The absorption onset corresponds to the bandgap (Eg) of 1.95 and 1.70 eV for **2c-1** and **2c-2**, respectively. The solar cell using donor **2c-1** shows a V_{OC} of 0.796 V, together with an FF of 0.34, and a resulted PCE of 1.17 %. However, **2c-2** has a higher efficiency of 1.34 % with a J_{SC} of 4.14 mA cm^{-2} and a V_{OC} of 0.753 [25]. Later, Roncali and coworkers connected a 5-hexyl-2,2'-bithienyl to the axial phenyl ring of compound **2c-2**, thus producing **2d**. This structural modification has little effect on the energy levels and absorptions of **2d** but improves its charge-transport properties, as supported by the electrochemical,

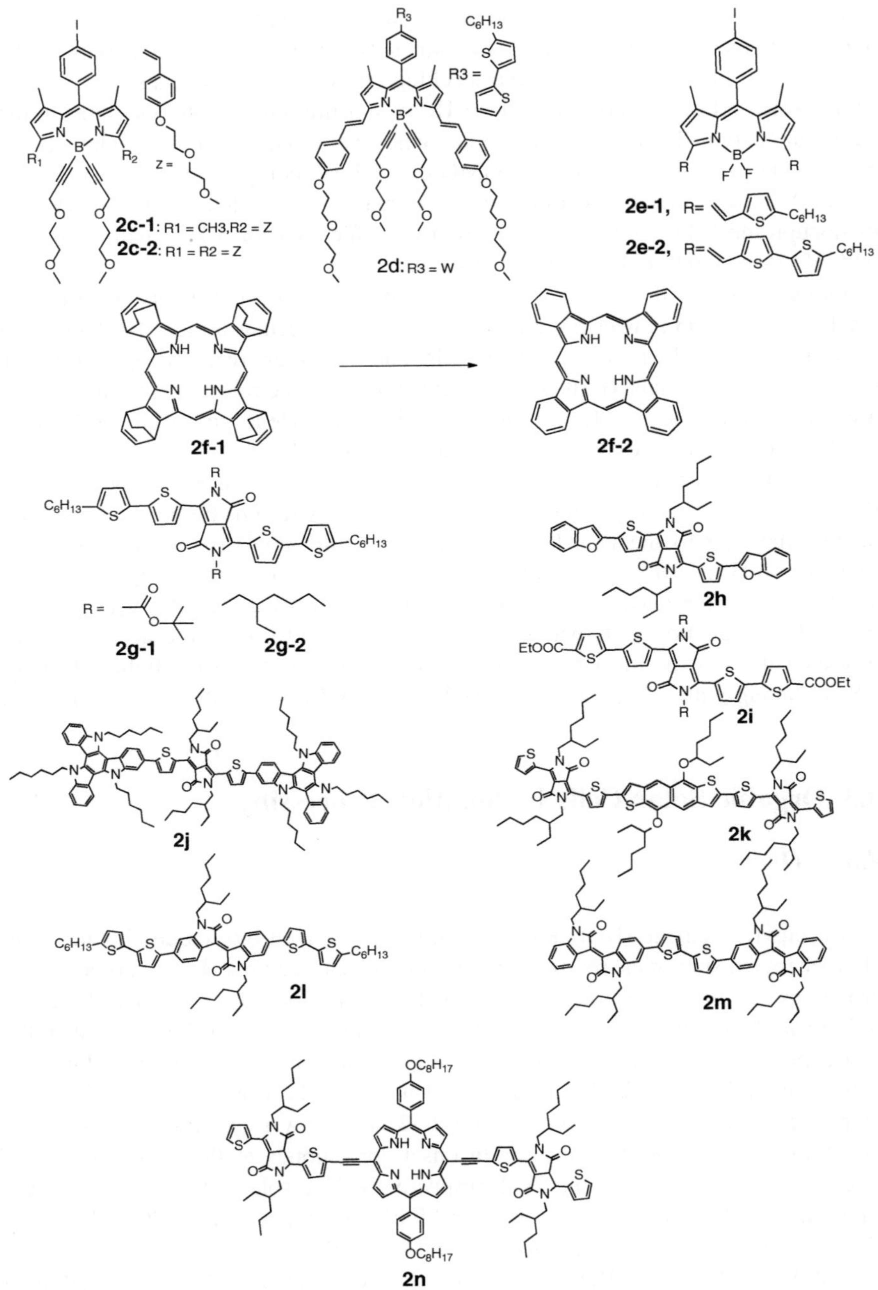

Fig. 4.2 Chemical structures of dyes

optical, and mobility data. Thus improved efficiency of 2.17 % was obtained with a V_{OC} of 0.75 V and J_{SC} of 7.0 mA cm^{-2} [26]. Ziessel and coworkers reported a new series of thienyl-BODIPY dyes (**2e-1** and **2e-2**) and studied their photophysical and electronic properties as well as their performances as donor component in OSCs, with a special focus on the influence of the solubilizing chains. In THF solution, **2e-1** and **2e-2** show two strong absorptions in the 430–450 and 560–750 nm ranges, with extinction coefficients of about 100000 M^{-1} cm^{-1}. In contrast to the mono-thiophene styryl units, red-shifted absorption by 47 nm can be found in **2e-2**. This could be explained by its more conjugated styryl bithiophene arms. With **2e-1** as donor and PCBM as acceptor, a PCE of 1.4 % was obtained with a J_{SC} of 5.84 mA cm^{-2} and V_{OC} of 0.76 V, whereas for **2e-2**, a higher PCE of 4.7 % was obtained with a J_{SC} of 14.2 mA cm^{-2} and a V_{OC} of 0.70 V [27] (Table 4.2).

Nakamura demonstrated new solution-processable three layered p-i-n OSCs (the i meaning an inter-layer contains donor and acceptor components), composed of tetrabenzoporphyrin (BP, **2f**) and silylmethyl [60] fullerene acceptor (SIMEF). The thermally transformable phthalocyanine enables formation of an "ordered" heterojunction rather than BHJ. A impressive PCE of 5.2 % was obtained [28]. The controllable trilayer cells deliver a large J_{SC} (10.3 mA cm^{-2}) and FF (65 %), leading to a respectable overall PCE of 5.2 %.

The diketopyrrolopyrrole (DPP) structure is an ideal aromatic building block for organic electronic materials because of its good photochemical stability, intense light absorption, and an extended π-conjugated framework. In addition, the accessibility of the lactam nitrogens for chemical modification provides a versatile handle for altering its physical properties, such as solid-state packing and solubility [37–40].

Table 4.2 Properties and device characteristics of dyes in Fig. 4.2

Molecules	HOMO/LUMO (eV/eV)	E_g (eV)	V_{oc} (V)	J_{sc} (mA cm^{-2})	FF (%)	PCE (%)	Reference
2c-1	−3.66/−5.69	1.95	0.796	4.43	34.0	1.17	[25]
2c-2	−3.75/−5.56	1.70	0.753	4.14	44.0	1.34	[25]
2d	−3.70/−5.61	1.70	0.750	7.70	0.38	2.17	[26]
2e-1	−3.81/−5.46	1.60	0.760	5.84	31.0	1.40	[27]
2e-2	−3.84/−5.34	1.45	0.700	14.30	47.0	4.70	[27]
2f-2	N.A.	N.A.	0.750	10.50	65.0	5.20	[28]
2g-1	−3.00/−5.03	1.51	0.670	8.42	45.0	2.33	[29]
2g-2	−3.70/−5.20	1.55	0.750	9.20	44.0	3.00	[30]
2h	−3.40/−5.20	N.A.	0.920	10.0	48.0	4.40	[31]
2i	−3.60/−5.33	1.65	0.940	8.55	0.50	4.02	[32]
2j	−3.60/−5.31	1.70	0.63	14.6	0.58	5.30	[33]
2k	−3.68/−5.40	1.72	0.84	11.27	0.42	4.06	[34]
2l	−3.90/−5.50	1.67	0.66	2.40	0.36	0.55	[35, 36]
2m	−3.80/−5.50	1.76	0.95	7.80	0.45	3.31	[35]

Nguyen et al. synthesized a series of SM (**2g-1**, **2g-2**, and **2h**). In molecule **2g-1**, the lactam nitrogens of the DPP unit were protected by *tert*-butyloxycarbonyl (Boc) groups to increase the solubility. This molecule delivers intense absorption in the visible and near-infrared regions and exhibits a hole mobility of $\sim 10^{-6}$ cm^2/V s, as determined by the SCLC (space-charge limited current) model. A 7:3 weight ratio of donor/PCBM ([6]-phenyl C_{61}-butyric acid methyl ester) provided a PCE of 2.3 % under simulated AM 1.5G solar irradiation of 100 mW/cm^2 [29]. To improve the thermal stability and solubility, the Boc groups on the DPP structure in **2g-1** was replaced by ethylhexyl chains, affording **2g-2**. Compared to the Boc derivative, this structure modification does increase its thermal stability and solubility, lowers the HOMO level, and further enhances film forming properties. The high degree of ordering in the pristine donor film is maintained in blended films, and thus good hole motilities were obtained. After thermal annealing (100 °C for 5 min), devices prepared from a **2g-2**: $PC_{70}BM$ (1:1, w/w) blend gave a PCE of 3.0 % with an FF of 0.45 [30]. To tune the electronic and optical properties further, the terminating group of hexylthiophene units in **2g-2** was replaced by benzofuran units, thus affording **2h**. After thermal annealing (100 °C for 10 min), devices prepared from a **2h** 5:$PC_{70}BM$ (6:4, w/w) blend gave a PCE of 4.4 % with an FF of 0.48 [31]. Chen and coworkers reported an ethyl thiophene-2-carboxylate group end-capped DPP molecule, **2i**. Benefiting from its narrow bandgap and lower-lying HOMO, the OSCs based on **2i** showed a broad photovoltaic response range extending to around 750 nm and a very high V_{OC} of 0.94 V, affording a PCE of 4.02 % [32]. Using triazatruxene as the end-capper group and DPP as the central group, Ziessel, Leclerc and coworkers reported the synthesis and photovoltaic properties of dumbbell-shaped solution-processable **2j**. The result suggested that the triazatruxene core is an effective electron-donating unit for OSCs, providing a good trade-off between planarity and solubility. It shows end-to-end π–π interactions in the solid phase and leads to favorable active layer morphologies with low ratios of fullerene acceptors. With a 1:0.75 **2j**:$PC_{70}BM$ weight ratio, solar cells demonstrated a high PCE of 5.3 % with a V_{OC} of 0.63 V and J_{SC} of 14.6 mA cm^{-2} after thermal annealing (110 ° C for 20 min) [33]. Using a naphthadithiophene donor group as the core and DPP as terminal arms, an alternative A-D-A (Acceptor-Donor-Acceptor) structure (**2K**) was employed by Marks et al. Solar cells prepared from a 2**K**:PCBM (1.5:1.02) deliver a high PCE of 4.06 %, with a V_{OC} of 0.84 V, J_{SC} of 11.27 mA cm^{-2}, and FF of 0.42 by annealing at 110 °C for 10 min [34].

Inspired by the prominent device performance of using diketopyrrolopyrrole dyes as photovoltaic donors in BHJ OSCs, isoindigo-based dyes have also been investigated as donor materials in OSCs in recent years. The unique structure of isoindigo delivers a strong electron-withdrawing character and this character benefits from the conjugation of the lactam rings in conjunction with an extended large π-system under the bis-oxindole framework. In 2010, Reynolds and coworkers first utilized isoindigo dyes in OSCs. Using donor-acceptor-donor (D-A-D) and acceptor-donor-acceptor (A-D-A) structures, isoindigo-based dyes (**2l** and **2m**) were synthesized in conjunction with bithiophene as an electron donor. Annealed photovoltaic devices (100 °C) of **2m** deliver modest PCEs of 0.55 %, with a V_{OC} of 0.66 V, J_{SC} of

2.4 mA cm^{-2}, and a low fill factor of 0.36. The BHJ cells made from **2l** performed significantly better than devices made from **2m**. After annealing at 100 °C, solar cells made from **2l** showed a PCE of up to 1.76 %, with a V_{OC} of 0.74 V, J_{SC} of 6.3 mA cm^{-2}, and fill factor of 0.38 [35]. By combining solvent additives with complementary effects, further refinement of the device processing conditions was achieved, and the PCE of **2l** was further increased to near 3.7 % [36].

Inspired by the natural photosynthetic systems that utilize chlorophylls to absorb light, Peng and coworker explored porphyrin derivatives as the donor materials for OSC studies [41–43]. To facilitate the intramolecular charge transport, they introduced ethynylene to link a porphyrin core with different acceptor units. High PCEs of 7.23 % for **2n** with a J_{SC} of 16 mA cm^{-2}, V_{OC} of 0.71 V, and FF of 0.63 were achieved for the solution-processed BHJ OSC [43]. This value is the highest PCE for solution-processed BHJ OSCs based on porphyrin dyes. This result also ranks as one of the best PCEs (over 7 %) for solution-processed BHJ OSCs.

4.3.2 Triphenylamine Derivatives

Organic molecules containing triphenylamine (TPA) units have good solubility because of the propeller structure of TPA units, and high hole mobility. Therefore, the studies of TPA-containing molecules for OSCs, pioneered by Roncali and coworkers [44, 45] attracted wide attention. Here we have divided the TPA-containing molecules into two categories, namely, linear (Fig. 4.3) and star-shaped (Fig. 4.4). With this classification, the structure-property relationships are discussed.

Li and coworkers reported D-A-D type molecules with TPA as donor, benzothiadiazole (BT) as acceptor, and vinylthiophene as the bridge [46]. Based on the blend of **3a** and PCBM (1:1), the PCE was only 0.26 % for the device with a Ba/Al

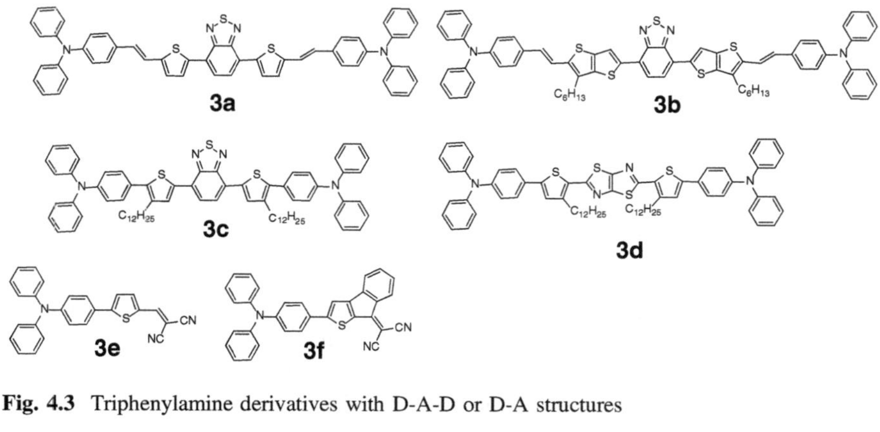

Fig. 4.3 Triphenylamine derivatives with D-A-D or D-A structures

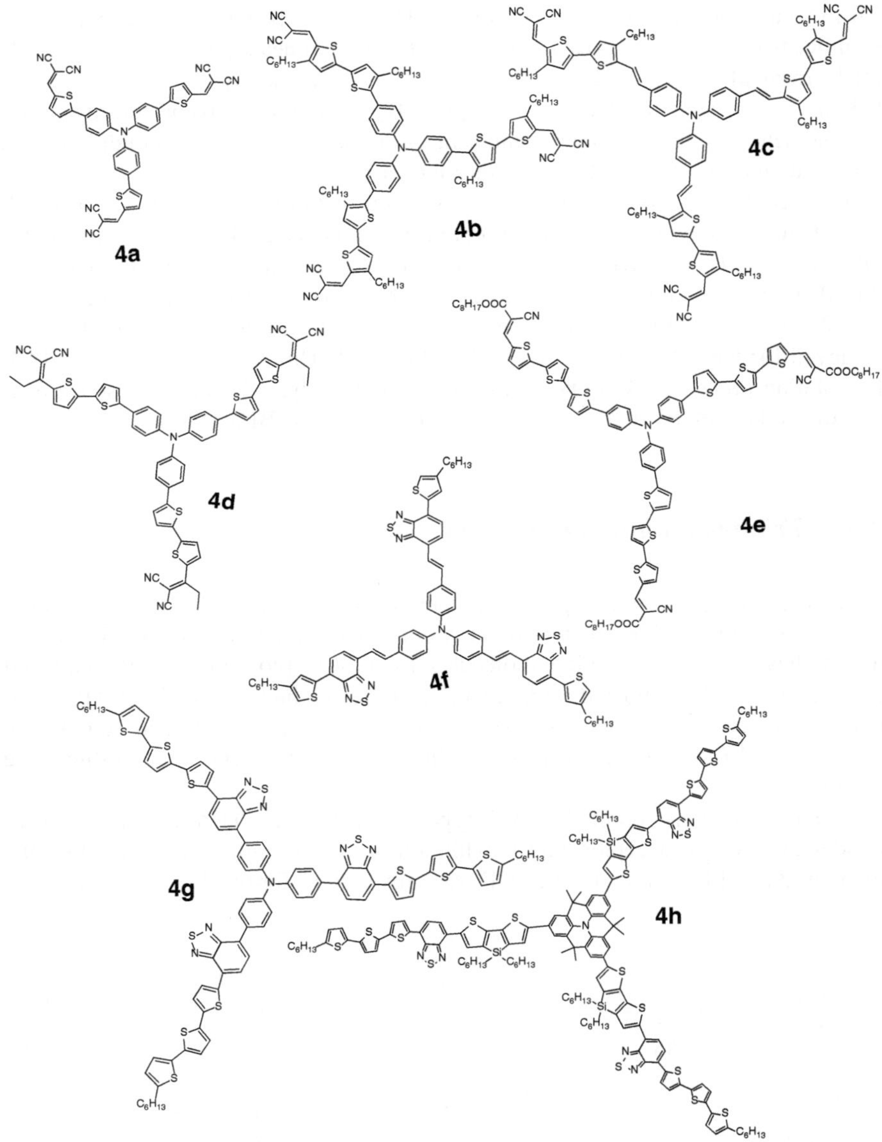

Fig. 4.4 Star-shaped organic molecules with TPA as central unit

cathode. Higher efficiency of 1.44 % with a V_{OC} of 0.74 V was realized by using a fused thiophene as bridge because of a larger conjugation in **3b** [47]. Under the molecular architecture of **3a**, removing of the vinyl group also affords **3c**. This chemical modification brought a larger V_{OC} of 0.86 V together with an improved efficiency of 1.06 % [48]. Replacing the acceptor unit of benzothiadiazole with thiazolothiazole (TTz), another D-A-D type molecule was obtained (**3d**). Because of the weak electron-accepting ability of the TTz unit, **3d** demonstrated a wide bandgap with an optical bandgap of 2.31 eV in film [49]. By tuning donor/acceptor blend ratio variation, OSCs with $PC_{70}BM$ as acceptor delivered perfect phase separation sizes of 10–20 nm and balanced carrier transport, leading to a PCE as high as 3.73 % under thermal annealing.

By combining a triphenylamine electron donor unit with a dicyanovinyl thiophene electron acceptor group, Roncali reported D-A type SM donors **3e** and **3f**. In **3f**, the dicyanovinyl group has been fused to the thiophene by a phenyl ring to modulate the absorption. In bilayer PHJ solar cells (with C_{70} as acceptor), the reference molecule **3e** demonstrated an efficiency of 2.53 % with a V_{OC} of 0.92 V, an FF of 0.42, and J_{SC} of 5.77 mA cm^{-2}, whereas for the molecule **3f** with larger conjugation, a higher efficiency of 2.97 % was obtained with a J_{SC} of 5.32 mA cm^{-2}, a V_{OC} of 0.97 V, and FF of 0.52 under the illuminaiton of AM 1.5G at 90 mW cm^{-2} [50].

Early in 2006, with TPA as core, a family of star-shaped SMs was first reported as donor materials by Roncali and coworkers. For the molecule with TPA as the core derivatized with dicyanovinyl group (**4a**), a PCE of 1.02 % together with a V_{OC} of 0.96 V was obtained [44]. Later, Li et al. reported two star-shaped D-π-A molecules (**4b** and **4c)** as donor components in solution-processed OSC [51]. These materials contain triphenylamine (TPA) as core and donor unit, dicyanovinyl (DCN) as end group and acceptor unit, and bithiophene or **4**, bithiophene vinylene as the conjugated bridge. PCE of the OSC based on **4b** as donor is 1.4 %. Benefitting from the vinylene bridge, the absorption profile of the **4c** film covers a wide wavelength ranging from 380 to 750 nm, and its absorption is red-shifted by ca. 40 nm compared to that of the **4b** film. With a J_{SC} of 7.76 mA cm^{-2} and a V_{OC} of 0.88 V, the PCE of the BHJ OSC based on a composite of **4c** and $PC_{70}BM$ (1:2, w/w) reached 3.0 %. The removing alkyl chain in the bithiophene bridge to the DCN acceptor end group afforded molecule **4d** with improved planarity [52]. By addition of a new additive 4-bromoanisole, devices based on a blend of **4d** and $PC_{70}BM$ demonstrated a PCE of 3.1 % without any post-treatment and further improved to 3.6 % under simulated AM 1.5G. Similar to the alkyl DCN group, the alkyl cyanoacetate acceptor group is another interesting electron-deficient group containing both cyano moiety and alkyl chain, Thus the greatest advantage of this alkyl cyanoacetate acceptor group is to induce internal charge transfer and improved solubility simultaneously [53]. With alkyl cyanoacetate acceptor group as the acceptor end group, Zhan reported another new star-shaped molecule **4e** [54]. Specifically, the use of alkyl cyanoacetate acceptor group can avoid tedious and time-consuming multi-step reactions. Without any post-treatment, the BHJ OSCs based on a **4e**:$PC_{70}BM$ (1:2, w/w) blend afforded a PCE of 3.60 % and an FF of 0.56 (Table 4.3).

Another star-shaped TPA-containing molecule is based on the D-A-D architecture arms. Recently, Li and coworkers reported such a type of molecule containing TPA as core and benzothiadiazole-(4-hexyl) thiophene as arms (**4f**), for application in solution-processed OSCs [55]. A thin film of **4f** shows broad and

Table 4.3 Properties and device characteristics of triphenylamine molecules in Figs. 4.3 and 4.4

Molecules	HOMO/LUMO (eV/eV)	E_g (eV)	V_{oc} (V)	J_{sc} (mA cm^{-2})	FF (%)	PCE (%)	Reference
3a	−3.30/−5.10	1.80	0.74	0.619	33.0	0.19	[46]
3b	−3.42/−5.10	1.64	0.74	5.71	34.0	1.44	[47]
3c	−2.99/−5.16	2.03	0.86	3.23	38.0	1.06	[48]
3d	−2.91/−5.39	2.31	0.91	9.39	43.7	3.73	[49]
3e	−3.79/−5.96	2.03	0.92	5.77	42.0	2.53	[50]
3f	−4.22/−5.96	1.59	0.97	5.32	52.0	2.97	[50]
4a	N.A.	1.78	0.96	3.65	29.0	1.02	[44]
4b	−3.34/−5.22	1.88	0.84	5.21	30.8	1.40	[51]
4c	−3.42/−5.03	1.61	0.88	7.76	43.9	3.00	[51]
4d	−3.41/−5.32	1.89	0.96	7.81	50.0	3.60	[52]
4e	−3.41/−5.32	1.95	0.88	7.30	56.0	3.60	[53]
4f	−3.08/−5.19	1.96	0.85	8.58	32.7	2.39	[55]
4g	−3.11/−5.28	1.90	0.87	9.51	52.0	4.30	[56]
4h	−3.66/−4.94	1.75	0.74	11.34	50.0	4.16	[57]

intense absorptions in the range 300–630 nm. The OSC device based on a blend of **4f** and $PC_{70}BM$ (1:3, w/w) exhibited a J_{SC} of 8.58 mA cm^{-2}, a V_{OC} of 0.85 V, and an FF of 0.327, leading to a PCE of 2.39 %, under the illumination of AM.1.5G, 100 mW cm^{-2}. Similar to **4f**, the removing of the vinyl group affords **4g** with higher efficiency. BHJ OSC cells based on **4g**:$PC_{70}BM$ (1:2, w:w) deliver a J_{SC} of 9.51 mA cm^{-2}, V_{OC} of 0.87 V, FF of 0.52, and PCE of 4.3 %. The PCE value of 4.3 % is among one of the highest reported values for solution-processed BHJ OSCs containing TPA-based small molecules [56]. In the above examples, the TPA unit plays an important roles as electron donor and hole transport mediator, but its propeller structure aroused the issue of intermolecular packing and resulted in lower hole mobility in BHJ active layer of the OSCs. To improve the planarity, Ko, Li, and coworkers introduced dimethylmethylene-bridged TPA core to replace TPA and synthesized **4h**. Relative to its counterpart with TPA as core, the planar star-shaped **4h** showed a slight redshift absorption and stronger absorbance. A note-worthy PCE value of 4.16 % was recorded with a J_{SC} of 11.34 mW cm^{-2} [57].

4.3.3 Oligothiophenes

For oligothiophenes, one synthetic advantage is to be relatively readily extended to longer oligomers. The other is to be appropriately functionalized at the terminal α-positions or the side β-positions. Under these chemical approaches, versatile structures can be obtained and their characteristics relevant to solar cells applications (such as mobilities, energy levels, and solid state packing) can be well controlled. To broaden absorption and improve solubility, branched oligothiophenes were developed [14].

Roncali et al. developed a tetrahedral oligothienyl silane derivative **5a** [58]. Compared with parent linear terthiophene, a 19-nm red-shifted absorption was observed. In blending with PCBM at 1:3 w/w ratio, a moderate PCE of 0.3 % was obtained under the simulated irradiation of AM 1.5G, 80 mW cm^{-2}. Although the device performances are mainly limited by the narrow absorption of the donor, results obtained with this short chain model compound confirmed the interest of the 3D approach. By changing the core to a phenyl group, a four-arm dendrimer with longer thiophene chains was developed [59]. The better conjugation in **5b** brought an improved PCE of 1.3 % using PCBM as acceptor under simulated AM 1.5G illumination. Under the 3D architecture, all-thiophene dendrimer (**5c** Fig. 4.5) with highly branched arms was reported by Peter Bäuerle [60] and used in BHJ solar cells as donor in combination with **PCBM** as acceptor. BHJ solar cells using dendrimer **5c** in a D-A ratio of 1:2 generated a PCE of 1.7 % with a high V_{OC} of 0.97 V. In order to improve the photon harvesting ability further, acceptor groups have been incorporated into the core or the terminal of the branched oligothiophenes. Under such an approach, Bäuerle et al. introduced an electron-deficient pyrazino[2,3 *g*]quinoxaline as core into the dendritic structure [61]. Dendrimer **5d** showed a broad absorption in the range 300–700 nm and a reduced bandgap of 1.7 eV, giving an efficiency of 1.3 %. Kopidakis and coworkers have developed such an oligothiophene (**5e**) with electron-deficient tricyanobenzene as core and electron-rich dendrons [62]. Compared to the control dendrimer with the same structure but without the electron-deficient core (**5e-1**), the appending electron-deficient core in **5e-2** planarizes the structure and lowers the bandgap, thus leading to enhanced structure order in bulk heterojunction films. The optical bandgap of **5e-2** was reduced to 1.8 eV, compared to 2.4 eV for **5e**. Oligomer **5e-2** showed a much higher PCE of 1.1 % compared to only 0.4 % for **5e-1**. Notably, despite its lower bandgap, an expected larger J_{SC} for **5e-2** as comparison to **5e-1** was not recorded. The electron trapping effect in the cyanobenzene core can account for its lower J_{SC}, impeding electron transfer to the acceptor [62]. Another approach for broadening the absorption is introducing acceptors into the terminal of oligothiophenes. Wong et al. incorporated dicyanovinyl units to the terminal branched oligothiophenes in thin films and, accompanied with a strong spectral broadening, the optical bandgap of the dendritic oligothiophene **5f** was reduced to 1.74 eV [63]. Together with a J_{SC} of 4.19 mA cm^{-2}, V_{OC} of 0.97 V, and FF of 0.42, device optimization produced a PCE of 1.72 %. Compared to linear D-A small molecules, however, this family of oligothiophenes usually delivers low FF values in solar cells because of the low mobility, and therefore more studies are needed to improve their FF values and solar cell performances.

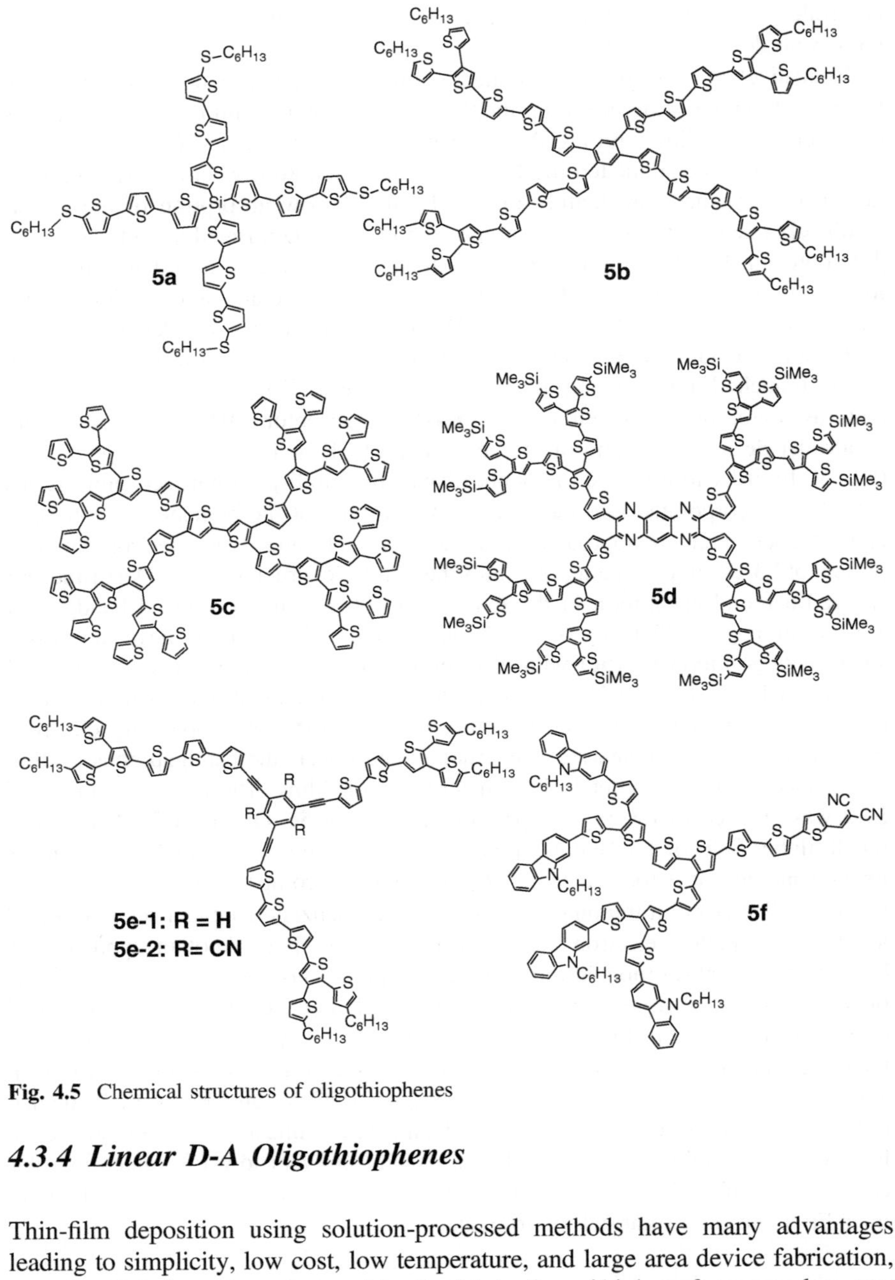

Fig. 4.5 Chemical structures of oligothiophenes

4.3.4 Linear D-A Oligothiophenes

Thin-film deposition using solution-processed methods have many advantages leading to simplicity, low cost, low temperature, and large area device fabrication, meaning high throughput that enables the fabrication of high-performance, low-cost electronics. Chen and coworker developed solution-processed linear oligothiophenes with terminal acceptors (**6a–6d**). They systematically varied the position and density of the alkyl side chain or the terminal acceptor group as well as the core in the as-developed oligothiophenes. Because of the efficient conjugation of the aromatic skeleton and intramolecular charge transfer (between the terminal acceptor unit and the central donor unit), the SMs obtained exhibited high hole mobilities and wide absorptions with high coefficients.

In 2010, Chen and coworkers reported DCV-substituted oligothiophene and used it as donor component for solution-processed OSCs. The BHJ solar cell with (**6a-1** (Fig. 4.6) and PCBM at the ratio of 1:1.4 exhibited a PCE as high as 3.7 % under AM 1.5G illumination in air [64]. Instead of a DCV terminal group, they incorporated the alkyl cyanoacetate terminal group in the oligothiophenes. In these oligothiophenes, by changing the ethyl group to longer octyl and 2-ethylhexyl groups, their absorption maxima in thin films are gradually blue shifted. These SMs all demonstrate high PCEs (4.46–5.08 %) for solution-processed BHJ OSCs. A high PCE of 5.08 % was recorded based on a blend of **6a-2** and PCBM without any special treatment [65]. However, this molecule contains a weak acceptor end unit of cyanoacetate group, which could not contribute too much to the overall light absorption. The change of cyanoacetate group to a stronger acceptor group of 3-ethylrhodanine, oligothiophene of **6a-3** with broader absorption was obtained [66].

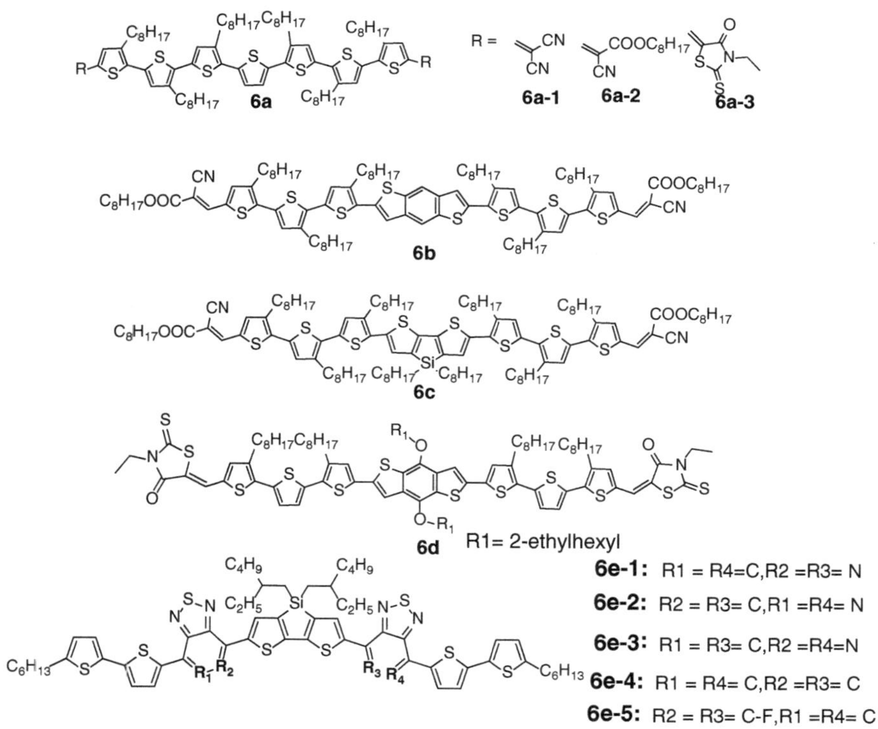

Fig. 4.6 Linear D-A oligothiophenes

The broad absorption resulted in a higher J_{SC} of 13.98 mA cm^{-2}. Together with a V_{OC} of 0.92 V, a high PCE of 6.10 % was obtained by using a blend of **6a-3**:PCBM as the active layer. Also, based on the molecular architecture of **6a-2**, Chen replaced the thiophene core with benzodithiophene (BDT) [67] and dithienosilole unit (DTS) [68] under the consideration of their large and rigid planar aromatic skeleton.

For the oligothiophene containing BDT units [67], this material (**6b**) showed high PCEs for solution-processed BHJ cells. With a high V_{OC} of 0.93 V and an FF of 0.599, a PCE of 5.44 % was recorded in a **6b**/PCBM blend with 1:0.5 weight ratio without any special treatment. For the DTS based oligothiophenes (**6c**), a higher PCE of 5.84 % was obtained along with a high fill factor of 0.64 [68].

With the success of the molecular architecture of **6b**, Chen and coworkers further modified the terminal acceptor group and the soluble side chains. Thus the small molecule, **6d**, was designed and synthesized [69], this molecule containing BDT as the central building block, 3-ethylrhodanine as terminal acceptor group, and dioctylterthiophene as the bridge. A PCE as high as 7.38 % (certified 7.10 %) for the device using **6d** as donor under the illumination of AM 1.5G irradiation, 100 mW cm^{-2} has been realized using the solution process. The efficiency is also comparable with that of the most highly efficient PSCs. This result demonstrated that highly efficient SM BHJ devices could indeed be realized through rational molecular design and device engineering (Table 4.4).

Table 4.4 Properties and device characteristics of oligothiophenes in Figs. 4.5 and 4.6

Molecules	HOMO/LUMO (eV/eV)	E_g (eV)	V_{oc} (V)	J_{sc} (mA cm^{-2})	FF (%)	PCE (%)	Reference
5a	N.A.	2.65	0.85	1.13	24.0	0.29	[58]
5b	N.A.	2.10	0.94	3.35	40.0	1.30	[59]
5c	−3.07/−5.28	2.28	0.97	4.19	42.0	1.70	[60]
5d	−3.90/−5.40	1.70	1.00	3.30	0.38	1.30	[61]
5e-1	−2.80/−5.40	2.40	0.75	2.00	28.0	0.40	[62]
5e-2	−3.30/−5.40	1.80	0.95	2.50	47.0	1.10	[62]
5f	−3.21/−5.06	1.74	0.97	4.19	42.0	1.72	[63]
6a-1	−3.40/−5.10	1.68	0.88	12.4	34.0	3.70	[64]
6a-2	−3.29/−5.13	1.74	0.86	10.74	55.0	5.08	[65]
6a-3	−3.68/−5.21	1.72	0.92	13.98	47.4	6.10	[66]
6b	−3.54/−5.11	1.83	0.93	9.77	59.9	5.44	[67]
6c	−3.26/−4.95	1.73	0.80	11.51	64.0	5.84	[68]
6d	−3.27/−5.02	1.74	0.93	12.21	65.0	7.38	[69]
6e-1	−3.72/−5.22	1.50	0.78	14.4	59.3	6.70	[70]
6e-2	−3.72/−5.26	1.50	0.73	12.70	60.0	5.56	[70]
6e-3	−3.78/−5.30	1.52	0.72	9.80	45.0	3.16	[70]
6e-4	−3.57/−5.15	1.58	0.83	0.90	25.8	0.19	[70]

Bazan et al. developed a series of isomorphic, solution-processable molecules, comprising electron rich 2-hexylbithiophene and dithienosilole as the donor [70, 71]. In these molecules, electron-deficient acceptor groups of 2,1,3-benzothiadiazole (BT) and [1, 2, 5] thiadiazolo[3,4-*c*]pyridine (PT) were employed as the building blocks (**6e**). In their study, by comparing BT- and PT-based SMs, the effect and significance of the pyridyl N-atom in the acceptors (BT or PT) was indentified, and the role of the regiochemistry on the device performance was disclosed. In solid film, when blended with fullerene acceptor ($PC_{70}BM$), the BT-based SM of **6e-4** shows no observable crystalline phase, and thus a poor efficiency of 0.18 % was obtained. For PT-based SMs, the PT regiochemistry within the aromatic skeleton was found to be close related to the solid state packing and BHJ cell efficiency.

When the pyridyl N-atoms in **6e-1** (**6e-2**) is symmetrically located at the proximal/proximal position (distal/distal position) relative to the central DTS donor unit, when blended with fullerene acceptors ($PC_{70}BM$), the composite film delivers a highly ordered nanomorphology, leading to high PCEs up of 7 % (5.6 %). However, relative to the central DTS donor unit, when the position of the pyridyl N-atoms (**6e-3**) changed from favored proximal/proximal configuration (distal/distal position) to less symmetric distal/proximal configuration, the self-assemble ability of the materials were reduced. Together with a low J_{SC} and FF value, the best PCE was only 3.2 % [70, 71].

Notably, the high efficiency of **6e-1** was achieved by employing metal oxides, i.e., MoO_X, as an anode buffer layer. Diminished performance was obtained using a PEDOT:PSS interlayer which is because of the protonation of the PT pyridyl nitrogen caused by the acidic nature of PEDOT:PSS. To remove this site sensitive group, 5-fluorobenzo[*c*] [1, 2, 5] thiadiazole (FBT) was used as the acceptor unit, affording **6e-5** [72]. With no lone pairs of electrons, the advantage of FBT is that of providing an electron-deficient functionality and being prone to participate in acid/base reactions. With PEDOT:PSS as anode buffer layer, a high efficiency of 7.0 % was obtained (V_{OC} = 0.81 V, J_{SC} = 12.8 mA cm^{-2}, and FF = 0.68), with the treatment of solution additive of DIO (0.4 vol. %), followed by thermal annealing at 70 °C [72]. By inserting an optical spacer (zinc oxide) between the active layer and the top Al electrode, its efficiency can be further improved to 8.9 %, which is comparable to that of polymer counterparts. The function of the ZnO buffer layer is to improve the light-harvesting of the active layer, increase the carrier collection efficiency, serve as a hole blocking layers, and reduce the carrier recombination at the interlayer [73]. To improve the efficiency further, Li and coworker extended the two-dimensional-conjugation concept to small molecule-based solar cells. Their motivation is inspired by the advantage of two-dimensional (2D)-conjugated polymers, such as main chain and side chain constructed broad absorptions (contributed by both the main chains and conjugated side chains) and 2D charge transport features [74–76] (Table 4.5).

Thus, with thiophene (**7c**) or bithiophene (**7a**) as conjugated-bridges, two solution-processable A-D-A-type SMs containing thienyl-substituted benzodithiophene (BDTT) as central and donor building block, and indenedione (ID) as

Table 4.5 Properties and device characteristics of oligothiophenes in Fig. 4.7

Molecules	HOMO/LUMO (eV/eV)	E_g (eV)	V_{oc} (V)	J_{sc} (mA cm^{-2})	FF (%)	PCE (%)	Reference
7a	−3.52/−5.16	1.60	0.92	11.05	66.4	6.75	[77]
7b	−3.52/−5.16	1.60	0.92	8.58	64.8	5.11	[77]
7c	−3.56/−5.19	1.61	1.03	10.07	54.7	5.67	[77]
7d	−3.56/−5.18	1.59	0.91	9.47	48.2	4.15	[77]
7e	−3.56/−5.18	1.83	0.92	6.89	63.0	4.00	[78]
7f-1	−3.27/−5.02	1.74	0.93	12.21	65.0	7.38	[7]
7f-2	−3.29/−5.06	1.77	0.96	11.92	59.4	6.79	[7]
7f-3	−3.27/−5.02	1.72	0.93	13.17	66.3	8.12	[7]
7f-4	−3.29/−5.07	1.76	0.92	12.09	72.1	8.02	[7]
7g	−3.45/−5.23	1.65	0.84	11.97	57.6	5.79	[79]

acceptor end groups, were designed and synthesized as photovoltaic donor materials in OSCs [77]. Their results show that solution absorptions of **7a** and **7c** with thiophene conjugated side chains are significantly enhanced in comparison with those of **7b** and **7d** without conjugated side chains. Further investigation shows that, the photovoltaic performance of **7a** and **7c** is also better than that of the corresponding molecules (**7b** and **7d**) with alkoxy side chains on BDT units. Solar cells were fabricated based on the organic molecules/$PC_{70}BM$ (1.5:1, w/w), the PCEs of the solar cells being 6.75 % for **7a**, 5.67 % for **7c**, 5.11 % for **7b**, and 4.15 % for **7d** under the illumination of AM 1.5G, 100 mW cm^{-2}. Instead of attaching the conjugated side-chain to the central BDT unit, Cui et al. [64] modified the thiophene linkage with conjugated side chains to synthesize the molecule **7e**. The BHJ OSCs based on **7e/PCBM** (1:0.5, w/w) delivers a PCE of 4.0 % together with a high V_{oc} of 0.92 V and a relatively high FF of up to 0.63 without any post-treatment [78].

Chen et al. also synthesized a series of SMs with conjugated side chains on the central BDT unit and investigated the effect of different conjugated side chains on the photovoltaic properties [7]. Among these molecules, **7f-3** gave the highest efficiency of 8.12 %. Notably, compared with the other three molecules, the highest J_{SC} is consistent with its most red-shifted absorption and lowest bandgap. Compared to **7f-3**, the bulkier bithiophene substituents in **7f-4** on BDT units side chains caused blue shift absorption. The OSC devices based on **7f-1** or the other three SMs all deliver high V_{OC} values over 0.9 V. For **7f-2** with long-alkyl-chain substituent on the thiophene units, the bulk effect of long alkyl chains can account for its especially high V_{OC} of 0.96 V. Also, based on the BDTT units, Lin et al. [79] and Huang et al. [21] independently reported a linear D-A SMs (**7g**) containing 5-alkylthiophene-2-yl-substituted BDT as core and DPP as arms. Although the device fabrication process for both sets of work is slightly different, a

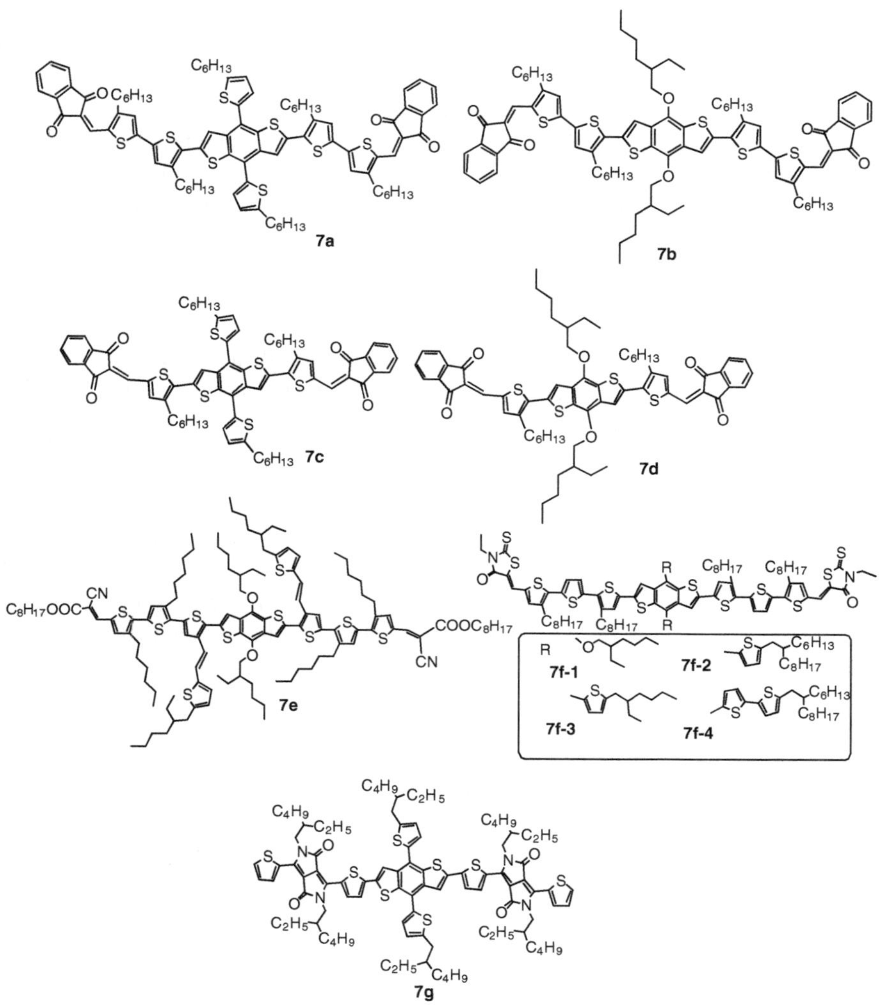

Fig. 4.7 A-D-A structured molecules with conjugated side chains

high efficiency of over 5 % was achieved for both cases, demonstrating that **7g** is also a promising photovoltaic donor material for solution-processed BHJ OSCs.

4.3.5 Organic Molecule Acceptors

The widely used photovoltaic acceptors are fullerene derivatives, such as PCBM and newly developed ICBA (fullerene bis-adducts) [80]. The unique features of fullerene acceptors include deep lower-lying LUMO (~3.70−4.2 eV), high electron mobility, reversible reduction with the ability to accept up to six electrons, and ultra-fast three-dimensional charge transfer. The aim of developing non-fullerene acceptors is to overcome their shortcomings associated with the fullerenes [81, 82]. Fullerene derivatives have disadvantages because of the weak absorption capacity in the visible range, high cost, and the limited variation range of LUMOs. Another motivation for developing n-type small molecules for photovoltaic acceptors is to overcome its V_{OC} limitation (ca. 1.1 V) [83]. In order to improve the PCEs of OSCs further, scientists have started to utilize n-type molecules with strong and broader absorptions and/or tunable LUMO energy levels as photovoltaic acceptors.

Current n-type organic semiconductors used for acceptors can be roughly classified into two types: (1) D-A structured molecules with a strong electron-withdrawing (A) unit and (2) naphthalenediimide (NDI) and perylene diimide (PDI) derivatives [84]. The first bilayer PHJ cells were reported by Tang, using Cu-phthalocyanine as the donor component and perylene-3,4,9,10-bis-benzimidazole (**8a**) as the acceptor component using the vacuum deposition method. Under AM2 conditions (75 mW cm^2), an efficiency of 0.95 % was obtained with an impressive FF of 65 % [85]. By incorporating soluble alkyl chain on the imide nitrogen atoms, the pentyl substituted PDI derivative (**8b**) was obtained. By blending with P3HT, corresponding OSC devices showed a J_{SC} of 1.65 mA cm^{-2}, V_{OC} of 0.45 V, FF of 0.34, and PCE of 0.25 % [86]. Notably, by changing the donor material with a narrow bandgap small molecule (**6e-5**), a high efficiency of 3.0 % was obtained with a V_{OC} of 0.78 V, a J_{SC} of 7.4 mA cm^{-2}, and an FF of 0.52 [87]. This efficiency value is one of the highest PCEs for a BHJ OSCs utilizing a non-fullerene acceptor at that time. The dramatic change in efficiency induced by the donor components indicated that the compatibility between donor and acceptor is very important.

To improve the performance of perylene diimides (PDI)-based acceptors further, one issue to be overcome is the over-strong aggregation behavior. This behavior leads to formation of phase separated acceptor domains (typically over 100 nm) in the active layer. This large domain size is largely beyond the efficient exciton diffusion length (normally 20–30 nm) [88, 89]. Thus, within these large domains, the excitons generated are strongly trapped, and this behavior leads to poor device performance. Reducing the aggregation size of the acceptor domains is thus very important to improve the efficiency of the PDI-based OSCs. To reduce the over-strong aggregation, Zhan et al. [90] recently developed twisting PDI derivative (**8c-2**). Compared to its monomeric counterpart **8c-1**, the twisting PDI of **8c-2** demonstrated significant reduction in its aggregation size. With 2D conjugated polymer PBDTTT-CT as donor, BHJ OSC was fabricated with 5 % DIO as additive. PDI acceptor of **8c-2** gave a best PCE of 4.03 %, whereas **8c-1** demonstrated a poor PCE of 0.13 %. This high PCE value is an outstanding result for non-fullerene photovoltaic acceptors. The result showed that, by just using n-type materials as photovoltaic acceptors, highly efficient OSC cells can also be fabricated by the modification of the structures of the PDIs (Table 4.6).

Table 4.6 Properties and device characteristics of oligothiophenes in Fig. 4.8

Molecules	HOMO/LUMO (eV/eV)	E_g (eV)	V_{oc} (V)	J_{sc} (mA cm^{-2})	FF (%)	PCE (%)	Reference
8b	N.A.	N.A.	0.78	7.40	52.0	3.0	[87]
8c-1	−3.74/−5.98	2.03	0.97	0.33	41.8	0.13	[90]
8c-2	−3.84/−5.65	1.69	0.85	8.86	54.1	4.03	[90]
8d	−4.10/−5.50	1.57	0.82	3.51	52.0	1.50	[91]
8e	−3.49/−5.87	N.A.	0.67	1.80	37.0	0.45	[92]
8f	−3.30/−5.8	N.A.	0.96	4.70	56.0	2.54	[93]
8g	−3.44/−6.27	2.83	0.95	6.35	0.48	2.90	[94]
8h	−3.70/−5.30	1.59	0.79	5.14	44.0	1.80	[95]
8i	−3.70/−5.40	N.A.	0.54	4.85	54.7	1.43	[96]
8j	−4.10/−5.90	N.A.	0.48	5.72	0.57	1.57	[97]
8k	−3.26/−5.26	1.85	1.18	2.68	37.9	1.20	[98]
8l	−3.28/−5.30	1.83	0.97	4.91	43.0	2.05	[99]
8m	−3.75/−5.95	2.10	0.95	3.92	67.0	2.43	[100]
8n	−3.79/−5.40	1.73	0.65	3.09	60.0	1.21	[101]

As a structural analogue to PDI, naphthalene diimide (NDI) is also attractive for constructing n-type molecular acceptors. One of the challenges for developing NDI-based acceptors is to overcome their absorption limitations. To broaden the absorption of NDI-based acceptors, Jenekhe and coworkers incorporated oligothiophene at the 2,6-positions of NDI to afford: **8d**. With this D-A motif, its absorption was extended to 790 nm. After thermal annealing of the cells at 100 °C for 10 min and using diiodooctane (0.2 %) as additive, the optimized cells with P3HT as donor and **8d** as acceptor showed a PCE of 1.5 % [91].

Besides the organic molecule acceptors based on PDI and NDI, some other n-type non-fullerene acceptors have also been designed in recent years. In order to function as n-type semiconductors, electron-deficient groups such as cyano and imide groups are introduced into the molecular architectures. Sellinger et al. reported non-fullerene acceptor based on aromatics (biphenyl, benzothiadiazole, fluorene) flanked with 2-vinyl-4,5-dicyanoimidazole [92]. Of the imidazole-based materials, **8e** has the most promising properties for OSC applications. With P3HT as donor, a PCE of 0.45 % was obtained with a V_{OC} of 0.67 V. Later, they modified the 2-vinyl-4,5-dicyanoimidazole group with vinylimides (phthalimide and naphthalimide). Of these two imide acceptors using P3HT as donor component, the phthalimide derivative of **8f** delivered a high V_{OC} of 0.96 V together with a maximum PCE of 2.54 %. This value is greater than the device (PCE = 0.1 %) using NI-BT (**8e**) as acceptor [93]. Pei and coworkers developed a series of fluoranthene-fused imide derivatives (**8g**) as non fullerene acceptors [94, 102]. With P3HT as donor, the PCEs vary from 2.14 to 2.89 %. The highest PCE of 2.89 % was achieved when **8g-2** served as the acceptor. Further studies show that the J_{SC} values of the inverted BHJ OSCs is determined by electron mobility of the blends

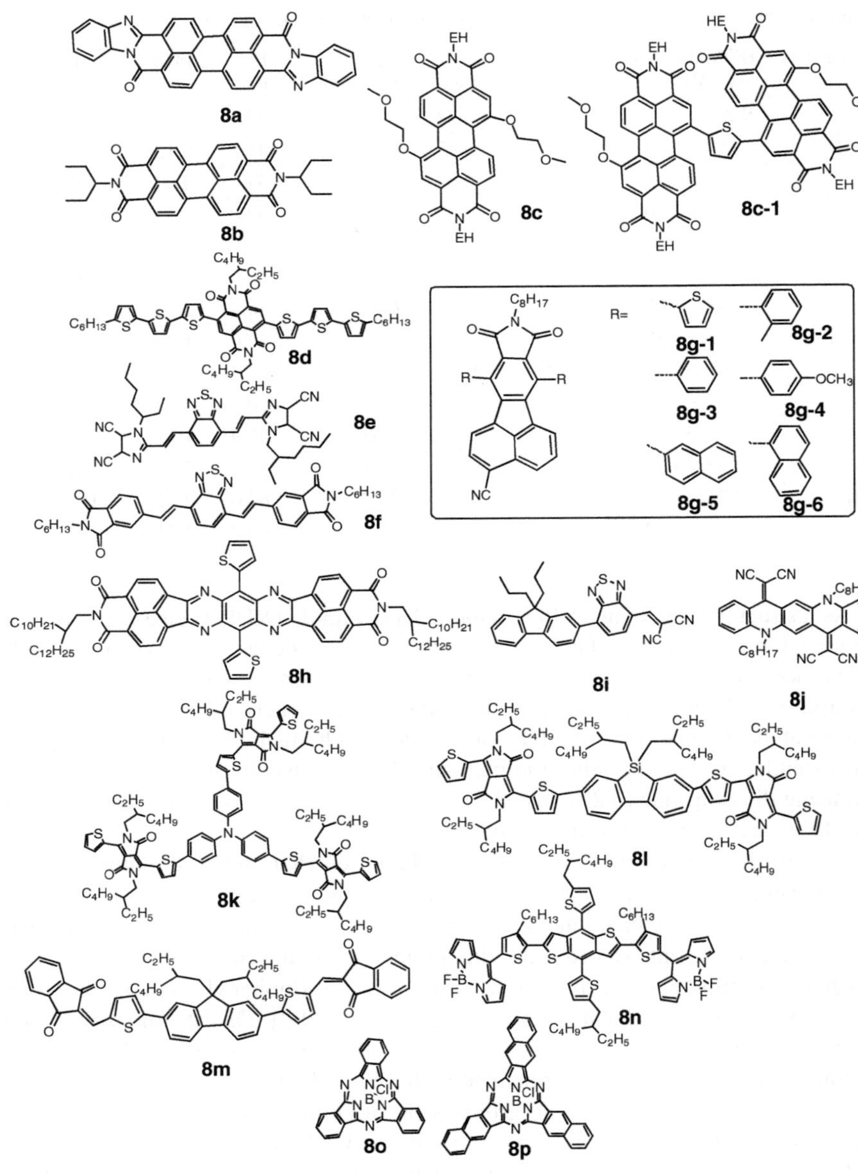

Fig. 4.8 Non-fullerene organic molecule acceptors

containing different acceptors [94]. Jenekhe and coworkers reported a series of tetraazabenzodifluoranthene diimides (BFIs), molecules containing a structure-tunable tetraazaanthracene core and two naphthalene imide units [95]. The 11-ring BFI system has an even larger aromatic skeleton than NDI and PDI. This larger aromatic skeleton is beneficial to extend the π conjugation, promote orbital overlapping, and improve carrier mobilities. By changing the substituents connected at the core position, the BFIs delivered large electron affinities (3.6–4.3 eV) with attractive low optical bandgaps of 2.5–1.6 eV. Using **8h** as acceptor and P3HT as donor, a PCE of 1.80 % was realized together with a V_{OC} of 0.79 V, a J_{SC} of 5.14 mA cm^{-2}, and an FF of 0.44.

Dicyanovinylene (DCV)-substituted arenes recently received focused attention as acceptors. Meredith and coworkers prepared the DCV-substituted fluorene-benzothiadiazole-based oligomer (**8i**) as a non-fullerene acceptor [96, 103]. **8i** has an optical gap of 1.7 eV and an electron affinity close to the standard fullerene acceptor (PCBM). The P3HT:**8i** bulk heterojunction solar cells give an efficiency of 1.43 % with a V_{OC} of 0.54 V, J_{SC} of 4.85 mA cm^{-2}, and FF of 0.547 [96].

Wang developed a class of DCV-substituted quinacridone derivatives [97]. This family of SMs showed intense absorption in the region from 650 to 700 nm where the donor component P3HT has a weak absorption. Among the quinacridone acceptors, **8j** demonstrated a best PCE of 1.57 % (J_{SC} = 5.7 mA cm^{-2}, V_{OC} = 0.48 V, and FF = 0.57).

In search of non-fullerene materials, chemical modified DPP units also produced promising acceptors. Under a 3D star-shaped structure, Lin et al. reported a new acceptor containing triphenylamine as core and diketopyrrolopyrrole as arm. Using P3HT as donor, solution-processed BHJ OSCs-based **8k** delivered a PCE of 1.20 %, and a high V_{OC} of 1.18 V, among the highest values reported for single junction organic solar cells [98]. Higher efficiency non-fullerene acceptor (**8l**) was realized by replacing the triphenylamine unit in **8k** with the dibenzosilole (DBS) unit. With the solvent annealing approach, solution-processed OSCs based on the P3HT:DBS-2DPP blend showed a PCE of 2.05 %, indicating that **8l** is a promising non-fullerene acceptor [99]. Watkins et al. presented an indandione-derived small molecule (**8m**) as acceptor [100]. With P3HT as donor, a PCE of 2.4 % with high V_{OC} of 0.95 V was realized. In particular, their findings show, besides the energy offset, that the electronic coupling should be considered, which is an extremely important parameter in the design of non-fullerene electron acceptors [100]. Thayumanavan et al. reported a series of A-D-A molecules containing 4,4-difluoro-4-bora-3*a*,4*a*-diaza-*s*-indacene as terminal acceptor moieties connected at the meso position. Deep LUMO energy levels and strong visible absorption make them good photovoltaic acceptors in BHJ OSCs. With P3HT as acceptor, inverted cells based on **8n** as acceptor shows a PCE of 1.51 %.

To use the multiple complementary absorbers better, Cnops et al. [104] developed multilayer cascade architecture. The three-layer structure contains two n-type non-fullerene acceptors (**8o** and **8p**) and a donor (**1d**). In the device, an efficient two-step exciton dissociation process was facilitated by this energy-relay cascade. All the three complementary absorbing materials contributed to the photocurrents, thus leading to a quantum efficiency above 75 % between 400 and 720 nm. For a fullerene-free device, the PCE of 8.4 % obtained (with a V_{OC} of 0.96 V, J_{SC} of 14.5 mA cm^{-2} and FF = 0.61) is extraordinarily high. This value even exceeds the

PCEs of previously reported fullerene-based OSCs. For improving the performance of non-fullerene-based devices, the new device architecture is an important alternative to conventional device structure.

4.4 Conclusion and Future Perspectives

In this chapter we have summarized the recent progress of OSCs involving small molecule donors, non-fullerene organic molecule acceptors, and donor–acceptor dyad systems for vacuum and solution-processed OSCs. The disclosed structure–property relationships of representative photovoltaic materials is beneficial for new material development. Some chemical strategies used for tailoring the photophysics properties, such as energy levels, the absorption, and charge mobilities, are also discussed. However, further challenges for developing SM OSCs are in-depth understanding of the relationship between their molecular structures, solid state packing, and photovoltaic properties. At the same time, the synthesis of photovoltaic material with high yields and large scale and good processability without using chlorinated solvents (such as chloroform, chlorobenzene, and dichlorobenzene) is also important for practical applications. Under the disclosed structure–property correlations, developing new small molecules in the context of device durability would certainly facilitate widespread application of this photovoltaic technology.

References

1. Lin Y, Li Y, Zhan X (2012) Small molecule semiconductors for high-efficiency organic photovoltaics. Chem Soc Rev 41:4245–4272
2. Mishra A, Baeuerle P (2012) Small molecule organic semiconductors on the move: promises for future solar energy technology. Angew Chem Int Ed 51:2020–2067
3. Li Y, Guo Q, Li Z, Pei J, Tian W (2010) Solution processable D-A small molecules for bulk-heterojunction solar cells. Energy Environ Sci 3:1427–1436
4. Walker B, Kim C, Nguyen T-Q (2011) Small molecule solution-processed bulk heterojunction solar cells. Chem Mater 23:470–482
5. Zhang F, Wu D, Xu Y, Feng X (2011) Thiophene-based conjugated oligomers for organic solar cells. J Mater Chem 21:17590–17600
6. Potscavage WJ, Sharma A, Kippelen B (2009) Critical interfaces in organic solar cells and their influence on the open-circuit voltage. Acc Chem Res 42:1758–1767
7. Zhou J, Zuo Y, Wan X, Long G, Zhang Q, Ni W, Liu Y, Li Z, He G, Li C, Kan B, Li M, Chen Y (2013) Solution-processed and high-performance organic solar cells using small molecules with a benzodithiophene unit. J Am Chem Soc 135:8484–8487
8. Liu Y, Chen C-C, Hong Z, Gao J, Yang Y, Zhou H, Dou L, Li G (2013) Solution-processed small-molecule solar cells: breaking the 10% power conversion efficiency. Sci Rep 3
9. Heliatek GmbH Press Release. http://www.heliatek.com
10. Sun Z, Ye Q, Chi C, Wu J (2012) Low band gap polycyclic hydrocarbons: from closed-shell near infrared dyes and semiconductors to open-shell radicals. Chem Soc Rev 41:7857–7889

11. Yoo S, Domercq B, Kippelen B (2004) Efficient thin-film organic solar cells based on pentacene/C[sub 60] heterojunctions. Appl Phys Lett 85:5427–5429
12. Wagner J, Gruber M, Hinderhofer A, Wilke A, Broeker B, Frisch J, Amsalem P, Vollmer A, Opitz A, Koch N, Schreiber F, Bruetting W (2010) High fill factor and open circuit voltage in organic photovoltaic cells with diindenoperylene as donor material. Adv Funct Mater 20:4295–4303
13. Hirade M, Adachi C (2011) Small molecular organic photovoltaic cells with exciton blocking layer at anode interface for improved device performance. Appl Phys Lett 99:153302
14. Sakai J, Taima T, Yamanari T, Saito K (2009) Annealing effect in the sexithiophene: C70 small molecule bulk heterojunction organic photovoltaic cells. Sol Energy Mater Sol Cells 93:1149–1153
15. Fitzner R, Reinold E, Mishra A, Mena-Osteritz E, Ziehlke H, Koerner C, Leo K, Riede M, Weil M, Tsaryova O, Weiss A, Uhrich C, Pfeiffer M, Baeuerle P (2011) Dicyanovinyl-substituted oligothiophenes: structure-property relationships and application in vacuum-processed small-molecule organic solar cells. Adv Funct Mater 21:897–910
16. Schulze K, Riede M, Brier E, Reinold E, Bauerle P, Leo K (2008) Dicyanovinyl-quinquethiophenes with varying alkyl chain lengths: investigation of their performance in organic devices. J Appl Phys 104:047511
17. Fitzner R, Mena-Osteritz E, Mishra A, Schulz G, Reinold E, Weil M, Koerner C, Ziehlke H, Elschner C, Leo K, Riede M, Pfeiffer M, Uhrich C, Baeuerle P (2012) Correlation of pi-conjugated oligomer structure with film morphology and organic solar cell performance. J Am Chem Soc 134:11064–11067
18. Wang S, Mayo EI, Perez MD, Griffe L, Wei G, Djurovich PI, Forrest SR, Thompson ME (2009) High efficiency organic photovoltaic cells based on a vapor deposited squaraine donor. Appl Phys Lett 94:233304
19. Wei G, Xiao X, Wang S, Zimmerman JD, Sun K, Diev VV, Thompson ME, Forrest SR (2011) Arylamine-based squaraine donors for use in organic solar cells. Nano Lett 11:4261–4264
20. Chen Y-H, Lin L-Y, Lu C-W, Lin F, Huang Z-Y, Lin H-W, Wang P-H, Liu Y-H, Wong K-T, Wen J, Miller DJ, Darling SB (2012) Vacuum-deposited small-molecule organic solar cells with high power conversion efficiencies by judicious molecular design and device optimization. J Am Chem Soc 134:13616–13623
21. Huang J, Zhan C, Zhang X, Zhao Y, Lu Z, Jia H, Jiang B, Ye J, Zhang S, Tang A, Liu Y, Pei Q, Yao J (2013) Solution-processed DPP-based small molecule that gives high photovoltaic efficiency with judicious device optimization. ACS Appl Mater Interfaces 5:2033–2039
22. Silvestri F, Irwin MD, Beverina L, Facchetti A, Pagani GA, Marks TJ (2008) Efficient squaraine-based solution processable bulk-heterojunction solar cells. J Am Chem Soc 130:17640–17641
23. Bagnis D, Beverina L, Huang H, Silvestri F, Yao Y, Yan H, Pagani GA, Marks TJ, Facchetti A (2010) Marked alkyl- vs alkenyl-substitutent effects on squaraine dye solid-state structure, carrier mobility, and bulk-heterojunction solar cell efficiency. J Am Chem Soc 132:4074–4075
24. Mayerhöffer U, Deing K, Gruß K, Braunschweig H, Meerholz K, Würthner F (2009) Outstanding short-circuit currents in BHJ solar cells based on NIR-absorbing acceptor-substituted squaraines. Angew Chem Int Ed 48:8776–8779
25. Rousseau T, Cravino A, Bura T, Ulrich G, Ziessel R, Roncali J (2009) BODIPY derivatives as donor materials for bulk heterojunction solar cells. Chem Commun 1673–1675
26. Rousseau T, Cravino A, Ripaud E, Leriche P, Rihn S, De Nicola A, Ziessel R, Roncali J (2010) A tailored hybrid BODIPY-oligothiophene donor for molecular bulk heterojunction solar cells with improved performances. Chem Commun 46:5082–5084
27. Bura T, Leclerc N, Fall S, Lévêque P, Heiser T, Retailleau P, Rihn S, Mirloup A, Ziessel R (2012) High-performance solution-processed solar cells and ambipolar behavior in organic field-effect transistors with thienyl-BODIPY scaffoldings. J Am Chem Soc 134:17404–17407
28. Matsuo Y, Sato Y, Niinomi T, Soga I, Tanaka H, Nakamura E (2009) Columnar structure in bulk heterojunction in solution-processable three-layered p-i-n organic photovoltaic devices

using tetrabenzoporphyrin precursor and silylmethyl 60 fullerene. J Am Chem Soc 131:16048–16050
29. Tamayo AB, Walker B, Nguyen T-Q (2008) A low band gap, solution processable oligothiophene with a diketopyrrolopyrrole core for use in organic solar cells. J Phys Chem C 112:11545–11551
30. Tamayo AB, Dang X-D, Walker B, Seo J, Kent T, Nguyen T-Q (2009) A low band gap, solution processable oligothiophene with a dialkylated diketopyrrolopyrrole chromophore for use in bulk heterojunction solar cells. Appl Phys Lett 94:103301
31. Walker B, Tomayo AB, Dang X-D, Zalar P, Seo JH, Garcia A, Tantiwiwat M, Nguyen T-Q (2009) Nanoscale phase separation and high photovoltaic efficiency in solution-processed, small-molecule bulk heterojunction solar cells. Adv Funct Mater 19:3063–3069
32. Chen M, Fu W, Shi M, Hu X, Pan J, Ling J, Li H, Chen H (2013) An ester-functionalized diketopyrrolopyrrole molecule with appropriate energy levels for application in solution-processed organic solar cells. J Mater Chem A 1:105–111
33. Bura T, Leclerc N, Bechara R, Lévêque P, Heiser T, Ziessel R (2013) Triazatruxene-diketopyrrolopyrrole dumbbell-shaped molecules as photoactive electron donor for high-efficiency solution processed organic solar cells. Adv Energ Mater 1118–1124
34. Loser S, Bruns CJ, Miyauchi H, Ortiz RP, Facchetti A, Stupp SI, Marks TJ (2011) A naphthodithiophene-diketopyrrolopyrrole donor molecule for efficient solution-processed solar cells. J Am Chem Soc 133:8142–8145
35. Mei J, Graham KR, Stalder R, Reynolds JR (2010) Synthesis of isoindigo-based oligothiophenes for molecular bulk heterojunction solar cells. Org Lett 12:660–663
36. Graham KR, Wieruszewski PM, Stalder R, Hartel MJ, Mei J, So F, Reynolds JR (2012) Improved performance of molecular bulk-heterojunction photovoltaic cells through predictable selection of solvent additives. Adv Funct Mater 22:4801–4813
37. Robb MJ, Ku SY, Brunetti FG, Hawker CJ (2013) A renaissance of color: new structures and building blocks for organic electronics. J Polym Sci Pol Chem 51:1263–1271
38. Walker B, Liu J, Kim C, Welch GC, Park JK, Lin J, Zalar P, Proctor CM, Seo JH, Bazan GC, Nguyen T-Q (2013) Optimization of energy levels by molecular design: evaluation of bis-diketopyrrolopyrrole molecular donor materials for bulk heterojunction solar cells. Energy Environ Sci 6:952–962
39. Qu S, Tian H (2012) Diketopyrrolopyrrole (DPP)-based materials for organic photovoltaics. Chem Commun 48:3039–3051
40. Liu J, Walker B, Tamayo A, Zhang Y, Nguyen T-Q (2013) Effects of heteroatom substitutions on the crystal structure, film formation, and optoelectronic properties of diketopyrrolopyrrole-based materials. Adv Funct Mater 23:47–56
41. Li L, Huang Y, Peng J, Cao Y, Peng X (2013) Enhanced performance of solution-processed solar cells based on porphyrin small molecules with a diketopyrrolopyrrole acceptor unit and a pyridine additive. J Mater Chem A 1:2144–2150
42. Huang Y, Li L, Peng X, Peng J, Cao Y (2012) Solution processed small molecule bulk heterojunction organic photovoltaics based on a conjugated donor-acceptor porphyrin. J Mater Chem 22:21841–21844
43. Qin H, Li L, Guo F, Su S-J, Peng J, Cao Y, Peng X (2014) Solution-processed bulk heterojunction solar cells based on a porphyrin small molecule with 7% power conversion efficiency. Energy Environ Sci. doi:10.1039/C1033EE43761B
44. Roquet S, Cravino A, Leriche P, Aleveque O, Frere P, Roncali J (2006) Triphenylamine-thienylenevinylene hybrid systems with internal charge transfer as donor materials for heterojunction solar cells. J Am Chem Soc 128:3459–3466
45. Leriche P, Frère P, Cravino A, Alévêque O, Roncali J (2007) Molecular engineering of the internal charge transfer in thiophene−triphenylamine hybrid π-conjugated systems. J Org Chem 72:8332–8336
46. He C, He QG, He YJ, Li YF, Bai FL, Yang CH, Ding YQ, Wang LX, Ye JP (2006) Organic solar cells based on the spin-coated blend films of TPA-th-TPA and PCBM. Sol Energy Mater Sol Cells 90:1815–1827

47. Deng D, Yang Y, Zhang J, He C, Zhang M, Zhang Z-G, Zhang Z, Li Y (2011) Triphenylamine-containing linear D-A-D molecules with benzothiadiazole as acceptor unit for bulk-heterojunction organic solar cells. Org Electron 12:614–622
48. Shang H, Fan H, Shi Q, Li S, Li Y, Zhan X (2010) Solution processable D-A-D molecules based on triphenylamine for efficient organic solar cells. Sol Energy Mater Sol Cells 94:457–464
49. Shi Q, Cheng P, Li Y, Zhan X (2012) A solution processable D-A-D molecule based on thiazolothiazole for high performance organic solar cells. Adv Energ Mater 2:63–67
50. Leliege A, Le Regent C-H, Allain M, Blanchard P, Roncali J (2012) Structural modulation of internal charge transfer in small molecular donors for organic solar cells. Chem Commun 48:8907–8909
51. Zhang J, Deng D, He C, He Y, Zhang M, Zhang Z-G, Zhang Z, Li Y (2011) Solution-processable star-shaped molecules with triphenylamine core and dicyanovinyl endgroups for organic solar cells. Chem Mater 23:817–822
52. Min J, Luponosov YN, Ameri T, Elschner A, Peregudova SM, Baran D, Heumueller T, Li N, Machui F, Ponomarenko S, Brabec CJ (2013) A solution-processable star-shaped molecule for high-performance organic solar cells via alkyl chain engineering and solvent additive. Org Electron 14:219–229
53. Zhang Z-G, Zhang K-L, Liu G, Zhu C-X, Neoh K-G, Kang E-T (2009) Triphenylamine −fluorene alternating conjugated copolymers with pendant acceptor groups: synthesis, structure−property relationship, and photovoltaic application. Macromolecules 42:3104–3111
54. Lin Y, Zhang Z-G, Bai H, Li Y, Zhan X (2012) A star-shaped oligothiophene end-capped with alkyl cyanoacetate groups for solution-processed organic solar cells. Chem Commun 48:9655–9657
55. Zhang J, Yang Y, He C, He Y, Zhao G, Li Y (2009) Solution-processable star-shaped photovoltaic organic molecule with triphenylamine core and benzothiadiazole−thiophene arms. Macromolecules 42:7619–7622
56. Shang H, Fan H, Liu Y, Hu W, Li Y, Zhan X (2011) A solution-processable star-shaped molecule for high-performance organic solar cells. Adv Mater 23:1554–1557
57. Paek S, Cho N, Cho S, Lee JK, Ko J (2012) Planar star-shaped organic semiconductor with fused triphenylamine core for solution-processed small-molecule organic solar cells and field-effect transistors. Org Lett 14:6326–6329
58. Roquet S, de Bettignies R, Leriche P, Cravino A, Roncali J (2006) Three-dimensional tetra (oligothienyl)silanes as donor material for organic solar cells. J Mater Chem 16:3040–3045
59. Kopidakis N, Mitchell WJ, van de Lagemaat J, Ginley DS, Rumbles G, Shaheen SE, Rance WL (2006) Bulk heterojunction organic photovoltaic devices based on phenyl-cored thiophene dendrimers. Appl Phys Lett 89:103524
60. Ma C-Q, Fonrodona M, Schikora MC, Wienk MM, Janssen RAJ, Baeuerle P (2008) Solution-processed bulk-heterojunction solar cells based on monodisperse dendritic oligothiophenes. Adv Funct Mater 18:3323–3331
61. Mastalerz M, Fischer V, Ma C-Q, Janssen RAJ, Bauerle P (2009) Conjugated oligothienyl dendrimers based on a pyrazino 2,3-g quinoxaline core. Org Lett 11:4500–4503
62. Rance WL, Rupert BL, Mitchell WJ, Koese ME, Ginley DS, Shaheen SE, Rumbles G, Kopidakis N (2010) Conjugated thiophene dendrimer with an electron-withdrawing core and electron-rich dendrons: how the molecular structure affects the morphology and performance of dendrimer: fullerene photovoltaic devices. J Phys Chem C 114:22269–22276
63. Zhang W, Ging Meng N, Hoi Lam T, Wong MS, Zhu F (2011) Synthesis and photovoltaic properties of functional dendritic oligothiophenes. J Polym Sci Pol Chem 49:1865–1873
64. Yin B, Yang L, Liu Y, Chen Y, Qi Q, Zhang F, Yin S (2010) Solution-processed bulk heterojunction organic solar cells based on an oligothiophene derivative. Appl Phys Lett 97
65. Liu Y, Wan X, Wang F, Zhou J, Long G, Tian J, You J, Yang Y, Chen Y (2011) Spin-coated small molecules for high performance solar cells. Adv Energ Mater 1:771–775

66. Li Z, He G, Wan X, Liu Y, Zhou J, Long G, Zuo Y, Zhang M, Chen Y (2012) Solution processable rhodanine-based small molecule organic photovoltaic cells with a power conversion efficiency of 6.1%. Adv Energ Mater 2:74–77
67. Liu Y, Wan X, Wang F, Zhou J, Long G, Tian J, Chen Y (2011) High-performance solar cells using a solution-processed small molecule containing benzodithiophene unit. Adv Mater 23:5387–5391
68. Zhou J, Wan X, Liu Y, Long G, Wang F, Li Z, Zuo Y, Li C, Chen Y (2011) A planar small molecule with dithienosilole core for high efficiency solution-processed organic photovoltaic cells. Chem Mater 23:4666–4668
69. Zhou J, Wan X, Liu Y, Zuo Y, Li Z, He G, Long G, Ni W, Li C, Su X, Chen Y (2012) Small molecules based on benzo 1,2-b:4,5-b' dithiophene unit for high-performance solution-processed organic solar cells. J Am Chem Soc 134:16345–16351
70. Takacs CJ, Sun Y, Welch GC, Perez LA, Liu X, Wen W, Bazan GC, Heeger AJ (2012) Solar cell efficiency, self-assembly, and dipole-dipole interactions of isomorphic narrow-band-gap molecules. J Am Chem Soc 134:16597–16606
71. Sun Y, Welch GC, Leong WL, Takacs CJ, Bazan GC, Heeger AJ (2012) Solution-processed small-molecule solar cells with 6.7% efficiency. Nat Mater 11:44–48
72. van der Poll TS, Love JA, Nguyen T-Q, Bazan GC (2012) Non-basic high-performance molecules for solution-processed organic solar cells. Adv Mater 24:3646–3649
73. Kyaw AKK, Wang DH, Wynands D, Zhang J, Nguyen T-Q, Bazan GC, Heeger AJ (2013) Improved light harvesting and improved efficiency by insertion of an optical spacer (ZnO) in solution-processed small-molecule solar cells. Nano Lett 13:3796–3801
74. Hou J, Za Tan, Yan Y, He Y, Yang C, Li Y (2006) Synthesis and photovoltaic properties of two-dimensional conjugated polythiophenes with Bi(thienylenevinylene) side chains. J Am Chem Soc 128:4911–4916
75. Zhang M, Fan H, Guo X, He Y, Zhang Z, Min J, Zhang J, Zhao G, Zhan X, Li Y (2010) Synthesis and photovoltaic properties of bithiazole-based donor−acceptor copolymers. Macromolecules 43:5706–5712
76. Zhang ZG, Zhang S, Min J, Cui C, Geng H, Shuai Z, Li Y (2012) Side chain engineering of polythiophene derivatives with a thienylene-vinylene conjugated side chain for application in polymer solar cells. Macromolecules 45:2312–2320
77. Shen S, Jiang P, He C, Zhang J, Shen P, Zhang Y, Yi Y, Zhang Z, Li Z, Li Y (2013) Solution-processable organic molecule photovoltaic materials with bithienyl-benzodithiophene central unit and indenedione end groups. Chem Mater 25:2274–2281
78. Cui C, Min J, Ho C-L, Ameri T, Yang P, Zhao J, Brabec CJ, Wong W-Y (2013) A new two-dimensional oligothiophene end-capped with alkyl cyanoacetate groups for highly efficient solution-processed organic solar cells. Chem Commun 49:4409–4411
79. Lin Y, Ma L, Li Y, Liu Y, Zhu D, Zhan X (2013) A solution-processable small molecule based on benzodithiophene and diketopyrrolopyrrole for high-performance organic solar cells. Adv Energ Mater 1166–1170
80. He Y, Chen H-Y, Hou J, Li Y (2010) Indene−C60 bisadduct: a new acceptor for high-performance polymer solar cells. J Am Chem Soc 132:1377–1382
81. He Y, Li Y (2011) Fullerene derivative acceptors for high performance polymer solar cells. Phys Chem Chem Phys 13:1970–1983
82. Li C-Z, Yip H-L, Jen AKY (2012) Functional fullerenes for organic photovoltaics. J Mater Chem 22:4161–4177
83. Faist MA, Kirchartz T, Gong W, Ashraf RS, McCulloch I, de Mello JC, Ekins-Daukes NJ, Bradley DDC, Nelson J (2012) Competition between the charge transfer state and the singlet states of donor or acceptor limiting the efficiency in polymer: fullerene solar cells. J Am Chem Soc 134:685–692
84. Sonar P, Fong Lim JP, Chan KL (2011) Organic non-fullerene acceptors for organic photovoltaics. Energy Environ Sci 4:1558–1574

85. Tang CW (1986) Two-layer organic photovoltaic cell. Appl Phys Lett 48:183–185
86. Guo X, Bu L, Zhao Y, Xie Z, Geng Y, Wang L (2009) Controlled phase separation for efficient energy conversion in dye/polymer blend bulk heterojunction photovoltaic cells. Thin Solid Films 517:4654–4657
87. Sharenko A, Proctor CM, van der Poll TS, Henson ZB, Nguyen T-Q, Bazan GC (2013) A high-performing solution-processed small molecule: perylene diimide bulk heterojunction solar cell. Adv Mater 4403–4406
88. Jiang B, Zhang X, Zhan C, Lu Z, Huang J, Ding X, He S, Yao J (2013) Benzodithiophene bridged dimeric perylene diimide amphiphiles as efficient solution-processed non-fullerene small molecules. Polym Chem 4:4631–4638
89. Lu Z, Zhang X, Zhan C, Jiang B, Zhang X, Chen L, Yao J (2013) Impact of molecular solvophobicity vs. solvophilicity on device performances of dimeric perylene diimide based solution-processed non-fullerene organic solar cells. Phys Chem Chem Phys 15:11375–11385
90. Zhang X, Lu Z, Ye L, Zhan C, Hou J, Zhang S, Jiang B, Zhao Y, Huang J, Zhang S, Liu Y, Shi Q, Liu Y, Yao J (2013) A potential perylene diimide dimer-based acceptor material for highly efficient solution-processed non-fullerene organic solar cells with 4.03% efficiency. Adv Mater 5791–5797
91. Ren G, Ahmed E, Jenekhe SA (2011) Non-fullerene acceptor-based bulk heterojunction polymer solar cells: engineering the nanomorphology via processing additives. Adv Energ Mater 1:946–953
92. Shin RYC, Kietzke T, Sudhakar S, Dodabalapur A, Chen Z-K, Sellinger A (2007) N-type conjugated materials based on 2-vinyl-4,5-dicyanoimidazoles and their use in solar cells. Chem Mater 19:1892–1894
93. Bloking JT, Han X, Higgs AT, Kastrop JP, Pandey L, Norton JE, Risko C, Chen CE, Bredas J-L, McGehee MD, Sellinger A (2011) Solution-processed organic solar cells with power conversion efficiencies of 2.5% using benzothiadiazole/imide-based acceptors. Chem Mater 23:5484–5490
94. Zhou Y, Dai Y-Z, Zheng Y-Q, Wang X-Y, Wang J-Y, Pei J (2013) Non-fullerene acceptors containing fluoranthene-fused imides for solution-processed inverted organic solar cells. Chem Commun 49:5802–5804
95. Li H, Kim FS, Ren G, Hollenbeck EC, Subramaniyan S, Jenekhe SA (2013) Tetraazabenzodifluoranthene diimides: building blocks for solution-processable n-type organic semiconductors. Angew Chem Int Ed 52:5513–5517
96. Fang Y, Pandey AK, Nardes AM, Kopidakis N, Burn PL, Meredith P (2013) A narrow optical gap small molecule acceptor for organic solar cells. Adv Energ Mater 3:54–59
97. Zhou T, Jia T, Kang B, Li F, Fahlman M, Wang Y (2011) Nitrile-substituted QA derivatives: new acceptor materials for solution-processable organic bulk heterojunction solar cells. Adv Energ Mater 1:431–439
98. Lin Y, Cheng P, Li Y, Zhan X (2012) A 3D star-shaped non-fullerene acceptor for solution-processed organic solar cells with a high open-circuit voltage of 1.18 V. Chem Commun 48:4773–4775
99. Lin Y, Li Y, Zhan X (2013) A solution-processable electron acceptor based on dibenzosilole and diketopyrrolopyrrole for organic solar cells. Adv Energ Mater 3:724–728
100. Winzenberg KN, Kemppinen P, Scholes FH, Collis GE, Shu Y, Singh TB, Bilic A, Forsyth CM, Watkins SE (2013) Indan-1,3-dioneelectron-acceptor small molecules for solution-processable solar cells: a structure-property correlation. Chem Commun 49:6307–6309
101. Poe AM, Della Pelle AM, Subrahmanyam AV, White W, Wantz G, Thayumanavan S (2014) Small molecule BODIPY dyes as non-fullerene acceptors in bulk heterojunction organic photovoltaics. Chem Commun 50:2913–2915
102. Zhou Y, Ding L, Shi K, Dai YZ, Ai N, Wang J, Pei J (2012) A non-fullerene small molecule as efficient electron acceptor in organic bulk heterojunction solar cells. Adv Mater 24:957

103. Schwenn PE, Gui K, Nardes AM, Krueger KB, Lee KH, Mutkins K, Rubinstein-Dunlop H, Shaw PE, Kopidakis N, Burn PL, Meredith P (2011) A small molecule non-fullerene electron acceptor for organic solar cells. Adv Energ Mater 1:73–81
104. Cnops K, Rand BP, Cheyns D, Verreet B, Empl MA, Heremans P (2014) 8.4% efficient fullerene-free organic solar cells exploiting long-range exciton energy transfer. Nat Commun 5:3406

Chapter 5
Conjugated Polymer Photovoltaic Materials

Long Ye and Jianhui Hou

5.1 Introduction

During the past few decades, conjugated polymers with various molecular structures have been explored for applications in polymer solar cells (PSCs). In order to obtain more efficient PSCs, great effort has been devoted to optimizing and synthesizing conjugated polymers with superior photovoltaic properties, which boost the power conversion efficiency (*PCE*) toward or even above 8–10 % in several research groups [1–13]. In this chapter, an overview of conjugated polymer photovoltaic materials is given to provide insights for molecular design and fine-tuning of high-performance photovoltaic polymers. First, we briefly summarize and provide design considerations of conjugated polymer photovoltaic materials. Second, representative photovoltaic polymers are introduced. Third, representative conjugated polymer acceptor materials are briefly introduced and discussed. Because photovoltaic properties of conjugated polymers are susceptible to device fabrication conditions and device structures, excluding the factors of optical and interfacial enhancement by device engineering, the '*initial PCE*' of the PSCs used in simple conventional devices of ITO/PEDOT: PSS/Polymer: PCBM/Ca (or Mg, Ba, LiF)/Al (or Ag, Au), as depicted in Fig. 5.1, is discussed.

L. Ye · J. Hou (✉)
State Key Laboratory of Polymer Physics and Chemistry, Beijing National Laboratory for Molecular Sciences, Institute of Chemistry, Chinese Academy of Sciences, Beijing 100190, People's Republic of China
e-mail: hjhzlz@iccas.ac.cn

Y. Li (ed.), *Organic Optoelectronic Materials*,
Lecture Notes in Chemistry 91, DOI 10.1007/978-3-319-16862-3_5

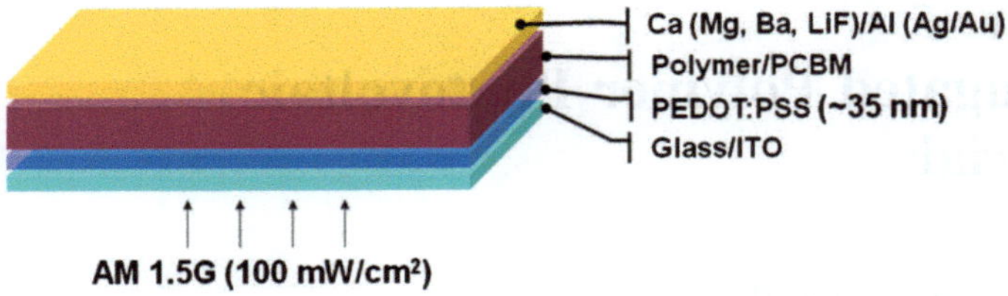

Fig. 5.1 Simple conventional device structure of a polymer solar cell

5.1.1 Brief Summary of Photovoltaic Polymers

As the key photovoltaic materials in PSCs, conjugated polymers can be classified into two types, namely donor and acceptor photovoltaic polymers. Following the pioneering works of Heeger et al. and Friend et al. in the 1990s [14, 15], thousands of polymer donors with different backbones and side groups have been developed, synthesized, and used in polymer solar cells during the past few decades, and have been deemed to be one of the driving forces of the development of PSCs. Conjugated polymer photovoltaic donor materials have been well discussed and reviewed recently [16–30]. However, to classify the polymer donors rationally is still quite difficult because of the rapid growth of the numbers of polymer donors. Herein, three types of the first used homopolymer donor materials, i.e., the derivatives of poly(1,4-phenylene vinylene) (PPVs), polythiophene (PTs), poly(thienylene vinylene) (PTVs), etc., are discussed as the first category of polymer donors (see Fig. 5.2). In recent years, the Donor–Acceptor copolymers (D–A copolymers) based on two or more conjugated building blocks have played important roles in promoting the development of the PSC field, so we provide more examples of conjugated D–A copolymers as the second category of polymer donors. For example, benzothiadiazole (BT)-based polymers, silole-containing polymers, diketopyrrolopyrrole (DPP)-based polymers, indacenodithiophene (IDT)-based polymers, Benzo[1,2-b:4,5-b′]dithiophene (BDT) based polymers, Thienopyrroledione (TPD) based polymers, etc., are introduced and discussed (see Fig. 5.2). In comparison with polymer donors, polymer acceptors have attracted less attention, although some recent work has shown that polymer acceptors have great potential in realizing highly efficient PSCs without using fullerene derivatives. Therefore, in the last section of this chapter, we give a brief introduction of polymer acceptors.

5.1.2 Design Considerations of Conjugated Polymer Photovoltaic Materials

The basic requirements for molecular design of high efficiency photovoltaic polymers have been discussed and summarized in several reviews [16–25]. It is known

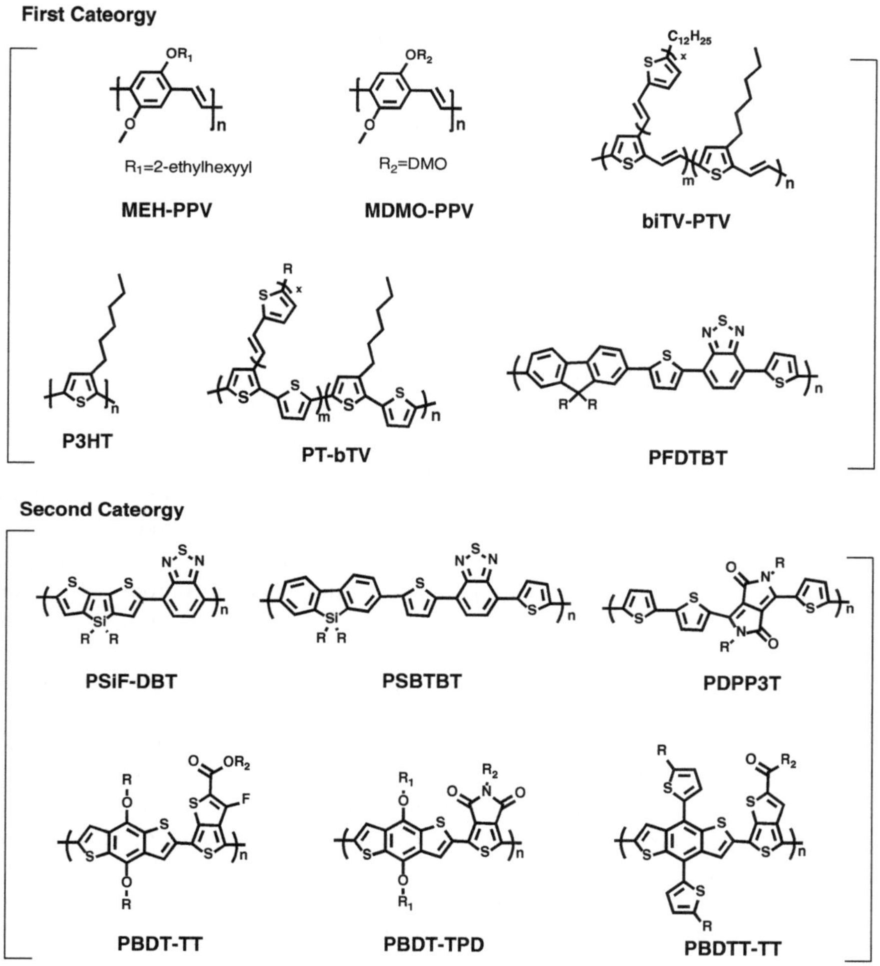

Fig. 5.2 Two categories of polymer donor materials and examples of representative polymers in each category

that the *PCE* of PSC can be calculated based on three photovoltaic parameters: the open circuit voltage (V_{oc}), the short circuit current (J_{sc}), and the fill factor (*FF*). For highly efficient photovoltaic polymers, important factors including solubility, light absorption, molecular energy level, mobility as well as morphology should be considered.

Appropriate solubility and good film-forming properties should first be considered in designing novel conjugated polymers. To get high quality thin films of conjugated polymers by solution coating processes, conjugated polymer donor materials must have good solubility in commonly used organic solvents such as

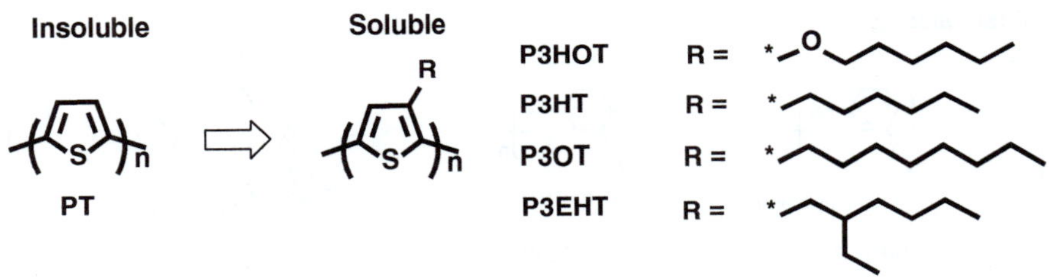

Fig. 5.3 Unsubstituted and substituted PTs

chloroform (CF), chlorobenzene (CB), toluene, and dichlorobenzene (DCB). Because conjugated polymers have rigid backbones and intermolecular π–π interaction provides strong driving force for aggregation, the unsubstituted conjugated polymers are all insoluble in organic solvents. Therefore, long, flexible and/or branched side groups, such as alkyls or alkoxys, are introduced as functional groups onto their backbones to overcome the strong aggregation effect and hence to afford solution processability for conjugated polymers. For example, as shown in Fig. 5.3, the unsubstituted polythiophene is an insoluble polymer; however, when alkyls, such as hexyl, octyl, or alkyls with more carbon numbers, are introduced as side groups, the derivatives can be readily dissolved in many types of organic solvents such as CF, toluene, CB, *o*-DCB, and so on.

For a polymer photovoltaic material, a broader and stronger absorption, matching well with the solar radiation spectrum, is necessary to achieve high J_{sc}. As shown in Fig. 5.4a, solar irradiation has a very broad spectrum, which is mainly distributed at the visible and infrared regions with a peak at ca. 700 nm, so to harvest solar light the photovoltaic polymer should absorb the majority of the wavelength region from 400 to 900 nm. Clearly, from the point of view of the absorption spectrum, the PSCs based on P3HT cannot make good use of solar light. Therefore, conjugated polymers with low band gaps (LBG polymers) were developed and used in PSCs, leading to the possibility of realizing high J_{sc}. In fact, the

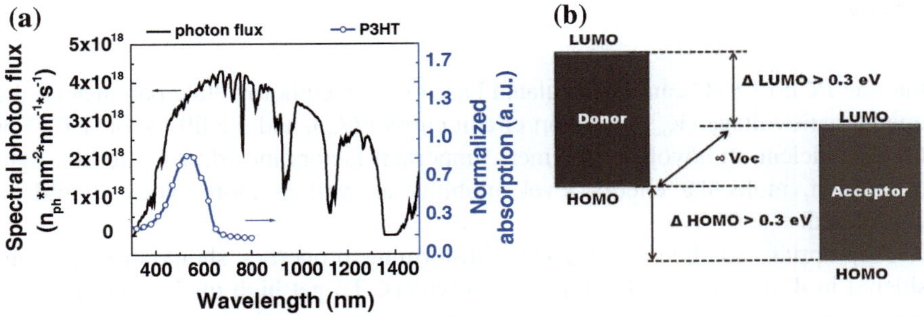

Fig. 5.4 **a** Spectral photon flux of AM 1.5 G and the UV-vis absorption of P3HT. **b** Schematic energy levels diagram of electron donors and acceptors in polymer solar cells

development of LBG polymers has played a very important role in the rapid progress in the PSC field.

The highest occupied molecular orbital (HOMO) and lowest unoccupied molecular orbital (LUMO) energy levels of photovoltaic polymers should be appropriate to maintain efficient charge separation and reduce energy loss. The schematic diagram of molecular energy and open-circuit voltage has been drawn in Fig. 5.4b. The value of V_{oc} for a PSC is directly proportional to the energy difference between the HOMO of the polymer donors and the LUMO of the acceptor materials [31]. As the LUMO of [6,6]-phenyl-C_{61} (or C_{71})-butyric acid methyl ester ($PC_{61}BM$ or $PC_{71}BM$) is generally considered to be −3.9 eV, the V_{oc} of polymer/PCBM-based polymer solar cells should rely on the HOMO level of polymers. Therefore, lowering the HOMO level of the polymer donors and thus reducing the energy loss during the charge separation process have been seen as key to achieving high V_{oc}.

Furthermore, in order to facilitate charge transport in the BHJ active layer, a high mobility should also be required for the polymer donors. The hole and electron mobility (μ_h and μ_e) are important parameters to evaluate the photovoltaic properties of the donor and acceptor photovoltaic materials. Moreover, in order to reduce or avoid germinate and bimolecular recombination in the BHJ layers, the donor and acceptor materials used in a BHJ active layer in a PSC device should have balanced mobilities for holes and electrons. For example, considering that μ_e of PCBM is ca. 10^{-3} $cm^2/(V\ s)$, when PCBM is used as the acceptor, μ_h of the polymer donors should be kept at the same level or higher than that of PCBM.

Morphological properties of polymer/PCBM blends are also of great importance for the photovoltaic performance of polymers [32–35]. For example, because of the low dielectric constants, the exciton diffusion length in the bulk of conjugated polymers is less than 10 nm, so nanoscale phase separation in the BHJ active layer is needed. If the aggregations of the donor and/or the acceptor are too big, the excitons are not diffused to the D/A interface efficiently, so that strong geminate recombination is observed. Besides of the size of phase separation in the BHJ blend, the crystallinity of conjugated polymers is also an important issue. As is well known, the transport of π-electrons in polymer aggregations is anisotropic, i.e., the intramolecular charge transport is along the conjugated backbones, whereas the intermolecular charge transport is along the overlapped π-orbits, which are perpendicular to the conjugated backbones. Therefore, various methods have been developed to modulate morphologies of conjugated polymers in the BHJ blends. For example, for the photovoltaic system based on a LBG polymer called PDPP3T [34] and PCBM, when the blend was processed with CF, a very low *PCE* (<2 %) was achieved; using CF and 1,8-diiodooctane as binary solvent, a moderate *PCE* of 4.7 % was achieved. Recently, Ye et al. [35] introduced a ternary solvent system of DCB/CF/DIO to optimize the morphology as well as the overall performance of the PDPP3T/PCBM system, which increased the *PCE* to 6.71 %. Morphological studies of the photovoltaic system of PDPP3T and PCBM clearly revealed that the phase separation size and the crystallinity of the polymer in the blend can be tuned effectively by modulating the processing solvents of the films. On the other hand, morphologies of conjugated polymers can also be tuned by changing their molecular structures, which have been deemed as one of the main tasks for molecular design of conjugated polymers.

To meet the above requirements for highly efficient polymer donors, various molecular design strategies have been developed to modulate the photovoltaic properties of conjugated polymers and many new conjugated polymers have been designed and used in PSCs. In the following sections, the design strategies and some of the representative conjugated polymer photovoltaic materials are discussed in detail.

5.2 Conjugated Polymer Donor Materials

5.2.1 Three Important Types of Homopolymer

5.2.1.1 Poly(1,4-Phenylene Vinylene) Derivatives

In 1995, Yu et al. [14] pioneered the concept of the bulk-heterojunction (BHJ) in MEH-PPV:fullerene systems, which is considered to be the best PSC device architecture to date. Two poly(1,4-phenylene vinylene) (PPV) derivatives including poly [(2-methoxy-5-(2-ethyl-hexyloxy))-1,4-phenylenevinylene] MEH-PPV and poly[(2-methoxy-5-(3′,7′-dimethyloctyloxy))-1,4-phenylenevinylene] (MDMO-PPV) are very important donor polymers, which have played a very important role in the early stage of the development of the field of PSCs. Their molecular structure and corresponding photovoltaic results are shown in Fig. 5.5 and Table 5.1, respectively.

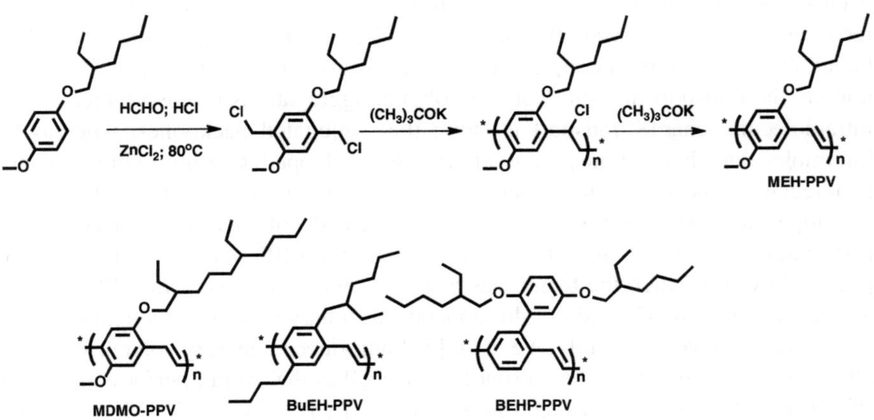

Fig. 5.5 Synthesis method and molecular structures of MEH-PPV and MDMO-PPV

Table 5.1 Photovoltaic results of some of the representative polymer donors of PPVs, PTVs, and the like

Materials	HOMO (eV)	V_{oc} (V)	J_{sc} (mA/cm^2)	FF (%)	PCE (%)	Ref.
MEH-PPV	−5.07	0.80	6.5	50	2.6	[42]
MDMO-PPV	−5.10	0.82	5.25	61	2.5	[40]
biTV-PTV	−4.77	0.48	2.27	30	0.32	[46]
P3CTV	−5.26	0.86	5.47	42.8	2.01	[47]
PBDTV	−5.16	0.71	6.46	57	2.63	[48]

MEH-PPV and MDMO-PPV show excellent solubility in commonly used solvents such as tetrahydrofuran, CF, toluene, xylenes, CB, and 1, 2-DCB. The solutions of these two polymers show very good film forming abilities, and the dilute solution of MEH-PPV or MDMO-PPV in CB (1–2 mg/mL) can still form high quality films by the spin coating process, which gives more opportunity to utilize these polymers to carry out studies in device physics and device engineering. Actually, these two polymers can also be used as electroluminescent materials in polymer light emitting devices.

PPV and its derivatives can be synthesized via various methods, including the Wessling precursor method [36], Gilch method [37], Heck coupling reaction [38], Knoevenagel polycondensation [39], etc. With the Wessling method, the PPV precursor can be synthesized by the reaction of bis-(sulfonium halide) salts of *p*-xylene with a base (NaOH) in water or alcohol solution, and then the precursor solution is spin-cast on ITO substrate, PPV film being formed by heat treatment at 180–300 °C under vacuum. The PPVs prepared by the Wessling precursor method have a lot of defects and impurities because of the oxidation of the precursor polymer, the residual precursor moieties, and undesired side reactions during the thermal conversion.

By Heck coupling reaction [38], organic halides and vinylbenzene compounds can be coupled to generate a carbon–carbon bond under the catalysis of Pd(0). Many functional groups, such as aldehyde, ester, nitryl, hydroxy and carboxy, have no obviously negative effects on the coupling reaction, so that this reaction has been widely used for preparation of PPVs. Cyano-substituted PPV can be readily prepared by the Knoevenagel polycondensation reaction between equimolar amounts of a terephthaldehyde derivative and a 1,4-diacetonnitrile-benzene derivative [39]. The condensation reaction takes place upon addition of excess potassium *tert*-butoxide or tetrabutylammonium hydroxide in THF/*tert*-butanol mixture at 50 °C. In the Knoevenagel condensation reaction, tetrabutylammonium hydroxide is used as catalyst, and the solvent for the reaction can be THF, toluene, or DMF.

As shown in Fig. 5.5, the Gilch method is very convenient to prepare PPV and its derivatives, and, by this method, 1,4-bis-chloromethyl-benzenes were treated with potassium *tert*-butoxide in non-hydroxylic solvents such as tetrahydrofuran [37]. The temperature of the reactant, the concentration of the monomer and the base, and the speed of the base addition are all crucial conditions for molecular weight and PDI of the polymer. The molecular weight can also be controlled by using a benzylchloride derivative as the end capping reagent. Many PPV derivatives can be prepared with high molecular weight and high purity. Therefore, the Gilch method is the most successful method for the synthesis of PPVs, and the classic materials of PPVs, MEH-PPV and MDMO-PPV, are synthesized by this method.

A breakthrough in the *PCE* of the PSC devices based on PPV derivatives were presented by Brabec, Sariciftci, and coworkers [40]. In 2001, Shaheen et al. [40] introduced LiF as the *n*-type buffer layer in MDMO-PPV-based PSC devices and investigated the effect of processing solvent on the photovoltaic performance. By replacing toluene with CB, optimal phase separation and increased interactions between conjugated polymers were observed in MDMO-PPV-based PSC devices. As a result, a dramatically improved *PCE* up to 2.5 % was achieved, which was a nearly a threefold enhancement over previously reported values and also the world record at that time. Afterwards, Brabec et al. [41] systematically investigated the effect of the thickness of LiF on the photovoltaic performances of MDMO-PPV/ $PC_{61}BM$-based PSC devices. Under optimal conditions, the PSC achieved an improved *PCE* of 3.3 %. Compared to PSCs without the LiF interfacial layer, the white light efficiencies of LiF-based PSC increased by over 20 %. Cao et al. [42] investigated the photovoltaic performance of MEH-PPV with a series of PCBM derivatives with different alkyl end groups on its side chain. The results revealed that the $PC_{61}BM$ derivative appending the butyl end group performs best as acceptor blended with MEH-PPV and achieved a high *PCE* of 2.6 %. Wienk et al. [43] incorporated $PC_{71}BM$ as the electron acceptor in thin-film polymer photovoltaic cells based on MDMO-PPV and provide a high J_{sc} because of the increased absorption in the visible region from $PC_{71}BM$ and an ultrafast charge transfer upon photoexcitation of MDMO-PPV or $PC_{71}BM$. In 2008, Tajima et al. [44] investigated the effect of regioregularity on the photovoltaic properties of the MDMO-PPV-based PSC devices. Fully regioregular MDMO-PPV was utilized for polymer photovoltaic devices, and their performance was compared with that of regiorandom MDMO-PPVs. A *PCE* up to 3.1 % was recorded in regioregular MDMO-PPV, whereas a moderate *PCE* of 1.7 % was achieved with regiorandom MDMO-PPV. The higher *PCE* of regioregular MDMO-PPV originated from both higher hole mobility and better nano-morphology. Mikroyannidis et al. [45] synthesized a novel LBG PPV derivative in 2010, which generated a record *PCE* for the PPV/PCBM system so far. However, although MEH-PPV and MDMO-PPV played vital roles in the early years (1995–2003) of PSC study, the narrow absorption range (400–560 nm) and poor hole mobility ($\sim 10^{-7}$ cm^2 V^{-1} s^{-1}) limited the photovoltaic performance of PPV derivatives.

5.2.1.2 Poly(thienylene vinylene) Derivatives and the Like

The likes of PPVs, poly(thienylene vinylene) derivatives (PTVs) were developed and applied in the PSC field because of their broad absorption and higher hole

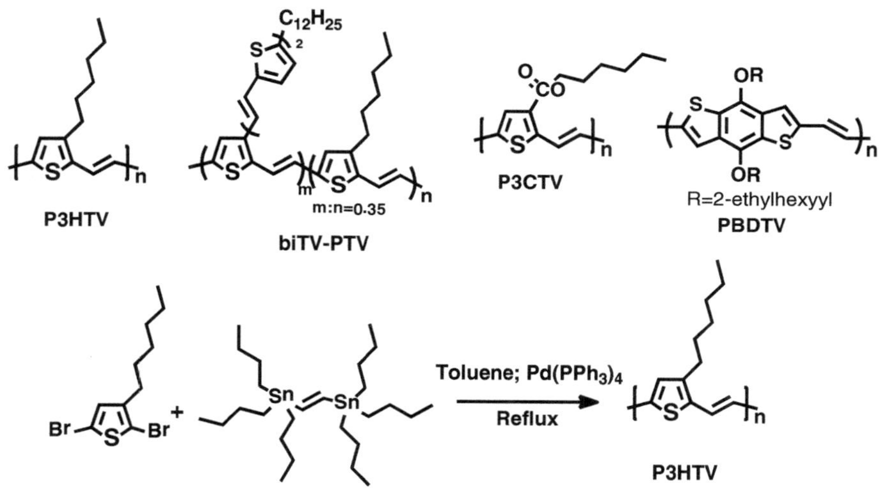

Fig. 5.6 Molecular structures and synthesis routes of PTVs

mobility. The molecular structure and corresponding photovoltaic results of a few PTV derivatives are provided in Fig. 5.6 and Table 5.1, respectively. Hexyl-substituted PTV, namely P3HTV, is the simplest soluble polymer among PTVs, although the photovoltaic performance of P3HTV is rather poor (PCE is ~0.2 %). The Stille coupling polycondensation reaction is widely used for synthesis of PTVs. As show in Fig. 5.6, PTVs can be easily prepared by using 2,5-dibromothiophenes and 1,2-bis-tributylstannylethylene as starting materials. In this reaction, aromatic solvents such as toluene, xylenes, and CB can be used as reaction solvents; the Pd(0) compound with appropriate ligand-like tetrakis(triphenylphosphine)palladium(0) can be used as the catalyst. Actually, the Stille coupling polycondensation reaction is also one of the most widely used methods for synthesis of other types of conjugated alternating copolymers. Therefore, although photovoltaic performance of the PSCs based on PTVs is quite low, the synthesis of PTVs paved the way for the study of molecular design and synthesis of highly efficient photovoltaic polymers.

Hou and Li et al. [46] introduced the conjugated side chains in the PTV backbone and synthesized a series of PTV derivatives with extended absorption in the UV-vis range from 350 to 740 nm. The *PCE* of the biTV-PTV-based PSC reached 0.32 %, which exhibited 52 % enhancement in comparison with that of the PSC device based on P3HTV under the same conditions. To overcome the intrinsic drawbacks of PTVs, for instance, high-lying HOMO levels and low V_{oc}, Li and coworkers [47] introduced an electron-deficient carboxylate group in P3HTV. The results indicated that the introduction of the carboxylate group in side chains can lower both LUMO and HOMO values of PTVs. Meanwhile, the photoluminescence of PTV can improved significantly. As a result, a higher V_{oc} up to 0.86 V was achieved. Under the D/A weight ratio of 1:2, a high *PCE* of 2.01 % was recorded. Notably, the *PCE* of the P3CTV-based PSC device is about 10 times higher than that of the device based on P3HTV. The *PCE* of 2.01 % is the highest efficiency for PSCs based on PTVs. The dramatic improvement of the photovoltaic performances of PTV by the carboxylate substitution demonstrated that PTVs might be promising photovoltaic polymers upon suitable structural modification.

In 2010, He et al. [48] reported a new vinylene-based polymer, PBDTV, which was copolymerized by vinylene and BDT via Pd-catalyzed Stille-coupling method. The PBDTV film depicted a broad absorption range covering from 350 to 618 nm and high hole mobility of 4.84×10^{-3} cm^2 V^{-1} s^{-1}. At the optimal conditions (the weight ratio of PBDTV:$PC_{71}BM$ of 1:4 and the active layer thickness of 65 nm), the *PCE* of the PBDTV-based PSC device reached 2.63 % with V_{oc} of 0.71 V, J_{sc} of 6.46 mA/cm^2, and *FF* of 57 % under the illumination of AM 1.5G, 100 mW/cm^2.

5.2.1.3 Polythiophene Derivatives

Polythiophenes (PTs), particularly regioregular poly(3-alkylthiophene)s (P3ATs), are a widely used class of polymer donors because of their excellent thermal and chemical stability as well as good charge transport properties. The photovoltaic properties of representative polythiophene derivatives (see Fig. 5.7) are listed in Table 5.2. Because PTs do not dissolve in most of the common solvents, alkyl substitution is a useful method to help with the solubility of P3ATs. The length of alkyl group in P3ATs plays an important role in determining the solubility, crystallinity, and morphology.

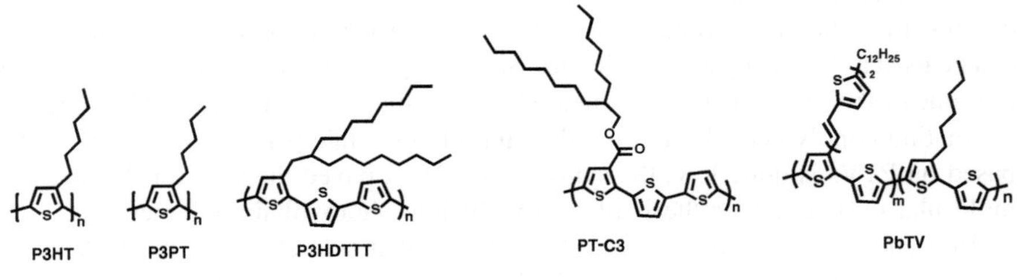

Fig. 5.7 Molecular structures of several PT derivatives

Table 5.2 Photovoltaic performances of some PT derivatives

Materials	HOMO (eV)	V_{oc} (V)	J_{sc} (mA/cm^2)	FF (%)	PCE (%)	Ref.
P3HT	−4.70	0.61	10.6	67.4	4.37	[61]
P3PT	−4.76	0.66	9.63	69	3.7	[69]
biTV-PT	−4.93	0.72	10.3	43	3.18	[73]
P3HDTTT	−5.30	0.82	6.33	66	3.4	[74]
PT-C3	−5.10	0.78	9.68	51.2	3.87	[75]

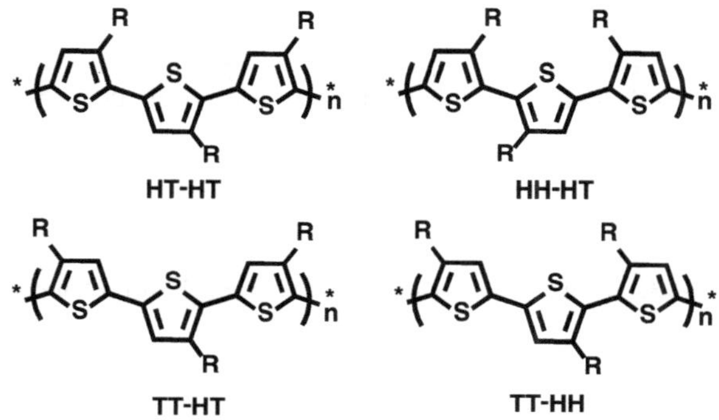

Fig. 5.8 Four regioisomers of poly(3-substituted thiophene)s

Generally, the polymerization of thiophenes is carried out at their 2- and 5-positions. For many PTs, such as poly(3-alkylthiophene)s, the repeated units are asymmetric, so there are three relative orientations available when two thiophene rings are coupled between the 2- and 5-positions. Usually, the 2-position is called the head, and the 5-position the tail. As shown in Fig. 5.8, this leads to a mixture of four regioisomers when 3-substituted (or asymmetric) thiophene monomers are employed [49]. The HT-HT structure of PTs is seen as regioregular polymers, and the HT-HT isomer proportion in the polymer is known as regioregularity. Regioregular poly(3-substituted thiophene) can easily access a low energy planar conformation, so the regioregularity is an important factor in characterization of poly(3-substituted thiophene).

^{1}H and ^{13}C NMR can be used to determine the structure and the regioregularity of PTs [50]. In a regioregular PT (HT-coupling ≈ 100 %), the proton at the 4-position of thiophene exhibits a neat peak at $\delta = 6.98$. There are four chemically distinct triad regioisomers in regioirregular PATs, as shown in Fig. 5.9. In ^{1}H NMR spectra, the TT-HT isomer has a peak at $\delta = 7.00$, HH-TT isomer has a peak at $\delta = 7.05$, and the HT-HH isomer has a peak at $\delta = 7.02$. Therefore, using the integral area of the peaks, the relative ratio of HT-HT couplings to non-HT-HT couplings can be determined. In the ^{13}C NMR spectrum, regioregular PTs exhibits four resonances in the aromatic region ($\delta = 128.5$, 130.5, 134.0, and 140.0 ppm), but regioirregular PTs show many resonances from 120 to 150 ppm.

PT derivatives can easily be synthesized by chemical oxidation methods [51]. In typical oxidation polymerization of PTs, 2,5-unsubstituted thiophenes, such as thiophene, 3-alkylthiophenes or 3-phenylthiophenes, etc., can be dissolved in CF, and, under an inert gas, the excessive oxidant ($FeCl_3$, $MoCl_5$, or $RuCl_3$) is added. The polymerization can take several hours. Generally, the $FeCl_3$ oxidation method has been widely used. The molecular weights of PTs prepared from this method range from 30 to 300 K, and the polydispersities range from 1.5 to 5.0. By this method, the regioregularities of poly(3-alkyl thiophene)s range from 70 to 80 %, whereas the regioregularity of poly(3-phenyl thiophene)s (P3PTs) can reach 90–95 %. The poor reproducibility is one of the major problems for oxidation polymerization reactions. As reported by Pomerantz et al. [52], the polymerization

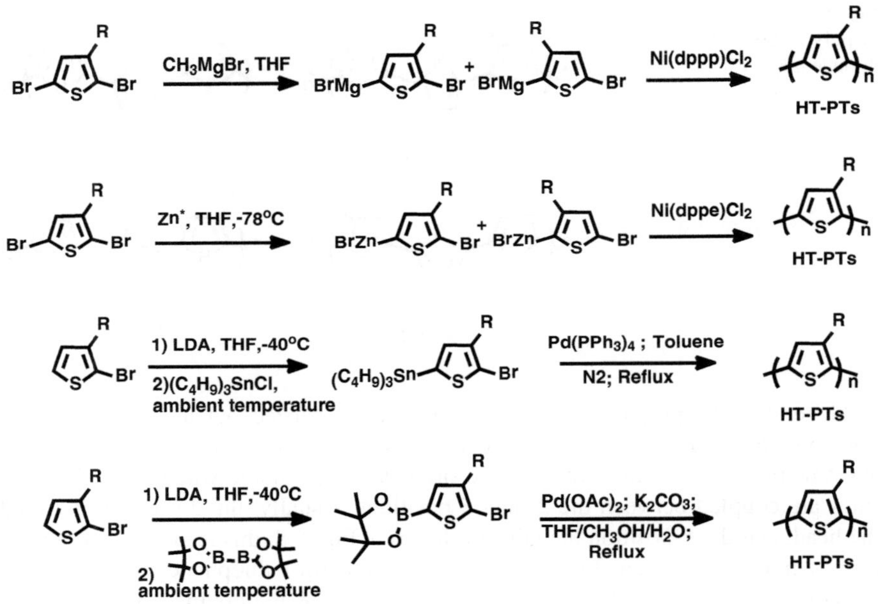

Fig. 5.9 Synthesis methods of regioregular poly(3-alkylthiophene)s by the McCullough and GRIM method, Rieke method, and the Stille and Suzuki coupling reactions

of 3-octylthiophene with $FeCl_3$ was repeated under identical reaction conditions five times, and the molecular weights of the five samples of poly(3-octylthiophene) ranged from 54 to 122 K with PDIs ranging from 1.6 to 2.7. Additionally, by using the identical preparation process, the polymer samples obtained from varied batches contain different levels of Fe impurities. Therefore, the oxidation method is seldom used to prepare PTs for applications in PSCs.

As shown in Fig. 5.9, regioregular poly(3-alkylthiophene)s can be prepared by various methods, such as the McCullough and GRIM method [53], the Rieke method [54], and the Stille [55] and Suzuki [56] coupling reactions. In 1992, McCullough et al. [53] reported the first synthesis method of head-to-tail coupled poly(3-alkylthiophene)s, and, from this method, close to 100 % HT-HT couplings. In this method, the monomer, 2-bromo-5-(bromomagnesio)-3-alkylthiophene, is obtained from 2-bromo-5-alkylthiophene at cryogenic temperature, and is then polymerized with catalytic amounts of Ni(dppp)Cl_2 (dppp is diphenylphosphino-propane). In this method, regioregular PTs were obtained in yields of 44–69 %. The Rieke method can also be used to synthesize regioregular poly(3-alkylthiophene) [54]. Regioselective control was realized on the basis of steric congestion at the reductive elimination step in the catalytic cycle. When the Ni(dppe)Cl_2 or Ni(dppp)Cl_2 was used as catalyst, the HT-couplings were more than 98.5 %. Other catalysts with less bulky, labile ligands such as PPh_3 combined with larger metal centers such as Pd lead to regiorandom poly(3-alkylthiophenes). The Stille [55] and Suzuki

[56] coupling methods are two convenient approaches to synthesize regioregular PTs, and these two methods exhibit great advantages in synthesis of multi-functional PTs because of the compatibilities of a large number of organic functional groups.

Regioregular poly(3-hexyl) thiophene (P3HT) is the most widely studied polymers in the field of PSC. The photovoltaic performance of P3HT/PCBM has been widely studied by numerous groups. It should be noted that the photovoltaic properties of P3HT/PCBM-based PSC devices are strongly dependent on the molecular weight and regioregularity of P3HT as well as the device fabrication methods [57–59]. Various strategies have therefore been developed to optimize the performance of P3HT-based polymer solar cells. In 2003, Sariciftci et al. [60] developed a postproduction treatment, namely, applying external voltage, which considerably improved the photovoltaic performance of solar cells based on the P3HT/PCBM system. Using this method, an enhancement of J_{sc} and an increase in external quantum efficiency (*EQE*) of 70 % are demonstrated. The breakthrough in the photovoltaic performance of P3HT/PCBM was realized by Li et al. [61] and Ma et al. [62] in 2005. Li et al. [61] developed a method of slow growth to optimize the morphology of P3HT/PCBM and an extremely high *PCE* of 4.4 % was realized, which was the highest value in the past decade. Alternatively, Ma et al. [62] applied post-production annealing at 150 °C, and P3HT/PCBM-based PSC devices with *PCE* approaching 5 % were achieved. In addition, these devices exhibited remarkable thermal stability. The improved performance was ascribed to the improved nanoscale morphologies, the increased crystallinities of P3HT, and the improved interfacial contact to the cathode, thereby enhancing the overall device efficiency. In 2010, He and Li et al. [63] introduced a novel fullerene acceptor, namely ICBA with a higher-lying LUMO level for the P3HT based polymers, which promoted the performance of P3HT to a new height by greatly increasing V_{oc}. After that, solvent additives and device engineering methods were utilized and increased the PCE of P3HT/ICBA to over 6.5 % [64, 65]. Clearly, the success of P3HT is largely associated with the active layer morphology [66, 67], such as using different casting solvents and film-forming speed, solvent and thermal annealing, etc.

Other P3HT-analogue polymers, for example poly(3-butylthiophene) (P3BT) and poly(3-pentylthiophene) (P3PT), were also explored in the device fabrications because of their similar crystalline and absorption characteristics. Nguyen et al. [68] investigated the effect of different alkyl lengths such as butyl, hexyl, octyl, decyl, and dodecyl on the photovoltaic properties of P3ATs-based PSC devices. Results revealed that longer alkyl chains (number of carbon atoms over 8) give poor efficiency and larger scale of phase separation. Jenekhe et al. [69] demonstrated that P3PT/PCBM-based BHJ PSC devices also give similar performances to those of P3HT/PCBM under the same conditions. Jenekhe and coworkers [70] also prepared P3BT nanowires and P3PT nanowires by solution-phase self-assembly, which were used to construct highly efficient P3AT/PCBM PSC devices. The fullerene/P3AT nanocomposite films showed an electrically bicontinuous nanoscale morphology and desirable *PCE* up to 3.3 %, which were identical with those of P3HT/PCBM-based photovoltaic cells. Afterwards, Gadisa et al. [71] optimized the performance of P3PT/PCBM and a high *PCE* of 4.6 % was observed. Later, Li et al. [72] optimized the performance of P3PT with different fullerene acceptors. A *PCE* of 3.1 % and a *PCE* up to 5.4 % were achieved in P3PT/PCBM- and P3PT/ICBA-based PSC devices, respectively.

Two structurally related polymers of P3HT with high performance were developed by Hou and coworkers [73, 74]. Hou and Li et al. introduced conjugated side chains in PTs and developed a novel class of two-dimension (2D) conjugated polymers. A typical example among these 2D polymers is PTs with bi(thienylenevinylene) side chains. Three 2D conjugated PTs with bi(thienylenevinylene) side chains biTV-PTs (see PbTV in Fig. 5.7) were designed and synthesized by Hou and Li et al. in 2006 [73]. Compared with the properties of P3HT, the biTV-PTs show broad absorption bands (350–650 nm), much stronger absorbance, and lower HOMO levels. The *PCE* of the biTV-PT-based PSC devices reached ~3.2 %, which is ~40 % increased relative to that (~2.4 %) of the devices based on P3HT under the same conditions. This discovery expanded the scope of designing 2D conjugated polymers for efficient PSC devices.

Although photovoltaic performance as well as the photocurrent of P3HT can be improved via morphological optimizations, the low voltage (~0.6 V) is still the limiting factor of P3HT. In 2009, Hou et al. [74] also developed an easy and effective way to promote the V_{oc} of a poly(3-alkylthiophene) by reducing the number of alkyl chains of P3HT, and a novel P3HT-analogue polymer, P3HDTTT (see Fig. 5.7) with a V_{oc} up to 0.82 V was obtained. In 2011, Li et al. [75] also introduced carboxylate substituent in polythiophene derivatives to tune downward the HOMO of polythiophene. A novel polythiophene derivative of PT-C3 (see Fig. 5.7) was also synthesized and characterized. The PSCs based on PT-C3/$PC_{71}BM$ exhibited relatively high V_{oc} of ~0.8 V. The *PCE* of the PSCs based on PT-C3 reached 3.87 % with V_{oc} = 0.78 V, J_{sc} of 9.68 mA cm^{-2}, and *FF* of 51.2 % under the illumination of AM1.5G, 100 mW/cm^2. These studies indicated that, in respect of further enhancing the J_{sc}, P3HDTTT would be a potential polymer to replace P3HT as blue absorber in tandem PSC devices.

5.2.2 *Donor–Acceptor Copolymers*

5.2.2.1 Benzothiadiazole (BT)-Based Polymers

D–A copolymers based on BT and its derivatives (see the molecular structures and photovoltaic parameters in Fig. 5.10 and Table 5.3) were explored because the HOMO and LUMO energy levels as well as band gaps can been effectively tuned by choosing donor (D) and acceptor (A) building blocks with appropriate electron-donating or -accepting natures. In 2003, the fluorene (FL) and benzothiadiazole (DTBT) copolymer, PFDTBT, was synthesized by Andersson and coworkers [76], which was generally considered as one of the first reports of donor–acceptor (D–A)

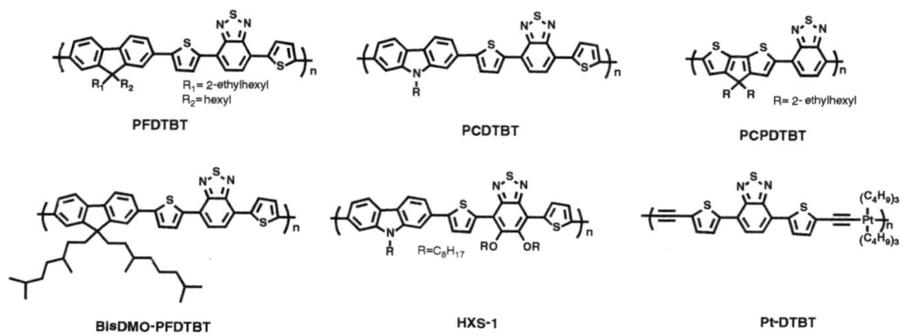

Fig. 5.10 Molecular structures of D–A copolymers based on benzothiadiazole

Table 5.3 Photovoltaic results of D–A copolymers based on benzothiadiazole

Materials	V_{oc} (V)	J_{sc} (mA/cm^2)	FF (%)	PCE (%)	μ_h [cm^2/(Vs)]	Ref.
BisDMO-PCDTBT	0.97	9.1	51	4.5	3×10^{-5}	[77]
PCPDTBT	0.62	16.2	55	5.5	–	[79]
PCDTBT	0.89	6.9	56	3.6	3×10^{-3}	[80]
HXS-1	0.81	9.6	69	5.4	–	[83]
Pt-DTBT	0.80	15.1	40	4.8	–	[84]

conjugated polymers. In this work, PFDTBT exhibited poor solubility but reached a moderate *PCE* of 2.2 %. Although the *PCE* was not high, the D–A copolymer exhibited potentials such as LBGs and tunable energy levels. Following the pioneer work of Andersson et al., numerous novel D–A copolymers based on different donor units and acceptor units were synthesized and utilized in PSC devices. It should be noted that D–A copolymers have been the most successful class of polymer photovoltaic materials for PSCs in recent years.

Considering the V_{oc} of this polymer is as high as 1 V, the performance of PFBTBT still has a large promotion space. In 2008, Hou et al. [77] altered the alkyl chains in the classical PFDTBT backbone. Two novel polymers, bisEH-PFDTBT and bisDMO-PFDTBT, employing the same polymer backbone as PFDTBT but different side chains, were studied to investigate the side-chain effects. After carefully optimizing the side chains, the *PCE* of bisDMO-PFDTBT was increased to 4.5 %. Notably, the saturated alkyl chains have little influence on the molecular energy levels of PFDTBT. From quantum-chemical calculations, both the ethyl groups are in the proximity of the conjugated backbone, and thus decreased the probability of π–π stacking in bisEH-PFDTBT originating from the steric effect. In contrast, there are two small methyl groups located on the third and seventh carbons of the bisDMO-PFDTBT which not only decrease the steric effect but also increase the solubility and hole mobility.

The cyclopentadithiophene and DTBT copolymer, PCPDTBT, is also a well-known LBG D–A copolymer, which was developed by Brabec et al. [78] in 2006. PCPDTBT is the first LBG polymer with high efficient photovoltaic activity in the IR spectral region. In 2008, Bazan et al. [79] introduced solvent additive 1,8-octanedithiol (OT) to optimize the nanomorphology of this polymer, and a twofold enhancement was observed. A high *PCE* of 5.5 % was recorded and regarded as a milestone of novel photovoltaic polymers. It should be noted that PCPDTBT is an amorphous polymer. The approach provided a feasible tool for modulating the heterojunction morphology in donor/acceptor systems where thermal annealing is not effective.

Another representative D–A copolymer is PCDTBT, which was copolymerized by carbazole (CZ) and DTBT by the Leclerc group [80]. PCDTBT had a low-lying HOMO level of −5.45 eV and a moderate band gap of 1.88 eV. A preliminary *PCE* of 3.6 % was measured at a donor/acceptor weight ratio of 1:4. Various methods were proposed to optimize the photovoltaic properties of PCDTBT. Notably, PCDTBT is among the most efficient, stable, and low-cost photovoltaic materials for PSC devices with a high V_{oc} (0.85–0.90 V), high *PCE* (6–7 %), and long lifetime (~7 years), which is now considered to be one of the new benchmarks for the development of highly efficient BHJ solar cells [81, 82]. Afterwards, Bo et al. [83] optimized the side chains of DTBT; a novel polymer, HXS-1 with planar configuration, achieved a high *PCE* over 5 % and excellent *FF* approaching 70 %.

A platinum metallopolyyne (herein called Pt-DTBT) with a LBG of 1.85 eV was also reported by Wong et al. [84]. The PSCs based on the metallated polymer exhibited an average *PCE* of 4.1 % without annealing. It is noteworthy that the devices based on the Pt-DTBT/PCBM system exhibited a very high J_{sc} of ~15 mA/cm^2 and relatively high V_{oc} of ~0.80 V. This is the first time that a metallated polymer was applied in PSC devices to get such a high *PCE*.

5.2.2.2 Silole-Containing Polymers

Fused coplanar thiophene-based heterocycles, such as dithieno[3,2-*b*:2′,3′-*d*]silole (DTS) and IDT, have been actively utilized as donor units to construct novel D–A copolymers. To optimize the photovoltaic performance of D–A copolymers as discussed above, such as PFDTBT and PCPDTBT, silole-containing building blocks, such as silafluorene (SiF) and dithienosilole (DTS), have been widely utilized in efficient D–A polymers. In the following we overview the representative polymers in the third generation. The majority of these polymers exhibited high *PCE* ranging from 5 to 7 %.

Cao and coworkers [85] successfully introduced the silicon atom into FL units and copolymerized the novel SiF unit and 4,7-di(2′-thienyl)-2,1,3-benzothiadiazole (DTBT). High-performance polymer solar cells composed of an alternating copolymer PSiF-DBT as the electron donor and $PC_{71}BM$ as the electron acceptor were investigated. A high *PCE* up to 5.4 % with a high V_{oc} of 0.90 V, a J_{sc} of 9.5 mA cm^{-2}, and a *FF* of 50.7 % was achieved under the illumination of AM

1.5*G* 80 mW/cm^2. Moreover, PSiF-DBT also showed a high hole mobility of ~1 × 10^{-3} cm^2 V^{-1} s^{-1}. Bo and coworkers [86] optimized the side chains in DTBT of the PSiFDBT, and obtained an improved *PCE* of up to 6.05 %. Driven by the high performance of PCPDTBT and PSiFDBT, Hou et al. [87] designed and synthesized a LBG DTS-containing polymer, PSBTBT. The preliminary *PCE* obtained with thermal annealing (140 °C, 5 min) was 5.1 % and then was increased to ~5.6 %. Chen et al. [88] further revealed the origin of the high performance of PSBTBT relative to PCPDTBT. Striking morphological changes were observed in polymer: fullerene bulk heterojunctions upon the substitution of the bridging atom. GIXRD investigations indicated increased π–π stacking in silole-based polymers compared to the carbon-bridged analogue [89, 90]. More importantly, the response range of the PSBTBT covered the whole visible range from 380 to 800 nm, which indicated that PSBTBT is an efficient red-absorbing polymer for tandem PSCs. Benefitting from the broad absorption (300–800 nm), *PCE* values up to 7 % were achieved in tandem PSC devices by the Yang group and others.

Afterwards, various units were copolymerized with these silole-containing building blocks and a class of low-band gap polymers was developed. Their molecular structure and photovoltaic parameters are depicted in Fig. 5.11 and Table 5.4. In particular, DTS-based copolymers show a broad absorption, relatively lower HOMO energy level, and higher hole mobility, which are attractive for

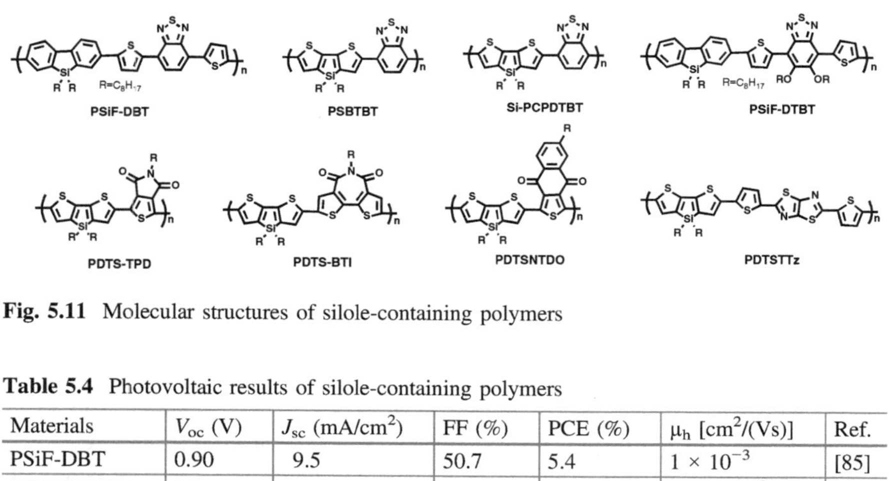

Fig. 5.11 Molecular structures of silole-containing polymers

Table 5.4 Photovoltaic results of silole-containing polymers

Materials	V_{oc} (V)	J_{sc} (mA/cm^2)	FF (%)	PCE (%)	μ_h [cm^2/(Vs)]	Ref.
PSiF-DBT	0.90	9.5	50.7	5.4	1 × 10^{-3}	[85]
PSiF-DTBT	0.91	11.5	58	6.05	3.2 × 10^{-3}	[86]
PSBTBT	0.68	12.7	55	5.1	3 × 10^{-3}	[87]
Si-PCPDTBT	0.58	14.92	61	5.24	1 × 10^{-3}	[89]
PDTSTTz	0.77	11.9	61	5.59	3.56 × 10^{-3}	[91]
PDTSNTDO	0.88	9.24	64	5.21	–	[92]
PDTS-BTI	0.80	12.81	62.3	6.41	–	[93]
PDTSTPD	0.88	12.2	68	7.3	1 × 10^{-4}	[94]

researchers. Considering that the thiazolothiazole (TTz) unit has a rigid and coplanar configuration and thereby ensures a highly extended π-electron system and strong π–π stacking, Zhang et al. [91] copolymerized DTS with the TTz acceptor unit. The *PCE* of the PSC based on PDTSTTz/$PC_{71}BM$ (1:1, wt/wt) reached 5.59 % with $V_{oc} = 0.77$ V, $J_{sc} = 11.9$ mA/cm^2, and $FF = 61$ % under optimized conditions (thermal annealing at 100 °C for 15 min). Cui et al. [92] designed a strong electron-withdrawing unit, naphtho[2,3-*c*]thiophene-4,9-dione, which was copolymerized with DTS to construct a D–A copolymer, PDTSNTDO, with a narrow band gap and lower lying HOMO level. The *PCE* of the PDTSNTDO-based device reached 5.21 %, with a high V_{oc} of 0.88 V.

Marks et al. [93] synthesized a new series of bithiopheneimide (BTI)-based donor–acceptor copolymers for efficient PSCs. Among these, PSC featuring BTI and DTS copolymer as donor and $PC_{71}BM$ as acceptor exhibited promising device performance with *PCE* up to 6.41 % and high V_{oc} over 0.80 V. The BTI analogue, TPD-based device exhibited 0.08 V higher V_{oc} with an enhanced *PCE* of 6.83 %, which is mainly attributed to the lower-lying HOMO induced by the higher imide group density in the backbone. Lu and Tao et al. [94] recently synthesized a new D–A copolymer PDTSTPD of DTS and thienopyrrole-4,6-dione (TPD) obtaining both a LBG (1.73 eV) and a deep HOMO level of −5.57 eV. When blended with $PC_{71}BM$, PDTSTPD exhibited an excellent *PCE* of 7.3 % on the photovoltaic devices with an active area of ~1 cm^2. These results demonstrate the great potential of DTS-based polymers for high-performance solar cells, and provide valuable insights into structure-property relationship of atom substitution.

5.2.2.3 Diketopyrrolopyrrole (DPP)-Based Polymers

As one of the high-performance pigments, the DPP unit was developed in the past few decades for constructing LBG (1.5 eV) conjugated polymers produced by its strong electron-withdrawing properties. Another attractive property of DPP is its excellent charge carrier mobility for both holes and electrons. Most of the photovoltaic performance of DPP-based polymers is likely to depend on its solvent-induced morphology [32].

Janssen et al. [95] introduced DPP unit in D–A copolymers in 2008. Dozens of newly designed DPP-based LBG polymers have been frequently reported since 2009. The molecular structure and photovoltaic parameters of these DPP based polymers are depicted in Fig. 5.12 and Table 5.5. Hou et al. [96] copolymerized DPP with various electron-donating monomers and obtained a series of new LBG polymers based on the DPP unit. Among these DPP-based polymers, benzodithiophene and DPP copolymer, PBDT-DPP showed a small band gap of 1.34 eV and a moderate *PCE* of 4.45 % was achieved.

When copolymerizing DPP with thiophene and the like, ultra-LBG (E_g ~1.3 eV) polymers such as PDPP3T [34] and PDPP3MT [97] can be obtained. The first highly performing DPP-containing polymer, PDPP3T, was synthesized [34] and applied in organic photovoltaic and field-effect transistor devices by Janssen et al. in 2009. After

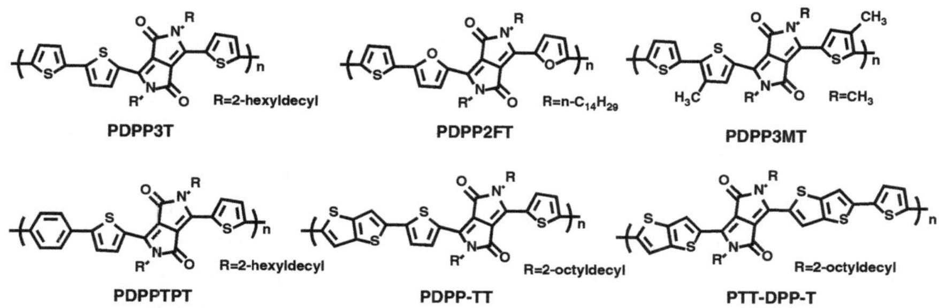

Fig. 5.12 Molecular structures of representative DPP-based efficient polymers

Table 5.5 Photovoltaic performances of the representative D–A copolymers based on DPP units

Copolymers	V_{oc} (V)	J_{sc} (mA/cm^2)	FF (%)	PCE (%)	μ_h [cm^2/(Vs)]	Ref.
PDPP3T	0.66	15.41	65.92	6.71	3.9×10^{-3}	[35]
PDPP3MT	0.60	17.8	66	6.8	–	[97]
PTT-DPP-T	0.58	15	61	5.4	–	[98]
PDPP2T-TT	0.66	14.8	70	6.9	–	[100]
PDPPTPT	0.80	10.8	65	5.5	–	[101]
PDPP2FT	0.65	14.8	64	6.5	7×10^{-4}	[102]

optimizing processing solvent and molecular weight, the moderate *PCE* of 4.7 % was increased to 6.7 % by Ye et al. [35] and 7.0 % by Janssen et al. [97]. To tune the coplanarity of PDPP3T, Janssen et al. [97] introduced methyl into the thiophene group of DPP unit, and a higher performance ultra-LBG polymer PDPP3MT was achieved with *PCE* of 6.8 % in the classic device configuration. Bronstein et al. [98, 99] reported the synthesis and polymerization of a novel thieno[3,2-*b*]thiophene-DPP-based building block in recent years. Copolymerization with thiophene afforded the resulting polymer, PTT-DPP-T, with a high hole mobility of 1.95 cm^2 V^{-1} s^{-1}. PSC devices comprised of PTT-DPP-T and $PC_{71}BM$ also exhibited an excellent *PCE* of 5.4 % and high J_{sc} up to 15 mA/cm^2 [98]. Later, Li et al. [100] designed a high-molecular-weight conjugated polymer based on an alternating electron-rich TT unit and an electron-deficient DPP unit, which also provided efficient polymer solar cells with *PCE* up to 6.9 %. The optimal morphology of the new polymer/PCBM blend reduced bimolecular recombination and thereby allowed a high *FF* up to 70 % and high *PCE*s over 6 % with film thickness up to 300 nm to be achieved. Because of the lower π-electron density of the benzene unit compared with thiophene, the DPP and benzene copolymer, PDPPTPT [101], showed a relatively higher band gap of 1.53 eV and lower HOMO level (−5.35 eV) together with a broad photo-response range up to 800 nm. When blended with $PC_{71}BM$, higher *PCE* of 5.5 % with $J_{sc} = 10.8$ mA/cm^2, $V_{oc} = 0.80$ V, and $FF = 65$ % was achieved. Clearly, PDPPTPT is a suitable photovoltaic polymer for multi-junction devices because of the balance between the broad absorption range and high voltage.

Frechet et al. [102, 103] developed a series of furan-containing DPP-based polymers, such as PDPP2FT and PDPP3F, with substantial power conversion efficiencies. Inserting furan into the backbone of the conjugated polymers enables the utility of relatively small solubilizing alkyl chains because of the significant contribution of the furan to overall polymer solubility in common organic solvents. PSC devices fabricated from PDPP2FT and $PC_{71}BM$ as active layers showed a high *PCE* reaching 6.5 % [102]. The design and synthesis of the successful examples of furan-containing LBG polymers paved the way to developing environmental photovoltaic polymers. Interestingly, it is also noted that furan-containing LBG polymers with high side-chain tunability also provide insights into molecular order in efficient PSCs. In recent years, the *PCE* of DPP based polymers was also increased to exceed 7 % by the Janssen group [104] and the Yang group [9–11] because of the molecular weight and side chain optimization. These highly efficient polymers are introduced in the fourth generation. Moreover, these red absorber materials exhibited potential applications in tandem and triple junction PSCs.

5.2.2.4 Indacenodithiophene (IDT)-Based Polymers

Ladder type units such as IDT, also known as thiophene-phenylene-thiophene (TPT), constitute a class of efficient D–A copolymers, which have been emerging as efficient building blocks since 2008 [105, 106]. Typically, the IDT unit is extremely versatile with a coplanar aromatic ring structure, and the electron density can be manipulated by the choice of the bridging group between the rings. The coplanarity of the IDT unit could enhance interchain interaction of the polymers and is expected to afford higher hole mobility. Ting et al. pioneered the IDT-based random and alternating copolymers for photovoltaic applications [107, 108]. In 2008, two LBG IDT-based polymers with high hole mobility (3.4×10^{-3} cm^2 V^{-1} s^{-1}) were designed and synthesized for application in PSCs by Ting and coworkers [106]. High-performance *PCE* of 4.4 % was obtained, which was superior to that of the analogous P3HT based PSC device under the same device fabrication condition. In 2010, Ting and collaborators [108] also synthesized an alternating copolymer (a-PTPTBT) based on IDT and BT and the highest *PCE* reached 6.4 % of the corresponding PSC under the optimal condition of solvent vapor annealing. The molecular structure and photovoltaic parameters of the representative IDT based polymers are depicted in Fig. 5.13 and Table 5.6.

Jen et al. [109] first applied aryl side chain substituted IDT in quinoxaline-based conjugated polymers and combined IDT and two quinoxaline derivatives to form novel polymers (PIDT-diphQ and PIDT-phanQ; Fig. 5.13). Because of the enhanced planarity of phenanthrenequinoxaline (phanQ), PIDTphanQ/$PC_{71}BM$-based PSC device exhibited an improved *PCE* of 6.24 % compared to the *PCE* of 5.69 % in PIDT-diphQ/$PC_{71}BM$-based device. Jen et al. also developed several high-performance IDT-based D–A copolymers such as PIDT-DFBT [110, 111]. Huang, Cao and coworkers [112] synthesized three low band-gap conjugated polymers via Stille copolymerization of IDT and naphtho[1,2-*c*:5,6-*c*]

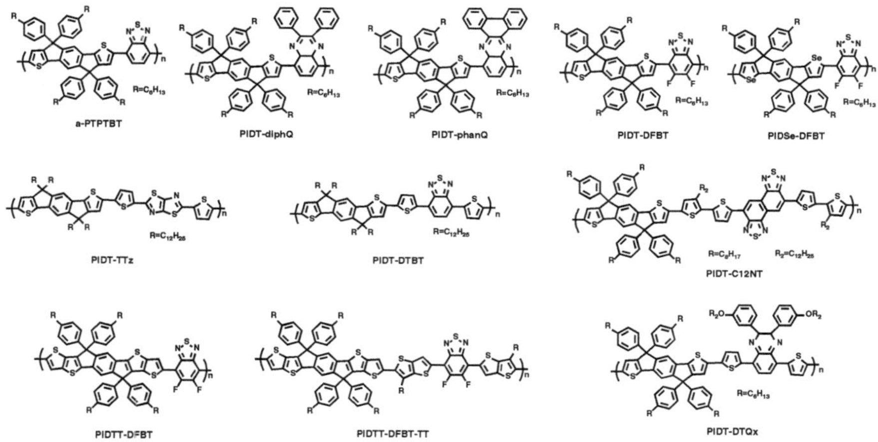

Fig. 5.13 Molecular structures of the representative IDT-based efficient polymers

Table 5.6 Photovoltaic results of the representative IDT-based polymers

Materials	V_{oc} (V)	J_{sc} (mA/cm^2)	FF (%)	PCE (%)	μ_h [cm^2/(Vs)]	Ref.
a-PTPTBT	0.85	11.2	67.2	6.41	2.4×10^{-5}	[108]
PIDT-diphQ	0.87	10.9	60	5.69	1.14×10^{-3}	[109]
PIDT-phanQ	0.87	11.2	64	6.2	2.06×10^{-3}	[109]
PIDT-DFBT	0.97	11.2	55	5.97	4×10^{-5}	[111]
PIDT-C12NT	0.90	10.21	55	5.05	2.42×10^{-4}	[112]
PIDT-TTz	0.89	13.3	48.9	5.79	4.99×10^{-3}	[113]
PIDT-DTBT	0.82	12.27	56.7	6.17	2.24×10^{-3}	[113]
PIDSe-DFBT	0.89	13.7	56.3	6.8	0.15	[114]
PIDTT-DFBT	0.95	12.21	61	7.03	3×10^{-4}	[111]
PIDTT-DFBT-TT	0.96	11.9	63	7.2	4×10^{-2}	[111]
PIDT-DTQx	0.87	12.34	70.23	7.51	1.18×10^{-4}	[115]

bis(1,2,5-thiadiazole) (NT) based monomers. The energy levels, absorption spectra, and band gaps of the target polymers were well tuned by utilizing different thiophene derivatives as spacer between IDT and NT units, and polymer PIDT-C12NT which employed bithiophene attached with dodecyl side chain as spacer exhibited superior properties compared with the other two copolymers. All polymers exhibited deep HOMO levels and subsequently led to high open circuit voltages of the fabricated PSC devices. The best performance (5.05 %) was achieved with PIDT-C12NT as donor polymer, which can be ascribed to its higher hole mobility, the optimal interpenetrating network, as well as enhanced absorption coefficient with respect to the other two polymers. The photovoltaic results demonstrated that the combination of IDT and NT with appropriate spacers might be a promising approach for the application of solar cells.

Considering potentials of the alkyl-substituted IDT, such as good planarity, good solubility, and high hole mobility, Zhang et al. [113] copolymerized alkyl-substituted IDT with different acceptor units including bithiazole (BTz), thiazolothiazole (TTz), tetrazine (TZ), and benzothiadiazole (DTBT). Among these copolymers, PIDTTTz has the highest hole mobility of 4.99×10^{-3} $cm^2\ V^{-1}\ s^{-1}$ and the *PCE* of the PSC based on PDTSTTz/$PC_{71}BM$ reached 5.79 % with a V_{oc} of 0.89 V, a J_{sc} of 13.3 mA/cm^2, and a *FF* of 48.9 %, under the donor/acceptor component ratio of 1:2 (wt/wt). In comparison with PIDT-TTz, PIDT-DTBT has a medium band gap of 1.68 eV and a similar hole mobility of 2.24×10^{-3} $cm^2\ v^{-1}\ s^{-1}$. The PSC based on PIDT-DTBT/$PC_{71}BM$ reached an even higher *PCE* of 6.17 %. These results indicated that the D–A copolymers based on the alkyl-substituted IDT unit are promising photovoltaic polymer materials because of the excellent hole mobility and solubility as well as deeper HOMO levels. The selenophene and other analogues of IDT were developed in recent years. For instance, Jen et al. [114] improved the molecular weight of IDSe-based polymers and a high *PCE* up to 6.8 % was recorded for a PIDSe-DFBT-based PSC device, which exhibited over 10 % enhancement compared to that of a PIDT-DFBT-based PSC device. This work demonstrated that selenium substitution on the IDT is an effective method to reduce the band gap and improve the photovoltaic performance of IDT-based polymers with higher molecular weight.

Recently, several IDT-based polymers realized high *PCE*s over 7 % and exhibited unique charge transport properties. For instance, Jen et al. [111] incorporated TT moiety in IDT unit and designed an IDTT unit. Utilizing IDTT unit in DFBT and TT bridged DFBT-based polymers, corresponding copolymers PIDTT-DFBT and PIDTT-DFBT-TT with *PCE* up to 7 % as well as V_{oc} over 0.95 V were observed. Hou et al. [115] incorporated IDT with DTQx and produced a highly efficient polymer PIDT-DTQx with a *PCE* up to 7.5 %, which was the highest value in IDT-based copolymers. Interestingly, PIDT-DTQx exhibited the best performance under the donor/acceptor weigh ratio of 1:4.

5.2.2.5 Benzodithiophene (BDT)-Based Polymers

Although various efficient photovoltaic polymers have been developed, the polymers discussed above (*PCE* < 7 %) still could not meet the need of commercial requirements. In this part, we overview the representative highly efficient photovoltaic polymers. In particular, the rapid progress of benzo[1,2-*b*:4,5-*b*′]dithiophene (BDT) and two dimensional BDT-based polymers are introduced. These polymers with *PCE* over 7 % draw lots of attention from both the industrial and the academic communities.

Benzodithiophene (BDT) has a large planar conjugated structure and easily forms π–π stacking, and thereby improves the hole mobility [116]. In 2008, Hou et al. [117] first introduced the BDT unit in the synthesis and application of photovoltaic polymers. In the work, alkoxy-substituted BDT was copolymerized with seven different units. The band gap and absorption spectrum of the BDT-based polymers could be effectively tuned within a wide range. In 2009, Yu and coworkers copolymerized BDT with thieno[3,4-*b*]thiophene (TT) and designed a series of BDT-TT copolymers, namely the PTB series [118–120]. Among these PTB series polymers, PTB7 [120] is the best-performing photovoltaic polymer because of the optimal side chain and functional substituents. Hou et al. also designed a series of BDT and TT copolymers, such as PBDTTT-C [121], PBDTTT-CF [122], and PBDTTT-S [123]. It should be mentioned that PBDTTT-CF and PTB7 are the first two polymers with *PCE* exceeding 7 %, which significantly pushes the PSC research to a new height. Following this pioneering work, *PCE*s of more than 7 % were frequently reported in various groups, as depicted in Fig. 5.14. The photovoltaic results of the corresponding polymers are listed in Table 5.7.

Besides TT, numerous building blocks including DPP, thieno[3,4-*c*]pyrrole-4,6-dione (TPD), benzothiadiazole (BT), and benzoxadiazole (BO) have been

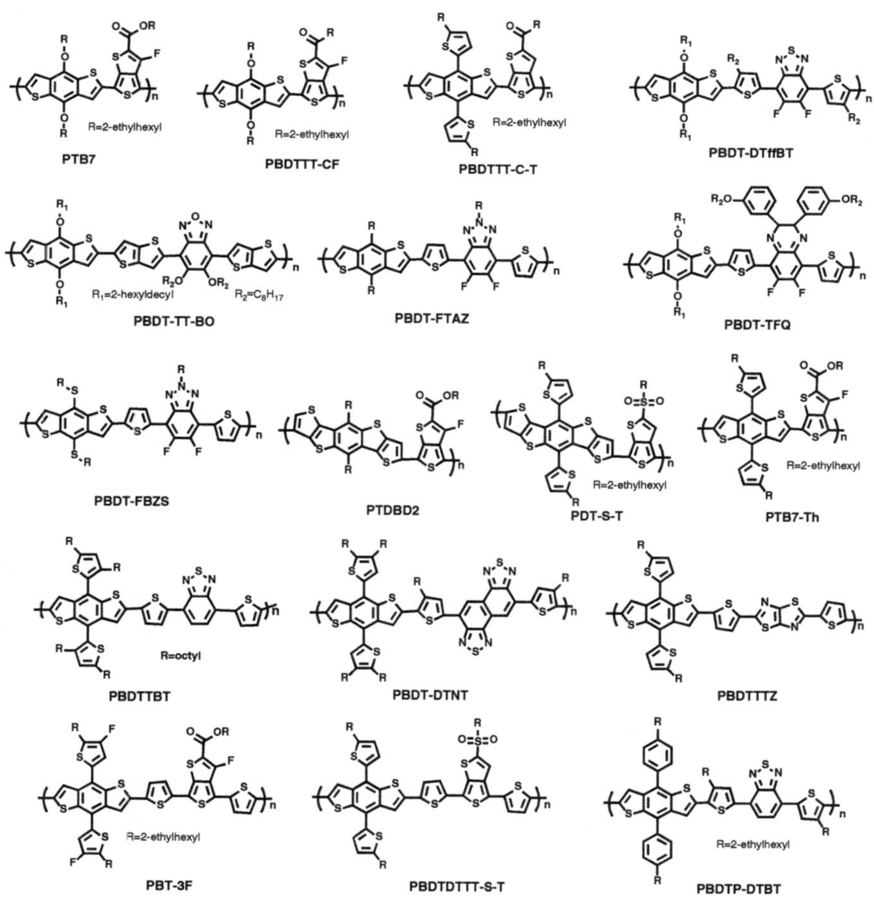

Fig. 5.14 Molecular structures of BDT-based highly efficient polymers

Table 5.7 Photovoltaic performance of the representative D–A copolymers based on BDT and BDT analogues

Materials	V_{oc} (V)	J_{sc} (mA/cm^2)	FF (%)	PCE (%)	μ_h [cm^2/(Vs)]	Ref.
PBDTTT-CF	0.76	15.2	66.9	7.73	7×10^{-4}	[122]
PTB7	0.74	14.5	68.97	7.4	5.8×10^{-4}	[120]
PBnDTDTffBT	0.91	12.91	61.2	7.2	8.3×10^{-3}	[124]
PBDT-FTAZ	0.79	12.45	72.2	7.1	1.03×10^{-3}	[125]
PBDTDTBTff-3	0.78	15.38	69.2	8.30	–	[126]
PBDT-TT-BO	0.76	13.87	66.6	7.05	0.023	[128]
PBDT-TFQ	0.76	18.2	58.1	8.0	–	[129]
PBDTFBZS	0.88	12.36	71.2	7.74	4.3×10^{-3}	[127]
PBDTTT-C-T	0.74	17.48	58.7	7.59	0.27	[131]
PBDTDTTT-S-T	0.70	17.07	66.3	7.81	2.76×10^{-3}	[132]
PBDTT-SeDPP	0.69	16.8	62	7.2	6.9×10^{-4}	[10]
PBDTP-DTBT	0.88	12.94	70.9	8.07	8.89×10^{-2}	[139]
PBDTTBT	0.92	10.7	57.5	5.66	–	[130]
PBDTTTZ	0.85	10.4	59.0	5.22	1.67×10^{-5}	[138]
PBDTDTNT	0.80	11.71	61.0	6.00	3×10^{-5}	[137]
PBT-3F	0.78	15.2	72.4	8.6	–	[140]
PDT-S-T	0.73	16.63	64.13	7.79	–	[143]
PTDBD2	0.89	13.0	65.3	7.6	–	[144]

successfully combined with the BDT unit to constitute highly efficient photovoltaic polymers.

You and collaborators [124] reported the first successful application of fluorinated benzothiadiazole in BDT-based polymers. The resulting polymer PBDTDTffBT exhibited down-shifted HOMO and LUMO energy levels and a similar band gap relative to its non-fluorinated analogue. As a result, the PSC device employing PBDTDTffBT achieved a greatly improved *PCE* of 7.2 %. Similarly, fluorine was incorporated into 2-alkyl-benzotriazoles (TAZ) by You et al. [125] and a novel polymer PBnDT–FTAZ was designed. Interestingly, although the band gap of fluorinated polymer PBnDT–FTAZ was ~2.0 eV, the copolymer exhibited a *PCE* above 7 % when mixed with $PC_{61}BM$, and a *PCE* above 6 % was still attained even for thickness up to 1 μm. The superior performance originated from their high hole mobility and low HOMO and LUMO levels. Very recently, Wang et al. [126] successfully optimized the similar fluorinated benzothiadiazole-based conjugated copolymers, PBDTDTBTff, with different side chains. A *PCE* up to 8.30 % was achieved in PBDTDTBTff-3-based PSC devices with ~100 nm thickness active layers without any processing additives or post-treatments, which was the highest value for the conventional single-junction polymer solar cells via simple fabrication architecture. In addition, it is noteworthy that PBDTDTBTff-3 could afford high *PCE*s of 7.27 % at ~200 nm thickness active layers and 6.56 %, even for thicknesses up to ~300 nm. Therefore, the results demonstrated that BDT and fluorinated benzothiadiazole polymers should be promising candidates for developing high-performance large-scale roll-to-roll fabrication of PSCs.

Peng et al. [127] further applied dialkylthiol-substituted BDT in the synthesis of conjugated copolymers based on monofluorinated benzotriazole (TAZ) acceptor block, and a high-performance PBDTFBZS was developed. As expected, wide band-gaps and deep HOMO and LUMO energy levels were observed in the resulting copolymers. A *PCE* up to 7.74 % was achieved from the regular single device based on PBDTFBZS with a $V_{oc} = 0.88$ V, a $J_{sc} = 12.36$ mA/cm^2 as well as a high *FF* of 71.2 %. The enhanced V_{oc} can be ascribed to a low-lying HOMO energy level by the introduction of dialkylthiol and fluorine substituents on the PBDT-DTBT polymer backbone. The improvements in J_{sc} *FF* are probably because of high hole mobility, suppressed charge recombination, and optimal blend morphology. Because of the excellent performance of polymers, tandem PSC devices featuring PBDTFBZS as blue absorber material and DPP-based polymer as red absorber material exhibited high *PCE* up to 9.40 %. Compared to BT-based polymers, benzoxadiazole (BO)-based polymers exhibited a rather disappointed performance. Very recently, to overcome the relatively poor performance and low solubility of BO-based polymers, Li and coworkers [128] designed a TT-bridged polymer, namely, PBDT-TT-BO, which was copolymerized by BDT unit and TT-bridged BO acceptor unit. The *PCE* of the PSC device featuring PBDT-TT-BO as donor polymer reached 7.05 %, which was the champion result in BO containing conjugated polymers and comparable to that of its BT counterparts. A medium-band gap fluorinated quinoxaline-based conjugated polymer of PBDT-TFQ was designed and synthesized by Chou and coworkers [129]. With an optimized blend ratio of PBDT-TFQ:$PC_{71}BM$ (1:1, wt/wt), a high *PCE* of 8.0 % was obtained, with a V_{oc} of 0.76 V, a J_{sc} of 18.2 mA/cm^2, and a *FF* of 58.1 %. The resulting copolymer achieved an extremely high J_{sc}, which was probably caused by the higher hole mobility of PBDT-TFQ together with the better morphology for efficient exciton dissociation and charge transport.

In recent years, a large class of donor–acceptor copolymers based on the two-dimensional conjugated BDT units (collectively called 2D-BDT units) was developed by Hou and coworkers [130–136]. In 2010, Huo et al. [130] copolymerized thiophene bridged 2,1,3-benzothiadiazole (BT) with an alkylthienyl-substituted BDT unit, and the first 2D-BDT-based polymer, PBDTTBT with V_{oc} up to 0.92 V and *PCE* over 5.6 % was achieved. By introducing the 2D-BDT units such as alkylthienyl substituted BDT, the *PCEs* of several novel polymers, including PBDTTT-C-T, PBDTTDTT-S-T, PBDTP-DTBT, and PBT-3F, have been increased to 8–9 % in Hou's group. A typical example is PBDTTT-C-T. Huo et al. [131] replaced the alkyl chain with an alkylthienyl side chain in the BDT unit of PBDTTT-C, which resulted in a high-performance polymer PBDTTT-C-T. Notably, PBDTTT-C-T has been widely utilized in versatile photovoltaic devices such as inverted devices because of its excellent properties. The 2D-BDT design rules were also successfully applied to more than three pairs of BDT-based systems [132–136]. Two donor–acceptor conjugated polymers, PBDT-DTBT and PBDT-DTNT, based on 2,1,3-benzothiadiazole (BT) and naphtho[1,2-*c*:5,6-*c*]bis [1, 2, 5] thiadiazole (NT), have been designed, synthesized, and characterized by Huang and collaborators [137]. Compared with BT, NT contains two fused 1,2,5-thiadiazole rings which narrow the band gap, enhance the interchain packing, and improve the charge mobility of the resulting polymer. Consequently, the NT-based polymer PBDT-DTNT exhibited considerably better photovoltaic performance with a *PCE* of 6.00 %, which is significantly higher than that of the BT-based polymer PBDT-DTBT under identical conditions. Huo et al. [138] also synthesized a wide band gap BDTT-based polymer, PBDTTTZ, and achieved a desirable *PCE* of 5.21 %, which is one of the highest among wide band gap (>2 eV) polymers. Considering that the absorption edge of the PBDTTTZ is 620 nm, which is ~20 nm shorter than that of P3HT, PBDTTTZ should be an excellent blue absorber for tandem devices relative to P3HT. Clearly, the impressive performance of BDT-based polymers has shown its obvious potential for achieving high performance in PSCs. The photovoltaic results demonstrated that photovoltaic polymers based on 2D-BDT units exhibited improved hole mobilities, and significantly improved photovoltaic performance relative to those of their corresponding alkoxy-substituted BDT-based photovoltaic polymers. Therefore, replacing BDT with 2D-BDT units should be a method to enhance the efficiency with wide applicability.

Similarly, Chen et al. [12] recently introduced the alkylthienyl side chain in the BDT unit of PTB7-Th, which produced a significantly improved *PCE* (~9.35 %) relative to PTB7 under the same conditions. Yang et al. [9–11] incorporated the 2D-BDT unit in the DPP based polymers, and high-performance low band-gap polymers such as PBDTT-DPP [9] and, PBDTT-SeDPP [10] were designed and synthesized. These PBDTT-DPP materials played vital roles in versatile highly efficient semi-transparent and tandem PSC devices. It is worth mentioning that the tandem devices incorporating PBDTT-DPP as LBG absorbing polymers have increased the *PCE* to a new height (PCE > 9 %) and have drawn worldwide attention.

Recently, alkylphenyl substituted BDT (BDT-P) has attracted much attention as a weak electron-donating unit with a large π-conjugated area and good planarity. Yang and coworkers [11] reported a series of copolymers based on BDT-P for photovoltaic application. Results revealed that BDT-P-based polymers exhibit similar photovoltaic performance and relatively higher V_{oc} in comparison with polymers based on alkylthienyl substituted BDT. Afterwards, Zhang et al. [139] designed and synthesized a novel copolymer PBDTP-DTBT based on benzothiadiazole and BDT-P. The best-performing PSC device based on PBDTP-DTBT/$PC_{71}BM$ (1:1.5, wt/wt) reached a *PCE* up to 8.07 % with a V_{oc} = 0.88 V, a J_{sc} = 12.94 mA/cm^2, and a *FF* = 70.9 % under the irradiation of AM 1.5G, 100 mW/cm^2. Interestingly, with only 0.5 vol.% DIO, the PBDTP-DTBT-based D/A blends exhibited the best performance because of the well-tuned morphology. Although the DIO volume was higher than 1 %, the *PCE* was greatly reduced to a moderate value (<7 %). To ameliorate the relatively low voltage in alkylthienyl-substituted BDT and thiophene-bridged TT-based copolymers, Zhang et al. [140] further demonstrated the synergistic effect of introducing fluorine (F) atoms of lowering the molecular energy levels, HOMO and LUMO levels of copolymers of alkylthienyl substituted BDT and thiophene-bridged TT. When three F atoms were introduced in both the donor and acceptor units, the PSC device based on the trifluorinated polymer (PBT-3F) showed an extremely high *PCE* of 8.6 % and a significantly improved V_{oc} of 0.78 V, which was the efficiency record for PSCs utilizing the classic device configuration. This work demonstrated that introducing F onto the appropriate positions of the donor units in D–A polymers, especially BDT-based polymers, is a promising method to modulate effectively the molecular energy levels for better applications in PSCs.

Driven by the great success of BDT, various BDT analogues were designed and synthesized for producing high-performance D–A copolymers [141–144]. For instance, Hou et al. [142] designed a novel BDT analogue, namely DTBDT, and synthesized a series of DTBDT-based copolymers. Among these polymers, PDT-S-T [143] exhibited the best performance with *PCE* up to 7.79 % because of favorable and ordered molecular packing originating from the linear conformation of the backbone. Yu et al. [144] copolymerized DTBDT and TT, and a highly efficient polymer, PTDBD2, with *PCE* over 7.5 % was obtained. The V_{oc} of the PTDBD2-based PSC device is as high as 0.89 V, which is remarkably higher than that (0.74 V) of the PTB7-based PSC device. These high results indicated that DTBDT should be a promising candidate building block for highly efficient photovoltaic materials.

From the photovoltaic results listed in Table 5.7, BDT and its analogues have been shown to be the most successful building block for highly efficient photovoltaic polymers over 7 %. Notably, the photovoltaic performance of BDT-based polymers such as PTB7, PBDTTT-C-T, and PBDT-DTNT was further increased to 9 % by several groups via device innovations [2–8]. These device optimizations are not presented in this chapter in detail.

5.2.2.6 Thienopyrroledione (TPD)-Based Polymers

The thieno[3,4-*c*]pyrrole-4,6-dione (TPD) family of conjugated polymers has shown promise for PSC applications (see Fig. 5.15 and Table 5.8). The advantage of TPD is relatively cheap and easy synthesis. Incorporation of appropriate alkyl chains on TPD not only enables the preparation of soluble polymers but also greatly tunes the molecular packing as well as blend morphology. TPD-based push-pull polymers for photovoltaic applications were proposed in 2010 by Leclerc and co-workers [145, 146].

In the search for novel polymers suitable for PSCs, Leclerc [146], Jen [147], Xie [148], and Frechet [149] independently synthesized PBDTTPD, which included the benzodithiophene (BDT) unit as donor moiety and TPD as acceptor moiety. The preliminary *PCE* of the PBDTTPD/$PC_{71}BM$-based PSC device was as high as 5.5 % without additive optimization [146]. After utilizing coadditives of chloronaphthalene (CN) and diiodooctane (DIO), a *PCE* up to 7.1 % was achieved by Tao and coworkers [150]. Very recently, alkyl chain engineering was successfully utilized in PBDT-TPD, which dramatically promoted the efficiency of PBDT-TPD by Frechet and coworkers [151]. In their work, replacing branched side chains by linear ones in the

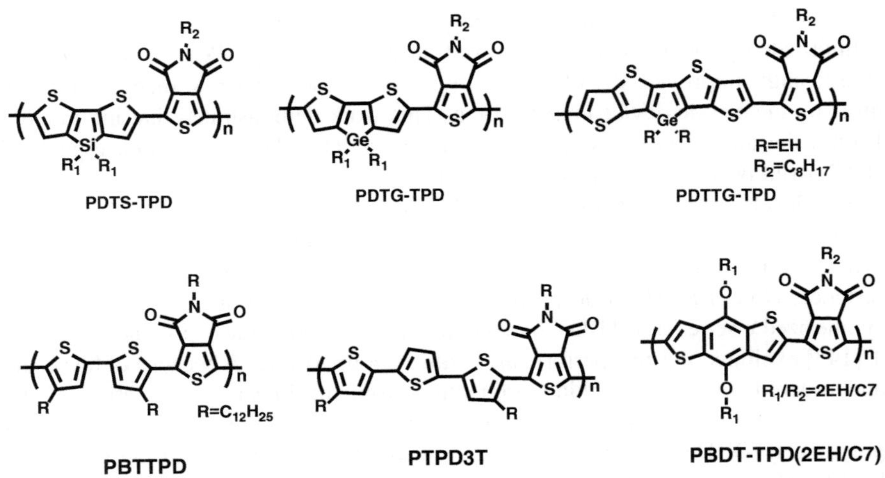

Fig. 5.15 Molecular structures of TPD-based highly efficient polymers

Table 5.8 Photovoltaic performance of the representative D–A copolymers based on BDT analogues

Materials	V_{oc} (V)	J_{sc} (mA/cm^2)	FF (%)	PCE (%)	μ_h [cm^2/(Vs)]	Ref.
PDTS-TPD	0.88	12.2	68	7.3	1×10^{-4}	[94]
PBDT-TPD	0.97	12.6	70	8.5	–	[151]
PDTG-TPD	0.85	12.6	68	7.3	–	[152]
PDTTG-TPD	0.81	13.85	64	7.2	–	[154]
PBTTPD	0.92	13.1	61	7.3	–	[155]
PTPD3T	0.80	12.5	79.6	7.9	1.2×10^{-3}	[157]

BDT motifs induced a critical change in polymer self-assembly and backbone orientation in thin films which correlates with a dramatic drop in solar cell efficiency. In contrast, for polymers with branched alkyl-substituted BDT motifs, controlling the number of carbon atoms in the linear alkyl-substituted TPD motifs could effectively improve material performance. Optimized through this approach, PBDTTPD polymer-based PSC devices acquired high *PCE* of 8.5 % and high V_{oc} of 0.97 V, making PBDTTPD one of the best polymer candidates for the wide band gap subcell of tandem PSCs. This report emphasized the determining role that linear side-chain substituents play in the device performance of PBDTTPD.

In the previous part, a TPD-silole copolymer, PDTS-TPD with high *PCE* up to 7 %, was introduced [94]. In order to improve the intermolecular interactions of silole-based polymers further, Reynolds and co-workers [152] substituted silicon by the larger germanium atom and prepared the first dithienogermole (DTG)-containing conjugated polymer. The dithienogermole-thienopyrrolodione copolymer,

PDTG-TPD, displayed an absorption shift to 735 nm, and a higher HOMO level than the analogous copolymer containing the commonly utilized DTS heterocycle. When PDTG-TPD was utilized in inverted PSCs, the cells displayed an average *PCE* of 7.3 %, relative to 6.6 % for the DTS-containing PSCs prepared under identical conditions. Notably, Reynolds et al. [153] also fabricated highly efficient PSC devices based on PDTG-TPD with a certified *PCE* up to 7.4 %, which is among the highest *PCE*s reported for photovoltaic polymers compatible with the roll-to-roll process. Followed by PDTS-TPD and PDTG-TPD, dithienogermolodithiophene (DTTG) was also incorporated as building blocks in TPD-based polymers because of the potentials of extended conjugation length and improved coplanarity. Very recently, Heeney et al. [154] reported the first synthesis of a novel ladder-type fused ring donor, DTG, in which two thieno[3,2-*b*]thiophene units are held coplanar by a bridging dialkyl germanium (Ge). Polymerization of DTTG with TPD afforded a polymer, PDTTG-TPD, with an optical band gap of 1.75 eV combined with a HOMO level of −5.68 eV. Bulk heterojunction PSC devices based on PDTTG-TPD/$PC_{71}BM$ afforded a *PCE* up to 7.2 % without the need for thermal annealing or processing additives. The preliminary results indicated that DTTG-based polymers are promising candidates for high-performance PSC devices. It should also be noted that the synthetic route provides considerable synthetic scope to promote the performance further by altering the bridging atoms.

Another interesting example of TPD-based copolymer was published by Wei and co-workers [155]. Wei et al. designed a crystalline polymer PDTTPD, which was constructed by bithiophene and TPD. A preliminary *PCE* of 5.0 % ($V_{oc} = 0.94$ V, $J_{sc} = 9.1$ mA/cm^2) was achieved for the PBTTPD/$PC_{71}BM$ system. Incorporating a small amount of diiodohexane (DIH) in the blend resulted in the formation of substantially enhanced polymer crystallinity and smaller as well as better dispersed $PC_{71}BM$ domains. As anticipated, an improved J_{sc} as high as 12.1 mA/cm^2 and a *PCE* of 7.3 % were recorded. Following the work of Wei et al., Marks and co-workers [156, 157] developed a series of TPD copolymers bearing thiophene, bithiophene, terthiophene, and quaterthiophene derivatives, respectively, as electron-donating moieties. Among these, a high-performance polymer PTPD3T [157] exhibited a high *PCE* of 7.9 % and extremely high *FF* up to 79.6 %. It is important to emphasize that the synthesis of monomers and polymers is both easy and versatile, wheres that of DTS or DTG derivatives is not.

5.2.2.7 Other Highly Efficient Polymers

As shown in Fig. 5.16 and Table 5.9, besides the above-mentioned polymers, other efficient polymers based on monomers such as quinoxaline (QX), thiazolo[5,4-*d*] thiazole (TTZ), thiophene bridged 2,1,3-benzothiadiazole, or naphtho[1,2-*c*:5,6-*c*] bis [1, 2, 5] thiadiazole also exhibited a high *PCE* over 6 %. QX has been widely implemented as an electron-deficient monomer of LBG polymers in PSCs. In 2010, Wang et al. [158] developed an easily synthesized donor–acceptor polymer (TQ-1) which showed *PCE* up to 6 %, with a high V_{oc} of 0.89 V, indicating that this

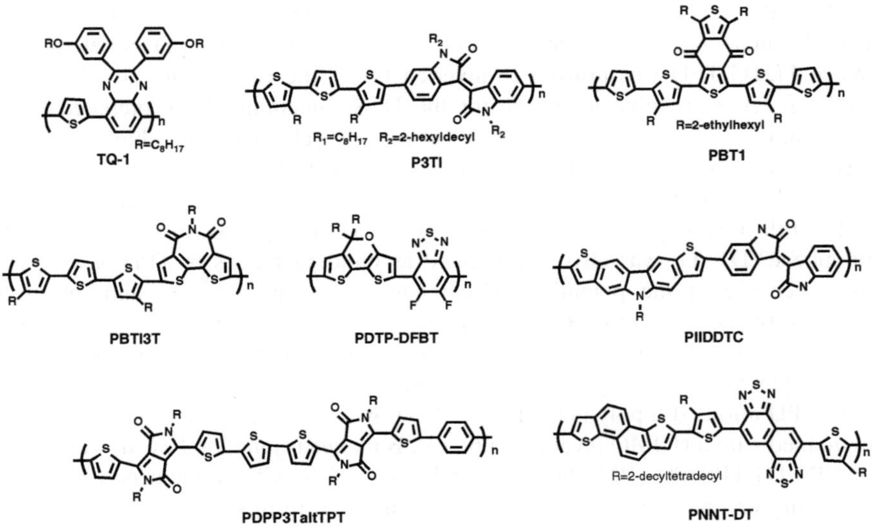

Fig. 5.16 Molecular structures of other highly efficient polymers with *PCE* over 6 %

Table 5.9 The photovoltaic parameters of other highly efficient polymers with *PCE* over 6 %

Materials	V_{oc} (V)	J_{sc} (mA/cm^2)	FF (%)	PCE (%)	μ_h [cm^2/(Vs)]	Ref.
TQ-1	0.89	10.5	64	6.0	–	[158]
P3TI	0.70	13.1	69	6.3	–	[159]
PBT-1	0.83	11.57	71	6.88	–	[160]
PIIDDTC	0.78	15.2	69	8.2	4.9×10^{-4}	[161]
PNNT-BT	0.82	15.6	64	8.2	1.7×10^{-3}	[162]
PDPP3TaltTPT	0.75	15.9	67	8.0	–	[163]
PBTI3T	0.86	12.9	77.8	8.66	1.5×10^{-3}	[153]
PDTP-DFBT	0.70	18.0	63	8.0	3×10^{-3}	[164]

polymer is a particularly promising candidate for high-efficiency low-cost polymer solar cells. Similar to DPP, isoindigo (IID) is also a high-performance pigment, which was developed in recent years for constructing LBG (<1.5 eV) conjugated polymers because of its strong electron-withdrawing character. By choosing the appropriate electron-rich unit terthiophene as the donor and IID as the acceptor, an easily accessible IID-based high-performance photovoltaic polymer (P3TI) was also reported by Wang and coworkers [159]. P3TI/PC$_{71}$BM-based PSC devices have an *IQE* of ~ 87 % and a V_{oc} of 0.7 V with an optimized device *PCE* up to 6.3 %. Qian et al. [160] designed a novel polymer, PBT1, which is copolymerized with benzodithiophene-4,8-dione (BDD) and α-quaterthiophene units. Interestingly, a *PCE* up to 6.88 % was recorded in a PBT1-based PSC under optimal condition of high donor:acceptor ratio (1.5:1, wt:wt) and small thickness (75 nm). This work provided a successful example of using molecular structure as a tool to realize optimal photovoltaic performance with high polymer content, thereby enabling the realization of efficient absorption in thin films.

A recent breakthrough in several building units was made. Geng and coworkers [161] designed a five-ring-fused aromatic unit, namely dithieno[3,2-*b*;6,7-*b*]carbazole (DTC), which is structurally related to FL. Optimized by inverted device structures, PIIDDTC achieved an extremely high *PCE* of 8.2 %. Notably, PIIDDTC is the first IID-based polymer with a *PCE* beyond 7 %, and amorphous polymer with a *PCE* beyond 8 %. Naphthodithiophene (NDT) also emerged as an efficient building block in highly efficient photovoltaic polymers. Osaka et al. [162] recently developed photovoltaic copolymers based on naphthodithiophene (NDT). Introducing linear alkyl chains improved the solubility as well as gave rise to a change in the orientation without any alteration of the energy levels, which resulted in quite an impressive *PCE* over 8 % in a conventional single-junction PSC device. Surprisingly, the introduction of linear alkyl chains led to a drastic change in polymer orientation into the face-on motif, which was beneficial for the charge transport in solar cells and promotion of the photovoltaic performance. In addition, PSC devices based on the NNT-BT yielded *PCE*s as high as 8.2 % with thickness up to 300 nm. These results indicated that this polymer platform is of particular interest in the understanding of molecular packing and carrier transport in high-performance photoelectronic polymers.

A regular alternating terpolymer design strategy was proposed and applied to produce a photovoltaic polymer with tailored energy levels and optical band gap by Janssen and collaborators [163]. High-molecular-weight photovoltaic materials the terpolymer PDPP3TaltDPP were obtained with high efficiencies up to 8.0 % in PSCs employing $PC_{71}BM$ as acceptors. Relative to the *PCE* of control copolymers PDPP3T (7.1 %) and PDPPTPT (7.4 %), the *PCE* of PDPP3TaltTPT exhibited more than 5 % improvement, which demonstrated that terpolymer could outperform the two parent copolymers when the design strategy was applied.

Similar to PTPD3T, Facchetti [26, 170] also developed a high-performance polymer PBTI3T, which demonstrated excellent *PCE* over 8.6 % and exceptionally high *FF* approaching 80 %. The high *FF* of PBTI3T is comparable to that of their inorganic counterparts. The extremely high *FFs* originated from the highly ordered, closely packed, and properly oriented active-layer microstructures with optimal horizontal phase separation and vertical phase gradation, which is beneficial for efficient charge collection and eliminated bulk as well as interfacial bimolecular recombination. This work depicted a comprehensive example to produce high *FF* in PSC device by integrating complementary materials design, synthesis, processing, and device engineering strategies.

Recently, a dithieno[3,2-*b*:2′,3′-*d*]pyran (DTPy)-containing polymer, PDTP-DFBT, was reported by Yang and coworkers [164]. The electron-donating property of the DTPy unit was found to be the strongest among the most frequently used donor units such as benzodithiophene (BDT) or cyclopentadithiophene (CPDT) units. When DTPy unit was polymerized with the strongly electron-deficient difluorobenzothiadiazole (DFBT) unit, a LBG (E_g = 1.38 eV) polymer PDTP-DFBT was obtained. In comparison with BDT or CPDT units, the DTP-based polymer

PDTP-DFBT showed significantly improved solubility and processability as well as *PCE*. Excellent performance in single and double junction solar cells was obtained with the *PCEs* reaching 8.0 and 10.6 %, respectively, which demonstrated that DTPy unit is a promising building block for high-performance photovoltaic materials.

5.3 Conjugated Polymer Acceptor Materials

Although fullerene acceptors, particularly the well-known PCBM, bis-PCBM, and ICBA have been widely used and performed well with photovoltaic donor polymers in BHJ PSC devices, these materials are still not the best choices for photovoltaic industry due to the high cost and narrow absorption coverage [165]. Because of the high absorption coefficient, broader absorption, and high electron mobilities as well as suitable affinities, non-fullerene acceptors such as perylene bisimide small molecules [166–168], as well as polymers [169, 170], have attracted substantial interest as alternative acceptor materials to fullerene acceptors.

LBG conjugated polymers can also be used as an acceptor in a PSC if its LUMO energy level is low enough. As an electron-deficient building block, PDI can be copolymerized easily with a wide variety of electron-rich units to tailor the molecular levels as well as absorption properties of the resulting D–A copolymers. The molecular structures and photovoltaic results of all-polymer solar cells with *PCE* over 1 % are summarized in Table 5.10 and Fig. 5.17. In 2007, Zhan et al. [171] reported the synthesis of the first soluble rylene containing polymer based on alternating perylene diimide (PDI) and dithienothiophene (DTT), which exhibited good solution processability, broad absorption, excellent thermal stability, and high electron affinity. Electron mobilities as high as 1.3×10^{-2} $cm^2\ V^{-1}\ s^{-1}$ have been measured by the OFET method. All-polymer solar cells using this polymer as acceptor polymer and a polythiophene derivative as donor polymer achieved a high *PCE* of over 1 % under AM 1.5G at 100 mW/cm^2. Similarly, a novel D–A copolymer of perylene diimide (PDI) and dithienothiophene, PPDIDTT-2, was also

Table 5.10 The photovoltaic results of representative acceptor polymer-based PSC devices

Acceptors	Donors	V_{oc} (V)	J_{sc} (mA/cm^2)	FF (%)	PCE (%)	Ref.
PPDIDTT	PT-1	0.63	4.2	39	1.03	[171]
PPDIDTT	PBDTTT-C-T	0.75	8.55	51.5	3.45	[180]
PPDIDTT-2	PT-2	0.69	5.02	43	1.48	[173]
PC-PDI	PT-2	0.70	6.35	50	2.23	[174]
F8TBT	P3HT	1.1	4.0	41	1.8	[176]
P(NDI2OD-T2)	PTB7	0.62	3.4	39	1.1	[178]
N2200	PTQ-1	0.84	8.40	56.0	4.00	[182]
PC-NDI	PTB7	0.88	4.07	38.0	1.34	[181]
PC-NDI	TTV7	0.88	7.71	54.0	3.68	[181]
PNDIT	PSEHTT	0.61	3.80	56	1.30	[179]
PNDIS-HD	PSEHTT	0.76	7.78	55	3.26	[179]

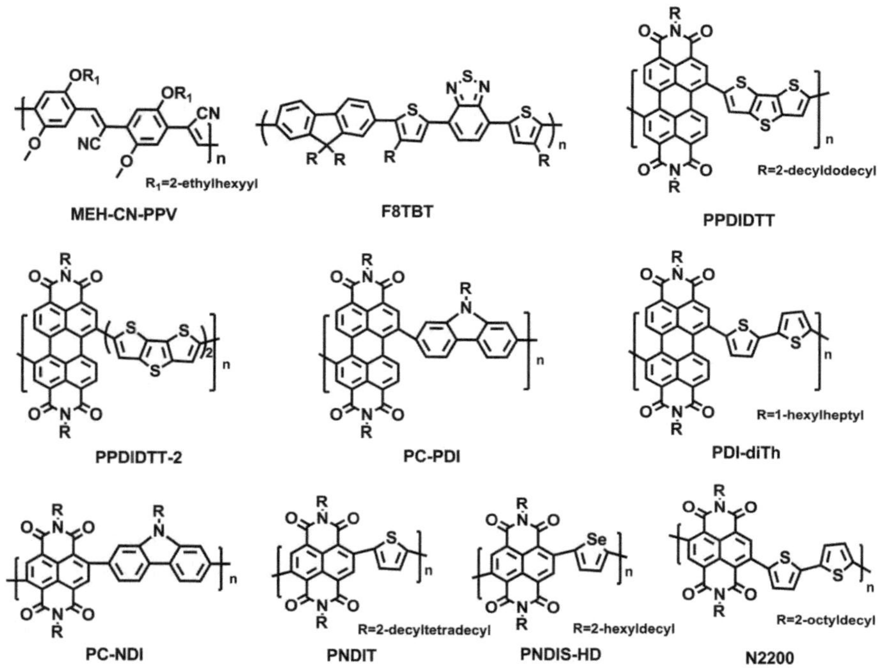

Fig. 5.17 Typical examples of highly efficient acceptor polymers with *PCE* over 1 %

synthesized by Zhan et al. [172], showing a LBG of 1.46 eV and high electron affinity with a LUMO level of 3.9 eV. The copolymer exhibited broad absorption throughout the visible and into the near-IR region. All PSCs were fabricated with the blend PPDIDTT-2 as acceptor and PT-2 as donor (1:1 wt/wt), showing a *PCE* over 1 %. Soon after, Tan et al. [173] optimized the all-polymer solar cells based on the narrow band gap alternating copolymer of perylene diimide and bis(dithieno-thiophene) (PPDIDTT-2). In their work, a polythiophene derivative substituted by a tris(thienylenevinylene) conjugated side chain (PT-2) are used as donor polymer. The optimized device based on the blend of PT-2 and PPDIDTT-2 in the ratio 3:1 (wt/wt) achieved a J_{sc} of 5.02mA cm^{-2} and a *PCE* of 1.48 %, under AM 1.5 G illumination at 100mW cm^{-2}. In 2010, Zhou et al. [174] systematically investigated all-polymer solar cells based on six perylene diimide-containing polymers (PX-PDI) as acceptor polymers and two polythiophene derivatives (P3HT and PT-2) as donor polymers. The highest *PCE* of 2.23 % was obtained in all-PSCs. The photovoltaic results also depicted that the PSC devices based on the PT1/PX-PDI systems have higher V_{oc} (0.58–0.76 V) than those of the devices based on the P3HT/PX-PDI blends (0.44–0.58 V). Owing to the application of solvent mixtures (toluene/CF, 9:1), the highest *PCE* of all-PSCs based on PT-2/PC-PDI reached 2.23 % because of the optimal phase separation. This work demonstrated that the solvent strategy and two-dimensional conjugated polymer donors are two effective methods to promote the performance of all-polymer solar cells. Pei et al. [175] developed two perylene diimide (PDI)-based acceptor polymers, r-PDI-diTh and i-PDI-diTh, which were synthesized by introducing a bulky side chain and thereby suppressing the π–π interactions between PDI units in the backbones of acceptor polymers. Therefore, more effective phase segregation of these acceptors with P3HT was realized. When regio-random polymer i-PDI-diTh was blended with P3HT, the PSC device gave a low *PCE* of 0.45 %. By using a similarly structured but regioregular polymer, r-PDI-diTh, the *PCE* further increased to 0.94 % because of the limited defect. By employing the inverted device configuration to match the vertical phase separation of donor polymer/acceptor polymer system better, a desirable *PCE* up to 2.17 % was achieved from the regioregular acceptor polymer, r-PDI-diTh-based PSC devices.

In 2007, McNeill and coworkers [176] reported efficient photovoltaic diodes which used ambipolar poly((9,9-dioctylfluorene)-2,7-diyl-*alt*-[4,7-bis(3-hexylthien-5-yl) -2,1,3-benzothiadiazole]-2′,2″-diyl) (F8TBT) both as acceptor polymer blending with P3HT and as donor polymer blending with $PC_{61}BM$. In both cases, external quantum efficiencies of over 25 % were achieved. In particular, a *PCE* up to 1.8 % and extremely high V_{oc} over 1 V were recorded for the optimized F8TBT/P3HT-based PSC device. Further studies by McNeill et al. [177] demonstrated that the relatively low efficiency of the P3HT/F8BT system can be attributed to poor charge generation and separation efficiencies which result from the failure of P3HT reorganization.

McNeill et al. [178] recently applied PTB7 as the donor polymer in the all-polymer solar cells. A relatively low efficiency of 1.1 % was observed. Clearly, the lack of suitable acceptor polymers has limited the photocurrent and efficiency of polymer/polymer bulk heterojunction solar cells. To overcome the problem, Jenekhe et al. [179] evaluated three naphthalene diimide (NDI) copolymers as acceptor materials in BHJ solar cells. Relatively poor performance ($\sim$1.3 %) was observed in the NDI and thiophene copolymer, PNDIT based all-polymer solar cells. PSCs based on an NDI-selenophene copolymer (PNDIS-HD) acceptor and a thiazolothiazole copolymer (PSEHTT) donor exhibited a high *PCE* of 3.3 %. The observed *FF* values of 55–60 % were impressively high among all-polymer solar cells and comparable to typical values observed in polymer/PCBM systems. Amazingly, this efficiency was comparable to the performance of PSEHTT/PCBM-based PSC devices. The lamellar crystalline morphology of PNDIS-HD, leading to balanced electron and hole transport in the polymer/polymer blend solar cells, should account for the good performance.

Cheng et al. [180] recently introduced two-dimensional polymer PBDTTT-C-T as donor polymer and proposed a binary additive approach to optimize the performance of PPDIDTT-based all-polymer solar cells, and a dramatically improved *PCE* up to 3.45 % was observed. Zhou et al. [181] recently reported a high *PCE* of 3.68 % for the all-polymer solar cell utilizing TTV as the donor polymer and PC-NDI as the acceptor polymer. By introducing a conjugated side chain in the TT unit of donor polymer PTB7, the miscibility of the polymer/polymer blend was greatly improved. In addition, adding a small amount (1 vol.%) of 1,8-diiodooctane (DIO) could increase the aggregation of acceptor polymer, PC-NDI. For the PTB7/PC-NDI-based all-PSCs, the *PCE* decreased to 1.12 % with V_{oc} = 0.86 V, J_{sc} = 3.83 mA/cm^2, and *FF* = 34 %. This work demonstrated that two-dimensional conjugated polymer might be ideal donor polymers for all-polymer solar cells because of the superior properties. More recently, Mori et al. [182] fabricated highly efficient LBG donor/acceptor polymer blend solar cells by utilizing TQ-1 as donor polymer and N2200 as acceptor polymer. The device performance was optimized at a donor/acceptor blending ratio of 7:3 (wt:wt) to result in a high J_{sc} of 8.85 mA/cm^2, a *FF* of 55 %, and a V_{oc} of 0.84 V. The high *PCE* exceeded 4 %, which was among the highest values reported for all-polymer solar cells. The high photovoltaic performance demonstrated the great potential of polymer/polymer blend solar cells as a promising alternative to polymer/fullerene solar cells.

Because the typical exciton diffusion length of polymer/polymer blends is approximately 10 nm, the larger phase separation length and smaller donor/acceptor interfacial area in the polymer/polymer systems might be the origin of inefficient exciton dissociation and relatively low performance in these systems. To improve the performance of all-polymer solar cells, two strategies could be utilized. One strategy is morphology tuning, for instance, altering the processing solvent to modulate the scale and degree of phase separation of D/A blends. The other strategy is to rationally select and design superior donor polymers.

5.4 Summary and Outlook

In this chapter, an overview of conjugated polymer photovoltaic materials developed during the past two decades are summarized and commented. The basic design considerations of the conjugated polymer photovoltaic materials for the application in polymer solar cells are also discussed along with examples. The rapid progress in novel conjugated photovoltaic polymers demonstrates that there is plenty of room to bring the cost of PSCs down by developing low-cost and highly efficient polymeric photovoltaic materials [183, 184]. Guided by computational calculations and screening [185], further improvements will be realized in the near future by fine optimization of integrated backbones, side chains, and molecular weight [186] based on well-defined highly efficient polymers.

References

1. Li XH, Choy WCH, Huo LJ, Xie FX, Sha WEI, Ding BF, Guo X, Li YF, Hou JH, You JB, Yang Y (2012) Dual plasmonic nanostructures for high performance inverted organic solar cells. Adv Mater 24:3046–3052
2. Chang C-Y, Zuo L, Yip H-L, Li Y, Li C-Z, Hsu C-S, Cheng Y-J, Chen H, Jen AKY (2013) A versatile fluoro-containing low-bandgap polymer for efficient semitransparent and tandem polymer solar cells. Adv Funct Mater 23:5084–5090

3. Liu S, Zhang K, Lu J, Zhang J, Yip H-L, Huang F, Cao Y (2013) High-efficiency polymer solar cells via the incorporation of an amino-functionalized conjugated metallopolymer as a cathode interlayer. J Am Chem Soc 135:15326–15329
4. Yang TB, Wang M, Duan CH, Hu XW, Huang L, Peng JB, Huang F, Gong X (2012) Inverted polymer solar cells with 8.4 % efficiency by conjugated polyelectrolyte. Energy Environ Sci 5:8208–8214
5. Tan Z, Li L, Wang F, Xu Q, Li S, Sun G, Tu X, Hou X, Hou J, Li Y (2014) Solution-processed rhenium oxide: A versatile anode buffer layer for high performance polymer solar cells with enhanced light harvest. Adv Energy Mater 4. doi:10.1002/aenm.201300884
6. Duan CH, Zhang K, Guan X, Zhong CM, Xie HM, Huang F, Chen JW, Peng JB, Cao Y (2013) Conjugated zwitterionic polyelectrolyte-based interface modification materials for high performance polymer optoelectronic devices. Chem Sci 4:1298–1307
7. He ZC, Zhong CM, Huang X, Wong WY, Wu HB, Chen LW, Su SJ, Cao Y (2011) Simultaneous enhancement of open-circuit voltage, short-circuit current density, and fill factor in polymer solar cells. Adv Mater 23:4636–4643
8. He ZC, Zhong CM, Su SJ, Xu M, Wu HB, Cao Y (2012) Enhanced power-conversion efficiency in polymer solar cells using an inverted device structure. Nat Photonics 6:591–595
9. Dou LT, You JB, Yang J, Chen CC, He YJ, Murase S, Moriarty T, Emery K, Li G, Yang Y (2012) Tandem polymer solar cells featuring a spectrally matched low-bandgap polymer. Nat Photonics 6:180–185
10. Dou LT, Chang WH, Gao J, Chen CC, You JB, Yang Y (2013) A selenium-substituted low-bandgap polymer with versatile photovoltaic applications. Adv Mater 25:825–831
11. Dou LT, Gao J, Richard E, You JB, Chen CC, Cha KC, He YJ, Li G, Yang Y (2012) Systematic investigation of benzodithiophene- and diketopyrrolopyrrole-based low-bandgap polymers designed for single junction and tandem polymer solar cells. J Am Chem Soc 134:10071–10079
12. Liao S-H, Jhuo H-J, Cheng Y-S, Chen S-A (2013) Fullerene derivative-doped zinc oxide nanofilm as the cathode of inverted polymer solar cells with low-bandgap polymer (PTB7-Th) for high performance. Adv Mater 25:4766–4771
13. You J, Dou L, Yoshimura K, Kato T, Ohya K, Moriarty T, Emery K, Chen C-C, Gao J, Li G, Yang Y (2013) A polymer tandem solar cell with 10.6 % power conversion efficiency. Nat Commun 4:1446
14. Yu G, Gao J, Hummelen JC, Wudl F, Heeger AJ (1995) Polymer photovoltaic cells—enhanced efficiencies via a network of internal donor–acceptor heterojunctions. Science 270:1789–1791
15. Halls JJM, Walsh CA, Greenham NC, Marseglia EA, Friend RH, Moratti SC, Holmes AB (1995) Efficient photodiodes from interpenetrating polymer networks. Nature 376:498–500
16. Li YF, Zou YP (2008) Conjugated polymer photovoltaic materials with broad absorption band and high charge carrier mobility. Adv Mater 20:2952–2958
17. Cheng YJ, Yang SH, Hsu CS (2009) Synthesis of conjugated polymers for organic solar cell applications. Chem Rev 109:5868–5923
18. Chen J, Cao Y (2009) Development of novel conjugated donor polymers for high-efficiency bulk-heterojunction photovoltaic devices. Acc Chem Res 42:1709–1718
19. Beaujuge PM, Fréchet JMJ (2011) Molecular design and ordering effects in π-functional materials for transistor and solar cell applications. J Am Chem Soc 133:20009–20029
20. He F, Yu LP (2011) How far can polymer solar cells go? In need of a synergistic approach. J Phys Chem Lett 2:3102–3113
21. Li YF (2012) Molecular design of photovoltaic materials for polymer solar cells: toward suitable electronic energy levels and broad absorption. Acc Chem Res 45:723–733
22. Ye L, Zhang SQ, Huo LJ, Zhang MJ, Hou JH (2014) Molecular design toward highly efficient photovoltaic polymers based on two-Dimensional conjugated benzodithiophene. Acc Chem Res 47:1595–1603
23. Zhou HX, Yang LQ, You W (2012) Rational design of high performance conjugated polymers for organic solar cells. Macromolecules 45:607–632

24. Henson ZB, Mullen K, Bazan GC (2012) Design strategies for organic semiconductors beyond the molecular formula. Nat Chem 4:699–704
25. Son HJ, Carsten B, Jung IH, Yu LP (2012) Overcoming efficiency challenges in organic solar cells: rational development of conjugated polymers. Energy Environ Sci 5:8158–8170
26. Facchetti A (2011) pi-conjugated polymers for organic electronics and photovoltaic cell applications. Chem Mater 23:733–758
27. Spanggaard H, Krebs FC (2004) A brief history of the development of organic and polymeric photovoltaics. Sol Energy Mater Sol C 83:125–146
28. Liang YY, Yu LP (2010) A new class of semiconducting polymers for bulk heterojunction solar cells with exceptionally high performance. Acc Chem Res 43:1227–1236
29. Zhan XW, Zhu DB (2010) Conjugated polymers for high-efficiency organic photovoltaics. Polym Chem 1:409–419
30. Helgesen M, Sondergaard R, Krebs FC (2010) Advanced materials and processes for polymer solar cell devices. J Mater Chem 20:36–60
31. Scharber MC, Wuhlbacher D, Koppe M, Denk P, Waldauf C, Heeger AJ, Brabec CL (2006) Design rules for donors in bulk-heterojunction solar cells—towards 10 % energy-conversion efficiency. Adv Mater 18:789–794
32. Liu F, Gu Y, Shen X, Ferdous S, Wang H-W, Russell TP (2013) Characterization of the morphology of solution-processed bulk heterojunction organic photovoltaics. Prog Polym Sci 38:1990–2052
33. Ye L, Jing Y, Guo X, Sun H, Zhang S, Zhang M, Huo L, Hou J (2013) Remove the residual additives toward enhanced efficiency with higher reproducibility in polymer solar cells. J Phys Chem C 117:14920–14928
34. Bijleveld JC, Zoombelt AP, Mathijssen SGJ, Wienk MM, Turbiez M, de Leeuw DM, Janssen RAJ (2009) Poly(diketopyrrolopyrrole-terthiophene) for ambipolar logic and photovoltaics. J Am Chem Soc 131:16616–16617
35. Ye L, Zhang S, Ma W, Fan B, Guo X, Huang Y, Ade H, Hou J (2012) From binary to ternary solvent: morphology fine-tuning of D/A blends in PDPP3T-based polymer solar cells. Adv Mater 24:6335–6341
36. Wessling RA (1985) The polymerization of xylylene bisdialkyl sulfonium salts. J Polym Sci Polym Symp 72:55–66
37. Gilch HG, Wheelwright WL (1966) Polymerization of α-halogenated p-xylenes with base. J Polym Sci A-1: Polym Chem 4:1337–1349
38. Hou J, Fan B, Huo L, He C, Yang C, Li Y (2006) Poly(alkylthio-p-phenylenevinylene): synthesis and electroluminescent and photovoltaic properties. J Polym Sci A: Polym Chem 44:1279–1290
39. Namazi H, Assadpour A, Pourabbas B, Entezami A (2001) Polycondensation of bis (cyanoacetate) and a,10bdihydrobenzofuro[2,3-b]benzofuran-2,9-dicarbaldehyde via knoevenagel reaction: synthesis of donor–acceptor polymers containing shoulder-to-shoulder main chains. J Appl Polym Sci 81:505–511
40. Shaheen SE, Brabec CJ, Sariciftci NS, Padinger F, Fromherz T, Hummelen JC (2001) 2.5 % efficient organic plastic solar cells. Appl Phys Lett 78:841–843
41. Brabec CJ, Shaheen SE, Winder C, Sariciftci NS, Denk P (2002) Effect of LiF/metal electrodes on the performance of plastic solar cells. Appl Phys Lett 80:1288–1290
42. Zhou QM, Zheng LP, Sun DK, Deng XY, Yu G, Cao Y (2003) Efficient polymer photovoltaic devices based on blend of MEH-PPV and C-60 derivatives. Synth Met 135:825–826
43. Wienk MM, Kroon JM, Verhees WJH, Knol J, Hummelen JC, van Hal PA, Janssen RAJ (2003) Efficient methano[70]fullerene/MDMO-PPV bulk heterojunction photovoltaic cells. Angew Chem Int Ed 42:3371–3375
44. Tajima K, Suzuki Y, Hashimoto K (2008) Polymer photovoltaic devices using fully regioregular poly[(2-methoxy-5-(3′,7′-dimethyloctyloxy))-1,4-phenylenevinylene]. J Phys Chem C 112:8507–8510

45. Mikroyannidis JA, Kabanakis AN, Balraju P, Sharma GD (2010) Enhanced performance of bulk heterojunction solar cells using novel alternating phenylenevinylene copolymers of low band gap with cyanovinylene 4-nitrophenyls. Macromolecules 43:5544–5553
46. Hou JH, Tan Z, He YJ, Yang CH, Li YF (2006) Branched poly(thienylene vinylene)s with absorption spectra covering the whole visible region. Macromolecules 39:4657–4662
47. Huo LJ, Chen TL, Zhou Y, Hou JH, Chen HY, Yang Y, Li YF (2009) Improvement of photoluminescent and photovoltaic properties of poly(thienylene vinylene) by carboxylate substitution. Macromolecules 42:4377–4380
48. He Y, Zhou Y, Zhao G, Min J, Guo X, Zhang B, Zhang M, Zhang J, Li Y, Zhang F, Inganäs O (2010) Poly(4,8-bis(2-ethylhexyloxy)benzo[1,2-b:4,5-b′]dithiophene vinylene): synthesis, optical and photovoltaic properties. J Polym Sci A: Polym Chem 48:1822–1829
49. Sato M, Morii H (1991) Configurational feature of electrochemically-prepared poly(3-dodecylthiophene). Polym Commun 32:42–44
50. McCullough RD (1998) The chemistry of conducting polythiophenes. Adv Mater 10:93–116
51. Sugimoto R, Takeda S, Gu HB, Yoshino K (1986) Preparation of soluble polythiophene derivatives utilizing transition metal halides as catalysts and their property. Chem Express 1:635–638
52. Pomerantz M, Tseng JJ, Zhu H, Sproull SJ, Reynolds JR, Uitz R, Arnott HJ, Haider MI (1991) Processable polymers and copolymers of 3-alkylthiophenes and their blends. Synth Met 41:825–830
53. Mccullough RD, Lowe RD (1992) Enhanced electrical-conductivity in regioselectively synthesized poly(3-alkylthiophenes). J Chem Soc Chem Comm 1:70–72
54. Chen TA, Rieke RD (1993) Polyalkylthiophenes with the smallest bandgap and the highest intrinsic conductivity. Synth Met 60:175–177
55. Iraqi A, Barker GW (1998) Synthesis and characterisation of telechelic regioregular head-to-tail poly(3-alkylthiophenes). J Mater Chem 8:25–29
56. Guillerez S, Bidan G (1998) New convenient synthesis of highly regioregular poly (3-octylthiophene) based on the Suzuki coupling reaction. Synth Met 93:123–126
57. Brabec CJ (2004) Organic photovoltaics: technology and market. Sol Energy Mat Sol C 83:273–292
58. Dang MT, Hirsch L, Wantz G (2011) P3HT:PCBM, best seller in polymer photovoltaic research. Adv Mater 23:3597–3602
59. Dang MT, Hirsch L, Wantz G, Wuest JD (2013) Controlling the morphology and performance of bulk heterojunctions in solar cells. Lessons learned from the benchmark poly (3-hexylthiophene):[6, 6]-phenyl-C_{61}-butyric acid methyl ester system. Chem Rev 113:3734–3765
60. Padinger F, Rittberger RS, Sariciftci NS (2003) Effects of postproduction treatment on plastic solar cells. Adv Funct Mater 13:85–88
61. Li G, Shrotriya V, Huang JS, Yao Y, Moriarty T, Emery K, Yang Y (2005) High-efficiency solution processable polymer photovoltaic cells by self-organization of polymer blends. Nat Mater 4:864–868
62. Ma WL, Yang CY, Gong X, Lee K, Heeger AJ (2005) Thermally stable, efficient polymer solar cells with nanoscale control of the interpenetrating network morphology. Adv Funct Mater 15:1617–1622
63. He YJ, Chen HY, Hou JH, Li YF (2010) Indene-C-60 bisadduct: a new acceptor for high-performance polymer solar cells. J Am Chem Soc 132:1377–1382
64. Zhao G, He Y, Li Y (2010) 6.5 % efficiency of polymer solar cells based on poly(3-hexylthiophene) and indene-C_{60} bisadduct by device optimization. Adv Mater 22:4355–4358
65. Guo X, Cui CH, Zhang MJ, Huo LJ, Huang Y, Hou JH, Li Y (2012) High efficiency polymer solar cells based on poly(3-hexylthiophene)/indene-C-70 bisadduct with solvent additive. Energy Environ Sci 5:7943–7949
66. Li G, Yao Y, Yang H, Shrotriya V, Yang G, Yang Y (2007) "Solvent annealing" effect in polymer solar cells based on poly(3-hexylthiophene) and methanofullerenes. Adv Funct Mater 17:1636–1644

67. Yao Y, Hou JH, Xu Z, Li G, Yang Y (2008) Effect of solvent mixture on the nanoscale phase separation in polymer solar cells. Adv Funct Mater 18:1783–1789
68. Nguyen LH, Hoppe H, Erb T, Günes S, Gobsch G, Sariciftci NS (2007) Effects of annealing on the nanomorphology and performance of poly(alkylthiophene): fullerene bulk-heterojunction solar cells. Adv Funct Mater 17:1071–1078
69. Wu PT, Xin H, Kim FS, Ren GQ, Jenekhe SA (2009) Regioregular poly(3-pentylthiophene): synthesis, self-assembly of nanowires, high-mobility field-effect transistors, and efficient photovoltaic cells. Macromolecules 42:8817–8826
70. Xin H, Kim FS, Jenekhe SA (2008) Highly efficient solar cells based on poly (3-butylthiophene) nanowires. J Am Chem Soc 130:5424–5425
71. Gadisa A, Oosterbaan WD, Vandewal K, Bolsée J-C, Bertho S, D'Haen J, Lutsen L, Vanderzande D, Manca JV (2009) Effect of alkyl side-chain length on photovoltaic properties of poly(3-alkylthiophene)/PCBM bulk heterojunctions. Adv Funct Mater 19:3300–3306
72. Sun Y, Cui C, Wang H, Li Y (2012) High-efficiency polymer solar cells based on poly (3-pentylthiophene) with indene-C_{70} bisadduct as an acceptor. Adv Energy Mater 2:966–969
73. Hou JH, Tan ZA, Yan Y, He YJ, Yang CH, Li YF (2006) Synthesis and photovoltaic properties of two-dimensional conjugated polythiophenes with bi(thienylenevinylene) side chains. J Am Chem Soc 128:4911–4916
74. Hou JH, Chen TL, Zhang SQ, Huo LJ, Sista S, Yang Y (2009) An easy and effective method to modulate molecular energy level of poly(3-alkylthiophene) for high-V-oc polymer solar cells. Macromolecules 42:9217–9219
75. Zhang MJ, Guo X, Yang Y, Zhang J, Zhang ZG, Li YF (2011) Downwards tuning the HOMO level of polythiophene by carboxylate substitution for high open-circuit-voltage polymer solar cells. Polym Chem 2:2900–2906
76. Svensson M, Zhang F, Veenstra SC, Verhees WJH, Hummelen JC, Kroon JM, Inganäs O, Andersson MR (2003) High-performance polymer solar cells of an alternating polyfluorene copolymer and a fullerene derivative. Adv Mater 15:988–991
77. Chen MH, Hou J, Hong Z, Yang G, Sista S, Chen LM, Yang Y (2009) Efficient polymer solar cells with thin active layers based on alternating polyfluorene copolymer/fullerene bulk heterojunctions. Adv Mater 21:4238–4242
78. Mühlbacher D, Scharber M, Morana M, Zhu Z, Waller D, Gaudiana R, Brabec C (2006) High photovoltaic performance of a low-bandgap polymer. Adv Mater 18:2884–2889
79. Peet J, Kim JY, Coates NE, Ma WL, Moses D, Heeger AJ, Bazan GC (2007) Efficiency enhancement in low-bandgap polymer solar cells by processing with alkane dithiols. Nat Mater 6:497–500
80. Blouin N, Michaud A, Leclerc M (2007) A low-bandgap poly(2,7-Carbazole) derivative for use in high-performance solar cells. Adv Mater 19:2295–2300
81. Park SH, Roy A, Beaupre S, Cho S, Coates N, Moon JS, Moses D, Leclerc M, Lee K, Heeger AJ (2009) Bulk heterojunction solar cells with internal quantum efficiency approaching 100 %. Nat Photonics 3:297–302
82. Beaupre S, Leclerc M (2013) PCDTBT: en route for low cost plastic solar cells. J Mater Chem A 1:11097–11105
83. Qin RP, Li WW, Li CH, Du C, Veit C, Schleiermacher HF, Andersson M, Bo ZS, Liu ZP, Inganas O, Wuerfel U, Zhang FL (2009) A planar copolymer for high efficiency polymer solar cells. J Am Chem Soc 131:14612–14613
84. Wong WY, Wang XZ, He Z, Djurisic AB, Yip CT, Cheung KY, Wang H, Mak CSK, Chan WK (2007) Metallated conjugated polymers as a new avenue towards high-efficiency polymer solar cells. Nat Mater 6:521–527
85. Wang EG, Wang L, Lan LF, Luo C, Zhuang WL, Peng JB, Cao Y (2008) High-performance polymer heterojunction solar cells of a polysilafluorene derivative. Appl Phys Lett 92:033307
86. Song J, Du C, Li C, Bo Z (2011) Silole-containing polymers for high-efficiency polymer solar cells. J Polym Sci A: Polym Chem 49:4267–4274

87. Hou JH, Chen HY, Zhang SQ, Li G, Yang Y (2008) Synthesis, characterization, and photovoltaic properties of a low band gap polymer based on silole-containing polythiophenes and 2,1,3-benzothiadiazole. J Am Chem Soc 130:16144–16145
88. Chen HY, Hou JH, Hayden AE, Yang H, Houk KN, Yang Y (2010) Silicon atom substitution enhances interchain packing in a thiophene-based polymer system. Adv Mater 22:371–375
89. Scharber MC, Koppe M, Gao J, Cordella F, Loi MA, Denk P, Morana M, Egelhaaf HJ, Forberich K, Dennler G, Gaudiana R, Waller D, Zhu ZG, Shi XB, Brabec CJ (2010) Influence of the bridging atom on the performance of a low-bandgap bulk heterojunction solar cell. Adv Mater 22:367–370
90. Morana M, Azimi H, Dennler G, Egelhaaf H-J, Scharber M, Forberich K, Hauch J, Gaudiana R, Waller D, Zhu Z, Hingerl K, van Bavel SS, Loos J, Brabec CJ (2010) Nanomorphology and charge generation in bulk heterojunctions based on low-bandgap dithiophene polymers with different bridging atoms. Adv Funct Mater 20:1180–1188
91. Zhang M, Guo X, Li Y (2011) Synthesis and characterization of a copolymer based on thiazolothiazole and dithienosilole for polymer solar cells. Adv Energy Mater 1:557–560
92. Cui C, Fan X, Zhang M, Zhang J, Min J, Li Y (2011) A D-A copolymer of dithienosilole and a new acceptor unit of naphtho[2,3-c]thiophene-4,9-dione for efficient polymer solar cells. Chem Commun 47:11345–11347
93. Guo X, Zhou N, Lou SJ, Hennek JW, Ponce Ortiz R, Butler MR, Boudreault P-LT, Strzalka J, Morin P-O, Leclerc M, López Navarrete JT, Ratner MA, Chen LX, Chang RPH, Facchetti A, Marks TJ (2012) Bithiopheneimide–dithienosilole/dithienogermole copolymers for efficient solar cells: information from structure–property–device performance correlations and comparison to thieno[3,4-c]pyrrole-4,6-dione analogues. J Am Chem Soc 134:18427–18439
94. Chu T-Y, Lu J, Beaupré S, Zhang Y, Pouliot J-R, Wakim S, Zhou J, Leclerc M, Li Z, Ding J, Tao Y (2011) Bulk heterojunction solar cells using thieno[3,4-c]pyrrole-4,6-dione and dithieno[3,2-b:2′,3′-d]silole copolymer with a power conversion efficiency of 7.3 %. J Am Chem Soc 133:4250–4253
95. Wienk MM, Turbiez M, Gilot J, Janssen RAJ (2008) Narrow-bandgap diketo-pyrrolo-pyrrole polymer solar cells: the effect of processing on the performance. Adv Mater 20:2556–2560
96. Huo LJ, Hou JH, Chen HY, Zhang SQ, Jiang Y, Chen TL, Yang Y (2009) Bandgap and molecular level control of the low-bandgap polymers based on 3,6-dithiophen-2-yl-2,5-dihydropyrrolo[3,4-c]pyrrole-1,4-dione toward highly efficient polymer solar cells. Macromolecules 42:6564–6571
97. Li W, Furlan A, Hendriks KH, Wienk MM, Janssen RAJ (2013) Efficient tandem and triple-junction polymer solar cells. J Am Chem Soc 135:5529–5532
98. Bronstein H, Chen ZY, Ashraf RS, Zhang WM, Du JP, Durrant JR, Tuladhar PS, Song K, Watkins SE, Geerts Y, Wienk MM, Janssen RAJ, Anthopoulos T, Sirringhaus H, Heeney M, McCulloch I (2011) Thieno[3,2-b]thiophene-diketopyrrolopyrrole-containing polymers for high-performance organic field-effect transistors and organic photovoltaic devices. J Am Chem Soc 133:3272–3275
99. Bronstein H, Collado-Fregoso E, Hadipour A, Soon YW, Huang Z, Dimitrov SD, Ashraf RS, Rand BP, Watkins SE, Tuladhar PS, Meager I, Durrant JR, McCulloch I (2013) Thieno[3,2-b]thiophene-diketopyrrolopyrrole containing polymers for inverted solar cells devices with high short circuit currents. Adv Funct Mater 23:5647–5654
100. Li W, Hendriks KH, Roelofs WSC, Kim Y, Wienk MM, Janssen RAJ (2013) Efficient small bandgap polymer solar cells with high fill factors for 300 nm thick films. Adv Mater 25:3182–3186
101. Bijleveld JC, Gevaerts VS, Di Nuzzo D, Turbiez M, Mathijssen SGJ, de Leeuw DM, Wienk MM, Janssen RAJ (2010) Efficient solar cells based on an easily accessible diketopyrrolopyrrole polymer. Adv Mater 22:E242–E246
102. Yiu AT, Beaujuge PM, Lee OP, Woo CH, Toney MF, Frechet JMJ (2012) Side-chain tunability of furan-containing low-band-gap polymers provides control of structural order in efficient solar cells. J Am Chem Soc 134:2180–2185

103. Woo CH, Beaujuge PM, Holcombe TW, Lee OP, Frechet JMJ (2010) Incorporation of furan into low band-gap polymers for efficient solar cells. J Am Chem Soc 132:15547–15549
104. Li WW, Hendriks KH, Furlan A, Roelofs WSC, Wienk MM, Janssen RAJ (2013) Universal correlation between fibril width and quantum efficiency in diketopyrrolopyrrole-based polymer solar cells. J Am Chem Soc 135:18942–18948
105. Mcculloch I, Ashraf RS, Biniek L, Bronstein H, Combe C, Donaghey JE, James DI, Nielsen CB, Schroeder BC, Zhang WM (2012) Design of semiconducting indacenodithiophene polymers for high performance transistors and solar cells. Acc Chem Res 45:714–722
106. Chen CP, Chan SH, Chao TC, Ting C, Ko BT (2008) Low-bandgap poly(thiophene-phenylene-thiophene) derivatives with broaden absorption spectra for use in high-performance bulk-heterojunction polymer solar cells. J Am Chem Soc 130:12828–12833
107. Yu CY, Chen CP, Chan SH, Hwang GW, Ting C (2009) Thiophene/phenylene/thiophene-based low-bandgap conjugated polymers for efficient near-infrared photovoltaic applications. Chem Mater 21:3262–3269
108. Chen YC, Yu CY, Fan YL, Hung LI, Chen CP, Ting C (2010) Low-bandgap conjugated polymer for high efficient photovoltaic applications. Chem Commun 46:6503–6505
109. Zhang Y, Zou JY, Yip HL, Chen KS, Zeigler DF, Sun Y, Jen AKY (2011) Indacenodithiophene and quinoxaline-based conjugated polymers for highly efficient polymer solar cells. Chem Mater 23:2289–2291
110. Zhang Y, Chien SC, Chen KS, Yip HL, Sun Y, Davies JA, Chen FC, Jen AKY (2011) Increased open circuit voltage in fluorinated benzothiadiazole-based alternating conjugated polymers. Chem Commun 47:11026–11028
111. Xu YX, Chueh CC, Yip HL, Ding FZ, Li YX, Li CZ, Li XS, Chen WC, Jen AKY (2012) Improved charge transport and absorption coefficient in indacenodithieno[3,2-b]thiophene-based ladder-type polymer leading to highly efficient polymer solar cells. Adv Mater 24:6356–6361
112. Wang M, Hu XW, Liu LQ, Duan CH, Liu P, Ying L, Huang F, Cao Y (2013) Design and synthesis of copolymers of indacenodithiophene and naphtho[1,2-c:5,6-c]bis(1,2,5-thiadiazole) for polymer solar cells. Macromolecules 46:3950–3958
113. Zhang MJ, Guo X, Wang XC, Wang HQ, Li YF (2011) Synthesis and photovoltaic properties of D-A copolymers based on alkyl-substituted indacenodithiophene donor unit. Chem Mater 23:4264–4270
114. Intemann JJ, Yao K, Yip HL, Xu YX, Li YX, Liang PW, Ding FZ, Li XS, Jen AKY (2013) Molecular weight effect on the absorption, charge carrier mobility, and photovoltaic performance of an indacenodiselenophene-based ladder-type polymer. Chem Mater 25:3188–3195
115. Guo X, Zhang MJ, Tan JH, Zhang SQ, Huo LJ, Hu WP, Li YF, Hou JH (2012) Influence of D/A ratio on photovoltaic performance of a highly efficient polymer solar cell system. Adv Mater 24:6536–6541
116. Huo LJ, Hou JH (2011) Benzo[1,2-b:4,5-b′]dithiophene-based conjugated polymers: band gap and energy level control and their application in polymer solar cells. Polym Chem 2:2453–2461
117. Hou JH, Park MH, Zhang SQ, Yao Y, Chen LM, Li JH, Yang Y (2008) Bandgap and molecular energy level control of conjugated polymer photovoltaic materials based on benzo [1,2-b : 4,5-b′]dithiophene. Macromolecules 41:6012–6018
118. Liang YY, Wu Y, Feng DQ, Tsai ST, Son HJ, Li G, Yu LP (2009) Development of new semiconducting polymers for high performance solar cells. J Am Chem Soc 131:56–57
119. Liang YY, Feng DQ, Wu Y, Tsai ST, Li G, Ray C, Yu LP (2009) Highly efficient solar cell polymers developed via fine-tuning of structural and electronic properties. J Am Chem Soc 131:7792–7799
120. Liang YY, Xu Z, Xia JB, Tsai ST, Wu Y, Li G, Ray C, Yu LP (2010) For the bright future-bulk heterojunction polymer solar cells with power conversion efficiency of 7.4 %. Adv Mater 22:E135–E138

121. Hou JH, Chen HY, Zhang SQ, Chen RI, Yang Y, Wu Y, Li G (2009) Synthesis of a low band gap polymer and its application in highly efficient polymer solar cells. J Am Chem Soc 131:15586–15587
122. Chen HY, Hou JH, Zhang SQ, Liang YY, Yang GW, Yang Y, Yu LP, Wu Y, Li G (2009) Polymer solar cells with enhanced open-circuit voltage and efficiency. Nat Photonics 3:649–653
123. Huang Y, Huo LJ, Zhang SQ, Guo X, Han CC, Li YF, Hou JH (2011) Sulfonyl: a new application of electron-withdrawing substituent in highly efficient photovoltaic polymer. Chem Commun 47:8904–8906
124. Zhou HX, Yang LQ, Stuart AC, Price SC, Liu SB, You W (2011) Development of fluorinated benzothiadiazole as a structural unit for a polymer solar cell of 7 % efficiency. Angew Chem Int Ed 50:2995–2998
125. Wang N, Chen Z, Wei W, Jiang ZH (2013) Fluorinated benzothiadiazole-based conjugated polymers for high-performance polymer solar cells without any processing additives or post-treatments. J Am Chem Soc 135:17060–17068
126. Price SC, Stuart AC, Yang LQ, Zhou HX, You W (2011) Fluorine substituted conjugated polymer of medium band gap yields 7 % efficiency in polymer-fullerene solar cells. J Am Chem Soc 133:4625–4631
127. Li K, Li Z, Feng K, Xu X, Wang L, Peng Q (2013) Development of large band-gap conjugated copolymers for efficient regular single and tandem organic solar cells. J Am Chem Soc 135:13549–13557
128. Wang XC, Jiang P, Chen Y, Luo H, Zhang ZG, Wang HQ, Li XY, Yu G, Li YF (2013) Thieno[3,2-b]thiophene-bridged D-pi-A polymer semiconductor based on benzo[1,2-b:4,5-b′]dithiophene and benzoxadiazole. Macromolecules 46:4805–4812
129. Chen HC, Chen YH, Liu CC, Chien YC, Chou SW, Chou PT (2012) Prominent short-circuit currents of fluorinated quinoxaline-based copolymer solar cells with a power conversion efficiency of 8.0 %. Chem Mater 24:4766–4772
130. Huo LJ, Hou JH, Zhang SQ, Chen HY, Yang Y (2010) A polybenzo[1,2-b:4,5-b′] dithiophene derivative with deep HOMO level and its application in high-performance polymer solar cells. Angew Chem Int Ed 49:1500–1503
131. Huo LJ, Zhang SQ, Guo X, Xu F, Li YF, Hou JH (2011) Replacing alkoxy groups with alkylthienyl groups: a feasible approach to improve the properties of photovoltaic polymers. Angew Chem Int Ed 50:9697–9702
132. Huang Y, Guo X, Liu F, Huo LJ, Chen YN, Russell TP, Han CC, Li YF, Hou JH (2012) Improving the ordering and photovoltaic properties by extending pi-conjugated area of electron-donating units in polymers with D-A structure. Adv Mater 24:3383–3389
133. Duan RM, Ye L, Guo X, Huang Y, Wang P, Zhang SQ, Zhang JP, Huo LJ, Hou JH (2012) Application of two-dimensional conjugated benzo[1,2-b:4,5-b′]dithiophene in quinoxaline-based photovoltaic polymers. Macromolecules 45:3032–3038
134. Guo X, Zhang MJ, Huo LJ, Xu F, Wu Y, Hou JH (2012) Design, synthesis and photovoltaic properties of a new D-pi-A polymer with extended pi-bridge units. J Mater Chem 22:21024–21031
135. Zhang SQ, Ye L, Wang Q, Li ZJ, Guo X, Huo LJ, Fan HL, Hou JH (2013) Enhanced photovoltaic performance of diketopyrrolopyrrole (DPP)-based polymers with extended pi conjugation. J Phys Chem C 117:9550–9557
136. Qian DP, Ye L, Zhang MJ, Liang YR, Li LJ, Huang Y, Guo X, Zhang SQ, Tan ZA, Hou JH (2012) Design, application, and morphology study of a new photovoltaic polymer with strong aggregation in solution state. Macromolecules 45:9611–9617
137. Wang M, Hu XW, Liu P, Li W, Gong X, Huang F, Cao Y (2011) Donor acceptor conjugated polymer based on naphtho[1,2-c:5,6-c]bis[1, 2, 5]thiadiazole for high-performance polymer solar cells. J Am Chem Soc 133:9638–9641
138. Huo LJ, Guo X, Zhang SQ, Li YF, Hou JH (2011) PBDTTTZ: a broad band gap conjugated polymer with high photovoltaic performance in polymer solar cells. Macromolecules 44:4035–4037

139. Zhang MJ, Gu Y, Guo X, Liu F, Zhang SQ, Huo LJ, Russell TP, Hou JH (2013) Efficient polymer solar cells based on benzothiadiazole and alkylphenyl substituted benzodithiophene with a power conversion efficiency over 8 %. Adv Mater 25:4944–4949
140. Zhang M, Guo X, Zhang S, Hou J (2014) Synergistic effect of fluorination on molecular energy level modulation in highly efficient photovoltaic polymers. Adv Mater 26:1118–1123
141. Huo LJ, Ye L, Wu Y, Li ZJ, Guo X, Zhang MJ, Zhang SQ, Hou JH (2012) Conjugated and nonconjugated substitution effect on photovoltaic properties of benzodifuran-based photovoltaic polymers. Macromolecules 45:6923–6929
142. Wu Y, Li ZJ, Guo X, Fan HL, Huo LJ, Hou JH (2012) Synthesis and application of dithieno[2,3-d:2′,3′-d′]benzo[1,2-b:4,5-b′] dithiophene in conjugated polymer. J Mater Chem 22:21362–21365
143. Wu Y, Li ZJ, Ma W, Huang Y, Huo LJ, Guo X, Zhang MJ, Ade H, Hou JH (2013) PDT-S-T: a new polymer with optimized molecular conformation for controlled aggregation and pi-pi stacking and its application in efficient photovoltaic devices. Adv Mater 25:3449–3455
144. Son HJ, Lu LY, Chen W, Xu T, Zheng TY, Carsten B, Strzalka J, Darling SB, Chen LX, Yu LP (2013) Synthesis and photovoltaic effect in dithieno[2,3-d:2′,3′-d′]Benzo[1,2-b:4,5-b′] dithiophene-based conjugated polymers. Adv Mater 25:838–843
145. Pron A, Berrouard P, Leclerc M (2013) Thieno[3,4-c]pyrrole-4,6-dione-based polymers for optoelectronic applications. Macromol Chem Physic 214:7–16
146. Zou YP, Najari A, Berrouard P, Beaupre S, Aich BR, Tao Y, Leclerc M (2010) A thieno[3,4-c] pyrrole-4,6-dione-based copolymer for efficient solar cells. J Am Chem Soc 132:5330–5331
147. Zhang Y, Hau SK, Yip HL, Sun Y, Acton O, Jen AKY (2010) Efficient polymer solar cells based on the copolymers of benzodithiophene and thienopyrroledione. Chem Mater 22:2696–2698
148. Zhang GB, Fu YY, Zhang Q, Xie ZY (2010) Benzo[1,2-b:4,5-b′]dithiophene-dioxopyrrolothiophen copolymers for high performance solar cells. Chem Commun 46:4997–4999
149. Piliego C, Holcombe TW, Douglas JD, Woo CH, Beaujuge PM, Frechet JMJ (2010) Synthetic control of structural order in N-alkylthieno[3,4-c]pyrrole-4,6-dione-based polymers for efficient solar cells. J Am Chem Soc 132:7595–7597
150. Aich BR, Lu JP, Beaupre S, Leclerc M, Tao Y (2012) Control of the active layer nanomorphology by using co-additives towards high-performance bulk heterojunction solar cells. Org Electron 13:1736–1741
151. Cabanetos C, El Labban A, Bartelt JA, Douglas JD, Mateker WR, Frechet JMJ, McGehee MD, Beaujuge PM (2013) Linear side chains in benzo[1,2-b:4,5-b′]dithiophene-thieno[3,4-c] pyrrole-4,6-dione polymers direct self-assembly and solar cell performance. J Am Chem Soc 135:4656–4659
152. Amb CM, Chen S, Graham KR, Subbiah J, Small CE, So F, Reynolds JR (2011) Dithienogermole as a fused electron donor in bulk heterojunction solar cells. J Am Chem Soc 133:10062–10065
153. Small CE, Chen S, Subbiah J, Amb CM, Tsang SW, Lai TH, Reynolds JR, So F (2012) High-efficiency inverted dithienogermole-thienopyrrolodione-based polymer solar cells. Nat Photonics 6:115–120
154. Zhong HL, Li Z, Deledalle F, Fregoso EC, Shahid M, Fei ZP, Nielsen CB, Yaacobi-Gross N, Rossbauer S, Anthopoulos TD, Durrant JR, Heeney M (2013) Fused dithienogermolodithiophene low band gap polymers for high-performance organic solar cells without processing additives. J Am Chem Soc 135:2040–2043
155. Su MS, Kuo CY, Yuan MC, Jeng US, Su CJ, Wei KH (2011) Improving device efficiency of polymer/fullerene bulk heterojunction solar cells through enhanced crystallinity and reduced grain boundaries induced by solvent additives. Adv Mater 23:3315–3319
156. Guo XG, Ortiz RP, Zheng Y, Kim MG, Zhang SM, Hu Y, Lu G, Facchetti A, Marks TJ (2011) Thieno[3,4-c]pyrrole-4,6-dione-based polymer semiconductors: toward high-performance, air-stable organic thin-film transistors. J Am Chem Soc 133:13685–13697

157. Guo XG, Zhou NJ, Lou SJ, Smith J, Tice DB, Hennek JW, Ortiz RP, Navarrete JTL, Li SY, Strzalka J, Chen LX, Chang RPH, Facchetti A, Marks TJ (2013) Polymer solar cells with enhanced fill factors. Nat Photonics 7:825–833
158. Wang EG, Hou LT, Wang ZQ, Hellstrom S, Zhang FL, Inganas O, Andersson MR (2010) An easily synthesized blue polymer for high-performance polymer solar cells. Adv Mater 22:5240–5244
159. Wang EG, Ma ZF, Zhang Z, Vandewal K, Henriksson P, Inganas O, Zhang FL, Andersson MR (2011) An easily accessible isoindigo-based polymer for high-performance polymer solar cells. J Am Chem Soc 133:14244–14247
160. Qian DP, Ma W, Li ZJ, Guo X, Zhang SQ, Ye L, Ade H, Tan ZA, Hou JH (2013) Molecular design toward efficient polymer solar cells with high polymer content. J Am Chem Soc 135:8464–8467
161. Deng Y, Liu J, Wang J, Liu L, Li W, Tian H, Zhang X, Xie Z, Geng Y, Wang F (2014) Dithienocarbazole and isoindigo based amorphous low bandgap conjugated polymers for efficient polymer solar cells. Adv Mater 26:471–476
162. Osaka I, Kakara T, Takemura N, Koganezawa T, Takimiya K (2013) Naphthodithiophene-naphthobisthiadiazole copolymers for solar cells: alkylation drives the polymer backbone flat and promotes efficiency. J Am Chem Soc 135:8834–8837
163. Hendriks KH, Heintges GHL, Gevaerts VS, Wienk MM, Janssen RAJ (2013) High-molecular-weight regular alternating diketopyrrolopyrrole-based terpolymers for efficient organic solar cells. Angew Chem Int Ed 52:8341–8344
164. Dou L, Chen C-C, Yoshimura K, Ohya K, Chang W-H, Gao J, Liu Y, Richard E, Yang Y (2013) Synthesis of 5H-dithieno[3,2-b:2′,3′-d]pyran as an electron-rich building block for donor–acceptor type low-bandgap polymers. Macromolecules 46:3384–3390
165. Ye L, Zhang SQ, Qian DP, Wang Q, Hou JH (2013) Application of Bis-PCBM in polymer solar cells with improved voltage. J Phys Chem C 117:25360–25366
166. Zhang X, Lu Z, Ye L, Zhan C, Hou J, Zhang S, Jiang B, Zhao Y, Huang J, Zhang S, Liu Y, Shi Q, Liu Y, Yao J (2013) A potential perylene diimide dimer-based acceptor material for highly efficient solution-processed non-fullerene organic solar cells with 4.03 % efficiency. Adv Mater 25:5791–5797
167. Jiang W, Ye L, Li XG, Xiao CY, Tan F, Zhao WC, Hou JH, Wang ZH (2014) Bay-linked perylene bisimides as promising non-fullerene acceptors for organic solar cells. Chem Commun 50:1024–1026
168. Sonar P, Lim JPF, Chan KL (2011) Organic non-fullerene acceptors for organic photovoltaics. Energy Environ Sci 4:1558–1574
169. Eftaiha AF, Sun JP, Hill IG, Welch GC (2014) Recent advances of non-fullerene, small molecular acceptors for solution processed bulk heterojunction solar cells. J Mater Chem A 2:1201–1213
170. Facchetti A (2013) Polymer donor-polymer acceptor (all-polymer) solar cells. Mater Today 16:123–132
171. Zhan XW, Tan ZA, Domercq B, An ZS, Zhang X, Barlow S, Li YF, Zhu DB, Kippelen B, Marder SR (2007) A high-mobility electron-transport polymer with broad absorption and its use in field-effect transistors and all-polymer solar cells. J Am Chem Soc 129:7246–7247
172. Zhan XW, Tan ZA, Zhou EJ, Li YF, Misra R, Grant A, Domercq B, Zhang XH, An ZS, Zhang X, Barlow S, Kippelen B, Marder SR (2009) Copolymers of perylene diimide with dithienothiophene and dithienopyrrole as electron-transport materials for all-polymer solar cells and field-effect transistors. J Mater Chem 19:5794–5803
173. Tan ZA, Zhou EJ, Zhan XW, Wang X, Li YF, Barlow S, Marder SR (2008) Efficient all-polymer solar cells based on blend of tris(thienylenevinylene)-substituted polythiophene and poly[perylene diimide-alt-bis(dithienothiophene)]. Appl Phys Lett 93:073309
174. Zhou EJ, Cong JZ, Wei QS, Tajima K, Yang CH, Hashimoto K (2011) All-polymer solar cells from perylene diimide based copolymers: material design and phase separation control. Angew Chem Int Ed 50:2799–2803

175. Zhou Y, Yan QF, Zheng YQ, Wang JY, Zhao DH, Pei J (2013) New polymer acceptors for organic solar cells: the effect of regio-regularity and device configuration. J Mater Chem A 1:6609–6613
176. McNeill CR, Abrusci A, Zaumseil J, Wilson R, McKiernan MJ, Burroughes JH, Halls JJM, Greenham NC, Friend RH (2007) Dual electron donor/electron acceptor character of a conjugated polymer in efficient photovoltaic diodes. Appl Phys Lett 90:193506
177. McNeill CR, Abrusci A, Hwang I, Ruderer MA, Muller-Buschbaum P, Greenham NC (2009) Photophysics and photocurrent generation in polythiophene/polyfluorene copolymer blends. Adv Funct Mater 19:3103–3111
178. Tang YQ, McNeill CR (2013) All-polymer solar cells utilizing low band gap polymers as donor and acceptor. J Polym Sci B: Polym Phys 51:403–409
179. Earmme T, Hwang YJ, Murari NM, Subramaniyan S, Jenekhe SA (2013) All-polymer solar cells with 3.3 % efficiency based on naphthalene diimide-selenophene copolymer acceptor. J Am Chem Soc 135:14960–14963
180. Cheng P, Ye L, Zhao X, Hou J, Li Y, Zhan X (2014) Binary additives synergistically boost the efficiency of all-polymer solar cells up to 3.45 %. Energy Environ Sci. doi:10.1039/C3EE43041C
181. Zhou E, Cong JZ, Hashimoto K, Tajima K (2013) Control of miscibility and aggregation via the material design and coating process for high-performance polymer blend solar cells. Adv Mater 25:6991–6996
182. Mori D, Benten H, Okada I, Ohkita H, Ito S (2013) Low-bandgap donor/acceptor polymer blend solar cells with efficiency exceeding 4 %. Adv Energy Mater. doi:10.1002/aenm.201301006
183. Chiechi RC, Hummelen JC (2012) Polymer electronics, quo vadis? Acs Macro Lett 1:1180–1183
184. Koster LJA, Shaheen SE, Hummelen JC (2012) Pathways to a new efficiency regime for organic solar cells. Adv Energy Mater 2:1246–1253
185. Kanal IY, Owens SG, Bechtel JS, Hutchison GR (2013) Efficient computational screening of organic polymer photovoltaics. J Phys Chem Lett 4:1613–1623
186. Liu C, Wang K, Hu XW, Yang YL, Hsu CH, Zhang W, Xiao S, Gong X, Cao Y (2013) Molecular weight effect on the efficiency of polymer solar cells. ACS Appl Mater Inter 5:12163–12167

Chapter 6
Organic Semiconductor Electroluminescent Materials

Gufeng He

Abstract This chapter reviews the important progress made on small molecule electroluminescent materials used in organic light-emitting diode (OLED). In many cases we describe not only the material structures but also the properties associated with these materials, such as energy level, absorption and photoluminescence (PL) peaks, PL quantum yield, exciton life time, and so on. The performances of related devices are covered as well if they are available.

Keywords Organic light-emitting diodes · Fluorescence · Phosphorescence · Efficiency

6.1 Introduction

The organic light-emitting diode (OLED) has attracted tremendous interest since the scientists at Kodak reported a high efficiency device with a double layer structure in 1987 [1]. Actually, as an electroluminescence (EL) phenomenon, it can be traced back to Pope's early work in the 1960s [2], where he used single crystals of anthracene as EL material and the driving voltage was around 400 V. To reduce the driving voltage and to improve the efficiency, much effort has been spent ever since. A major step forward in this field was the development of thin-film organic electroluminescent devices with relatively low driving voltages (below 30 V) by Vincett in 1982 [3].

The first practical OLED demonstrated by Tang in 1987 used a heterojunction structure of organic amorphous thin films deposited by vacuum thermal evaporation of small molecules. One of these layers was a hole-transporting aromatic diamine, whereas the other was an electron-transporting and emissive layer of tris (8-hydroxyquinoline) aluminum (Alq_3). This was the first time that an organic

G. He (✉)
Shanghai Jiao Tong University, Shanghai, China
e-mail: gufenghe@sjtu.edu.cn

Y. Li (ed.), *Organic Optoelectronic Materials*,
Lecture Notes in Chemistry 91, DOI 10.1007/978-3-319-16862-3_6

electroluminescent device could reach a high brightness (>1000 cd m^{-2}) with a relatively low operating voltage (<10 V) and an external quantum efficiency (EQE) of about 1 %. To approach this, each material has its own function, and this strategy has become a basic rule for designing high-performance OLED structures. Soon after, in 1990, Richard Friend et al. [4] successfully discovered an OLED using a conjugated polymer as emitter by spin-coating a soluble precursor and then thermal treating it to form a conjugated polymer, which is usually called a polymer light-emitting diode (PLED). In this device, a single layer of poly(*p*-phenylenevinylene) (PPV), sandwiched between indium tin oxide (ITO) and Al electrodes, emitted green-yellow light under applied voltage. The device efficiency and relatively low turn-on voltage held promise for a large-area, low-cost solution-processable commercial application. In 1998, the Thompson and Forrest groups [5] boosted OLED internal quantum efficiency limit from 25 to 100 % by using phosphorescent electroluminescent emissive materials. Heavy metal atoms were widely used in phosphorescent materials to mix the triplet excited states and the singlet excited states to make the radiative decay possible from triplet state to ground state. Recently, Adachi group [6, 7] from Kyushu University realized 100 % internal quantum efficiency via Thermally Activated Delayed Fluorescence (TADF), without using heavy metal chelate. This is another breakthrough in the development of organic electroluminescent materials.

The performance of an OLED depends very much on the electroluminescent materials used in the device. In this chapter, we attempt to review the important small molecule materials used in OLEDs, especially electroluminescent materials, and have tried to present a simple unbiased selection of representative data from the original authors' works.

6.2 Working Mechanism of OLEDs

6.2.1 Working Mechanism

A simple OLED structure is composed of an extremely thin organic electroluminescent film, sandwiched between two electrodes. When an external bias is applied to the device, holes and electrons are injected from the anode and the cathode, respectively, into the organic semiconductor layer. The holes and electrons migrate in the organic layer, then meet and recombine to form excitons. Finally, the excitons decay to the ground states radiatively and emit light.

However, to improve the charge carrier injection and transport and to confine excitons to reach high performance, a multi-layer structure is commonly applied in OLEDs. A typical high efficiency OLED consists of five organic semiconductor layers—hole injection layer (HIL), hole transport layer (HTL), emissive layer (EML), electron transport layer (ETL), and electron injection layer (EIL)—where the HTL is often used as electron blocking layer (EBL), whereas ETL is used as hole blocking layer (HBL) (Fig. 6.1).

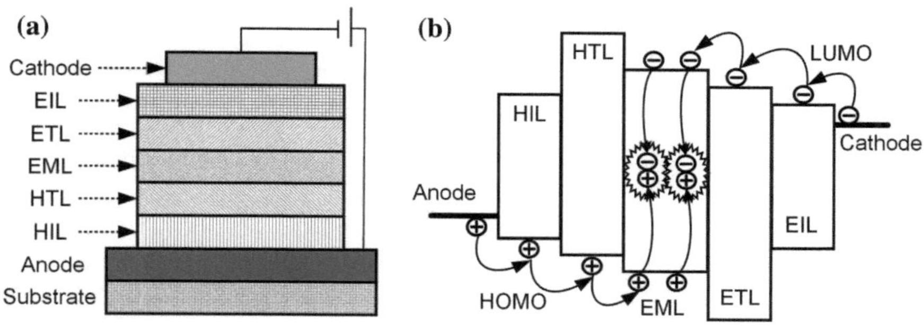

Fig. 6.1 Device structure and working mechanism

6.2.2 Anode and Hole Injection Material

As a perfect anode of an OLED, it must have high conductivity, good stability, and high transparency in the visible range when used in a bottom-emission or transparent OLED. To ensure an effective hole injection into the organic semiconductor, the anode should have high work function, which is close to the highest occupied molecular orbital (HOMO) of the adjacent HIL to minimize the hole injection barrier. The commonly used anode materials are transparent conducting oxide (TCO) and high work function metals such as Ni, Au, and Pt.

ITO is the most widely used TCO for anodes because of its fairly high electrical conductivity and outstanding optical transparency. The work function of ITO is quite sensitive to the cleaning procedure. Untreated ITO has a work function of 4.5~4.8 eV, and it can be increased up to 5 eV with oxygen plasma [8] or UV ozone treatment [9].

Although the work function of ITO can be up to 5 eV [10], about 0.4 eV hole injection barrier still exists between ITO and most of the hole transport materials (HTM). To minimize the energy barrier for injection and thus lower the operating voltage, a HIL is often employed. The HIL not only acts as an interfacial layer between the anode and the HTL to facilitate efficient hole injection, but also enhances the device stability by smoothing the surface and improving the film forming property of the subsequent organic layers. The most common organic HIL materials include porphyrinic metal complexes [11], star arylamines [12], and conducting polymer such as PEDOT:PSS(poly-3,4-ethylenedioxythiophene doped with polystyrene sulfonic acid; Fig. 6.2) [13].

Copper phthalocyanine (CuPc; Fig. 6.2) is a widely used pigment with very high thermal stability. Van Slyke et al. [11] inserted a thin CuPc layer between the anode (ITO) and the HTL (*N*,*N*′-diphenyl-*N*,*N*′-bis(1-naphthyl phenyl)-1,1′-biphenyl-4,4′-diamine, NPB), and achieved a low voltage and highly stable OLED with optimized thickness. This was attributed to a lower energy injection barrier from ITO to CuPc. Besides CuPc, other phthalocyanines and porphyrin materials with similar structure have also been used for HIL applications [14].

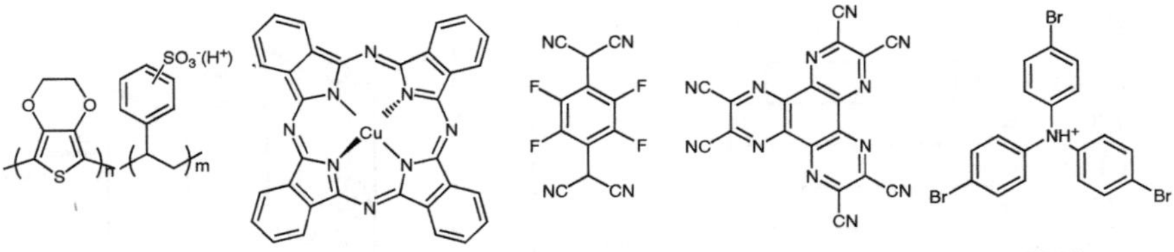

Fig. 6.2 Organic hole injection materials

In 2004, LG Chem, Ltd. [15] developed an effective material for a HIL, 1,4,5,8,9,11-hexaazatriphenylene-hexacarbonitrile (HAT-CN; Fig. 6.2). It is a strongly electron-withdrawing molecule because of its six nitrile groups, and a material with such a property is usually applied as an ETL. However, it was found that it could significantly improve hole injection even with low work function metals such as Al or Ag, which is critical for top-emission OLEDs. It has some interesting properties, including a very large value of HOMO level, a LUMO level very close to the Fermi level, and a high work function of over 6.0 eV.

Another category of HIL is conductive polymer such as PEDOT:PSS [13], PANI (polyaniline) [16], or PPy (polypyrrole) [17]. The most widely used is PEDOT:PSS, which is water soluble and can be deposited by solution processes such as spin coating, ink-jet printing, etc. The spin-coated PEDOT:PSS layer can smooth the ITO surface and provide better hole injection because of its relatively high work function (>5.0 eV). However, most of the available conductive polymer materials are strongly acidic because of the doping necessary to improve the solubility and conductivity. The acidity may destroy the underlying ITO surface by dissolving and even reacting with amine-containing materials, leading to a short device life time and poor shelf life.

The HILs described above are used as energy compatible materials to reduce the injection energy barrier. A thin insulating layer inserted between ITO and HTL has also been discovered to enhance the hole injection effectively, which is often called a buffer layer such as SiO_2 [18], SiO_xN_y [19], Teflon [20], LiF [21], and TiO_2 [22]. With an optimum thickness, such a buffer layer can reduce the operating voltage and improve the performance. Hou et al. [21] explained it with a tunneling model; the introduced insulating layer has two effects: the voltage drop across it lowers the ITO E_F which reduces the injection barrier, and the buffer layer itself adds an additional tunneling barrier. If the sum of these two barriers is lower than the initial barrier, the hole injection capability via tunneling is enhanced and the operating voltage decreases. However, if this buffer layer is too thick, the latter effect is more pronounced, i.e., the injection barrier reduced is smaller than the barrier introduced, and the buffer layer has negative effect. Some researchers attribute the enhanced performance to more homogeneous adhesion of the following HTL resulting from the improved smoothness of the ITO surface.

Some HTM can be mixed with oxidizer, Lewis acid, or strong electron-acceptors such as $FeCl_3$ [23], $SbCl_5$, iodine [24], tetrafluorotetracyanoquinodimethane

(F4-TCNQ; Fig. 6.2) [25], or tris(4-bromophenyl)aminiumhexachloroantimonate (TBAHA; Fig. 6.2) [26] to form effective p-type doping and obtain fantastic hole injection properties. The organic p-type doping, together with n-type doping is further discussed in detail later.

6.2.3 Cathode and Electron Injection Material

Opposite the anode material, a satisfactory cathode should have low work function to facilitate electron injection into the LUMO of the adjacent organic layer, as well as high conductivity. A low work function metal, such as Li, Mg, Ca, and Ba, is a good candidate. However, low work function usually means high chemical reactivity and poor stability. The reactive metals are sensitive to moisture and oxygen, making them difficult to process. A general solution is to use an alloy composed of low work function metal and noncorroding metal, and this usually provides good film-forming properties and stability.

The Mg:Ag (10:1 by volume) alloy was the most popular cathode material during the early stage of OLED development [27]. The addition of metal Ag not only improves the cathode stability but also enhances the adhesion of the cathode on the organic layer. Al is quite a stable metal, but the relatively high work function makes it difficult to inject electrons to the LUMO of most electron transport materials (ETM). Naka et al. [28] systematically investigated the influence of doping Al with various metals (Li, Ca, and Mg). They found that the lower the work function of the doped metal, the less the injection Schottky barrier.

Although the metal alloy as cathode provides good electron injection and relatively good stability, Li and Mg atoms may diffuse into the organic layer and form quenching sites for excitons, which influences the long-term stability of an OLED. Therefore, to use a high work function stable metal such as Al and Ag as cathode, a common approach is to insert an EIL between the cathode and the organic layer to enhance the electron injection.

The typical electron injection materials are alkali metal compounds and alkaline earth metal compound [29–31], such as Li_2O, Cs_2CO_3, CH_3COOM (M=Li, Na, K, Rb, Cs), LiF, CsF, and so on. The most popular electron injection material is LiF [32, 33], which is still widely used at the present day. The LiF/Al bilayer can reduce the injection barrier significantly, and the current-voltage characteristics of the OLED are greatly improved in the presence of a 0.5–1-nm LiF layer. Although the exact mechanism of such a buffer layer is unclear [34–36], one of the commonly accepted explanations is that LiF reacts with evaporated Al and liberates Li atoms, and Li dopes into the electron transport material at the interface and forms *n*-type doping [37, 38]. Because the thermal evaporation of LiF is a bit difficult, lithium-quinolate complexes (Fig. 6.3) 8-hydroxyquinolinolatolithium (Liq), 2-methyl-8-hydroxyquinolinolatolithium (LiMeq), 4-phenanthridinolatolithium (Liph), and 2-(5-phenyl-1,3,4-oxadiazolyl)phenolatolithium (LiOXD) have been synthesized and investigated as electron injection materials [39–43]. The results confirm that the

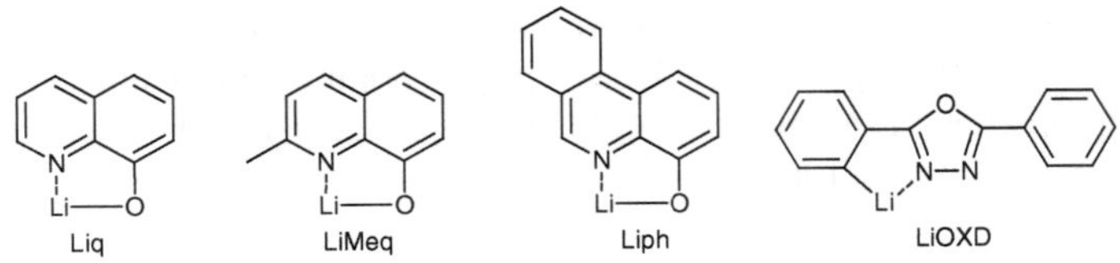

Fig. 6.3 Organic electron injection materials

function of these lithium-quinolate complexes is the same as that of LiF. A very thin layer (0.5–5.0 nm) of the complex deposited on the Alq_3 (8-hydroxyquinoline aluminum) layer enhances the electron injection and reduces the driving voltage. In addition, another advantage of using Liq over LiF as an injection layer is that the efficiency is less sensitive to the Liq thickness. Liq is recently more and more replacing LiF as EIL both in academia and industry for its superior properties.

6.2.4 Hole and Electron Transport Materials

HTM are very common in OLEDs, providing a hole-conductive pathway for positive charge carriers to migrate from the anode into the EML. It is usually required that they possess thermal stability with high glass transition temperature (T_g), high hole-mobility, and the capability of forming stable pin-hole-free thin-films upon evaporation. Hence it is important that the HTL materials are easily oxidized by removing an electron and the redox processes are reversible. To facilitate the hole transport into the EML, the HTL should have HOMO levels comparable to EML. As a good HTL, it is appreciated that it can confine the charge carriers and excitons in the EML effectively. Therefore a shallow LUMO level to block the electrons from the cathode side and large band gap to prevent exciton diffusion and quenching are desirable. With these properties in mind, some of the commonly employed HTL materials are triaryl amines (e.g., 4,4′,4″-tri(*N*-carbazolyl) triphenylamine, TCTA and 4,4′,4″-tris(3-methylphenylphenylamino)-triphenylamine, m-MTDATA) [44, 45], biphenyl diamine derivatives (e.g., *N*,*N*′-(3-methylphenyl)-1,1′-biphenyl-4,4′-diamine, TPD and NPB), and so on (Fig. 6.4).

For organic materials, the charge transport property of electrons is usually inferior to that of holes. Therefore, it is more important to select ETM with high mobility. Besides the mobility, the strategy of selecting a good ETM is similar to HTM. As such, the material needs to have a LUMO level close to the work function of the cathode to aid charge injection. Its HOMO needs to be deep to block the holes from the anode side and the band gap should be large. Furthermore, as all organic layers, it should have a high Tg for good stability and form smooth amorphous thin films. On the basis of these criteria, typical examples are: oxadiazole derivatives (5(4-biphenyl)-2-(4-*tert*-butyphenyl)-1,3,4-oxadiazole, PBD) [46], azole-based materials (1,3,5-tris(*N*-phenylbenzimidizole-2-yl)benzene, TPBI, and 3-phenyl-4-(1′-naphthyl)-5-phenyl-1,2,4-

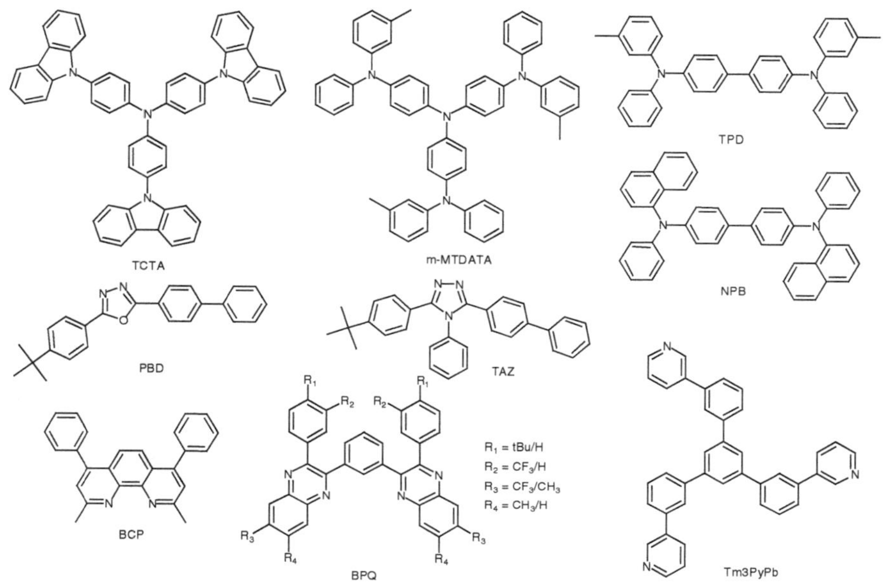

Fig. 6.4 Hole and electron transporting materials

triazoles, TAZ) [47, 48], quinolone derivatives, quinoxaline derivatives (bis(phenylquinoxaline, BPQ) [49], anthrazoline derivatives, phenanthroline derivatives (4,7-diphenyl-1,10-phenanthroline, BPhen and 2,9-dimethyl-4,7-diphenyl-1,10-phenanthroline, BCP) [50–54], silole [55, 56], pyridine-based materials (1,3,5-tri(*m*-pyrid-3-yl-phenyl)benzene, Tm_3PyPB) [57, 58], cyano and F-substituted compounds, metal chelates, and others (Fig. 6.4).

6.2.5 p- *and* n-*Type Doping Materials*

Similar to inorganic materials, some organic semiconductor materials can be doped by electron donors or acceptors to obtain excellent charge injection and transport properties. In organic semiconductors, the dopants either extract electrons from the HOMO states of the hosts to generate holes (*p*-type doping), or donate electrons to the LUMO states (*n*-type doping). Large aromatic molecules with strong π-electron donating or withdrawing properties are often used in organic electrical doping to avoid easy diffusion which happens with small dopants.

As for *p*-type doping, strong electron acceptor molecules such as orthochloranil [59], tetracyano-quinodimethane (TCNQ) [60], or dicyano-dichloroquinone (DDQ) [61] are required as dopants (Fig. 6.5). Among them, 2,3,5,6-tetrafluoro-7,7,8,8-tetracyanoquinodimethane (F4-TCNQ), capable of being doped in a variety of hole transport matrices [62, 63], is considered as the most successful dopant material in

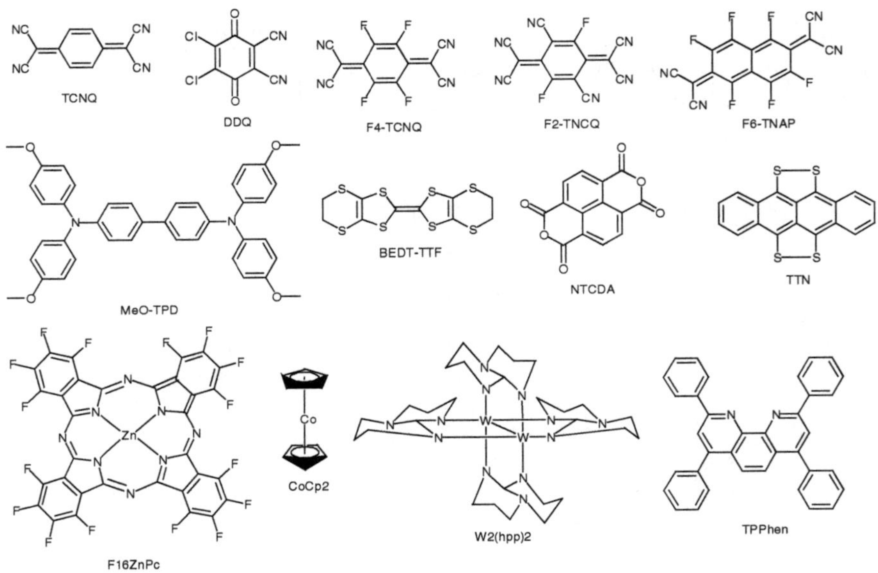

Fig. 6.5 *p*- and *n*-type doping materials

the application of optoelectronic devices. When it is doped into hole transport material *N,N,N′,N′*-tetrakis(4-methoxyphenyl)-benzidine (MeO-TPD, Fig. 6.5), a conductivity of 5×10^{-6}S/cm at 2 % doping ratio can be achieved, which is four orders higher than that of the undoped one. With such high conductivity, the voltage drop over a thick doped HTL at the normal driving condition is negligible, which enables a more freely OLED structure design.

More importantly, doped HTL provides much easier electron injection than undoped. In a model doped system, ZnPc:F4-TCNQ (30:1) doped on ITO substrate, a much stronger level bending of 0.9 eV is observed, and the Fermi level is only 0.23 eV away from the HOMO level [64]. The space charge layer is very thin, and the holes are easily injected via tunneling and form a quasi-ohmic contact.

F4-TCNQ has great successes in improving the hole injection and conductivity of HTLs. However, its low glass transition temperature and high volatility raise serious concerns over issues of cross-contamination, reproducibility, and thermal stability of the OLED devices. Some other *p*-type dopants with better thermal stability have been synthesized (Fig. 6.5), e.g., 3,6-difluoro-2,5,7,7,8,8-hexacyanoquinodimethane (F2-HCNQ) [65], 1,3,4,5,7,8-hexafluorotetracyanonaphthoquinodimethane (F6-TNAP) [66], and 2,2-(perfluoronaphthalene-2,6-diylidene) dimalononitrile (F6-TCNNQ) [67–69]. Because this field is now mainly of commercial interest and most of the improved dopants are proprietary materials [70, 71], they are not discussed in detail here.

In contrast to *p*-type doping, *n*-type molecular doping is intrinsically more difficult. For efficient doping, the HOMO level of the dopant must be energetically aligned with the LUMO level of the host material, which makes such materials unstable against oxygen in air. With increasing LUMO energy, the difficulty in finding suitable materials is increased. Hence, the most widely used *n*-type dopant is actually alkali metals, such as Li and Cs. In 1998, Kido [72] reported efficient OLEDs using electron transport material of bathophenanthroline (BPhen) doped with Li. Parthasarathy et al. [73] reached a conductivity of 3×10^{-5}S/cm for a 100 nm-thick BCP:Li film. Such bulk doping can also be reached by Li deposition on top of the organic layers, if it is followed by metal deposition. The high temperature during metal deposition may cause the Li to diffuse into the organic layer. The alkali metal acts as an efficient electron donor to the ETM to produce *n*-type doping. An analysis by secondary ion mass spectrometry (SIMS) depth profiling shows that a strong Li diffusion into the organic material can be up to 80 nm. However, the propensity to diffuse into the emission layer (EML) because of the alkali metal's small atom size may impact on light emission as well as stability [72]. Organic dopants are more robust and more stable because of their large aromatic ring. Bis(ethylenedithio)tetrathiafulvalene (BEDT-TTF) is the first *n*-type donor molecule coevaporated into naphthalene tetracarboxylic dianhydride (NTCDA) [74], but the conductivity is only one to two orders above that of nominally undoped NTCDA. Tetrathianaphthacene (TTN) is found to be an efficient donor in hexadecafluorophthalocyaninatozinc ($F_{16}ZnPc$) but not in Alq_3 which has a higher lying LUMO so that an energy transfer from the dopant to Alq_3 LUMO is not possible [75]. The strongly reducing molecule bis(cyclo-pentadienyl)cobalt(II) (cobaltocene, $CoCp_2$) [76] has an ionization energy of 4 eV, making it a promising material for molecular *n*-type doping. When it is doped into an electron transport material, a tris(thieno)hexaazatriphenylene derivative, it shows a 0.56-eV shift of the Fermi level toward the unoccupied states of the host.

Today the most successfully commercial *p*- and *n*-type dopants and corresponding host materials are produced by Novaled AG, which claims to have high stability and ease of handling in air and control of the evaporation. They reported the first practical molecular *n*-type dopant $W_2(hpp)_4$ [77–79], which could be doped into ETL material 2,4,7,9-tetraphenyl-1,10-phenanthroline (TPPhen; Fig. 6.5). The doped ETL significantly reduced the operating voltages in OLED. However, the dopant is sensitive to air, and can only be handled in an inert atmosphere. Later, they announced an air stable *n*-type dopant NDN26 [80, 81], although the material structure has not been disclosed.

When the electrically intrinsic EML is sandwiched between *p*-doped and *n*-doped transport layers, it forms a *p-i-n* OLED structure. A typical *p-i-n* bottom emission OLED consists of five organic layers with distinct functionalities: two doped transport layers, two undoped blocking layers, and an EML [82]. The doped charge transport layers (HTL and ETL) are responsible for efficient charge injection from the contacts into the devices, and for efficient charge transport to the EML. Although this three-layer structure already works rather well, one often faces problems with lacking charge balance, interface exciplexes, or exciton quenching by excess charge carriers. For this reason, the introduction of additional blocking layers (HBL and EBL) between the charge transport layers and the EML to confine the injected charges and generated excitons within the EML and to ensure a good charge balance is sometimes necessary to reach really high efficiency and high stability [83].

Because of electric doping in HTL and ETL, the conductivity of the organic stack increases to induce band bending at the interface, and thus it also reduces carrier injection energy barrier as well as operating voltage. For example, using CBP:Ir(ppy)$_3$ as the emitter system, green phosphorescent OLEDs with extremely low operating voltages and high quantum efficiency have been demonstrated [84]. These *p-i-n* devices attain a luminance of 1000 cd/m^2 at only 3 V, with an EQE of 9 % and a power efficiency of 28 lm/W. Remarkable for these *p-i-n* OLEDs is their low driving voltage: 100 cd/m^2 are reached at 2.6 V, which is close to the equivalent of the photon energy (2.4 eV, corresponding to the triplet energy in Ir(ppy)$_3$). This result confirms that the *p-i-n* structure significantly reduces the operating voltage while improving the current efficiency of organic EL devices.

The question of stability and operational life time is a very important issue for *p-i-n* OLEDs as dopants may diffuse to the emitting layer to produce potential luminance quencher, particularly for Li-doped device. Using a larger size of atomic dopant such as Cs or organic dopant can overcome most of the problems and the insertion of a stable hole blocking interlayer also plays an important role in improving life time. Meerheim et al. [85] demonstrated extremely stable and highly efficient red *p-i-n* OLEDs based on an iridium-based phosphorescent dye. Ten million hours of life time at initial luminance of 100 cd/m^2 is reached with 12.4 % EQE, which is attributed to low current density and the highly stable materials against both charge carriers and excitons.

The advantage of low operating voltage and high luminous efficiency of *p-i-n* structure are very important for the development of both AMOLED displays and lighting applications. As the increase of voltage drop over the entire doped transport layer with respect to thickness is very small, it makes it possible to optimize resonant light outcoupling with strong microcavity without impacting on charge carrier balance [81].

6.3 Fluorescent Electroluminescent Materials

The electroluminescent materials are definitely the most important materials in the OLED because they generate light. The first efficient OLED reported by the Kodak group used a pure Alq$_3$ film as EML [1]. Two years later, the same group developed a host-guest doped emitter system which is considered the key technology to realize high efficiency, long term stability OLED devices and make it possible to use for full color display and lighting application [86]. The energy of excitons electrogenerated on the host materials can be efficiently transferred to the red, green, and blue dopants, and then radiatively decay to the ground states and give out efficient EL. Because the dopant concentration is usually below 1 %, the self-quenching effect is significantly suppressed, high efficiency and pure color can be achieved [87].

The functions of host materials are mainly to act as transport charge carriers and to generate excitons, hence they should have the following properties: good hole or electron transport capability with thermal, chemical, and electrochemical stabilities, matching HOMO and LUMO energy levels with the dopant materials, and large overlap between photoluminescence (PL) spectrum of the host and the absorption of the dopant materials. Typical host materials can be either electron transport hosts, hole transport hosts, or bipolar hosts.

In a perfect host-guest emitter system, the emission should be exclusively from the dopant to achieve high efficiency and pure color. The dopant materials must have high PL yield and a satisfying color suitable for the applications. The color is often described with color chromaticity coordinates of Commission Internationale de l'Eclairage (CIE) 1931. For example, the color chromaticity coordinates of three primary red, green, and blue colors of NTSC standard in displays are (0.67, 0.33), (0.21, 0.71), and (0.14, 0.08), respectively. In the following sections, the development of some typical fluorescent red, green, and blue electroluminescent materials is discussed.

6.3.1 Red Fluorescent Materials

Red emitter 4-(dicyanomethylene)-2-methyl-6-[*p*-(dimethylamino)styryl]-4*H*-pyran (DCM) was the first dopant used in the host-guest doped emitter system introduced by Kodak in 1989 [27]. The PL quantum yield is 78 % with a peak maximum at 596 nm. The emission is quite broad, and the full width at half maximum (FWHM) is as large as 100 nm. A simple structure of ITO/HTL/Alq_3:DCM/Mg:Ag reaches an EQE of 2.3 %, which is twice as high as that without doping. However, the color is yellowish and the color coordinates are (0.56, 0.44). High doping concentration results in more saturated red emission, but the efficiency gets lower because of concentration quenching.

The DCM molecule has a donor-acceptor (so-called push-pull) structure. To shift the chromophore's color more reddish in the organic molecule design, one of the most effective ways is introducing regidization structure into the donor moiety. The researchers in Kodak introduced a more rigid julolidine ring into the DCM molecule; the 4-(dicyanomethylene)-2-methyl-6-(julolidyl-9-enyl)-4*H*-pyran (DCJ) molecule obtained has a peak wavelength of 630 nm, which is about 30 nm longer in wavelength. With the same structure as DCM, the device with DCJ shows a more saturated red emission compared with DCM. However, such red emission (0.64, 0.36) is only obtained when the dopant concentration is as high as 3 %, and the efficiency is only half of the maximum value because of self-quenching. At lower concentration, a green emission from the host material Alq_3 is observed, resulting from incomplete energy transfer. With four additional methyl groups on julolidine, the interaction between DCJT molecules is significantly suppressed, and the concentration quenching is less pronounced. All the DCM series molecules have an active methyl on pyran group, which may cause condensation reactions to form undesired bis-condensation by-products. This by-product shows a broad, very weak fluorescence [88].

To solve this problem, the active methyl group is substituted by bulky and sterically significant *tert*-butyl group to form DCJTB. The bulky *tert*-butyl group on the pyran ring avoids further condensation and improves the purity and thermal stability of the materials. Because of the inefficient energy transfer processes, emission from the Alq_3 is often observed when it is doped into Alq_3 host. To overcome this, Hamada et al. [89] reported improved red emission by using an emitter assist dopant such as rubrene as a sensitizer to assist the energy transfer processes between the host and the red dopant. Using the same strategy, Chen et al. [90, 91] developed a so-called co-hosted emitter system. A high efficiency of 4.5 cd/A is obtained and maintains a driving current density of 700 mA/cm^2 by optimization of device structures. The optimized OLED structure is ITO/CFx/NPB (120 nm)/Rubrene:Alq_3 (6:4):2 % DCJTB (30 nm)/Alq_3 (50 nm)/LiF(1 nm)/Al (200 nm). More importantly, the half-life is longer than 30,000 h at an initial luminance of 100 cd/m^2 (Table 6.1).

6.3.2 Green Fluorescent Materials

The luminous efficiency of a green OLED is much higher than the other two primary colors, red and blue, because the human eye is more sensitive to green. The most widely investigated green fluorescence dopant was the coumarin family because of its high PL quantum yield (up to 90 %) and saturated green color. One of the best green fluorescent dopants is (10-(2-benzothiazolyl)-1,1,7,7-tetramethyl-2,3,6,7-tetrahydro-1*H*,5*H*,11*H*-[l]benzo-pyrano[6,7,8-*ij*]quinolizin-11-one) (C-545T) [95]. With the julolidine group at the C-7 position, the molecule has better structural coplanarity, and the *p*-orbital of nitrogen overlaps with the *p*-orbitals of the phenyl ring for more effective conjugation. The relative movements between the molecular bonds become less active and the probability of non-radiative decay is getting smaller, which results in increased PL quantum yield to more than 90 %. The steric effects of the four methyl groups on the julolidyl ring significantly reduce the interaction between the molecules [96]. Further improvement of the coumarin dyes has been achieved by substituting *tert*-butyl groups at the benzothiazolyl ring as in C-545TB, or by adding a methyl group at the C-4 position as in C-545MT [97]. The concentration quenching problem could be further suppressed over a wide range of doping concentration from 2 to 12 % and the thermal property was also greatly improved without compromising its emissive color. In the device structure of ITO/CHF_3 plasma/NPB/Alq_3:1 % C-545TB/Alq_3/Mg:Ag, a saturated green emission CIE (0.30, 0.64) with a luminescent efficiency of 12.9 cd/A at driving current density of 20 mA/cm^2 was obtained.

Chen's group [98] has synthesized another green dopant C-545P by introducing asymmetric 4 methyl steric groups beside the nitrogen atom on the julolidyl ring. C-545P has good properties and photostability, and when used as a dopant in an

Table 6.1 Chemical structures and properties of red fluorescent materials

Material	Structure	HOMO (eV)	LUMO (eV)	Abs (nm)	λ_{max} (nm)	PL QE (%)	CIE (x, y)	Efficiency	Refs.
DCM	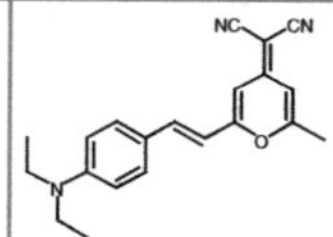	5.46	3.21		596	78	(0.56, 0.44)	2.3 %	[27, 92]
DCJ	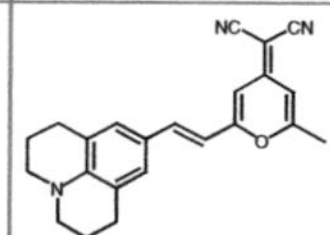	5.26	3.11		630		(0.64, 0.36)	2.3 %	[27, 89]
DCJT	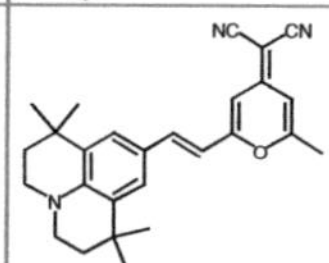	5.3	3.1	520	615		(0.64, 0.36)		[88, 93]
DCJTB	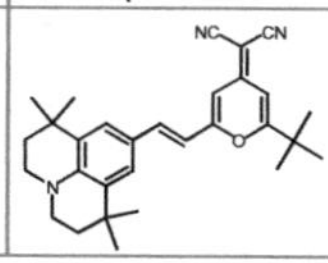	5.45[4]	3.23[4]	520	615	>90	(0.65, 0.35)	4.5 cd/A	[88, 90, 94]

Alq_3-hosted OLED, it shows 10 % higher luminous efficiency than that of C-545T while keeping the same color coordinates at (0.31, 0.65). This is attributed to the asymmetric substituents, which minimize aggregation at high concentration.

The coumarin dyes were usually doped in Alq_3 as the emitter system in the OLED research, which resulted in an unstable Alq_3^+ cation with excessive holes. Recently, more and more blue emissive materials have been employed as the hosts for green fluorescent dopants. In 2006, researchers in Sanyo [95] published an ultra-high green fluorescent green OLED using C545T as the dopant. The host material is 9,9′,10,10′-tetraphenyl-2,2′-bianthracene (TPBA), whose PL spectrum has excellent overlap with the absorption of C545T. An efficient energy transfer from the host to the dopant is expected. By using a novel electron transport material 9,10-bis[4-(6-methylbenzothiazol-2-yl)phenyl]anthracene (DBzA) in the device, a luminous efficiency as high as 29.8 cd/A has been reached at a current density of 20 mA/cm^2, which translates to an EQE of about 10 % (Table 6.2).

The second class of green dopants is the family of quinacridone (QA). QA was patented by Pioneer [99], although the patents on its derivatives have been filed by Kodak [102] for use in OLEDs. The imino and carbonyl groups on the QA molecule easily form intermolecular hydrogen bonding, which results in QA excimer or exciplex with Alq_3, leading to quenching of the fluorescence. Wakimoto et al. [103] substituted steric isopropyl groups beside the imino group to avoid the formation of hydrogen bonds and found that the steric hindrance invoked by the bulky substituents of QA prevents excimer formation and prolongs the life time of the devices. Another strategy to avoid forming excimer or exciplex by the hydrogen bonds is replacing the hydrogen on an imino group with an alkyl group, such as *N,N*-dimethylquinacridone (DMQA) developed by Kodak [87]. In a simple structure of ITO/Teflon/Alq_3:0.7 % DMQA/BAlq/Alq_3/Mg:Ag, Qiu et al. [104] achieved a luminous efficiency of 21.1 cd/A (EQE of 5.4 %) at current density of 418 mA/cm^2. Murata et al. [100] used *N,N′*-diethylquinacridone (DEQ) as green dopants in structure ITO/1-TNATA/NPB/Alq_3:DEQ/Alq_3/Mg:Ag, and found that the device was thermally durable and the quantum efficiency was temperature-independent.

6-*N,N*-Dimethylamino-1-methyl-3-phenyl-1*H*-pyrazolo[3,4-*b*]-quinoline (PAQ-NEt) is another green fluorescent dopant. Using a high doping level of ca. 16 % of PAQ-Net as a dopant in the device ITO/NPB/NPB:16 % PAQ-NEt/TPBI/Mg:Ag, Tao et al. [101] fabricated a device which gave a sharp, bright, and efficient green EL peaked at 530 nm with an FWHM of 60 nm. Interestingly, whereas the PL of the doped film showed emission of the host NPB material even at the high doping concentration of 20 %, the host emission was not observed in the EL spectrum. This is presumed to be because of charge-trapping processes, which occur in competition with the energy transfer process.

Other useful green dopants with very good EL performance belong to the class of bis(amine)-substituted anthracene derivatives which were reported by Idemitsu. One of the best green fluorescence dopants with high efficiency, good color and device life time was GD-403 [105] which had a luminous efficiency of 30 cd/A (EQE 8.5 %) at 20 mA cm^{-2} (3.7 V) with CIEx,y (0.25, 0.65). The T_{50} life time at an initial luminance of 1000 cd/m^2 was about 50,000 h.

Table 6.2 Chemical structures and properties of green fluorescent materials

Material	Structure	HOMO (eV)	LUMO (eV)	Abs (nm)	λ_{max} (nm)	PL QE (%)	CIE (x, y)	Efficiency	Refs.
Coumarin	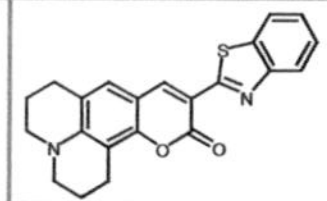				519	90		9 cd/A	[95]
C-545T	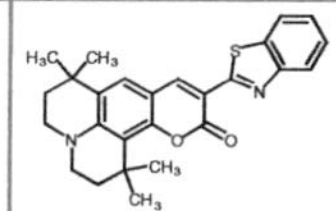	5.6	3.0	480	512	99	(0.24, 0.62)	29.8 cd/A	[95]
C-545 TB	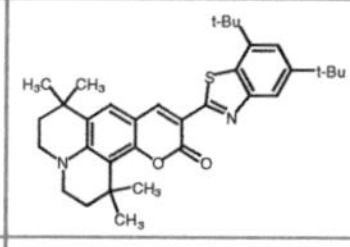						(0.30, 0.64)	12.9 cd/A	[97]
C-545MT	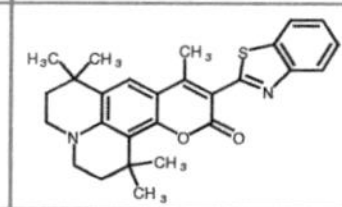							7.8 cd/A	[97]
C-545P	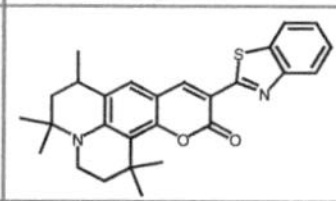	5.3	2.9	480	514	100	(0.31, 0.65)	11.3 cd/A	[98]
TPBA	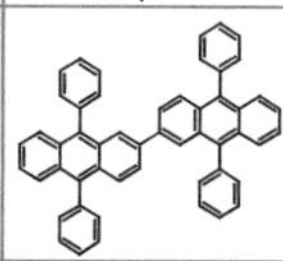	5.8	2.8		450				[95]
QA	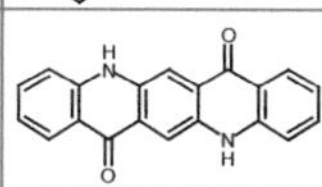				544		(0.35, 0.62)	3.2 lm/W	[99]

(continued)

Table 6.2 (continued)

Material	Structure	HOMO (eV)	LUMO (eV)	Abs (nm)	λ_{max} (nm)	PL QE (%)	CIE (x, y)	Efficiency	Refs.
DMQA	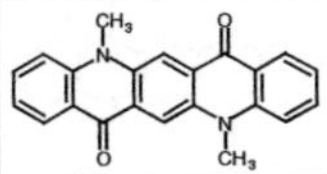				544		(0.39, 0.59)	7.3 cd/A	[87]
DEQ	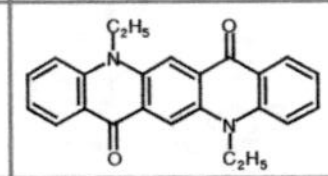				544			2.05 %	[100]
PAQ-Net	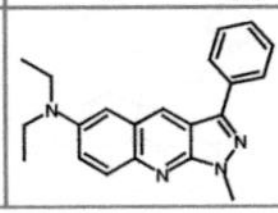	5.2	2.7	445	530	20	(0.22, 0.60)	5.99 cd/A	[101]

6.3.3 Blue Fluorescent Materials

High efficiency and long living phosphorescent red and green emitter are now commercially available. However, a high efficiency deep blue color as well as long life time is still a big challenge for phosphorescent emitters. As a result, the blue fluorescent materials are of particular importance in display and lighting applications.

The band gap of blue emitter is larger than red and green ones, and even larger band gaps are required for host materials to satisfy energy transfer requirements. Many large band gap organic materials have been explored for blue emission. They can be roughly categorized as follows: anthracenes, distyrylarylene, perylenes, fluorenes, heterocyclic compounds, and metal complexes.

Anthracene has a very high PL quantum yield because of its rigid ring structure which minimizes vibronic energy levels. Actually the first organic EL observed by Pope in 1963 was based on anthracene. Since then, anthracene derivatives have been intensively studied as an attractive building block and starting material because the chemical modification of anthracene is relatively easier than for most other rigid aromatic materials because of its better solubility in common organic solvents.

Anthracene itself is highly crystalline and tends to crystallize in thin films, which greatly limits its application in OLEDs. Fortunately, anthracene can be easily modified by directly introducing different functionalized blocks at the 9,10-positions or bulky steric substituents on the 2,6-positions to prevent its crystallization and improve its film-forming property.

Adachi et al. [106] introduced phenyl groups at 9,10-positions of anthracene, the obtained compound DPA having a PL quantum yield of almost unity. However the stability and OLED device efficiency were really poor. Substituting *tert*-butyl groups at various positions of DPA improves the thermal stabilities.

In 2002, Shi and Tang [107] at Kodak reported the first stable blue OLED emitter using 9,10-di(2-naphthyl)anthracene (ADN) as the host material, which showed high PL quantum yield both in solution and solid state. When it was doped with 2,5,8,11-tetra(*tert*-butyl)perylene (TBP) blue dopant emitter, the device achieved a luminous efficiency of 3.5 cd/A with CIEx,y coordinates of (0.15, 0.23) and a half-life of about 4000 h at an initial luminance of 636 cd/m^2. In further investigation it was observed that the ADN thin film was unstable and tended to crystallize under prolonged electrical stress or annealing at elevated temperatures (95 °C).The blue color was a bit greenish with CIEx,y of (0.20, 0.26). In attempts to solve these issues, the Kodak group [108] substituted *tert*-butyl group at C-2 position of anthracene, namely with 2-(*tert*-butyl)-9,10-di(2-naphthyl)anthracene (TBADN). Using the same dopant and device configuration, TBADN was able to generate a deeper blue emission of CIEx,y (0.13, 0.19), but its efficiency was lower than that of ADN (Table 6.3).

Chen and co-workers [113] systematically investigated the effects of alkyl substitution on the 2-position of ADN and found that the best way to stabilize the

Table 6.3 Chemical structures and properties of blue fluorescent materials

Material	Structure	HOMO (eV)	LUMO (eV)	Abs (nm)	λ_{max} (nm)	PL QE (%)	CIE (x,y)	Efficiency	Refs.
Anthracene	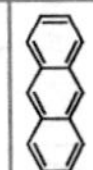	5.25	1.63	375, 357, 341, 329	420	46	(0.15, 0.15)	0.05 %	[106, 109]
DPA	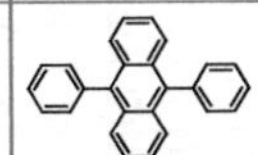	6.4	3.0	396, 374, 356, 340	414, 432	90	(0.145, 0.195)	1.8 %	[106, 110, 111]
ADN	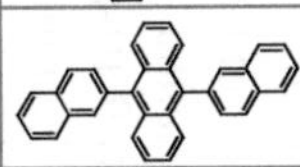	5.5	2.5	377	426	79	(0.15, 0.10)	1.8 cd/A	[112, 113]
TBP	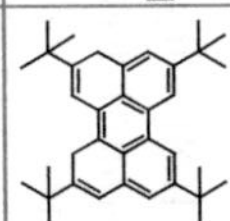	5.55	2.84	415, 440	460		(0.15, 0.23)	3.5 cd/A	[107, 114]
TBADN	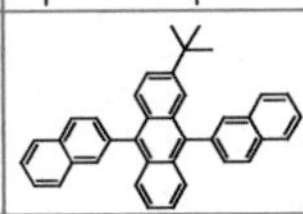				430		(0.13, 0.19)	1.0 cd/A	[108]
MADN	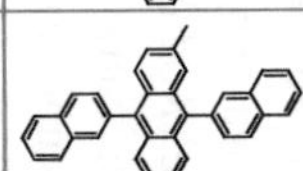	5.5	2.5	380	430		(0.14, 0.20)	4.4 cd/A	[113]
DSA-Ph	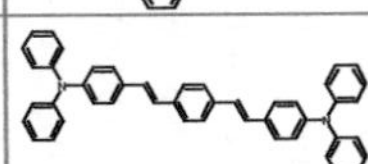	5.4	2.7	411	464, 490		(0.16, 0.32)	9.7 cd/A	[115, 116]
α,α-MADN	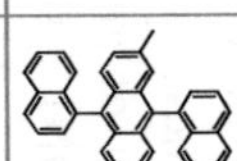	5.8	2.8	415	413		(0.15, 0.08)	0.7 cd/A	[117, 118]

(continued)

Table 6.3 (continued)

Material	Structure	HOMO (eV)	LUMO (eV)	Abs (nm)	λ_{max} (nm)	PL QE (%)	CIE (x,y)	Efficiency	Refs.
BD-1	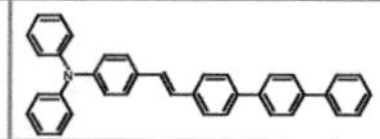	5.44	2.57	383	454, 472	71	(0.15, 0.13)	3.3 cd/A	[117, 119]
DPVBi	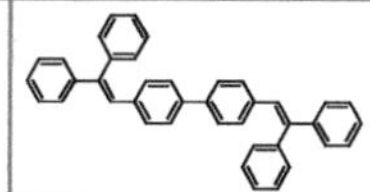	5.9	2.8	360	464	38	(0.15, 0.15)	2.0 cd/A	[120–123]
BCzVB	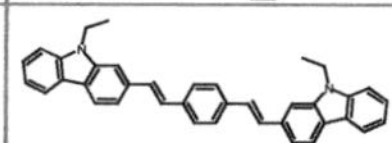	5.5	2.5	403	480		(0.15, 0.16)	3.4 cd/A	[120, 124–126]
BCzVBi	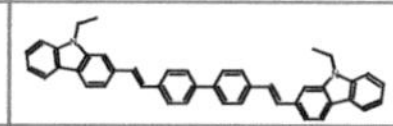	5.4	2.4	380	445, 475		(0.155, 0.157)	2.8 cd/A	[120, 125, 127]
DPVPA	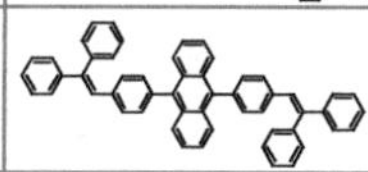	5.4	2.7	410	448	98.8	(0.14, 0.17)	3 %	[128–130]

morphology was to place a relatively small methyl group. It is well known as 2-methyl-9,10-di(2-napthyl)anthracene (MADN). The methyl substituent slightly disrupts the symmetry and increases the intermolecular distance, which can effectively suppress the problematic crystallization to fabricate the robust and amorphous film, but still keeping the HOMO and LUMO energy of ADN at 2.5 and 5.5 eV, respectively. The optimized MADN-based blue OLED (ITO/CFx/NPB/MADN/Alq_3/LiF/Al) can reach an efficiency of 1.4 cd/A with deep blue color coordinates of (0.15, 0.10), which is higher than that of the equivalent TBADN device (1.0 cd/A).

When using *p*-bis(*p*-*N*,*N*-diphenyl-aminostyryl)benzene (DSA-Ph) as a dopant, an optimized device with structure ITO/CFx/NPB/MADN:DSA-Ph/Alq_3/LiF/Al achieved a very high efficiency of 9.7 cd/A at 20 mA/cm^2 (5.7 V) [115]. The projected half-life is 46,000 h at an initial luminance of 100 cd/m^2, although the color purity is sacrificed with a major peak at 464 nm with a shoulder centered at 490 nm with CIE (0.16, 0.32).

In 2006, an isomer of MADN, 2-methyl-9,10-di(1-naphthyl)anthracene(α,α-MADN) was synthesized by replacing the 2-naphthysubstituents at C-9 and C-10 positions with the sterically more demanding 1-naphthyl substituent [117]. The peak of PL wavelength is 413 nm, around 17 nm blue-shift from that of MADN. The blue-shifted emission spectrum of α,α-MADN provides better overlap with the absorption of the deep blue emitter which is essential for efficient energy transfer. The device using di(4,4′-biphenyl)-[4-(2-[1,1′,4′,1″]terphenyl-4-yl-vinyl)phenyl] amine (BD-1) as blue dopant and α,α-MADN as the host produced a current efficiency of 3.3 cd/A with color coordinates of (0.15, 0.13), which is 1.5 times as high as that of the MADN:BD-1 system of 2.2 cd/A, although the life time is similar.

The introduction of bulky substituent leads to larger intermolecular distances and is a hindrance in packing, and shows amorphous behavior. A slight change of the unsymmetrical substituents at the 9 and 10 positions of anthracene not only efficiently improved the efficiency, but also strengthened amorphous morphological stability. Kang et al. [131] designed and synthesized a series of unsymmetric 9-(2-naphthyl)-10-arylanthracene derivatives with high thermal stability. By doping 6,12-bis(di(3,4-dimethylphenyl)amino)chrysene in such a host, the highest luminous efficiency is 9.9 cd/A at 20 mA/cm^2 with color coordinates of (0.14, 0.18).

The anthracene derivatives are often used as the host materials because of their high carrier mobility, good film morphology, and large band gap. However, the most efficient fluorescent blue emitters are the distyrylarylene (DSA) series patented by Idemitsu Kosan Co. Ltd., Japan. The basic structure of DSA is $Ar_2C{=}CH\text{-}(Ar')\text{-}CH{=}CAr_2$.

In 1995, Hosokawa et al. [132] first reported DSA-based host material 4,4′-bis (2,2-diphenylvinyl) biphenyl (DPVBi) and amino-substituted DSA dopants such as BCzVB and BCzVBi [120]. The nonplanar host DPVBi has nice film-forming properties and the LUMO and HOMO levels are 2.8 and 5.9 eV, respectively. The band gap of 3.1 eV is similar to that of ADN. With OLED structure of ITO/CuPc/TPD/DPVBi:DSA-amine/Alq_3/Mg:Ag, both BCzVB and BCzVBi gave almost identical EL emission with a peak at 468 nm and two shoulders at 445 and 510 nm. The EQE of 2.4 % was obtained at a current density of 8.28 mA/cm^2. The initial half-decay life time of the above device was measured to be 500 h at an initial luminance of 100 cd/m^2. Later, the same group used an improved HTL with a DSA host and a DSA–amine dopant that gave a life time of over 5000 h.

In 2004, eMagin disclosed a blue host material DPVPA [128], which replaces the biphenyl core of DPVBi with diphenylanthracene. The PL quantum yield is 2.6 times as high as that of DPVBi with a peak wavelength at 448 nm, which is around 20 nm red-shift caused by the extended conjugation chain. The EQE of DPVPA-based OLED is 3 % with CIE of (0.14, 0.17), which is much higher than that of MADN (1.5 %).

The perylene family is probably one of the most stable blue dopants because it doesn't have any functional group sensitive to chemistry, heat, or light. With 0.5 % tetra (*tert*-butyl) perylene (TBP) doped into MADN, the device can reach a luminous efficiency of 3.4 cd/A with CIE of (0.13, 0.20), and a half-life of 5000 h can be obtained. Jiang et al. [133] reported a stable blue OLED based on anthracene derivative JBEM host. With a structure of ITO/CuPc/NPB/JBEM:perylene/Alq/Mg:Ag, the device shows a maximum luminance of 7526 cd/m^{2}with CIE (0.14, 0.21). The device life time is over 1000 h at initial luminance of 100 cd/m^2, which is relatively higher longer than that of the device using DPVBi as the host.

6.3.4 Advanced Delayed Fluorescent Materials

In OLED devices, the injected electrons and holes form singlet or triplet excitons in the ratio of 1:3 according to the statistics of spin multiplicity. Therefore the upper limit of internal quantum efficiency is 25 % for fluorescent emitters, because only singlet excitons can decay radiatively to the ground state. The measured EQE cannot exceed 5 % for fluorescent OLED when an out-coupling efficiency of 20 % is accounted for. However, more and more ultra-high efficiencies of fluorescent OLEDs exceeding 10 % have been reported recently, which are attracting wide research interests in both academic institutions and industry.

Most of these studies focus on delayed fluorescent materials which can convert triplet excitons to singlet excitons through reverse intersystem crossing (ISC). There are two main methods to achieve efficient ISC: triplet-triplet annihilation (TTA) and TADF.

Since 2008, a few ultra-high efficiency red fluorescent OLEDs have been reported by companies such as Kodak, Idemitsu, and Novaled. For example, Kodak reported a high efficiency of 6–9 % EQE with fluorescent material 4,4′-bis[4-(di-*p*-tolylamino)styryl] biphenyl doped in 9,10-bis(2-naphthyl)-2-phenylanthracene (PADN), which is attributed to TTA contribution [134]. Based on the assumption that spin statistics dictate 20 % probability of generating a singlet exciton through TTA, they proposed a formula to estimate the maximum EQE of such a system: $0.2 \times (25\,\% + 0.2 \times 75\,\%) = 8\,\%$. Soon they [135] obtained an EQE of >11 % at

3 V using rubrene as emitter. This value is far beyond the theoretical limit of the fluorescent emitter imposed by spin statistics, and cannot be satisfactorily explained by previous equation. They surmise that the fluorescent OLED devices are capable of using a considerably larger fraction of triplet states than was previously believed. With the appropriate emissive material and host material, every two triplet excitons can form one singlet exciton to contribute to the light emission, i.e., the upper limit for the singlet excited state yield in the TTA process is 0.5. Therefore the maximum internal quantum efficiency of fluorescent OLEDs is to be 25 % + 0.5 × 75 % = 62.5 %. The estimated maximum EQE of the fluorescent OLEDs should be revised to at least 0.2 × 62.5 % = 12.5 % when assuming the optical outcoupling efficiency of 0.2.

TADF materials have a very small energy gap between singlet and triplet states, which allows the upconversion from triplet to singlet excited states. However, the reverse ISC rate ($\sim 10^6$ s^{-1}) is usually lower than the radiative rate ($\sim 10^9$ s^{-1}), resulting in an inefficient upconversion from triplet to singlet excited states. In addition, such materials always exhibit low fluorescence efficiencies. Thus it is really difficult to verify the effectiveness of TADF materials for emissive layers in OLED. It was not until 2009 that Adachi group at Kyushu University first introduced tin(IV) fluoride–porphyrin complexes into OLED [6]. It is demonstrated that the PL efficiency of such TADF material can be increased from ~1 to 3 % with an increase in temperature caused by reverse ISC. However, the activation energy of the reverse ISC (ΔE_{ST}) was still rather high at 0.24 eV, leading to inefficient reverse ISC, and the overall EL efficiency is still very low. In 2011, they designed a donor-acceptor system with steric hindrance incorporated between them to form intramolecular excited states, and the resulting molecule 2-biphenyl-4,6-bis(1,2-phenylindolo[2,3-*a*] carbazol-11-yl)-1,3,5-triazine (PIC-TRZ) has ΔE_{ST} of only 0.11 eV. With this molecule, a significant contribution of 30 % reverse ISC efficiency was realized under both PL and EL processes [136]. Later in 2012, the Adachi group [7] proposed a strategy to realize high reverse ISC efficiency by exciplex of electron-donating and electron-accepting molecules. A high reverse ISC of 86.5 % has been demonstrated using an exciplex system of 4,4′,4″-tris[3-methylphenyl(phenyl)amino]triphenylamine (m-MTDATA) as a donor and tris-[3-(3-pyridyl)mesityl]borane (3TPYMB) as acceptor. Using this emission system, the OLED achieves an EQE greater than 5 %, even with the rather low PL efficiency of 26 %, showing the contribution of efficient ISC significantly. For such a donor-acceptor system, the HOMO and the LUMO in exciplexes are mainly located on the donor and acceptor molecules, respectively. It is possible that the hole and electron wavefunctions are spatially separated and electronic exchange energy is very small, resulting in the triplet levels being very close to the singlet levels. However, the limited overlap of hole and electron wave functions generally results in very low fluorescence efficiency of exciplex materials according to the Franck–Condon principle. Therefore, the researchers focus more on intramolecular donor-acceptor system for both high fluorescence efficiency and high reverse ISC efficiency. Adachi's group further developed a series of highly efficient TADF emitters [137–144], which contains carbazole group as a donor and dicyanobenzene group as an electron acceptor. A small ΔE_{ST} is realized because the HOMO and the LUMO of these emitters are localized on the donor and acceptor moieties, respectively. Furthermore, most of these materials containing carbazolyl dicyanobenzene (CDCB) have PL quantum yields over 79 %, especially 1,2,3,5-tetrakis(carbazol-9-yl)-4,6-dicyanobenzene (4CzIPN) (94 ± 2 %). Using these TADF emitters in OLED, an EQE of 19.3 % for green emission is achieved, which is much higher than those of conventional fluorescent OLED, and comparable to high-efficiency phosphorescent OLEDs.

Although TADF materials have the potential to harvest ~100 % excitons, the mechanism of TADF still requires further clarification. The efficiency roll-off is quite serious for most TADF emitter based OLEDs [140, 145], which could be attributed to pronounced triplet-triplet or singlet-triplet quenching because of slow reverse ISC process. It is necessary to improve further the performance of TADF materials to meet the requirements of real applications [146].

6.4 Phosphorescent Electroluminescent Materials

Although fluorescent OLEDs have been extensively studied since the first practical OLED was invented by Kodak in 1987, a breakthrough of its development is the discovery of electrophosphorescence by Forrest and Thompson in 1998 [5]. Compared with fluorescent OLEDs using only singlet excitons for light emission, phosphorescent OLEDs can utilize both singlet and triplet excitons, which boosts the internal quantum efficiency (IQE) of OLEDs by a factor of four theoretically. Hence, phosphorescence has attracted tremendous research interest in both academia and industry for its significant performance potential. In fact, red phosphorescent electroluminescent material has already been used in commercial mobile phones since 2003. Highly efficient and long-living green phosphorescent material is also undergoing commercialization. Compared with the great achievements in red and green phosphorescent materials, blue phosphorescence seems to be the last obstacle against commercialization because of the lack of deep color and poor life time, although the EQE of blue phosphorescent OLED has already been over 20 % [57, 147–149].

Phosphorescence is the radiative decay from triplet excited state to the ground state, whereas fluorescence is the emission from singlet excited state in photophysics. Most organic material is singlet in the ground state (S0) because of the covalent bonding nature. The electrons in organic materials are all paired up, and the paired-up electrons must have opposite electron spins because of the Pauli Exclusion Principle. When receiving enough energy, such as from light, the molecules can be excited from ground state to excited state. The two electrons have either the opposite spin orientation called singlet or the same spin orientation called triplet. Each electron spin is denoted 's' and it equals either +1/2 or −1/2. The spin multiplicity (total spin number, S) is calculated by $S = 2\sum s + 1$, i.e., a spin multiplicity of 1 for singlet and 3 for triplet. If an organic material is excited from the ground state, the excited state usually remains as singlet because of the Wigner–Witmer selection rule in quantum mechanics for the process of electronic transition. Similarly, the phosphorescence of transition from the triplet excited state to ground state is prohibited.

In EL, the excited states are formed by the recombination of injected electrons and holes from the electrodes. The spin orientation of electrons and holes can be either positive or negative, and the generated excitons by electron and hole recombination can be either singlet or triplet. The quantum mechanical spin statistics dictates that for every singlet exciton created there are three triplet excitons. However, the attained triplet excitons in OLEDs radiatively decay very little to ground state because of the limitation of aforementioned selection rule.

The typical life time of the triplet excited state of organic material is usually longer than milliseconds (ms), which is long enough for releasing its energy via non-radiative decay such as the thermal vibration motion. The way to enable phosphorescence from organic materials is to suppress the thermal vibration motion by lowering temperatures down to liquid nitrogen temperature (77 K). However it is surely not practical for OLEDs, which should usually operate at room temperature. A practical and effective way is to reduce the life time of the triplet excited state so that the transition process can compete with the thermal vibration motion.

To enhance the phosphorescence, the most effective way is to use heavy metal atoms in organic molecules, which enhances spin-orbital coupling, the interaction between spin magnetic moment and orbital magnetic moment. The strength of such coupling is proportional to the proton number in the atomic nucleus, which is in turn related to the electromagnetic field generated by atomic electrons circling around. This is called the "heavy atom effect". The difference between triplet and singlet state become diminished when the spin-orbital coupling constant is large. The selection rule loses its restriction and radiative decay from the triplet excited state to the singlet ground state becomes feasible and the emission life time shortens, faster than the molecular thermal vibration motion. Up to now, the heavy atoms used in phosphorescent materials are all transition metals such as Ru, Re, Os, Ir, Pt, Au, and Hg. Among them, most phosphorescent materials focus on group VIIIB elements, Os, Ir, and Pt. Ir-containing phosphorescent materials in particular attract the most attention.

6.4.1 Red Phosphorescent Materials

The first phosphorescent material reported by Forrest and Thompson in 1998 was 2,3,7,8,12,13,17,18-octaethyl-21*H*,23*H*-porphine platinum(II) (PtOEP) [5], a red phosphorescent dye based on the heavy metal Pt. When doped into Alq_3 with co-evaporation, a peak EQE of 4 % was obtained with saturated red emission of (0.7, 0.3). The EQE can be further improved to 5.6 % when CBP is used as the host [50]. The efficiency was considerably higher than other fluorescent emitters, hence Pt complexes have attracted strong attention because of their phosphorescence nature

—phosphorescence instead of fluorescence mainly because of the following factors. First, PtOEP has an absorption peak at around 540 nm, and its fluorescence emission peak is at around 580 nm from the literature. In contrast, the observed EL peak is at 650 nm with a very nice narrow red emission band, which is much longer than the fluorescence emission of PtOEP and the Stokes shift is more than 100 nm. This is direct proof that the observed EL emission is phosphorescence instead of fluorescence. Furthermore, the life time of the emission at ~650 nm in solution was measured to be about ~50 μs, which is significantly longer than the typical fluorescence life time of 0.1–10 ns. All these spectroscopic data infer that the EL emission at 650 nm is phosphorescence rather than fluorescence. However, this PtOEP phosphorescent OLED shows serious efficiency roll-off, a typical EL behavior for phosphorescent OLEDs at elevated current density, which can be ascribed to their relatively long triplet life time. More Pt-based phosphorescent emitters have been revealed by Kwong et al. [150], such as PtOX and PtDPP, which show similar performance to PtOEP and the device performances are sensitive to driving current density. Bright saturated red emission with high efficiency at low current density and blue shifted to orange emission color with reduced efficiency are observed, which are caused by the saturation of triplet emissive sites because of the long-lived phosphorescence state of the platinum–porphyrin complex. The phosphorescence life time of Pt(II)–porphyrin complexes doped into Alq_3 with a concentration of 6 mol% are in the tens of μs range, for example, 39 μs for PtOX, 21 μs for PtDPP, and 37 μs for PtOEP. In order to achieve a stable red PHOLED, a phosphor of bis(2-(2′-benzo [4,5-*a*] thienyl)pyridinato-*N*, C3′)Pt(acetylacetonate) [btpPt(acac)] [151] with short life time of triplet state was synthesized. The life time of btpPt(acac) in TPBI host is only 5.6 μs, minimizing the saturation of triplet emissive states and TTA.

When the central metal atom Pt is replaced by iridium, the obtained red phosphor bis(2-(2′-benzo[4,5-*a*]thienyl)pyridinato-N,C3′) iridium(acetylacetonate) (btp2Ir (acac)) [151] has an even shorter phosphorescence life time of 4 μs. The maximum emission peak is at a wavelength of 616 nm with additional intensity peaks at 670 and 745 nm with CIE coordinates of (0.68, 0.32), which are close to the video display standards. A maximum EQE of 7.0 % is achieved at a current density of 0.01 mA/cm^2. At a higher current density of 100 mA/cm^2, an EQE of 2.5 % is still obtained. The reduced efficiency roll-off of btp2Ir(acac)doped PHOLED at high currents is mainly ascribed to short triplet life time.

The HOMO level of heavy metal complex is a mixture of the highest energy orbitals of the metal (e.g., 5d orbital for Ir, Pt, Os, or Re, 4f orbital for Tb) and the π-orbitals of the ligand, whereas the π^*-orbitals of the ligand make a contribution to the LUMO level [152, 153]. Ligands therefore play an important role in color emission and quantum efficiency. $Ir(piq)_3$ [154] and $(piq)_2Ir(acac)$ [155] have been reported by two different research groups about the same time, respectively. The common feature of these two Ir complexes is the cyclometalate ligand 1-phenyl-isoquinoline (piq). The difference between the two Ir complexes is the homoleptic one for $Ir(piq)_3$ and the heteroleptic one for $(piq)_2Ir(acac)$. In fact, cyclometalate ligand piq is very powerful for red phosphorescence. Most of the red phosphorescence materials developed later are based on derivatives of 1-phenylisoquinoline ligand. Both $Ir(piq)_3$ and $(piq)_2Ir(acac)$ show similar phosphorescence quantum yields of about 0.2 and resulting OLEDs have the same color coordinates of (0.68, 0.32). $Ir(piq)_3$ has a shorter phosphorescence life time of 3.5 μs and higher thermal stability. An EQE of 10.3 % and a power efficiency of 8.0 lm/W are obtained at 100 cd/m^2. $(piq)_2Ir(acac)$ has an even shorter life time of 1.2 μs, and an EQE of 8.46 % is achieved even at 20 mA/cm^2.

The performance of the OLED depends not only on the phosphorescent emitters but also on the host materials, as well as the HTL and ETL materials surrounding the EML. When $(piq)_2Ir(acac)$ was doped into four different host materials, Alq_3, BAlq, BCP, and OXD-7 (ITO/NPB/host:$(piq)_2Ir(acac)/Alq_3/Li_2O$/Al), it was confirmed that the electron-transporting host BAlq outperforms the rest of the host materials [156]. $(piq)_2Ir(acac)$ can also be employed in solution processed [157] device ITO/PEDOT:PSS/PVK/blends/Ba/Al, where the active layer contains a blend of electron-transporting 2-(4-biphenyl)-5-(4-*tert*-butylphenyl)-1,3,4-oxadiazole (PBD) or 3-phenyl-4-(1′-naphyl-5-phenyl-1,2,4-triazole) (TAZ), and one of three kinds of polymeric host material: poly-9,9-dioctylfluorene (PFO), polyvinylcarbazole (PVK), and polyfluorene-*p*-substituted triphenylamine (PFTA). Among them, PFTA blended with PBD is the best device, having EQE near 12 %, which is higher than that of $(piq)_2Ir(acac)$ OLED with the thermal evaporation process. Shortly after, the EQE of a similar device ITO/PEDOT:PSS/PVK:PBD: $(piq)_2Ir(acac)$/cathode was pushed up to 13 % by inserting a hole-transporting layer between PODOT:PSS and $(piq)_2Ir(acac)$ EML [158]. More recently, a new host material 2,5-bis(2-*N*-carbazolylphenyl)-1,3,4-oxadiazole (o-CzOXD) [159] has been employed in a $(piq)_2Ir(acac)$-based device, and an EQE of 18.5 % was obtained.

Liu's group [155] substituted a fluorine on 1-(phenyl)isoquinoline ligand; the obtained Ir(piq-F)$_2$(acac) has a slightly blue-shifted emission with CIEx, y of (0.61, 0.32) although the EQE remains 8.67 % at a current density of 20 mA/cm^2. Yang et al. [160] developed Ir(m-piq)$_2$(acac) by adding a methyl group on position 5 of 1-(phenyl)isoquioline, whose EL emission is around 623 nm. When the phenyl is replaced by a naphthyl ring, the Ir(1-niq)$_2$(acac) or Ir(2-niq)$_2$(acac) formed have emission peaks at 664 and 633 nm, respectively. $Ir(4F5Mpiq)_3$ was reported in 2005 [161] by the same research team who developed $Ir(piq)_3$. The maximum EQE is as high as 15.5 % at a more reasonable current density of 1.23 mA/cm^2, and the EQE remains at 7.9 % at a current density of 120 mA/cm^2 with luminance of 10,000 cd/m^2. However, the color is a bit worse with CIEx, y of (0.66, 0.34). Cheng and co-workers [162] have synthesized Ir complex of $(RDQ)_2Ir(acac)$ with 2-R-dibenzo[*f,h*]quinoxaline (RDQ) ligands, where R=H or methyl. With the devices of ITO/NPB/CBP:$(RDQ)_2Ir(acac)$/TPBI or BCP/Alq_3/Mg:Ag, a maximum EQE of 12.4 % has been reached, and the color coordinates are (x = 0.60–0.63, y = 0.37–0.40). When a *p-i-n* structure is applied, 20 % of EQE is achieved for the red emitter of $Ir(MDQ)_2(acac)$, and the operating voltage for 100 cd/m^2 is less than 2.4 V, close to the thermodynamic limit for the red emission. To get more red-shifted emission, a ligand with larger π-conjugation, such as benzo[*c*]acridine (BA) is desired. The

PHOLED with $(BA)_2Ir(acac)$ as the red phosphor shows a wavelength of 666 nm and the color coordinates are (0.64, 0.33) [163].

In addition to the larger π-conjugation space, the intramolecular donor-acceptor (D-A) systems can also shift the emission spectra to a longer wavelength [154]. Thiophene is an electron-donor group and pyridine an electron-acceptor group, thus the ligand of 2-thiophen-2-yl-pyridine (thpy) itself has a D-A character. To strengthen the D-A character, Tsuboyama et al. introduced a methyl group (an electron donor) into the thiophene moiety and a $-CF_3$ group (an electron acceptor) into the pyridine moiety of thpy, respectively. Visible red shifts in the PL spectra are found in the Ir(III) complex based on the modified ligands. Xu and co-workers [164] introduced the substituents of CH_3 and CF_3 into the pyridyl ring of btp ligand to tune Ir(III) complexes into the red region. HOMO levels are a mixture of Ir and 2-benzo[*b*]thiophen-2-yl-4-methyl-pyridyl (btmp) ligand orbitals, while the LUMO is predominantly btmp ligand based. $(btmp)_2Ir(acac)$ with the CH_3 group has a slightly lower oxidation potential, but $(btfmp)_2Ir(acac)$ (btfmp = 2-benzo[*b*]thiophen-2-yl-5-trifluoromethyl-pyridine) containing the CF_3 group is much more difficult to oxidate than $(btp)_2Ir(acac)$. The emission characteristics of these complexes can be tuned by either changing the substituents and their position on 2-benzo[*b*]thiophen-2-yl-pyridine or using different monoanionic ligands, showing emission λ_{max} values from 604 to 638 nm in CH_2Cl_2 solution at room temperature (Table 6.4).

From the color point of view, the above-mentioned iridium-based red phosphorescent materials are not as good as PtOEP, which has CIEx, y of (0.7, 0.3). In 2005, Sanyo [169] published on deep red phosphorescent emitters, diphenylquinoxaline-iridium compounds including Q3Ir and $(QR)_2Ir(acac)$, which show EL emission from 653 to 675 nm, with color coordinates of (0.70, 0.28). These materials have phosphorescence life time quantum yields of 50–79 %, and the life time of the triplet excited state of Q3Ir is only 1 μs. Zhou et al. [181] reported Ir(III) complexes by introducing triphenylamine dendrons to the key ligands, where the triphenylamine group lifted the HOMO level of the Ir(III) complexes. Compared with the tri(phenylisoquinoline) Ir(III), the HOMO of the Ir(III) complexes tailoring phenyl isoquinoline with triphenylamine groups was increased from −5.11 to −4.96 eV. The OLED devices emit pure red light with an EL maximum at around 640 nm and excellent CIE color coordinates of (0.70, 0.30). Some other red phosphorescent dyes, such as $(dpq)_2Ir(acac)$ [170], $(NAPQ)_2Ir(acac)$ [171], and $(Mpnq)_2Ir(acac)$ [172], have even larger x-coordinates of 0.71 and comparable y-coordinates below 0.30.

In addition to Ir(III)- and Pt(II)-based red phosphorescent emitters, as discussed, there are many organometallic phosphors based on other heavy metal ions, e.g., Os (II), Eu(IV), Ru(II), and Re(I).

Os is one of heavy metal atoms often used in phosphorescent materials. Jiang et al. [182] have designed a series of Os complexes, and the emission color can be tuned by changing the ligand. The phosphorescence life time is 0.6–1.9 μs, and the EL emission peak is from 620 to 650 nm, with color coordinates around (0.65, 0.33). $(fppz)_2Os(PPhMe2)_2$ [173], $(fppz)_2Os(PPh2Me)_2$ [174], and $(tptz)_2Os$

Table 6.4 Chemical structures and properties of red phosphorescent materials

Material	Structure	HOMO (eV)	LUMO (eV)	Abs (nm)	λ_{max} (nm)	PL QE (%)	E_T(eV) τ (μs)	CIE (x, y)	Efficiency	Refs.
PtOEP	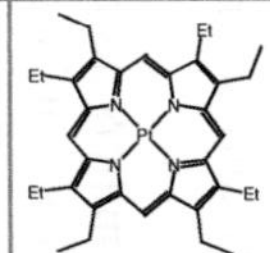	5.3	3.2	540	650	45	1.91 eV, 83 μs	(0.70, 0.30)	5.6 %	[150, 165]
PtOX	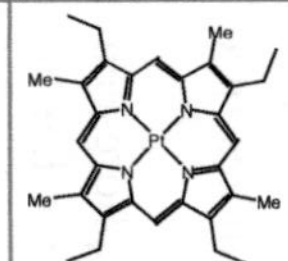	5.3	2.9		648	44	1.92 eV, 76 μs	(0.69, 0.30)	1.5 %	[150, 166]
PtDPP	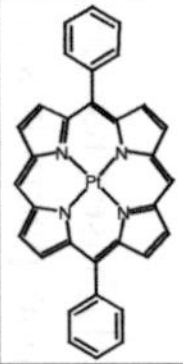	5.2	3.2		630, 695	16	1.97 eV, 34 μs	(0.67, 0.31)	0.25 %	[150]
btpPt(acac)	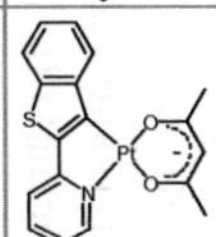	5.35	2.55	265, 318, 344, 427, 444	612	8	2.02 eV, 5.6 μs	(0.67, 0.33)	2.7 %	[151, 167]
$btp_2Ir(acac)$	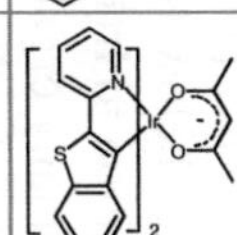				616, 670, 745		2.0 eV, 4.0 μs	(0.67, 0.33)	7.0 %	[151]

(continued)

Table 6.4 (continued)

Material	Structure	HOMO (eV)	LUMO (eV)	Abs (nm)	λ_{max} (nm)	PL QE (%)	E_T(eV) τ (μs)	CIE (x, y)	Efficiency	Refs.
$Ir(piq)_3$	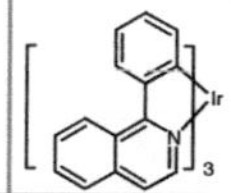	5.03	3.10	550, 600	620	26	2.05 eV, 2.8 μs	(0.68, 0.32)	10.3 %	[154]
$(piq)_2Ir(acac)$	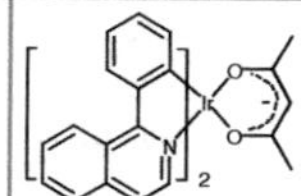	5.3	2.8	226, 286, 336, 350, 377, 477	622	20	2.0 eV, 1.67 μs	(0.68, 0.32)	18.5 %	[155, 159]
$Ir(piq\text{-}F)_2(acac)$	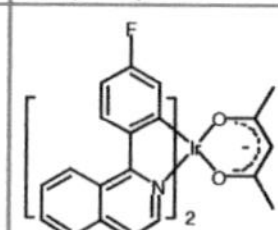			229, 289, 335, 352, 398, 452	600	33	2.07 eV, 1.20 μs	(0.61, 0.36)	8.67 %	[155]
$Ir(m\text{-}piq)_2(acac)$	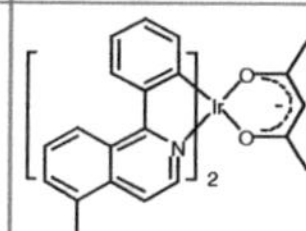	5.35			623		1.99 eV	(0.68, 0.32)	8.91 cd/A	[160]
$Ir(1\text{-}niq)_2(acac)$	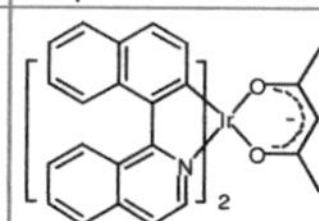	5.11			665		1.87 eV	(0.70, 0.27)	0.24 cd/A	[160]
$Ir(2\text{-}niq)_2(acac)$	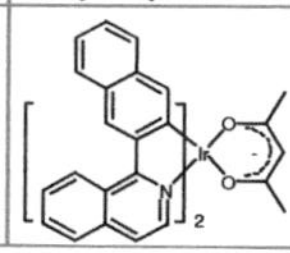	5.14			633		1.96 eV	(0.70, 0.30)	3.79 cd/A	[160]

(continued)

Table 6.4 (continued)

Material	Structure	HOMO (eV)	LUMO (eV)	Abs (nm)	λ_{max} (nm)	PL QE (%)	E_T(eV) τ (μs)	CIE (x, y)	Efficiency	Refs.
$Ir(4F5Mpiq)_3$	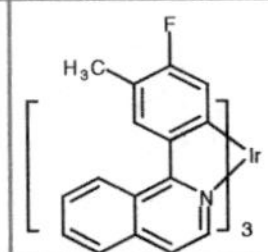	5.34	2.63	419, 471, 518	607	26	2.04 eV, 1.65 μs	(0.66, 0.34)	15.5 %	[161]
$(RDQ)_2Ir(acac)$	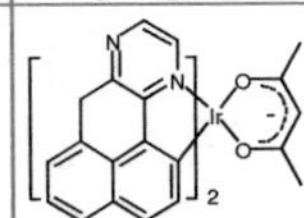			250, 370, 440, 530	618	53	2.01 eV	(0.62, 0.38)	11.9 %	[162]
$Ir(MDQ)_2(acac)$	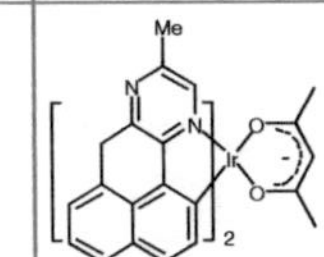	5.35	2.75	250, 370, 440, 530	608	48	2.04 eV	(0.63, 0.37)	20 %	[162]
$(BA)_2Ir(acac)$	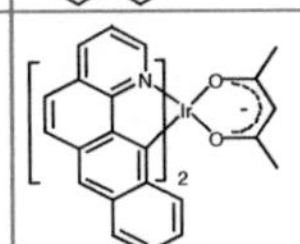			275, 346, 363, 463, 509	665		1.87 eV	(0.64, 0.33)	4.66 %	[163]
$(btmp)_2Ir(acac)$	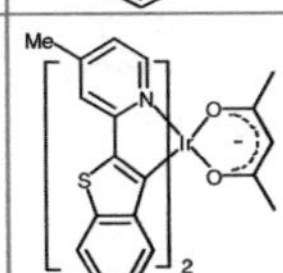	5.48	2.88	280, 322, 334, 449, 477	604	2.5	2.05 eV			[164]

(continued)

Table 6.4 (continued)

Material	Structure	HOMO (eV)	LUMO (eV)	Abs (nm)	λ_{max} (nm)	PL QE (%)	E_T(eV) τ (μs)	CIE (x, y)	Efficiency	Refs.
$(btfmp)_2Ir(acac)$	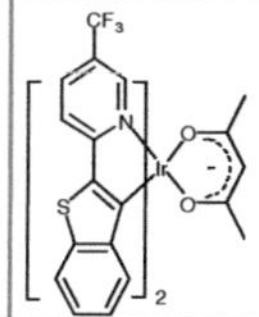	5.75	3.33	290, 335, 352, 369, 482, 513	638	4	1.95 eV	(0.69, 0.29)	9.6 %	[164, 168]
Q3Ir	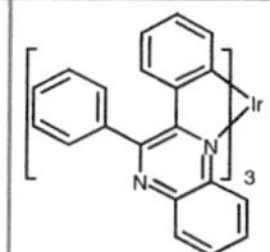				675		1.84 eV, 1.1 μs	(0.70, 0.28)		[169]
$(QR)_2Ir(acac)$	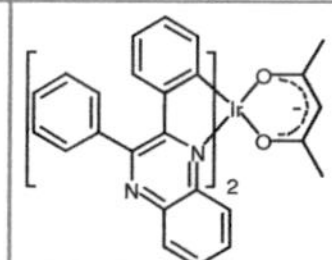			488	670	50	1.85 eV			[169]
$(dpq)_2Ir(acac)$	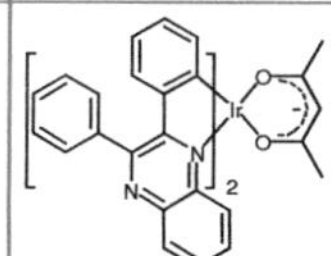			350, 380, 460	675		1.84 eV	(0.71, 0.27)	5.5 %	[170]
$(NAPQ)_2Ir(acac)$	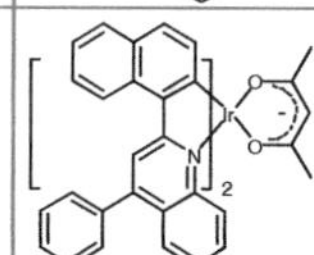	5.16	2.56	256, 282, 314, 382, 493, 539	640	8.5	1.94 eV, 1.22 μs	(0.71, 0.29)	2.1 %	[171]

(continued)

Table 6.4 (continued)

Material	Structure	HOMO (eV)	LUMO (eV)	Abs (nm)	λ_{max} (nm)	PL QE (%)	E_T(eV) τ (μs)	CIE (x, y)	Efficiency	Refs.
$(Mpnq)_2Ir(acac)$	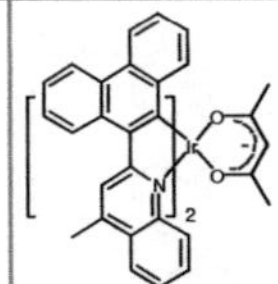	5.00	2.90	310, 377, 508, 535	663	29	1.87 eV, 3.3 μs	(0.71, 0.29)	9.78 %	[172]
$(fppz)_2Os$ $(PPhMe2)_2$	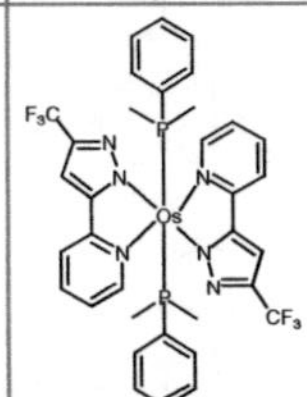	5.02	2.73	410, 465, 550	648		1.92 eV, 0.61 μs	(0.67, 0.33)	12.8 %	[173]
$(fppz)_2Os$ $(PPh2Me)_2$	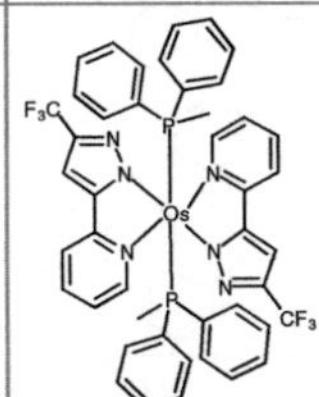	4.95	2.88	405, 454, 542	617	50	2.01 eV, 0.85 μs		4 cd/A	[174]
$(tptz)_2Os$ $(PPh2Me)_2$	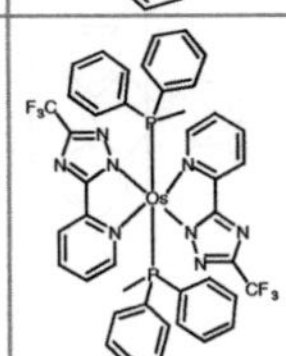	4.5	2.2		620		2.00 eV, 0.7 μs	(0.65, 0.34)	8.37 %	[175]

(continued)

Table 6.4 (continued)

Material	Structure	HOMO (eV)	LUMO (eV)	Abs (nm)	λ_{max} (nm)	PL QE (%)	E_T(eV) τ (μs)	CIE (x, y)	Efficiency	Refs.
$(fptz)_2Os$ $(PPh2Me)_2$	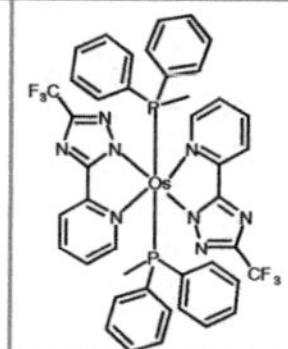	4.9	2.2				0.096 μs	(0.64, 0.36)	19.9 %	[176, 177]
$Eu(TFacac)_3phen$	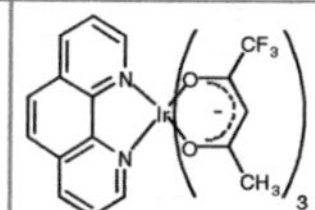				612		2.03 eV			[178]
$(ifpz)_2Ru$ $(PPh2Me)_2$	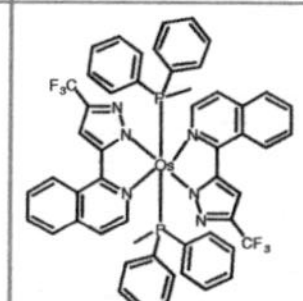	4.93	2.3	320, 353, 446, 523	636	24	1.95 eV, 1.82 μs	(0.67, 0.33)	7.03 %	[179]
dbufbpy-Re	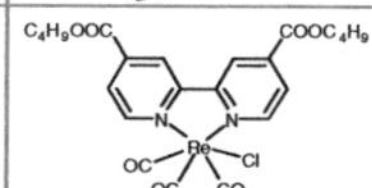			314, 429, 439	605, 635				2.5 cd/A	[180]

$(PPh2Me)_2$ [175] are all charge neutral red phosphorescent emitters. When $(fptz)_2Os(PPh2Me)_2$ was reported for the first time, NPB or 9,9-bis[4-(*N*,*N*-bis-biphenyl-4-yl-amino)phenyl]-9*H*-fluorene (BPAPF) were used as HTL and bathocuproine (BCP) was used as HBL. CBP was used for the host material in the device. Device emits orange instead of red EL if the dopant concentration is less than 20 wt %. At current density of 20 mA/cm^2, EQE is 11.5 % (NPB as HTL) and 13.3 % (BPAPF as HTL), respectively. When the same red phosphorescence emitter was doped into tris[4,9-phenylfluoren-9-ylphenyl]amine (TFTPA) and used 13,5-tris(*N*-phenylbenimidazol-2-yl)benzene (TPBI) as HBL, the device exhibits a maximum luminous efficiency of 29.9 cd/A and a power efficiency of 25 lm/W [176]. It is quite remarkable that the device has minor efficiency roll-off, and the efficiencies remain 29.2 cd/A and 22.2 lm/W at 1000 cd/m^2.

Recently, the EQE of $(fptz)_2Os(PPh2Me)_2$-based OLED has been further improved to 19.9 % at a luminance of 100 cd/m^2 [177]. For such high EL efficiency performance, it was attributed to the use of the new host material 2,7-bis(diphenylphosphoryl)-9-[4-(*N*,*N*-diphenylamino)phenyl]-9-phenylfluorene (POAPF). The bipolar property of the host POAPF guarantees the charge balance of the device. $(fptz)_2Os(PPh2Me)_2$ has a sufficiently high phosphorescence quantum yield of 0.62, although its emission peak is relatively short (617 nm). In order to reach a longer phosphorescence wavelength (more saturated red color), a relatively high dopant concentration ($\sim$20 wt%) is required, which normally results in severe TTA. The observed slight efficiency roll-off (EQE of 18.6 % at a luminance of 1000 cd/m^2) is mainly attributed to the short phosphorescence life time ($\sim$0.7 μsec) of $(fptz)_2Os$ $(PPh2Me)_2$. On the other hand, the symmetrical structure of $(fptz)_2Os(PPh2Me)_2$ induces a small net dipole moment and reduces dipole–dipole interaction and quench emission.

Zheng et al. [183] obtained $Eu(TFacac)_3phen$ using 1,1,1-trifluoroacetylacetone as ligand, and Male et al. [178] demonstrated $Eu(TTFA)_3phen$ using tris-(thiophenyltrifluoromethylacetylacetonate)(phenanthroline) as the ligand. Both emitters can be solution processed and mixed with PVK and PBD to fabricate the OLED devices. The EL emission from Eu complexes is extremely sharp because the emission is from the f-f transition of the europium atom.

Ru(II) complexes have attracted great attention owing to their high quantum efficiencies. The inadequate evaporation and serious thermal degradation resulting from the ionic nature makes cationic Ru(II) complexes unsuitable for wide application in PHOLEDs. However, a breakthrough has been achieved by Tung et al. [179, 184], who synthesized charge-neutral Ru(II) complexes which can be used in PHOLED by the vacuum-deposition method. An optimized device using $(ifpz)_2Ru$ $(PPh2Me)_2$ (ifpz = 3-trifluoromethyl-5-(1-isoquinolyl) pyrazolate) doped in CBP as EML reaches an EQE of 7.03 %, luminous efficiency of 8.02 cd/A, power efficiency of 2.74 lm/W at 20 mA/cm^2, and CIE coordinates of (0.67, 0.33).

Re(I) complexes are hardly used in PHOLEDs owing to the disadvantage of the saturation of emission states resulting from TTA, which leads to low efficiency at high current density. Li et al. [180] incorporated two butylformate groups at the 4 and 4′ positions of 2,2′-bipyridine, increased the steric hindrance effect, and reduced the TTA of (4,4′-dibutyl formate-2,2-bipyridine)Re$(CO)_3$Cl (dbufbpy-Re). A red PHOLED with a peak wavelength of ~610 nm, a maximum efficiency of 2.5 cd/A, and a maximum brightness of 1852 cd/m^2 is obtained and the efficiency was maintained at a value of 1.4 cd/A, even at a current density of 100 mA/cm^2.

6.4.2 Green Phosphorescent Materials

Ir$(ppy)_3$ was the first iridium-based green phosphorescent material for OLEDs, which was shortly followed by $(ppy)_2$Ir(acac) with a similar structure by the same research group. Both of them are transition metal coordinate complexes with simple molecular structure. They stand for two types of Ir complex phosphorescent materials: homoleptic complexes $[C\text{^}N]_3$Ir and heteroleptic complexes $[C\text{^}N]_2$Ir (LX). Their cyclometalation ligand 2-phehylpyridine (ppy) is the parent structure for many other cyclometalation ligands of phosphorescent materials. The photophysical properties of Ir$(ppy)_3$ and $(ppy)_2$Ir(acac) are quite similar, although Ir $(ppy)_3$ is homoleptic and $(ppy)_2$Ir(acac) is heteroleptic. The phosphorescence quantum yield of Ir$(ppy)_3$ is 0.40 and that of $(ppy)_2$Ir(acac) is 0.34. The triplet life time of Ir$(ppy)_3$ is 1.9 μs and that of $(ppy)_2$Ir(acac) is 1.6 μs. The emission peak of Ir$(ppy)_3$ is 514 nm with CIEx,y of (0.27, 0.63), which is a bit shorter than the 525 nm of $(ppy)_2$Ir(acac) with CIEx,y of (0.31, 0.64). The smaller CIE x-coordinate means that the emission of Ir$(ppy)_3$ is greener than that of $(ppy)_2$Ir(acac), whereas a slightly smaller CIE y-coordinate indicates a less saturated color of Ir$(ppy)_3$. Such difference of color chromaticity is not large but substantial and very similar trends have been observed for other homoleptic and heteroleptic coordination complexes as well. First, acac ancillary ligand has a weaker ligand-field strength than ppy. Weaker ligand-field strength causes less *d*-orbitals splitting, which results in lower energy of MLCT and hence longer phosphorescence wavelengths. Second, experimental evidence indicates that the molecular dipole quenches phosphorescence of the material. In terms of molecular dipole moment, facial homoleptic coordination complexes are smaller than heteroleptic coordination complexes, which in turn are smaller than meridional homoleptic coordination complexes.

When Ir$(ppy)_3$ was first reported, the maximum EQE had reached 8 %, which was the first time that the EQE of OLEDs surpassed 5 %, which was considered then the theoretical upper limit for fluorescent OLEDs. The $(ppy)_2$Ir(acac) OLEDs reported thereafter achieved even higher EQE up to 12 %. The Kido group [185] used a polymer buffer layer, tetraphenyldiamine-containing poly(arylene ether sulfone) (PTPDES) doped with tris(4-bromophenyl) aminium hexachloroantimonate (TBPAH) as an electron acceptor, and the Ir$(ppy)_3$-based OLED obtains an EQE of 21.6 % and a power efficiency of 82 lm/W (77 cd/A) at 3.0 V. In 2004, He et al. [186] employed double-emission layers with one predominantly hole transporting EML at anode side and one predominantly electron transporting EML at the cathode side in a *p-i-n* structure, and the optimized Ir$(ppy)_3$ OLED reaches a maximum EQE of 19.5 % and a power efficiency of 64 lm/W at 1000 cd/m^2 with smaller efficiency roll-off in comparison with the single EML. Mikami [187] used highly refractive index glass substrate (n = 2.0) and micro lens array as the extraction method; the Ir(ppy)$_3$-based OLED demonstrated a power efficiency of 210 lm/W at 10 cd/m^2 (an EQE of 50 %), and 100 lm/W at 1000 cd/m^2.

CBP is a widely used host material for green phosphorescent emitters; however, its thermal stability is relatively poor, which influences the operational life time of the OLED. Chen's group [188] synthesized 4,4′-*N,N*′-[di(3,6-di(*tert*-butyl)carbazole)biphenyl (ttbCBP) which possesses both a wide band gap (3.2 eV) and high Tg (175 °C) in 2008. Tsuzuki and Tokito [189] designed 4,4-bis-*N*-carbazolyl-9,9-spirobifluorene (CFL) host with spirobifluorene and two carbazole moieties. The Tg of CFL is 151 °C and the life time of the OLED using CFL as the host is longer than that of the OLED using CBP. A long–living Ir(ppy)$_3$ OLED is achieved by using a silane host, 9-(4-triphenylsilanyl-(1,1′,4,1″-yl)-9*H*-carbazole (TSTC) as the host material [190]. The tetraphenylsilane moiety is introduced to provide a high triplet energy level (2.4 eV), thermal and chemical stability, and glassy properties leading to high efficiency and operational stability of the devices. Ir(ppy)$_3$-based OLEDs using the TSTC host result in a maximum EQE of 19.8 % and power efficiency of 59.4 lm/W. High operational stability with a half-life of 160,000 h is also obtained at an initial luminance of 100 cd/m^2. High morphological stability of the host is desired for high efficiency and operational stability for electrophosphorescent devices. Yang's group [159] has developed a bipolar host *o*-CzOXD with twisted structure by linking the 9-position of carbazole with the ortho position of 2,5-diphenyl-1,3,4-oxadiazole, which results in good thermal and morphological stabilities and a high triplet energy level. Devices containing Ir(ppy)$_3$ dopant and o-CzOXD host show maximum external quantum efficiencies as high as 20.2 %, which is much higher than that of identical devices with the CBP host. The outstanding EL performances are attributed to the well-matched energy levels, which consequently leads to a more balanced injection and recombination of charge carriers.

Kido and co-workers reported a series of phenylpyridines and phenylpyrimidines as the combinational ET and HB materials with very high electron mobilities and higher triplet energy level than Ir(ppy)$_3$, e.g., bis-4,6-(3,5-di-3-pyridylphenyl)-2-methylpyrimidine (B3PYMPM) [191], 2-phenyl-4,6-bis(3,5-dipyridylphenyl) pyrimidine (BPyPPM) [192], 3,5,3′,5′-tetra-(*p*-pyrid-4-yl)phenyl[1,1′]biphenyl (p4PPP) [193], 3,5,3′,5′-tetra(*m*-pyrid-3-yl)phenyl[1,1′]biphenyl (m3PPP) [193], 3,5,3′,5′-tetra-(*m*-pyrid-4-yl)phenyl-1,1′-biphenyl (m4PPP) [193], 3,5,3′,5′-tetra(*p*-pyrid-3-yl)phenyl-1, 1′-biphenyl (p3PPP) [193], TmPyPB [194], and TpPyPB [194]. The electron mobilities of TmPyPb and TpPyPb are 1.0 × 10^{-3} and 7.9 × 10^{-3} cm^2/Vs, respectively, which are even higher than the hole mobilities of most HTM. By using B3PYMPM as the hole block and ETL [191], the EQE of Ir(ppy)$_3$-based OLED can be increased to 29 % at 100 cd/m^2, corresponding to a power efficiency of 133 lm/W (Table 6.5).

As the first (C^N)$_2$Ir(acac) type green phosphorescent emitter, a lot of device optimization efforts have been applied to (ppy)$_2$Ir(acac)-based OLEDs. Qin and co-workers [207] fabricated a (ppy)$_2$Ir(acac)-based OLED on p-silicon anode using

Table 6.5 Chemical structures and properties of green phosphorescent materials

Material	Structure	HOMO (eV)	LUMO (eV)	Abs (nm)	λ_{max} (nm)	PL QE (%)	E_T(eV) τ (μs)	CIE (x, y)	Efficiency	Refs.
Ir(ppy)$_3$	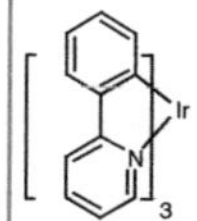	5.2	2.8	244, 283, 341, 377, 405, 455, 488	510	40	2.43 eV, 1.9 μs	(0.27, 0.63)	29 %	[191]
(ppy)$_2$Ir(acac)	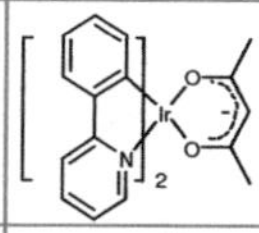	5.6	3.0	260, 345, 412, 460, 497	516	34	2.41 eV, 1.6 μs	(0.31, 0.64)	29.1 %	[195]
Ir(Czppy)$_3$	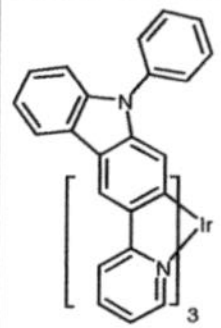	4.87	2.24	291, 316	515	43	2.41 eV, 0.46 μs	(0.24, 0.63)	11.56 %	[196]
Ir(FCzppy)$_3$	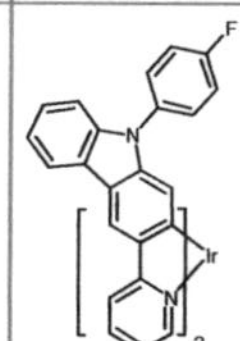	4.96	1.96	281, 315	513	43	2.42 eV, 0.18 μs	(0.27, 0.60)	6.69 %	[197]
Ir(CCzppy)$_3$	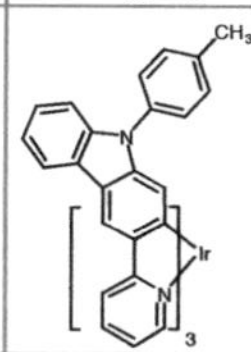	4.93	1.95	286, 318, 399	514	43	2.41 eV, 0.21 μs	(0.25, 0.62)	12.93 %	[197]

(continued)

Table 6.5 (continued)

Material	Structure	HOMO (eV)	LUMO (eV)	Abs (nm)	λ_{max} (nm)	PL QE (%)	E_T(eV) τ (μs)	CIE (x, y)	Efficiency	Refs.
$Ir(CzppyF)_3$	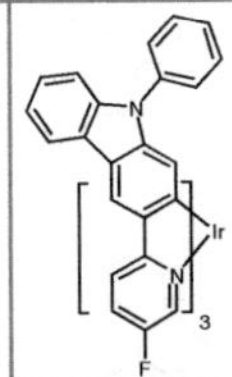	5.04	2.08	288, 312, 402	528	30	2.35 eV, 0.16 μs	(0.26, 0.57)	6.85 %	[197]
$Ir(CzppyC)_3$	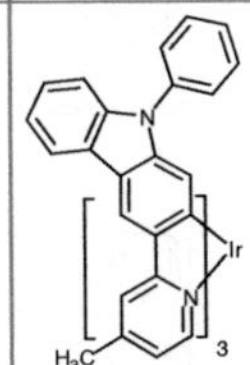	4.88	1.84	293, 315, 396	506	47	2.45 eV, 0.18 μs	(0.22, 0.58)	11.50 %	[197]
$Ir(CzppyF)_2Ir$ (acac)	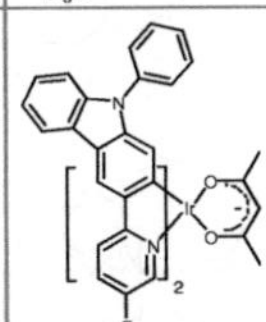	5.10	2.15	308, 356, 398, 440	531	35	2.34 eV, 0.22 μs	(0.31, 0.62)	10.23 %	[197]
$Ir(CzppyC)_2Ir$ (acac)	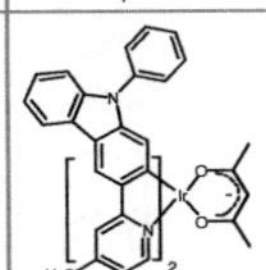	5.01	2.88	297, 325, 359, 424	506	41	2.45 eV, 0.20 μs	(0.23, 0.63)	10.77 %	[197]

(continued)

Table 6.5 (continued)

Material	Structure	HOMO (eV)	LUMO (eV)	Abs (nm)	λ_{max} (nm)	PL QE (%)	E_T(eV) τ (μs)	CIE (x, y)	Efficiency	Refs.
$Ir(mCP)_3$	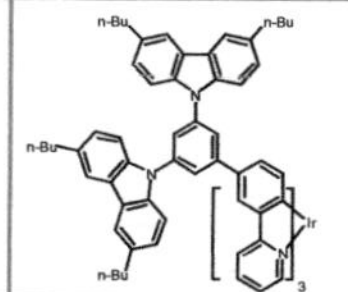	5.5	2.7			91			8.3 %	[198]
$(mCP)_2Ir(bpp)$	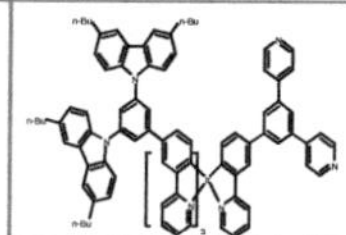	5.6	2.8			84			5.4 %	[198]
$Ir(Oppy)_3$	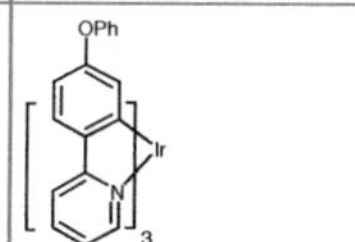	5.15	1.97	243, 290, 364, 400, 476	497	57	2.50 eV, 0.10 μs	(0.20, 0.60)	8.92 %	[199]
$(Oppy)_2Ir(acac)$	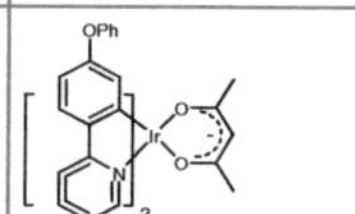	5.22	2.00	282, 403, 430, 481	505	40	2.49 eV, 1.66 μs	(0.26, 0.59)	11.05 %	[200]
$Ir(dsippy)_3$	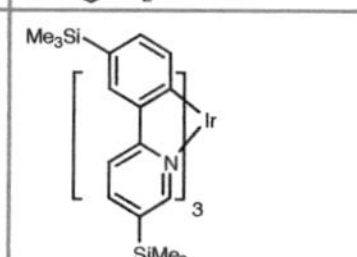	5.20	2.71	291, 394, 419, 460, 460	519		2.31 eV, 1.45 μs		39.2 cd/A	[201]

(continued)

Table 6.5 (continued)

Material	Structure	HOMO (eV)	LUMO (eV)	Abs (nm)	λ_{max} (nm)	PL QE (%)	E_T(eV) τ (μs)	CIE (x, y)	Efficiency	Refs.
$Ir(ppy)_2(CBDK)$	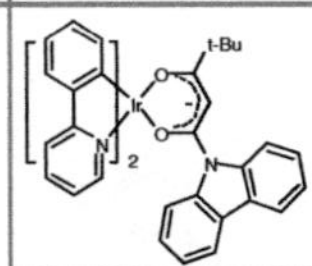	5.03	2.78	237, 283, 283, 293, 329, 343	487, 521		0.9 μs	(0.36, 0.60)	6.34 cd/A	[202]
Ir $(dfppy)_2(CBDK)$	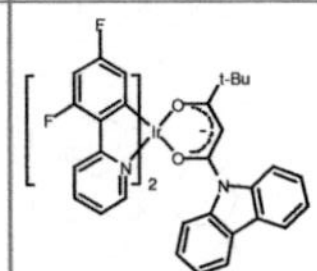	5.34	2.95	237, 283, 283, 293, 329, 343	487, 521		1.0 μs	(0.29, 0.50)	1.35 cd/A	[202]
$Ir(bim)_2acac$	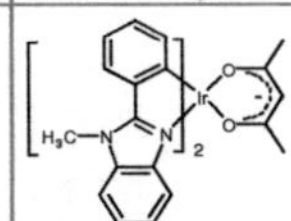				509	40		(0.32, 0.60)	22.5 cd/A	[203]
(dmbipy)Pt $(CCPh)_2$	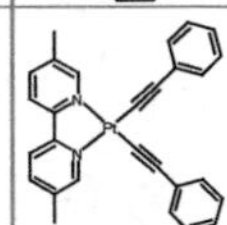			284, 395	554	50	1.2 μs			[204]
Pt(dpt)(oph)	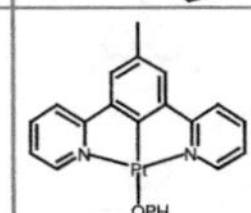				525		2.36 eV		17.56 %	[205]
Pt(dpp)Cl	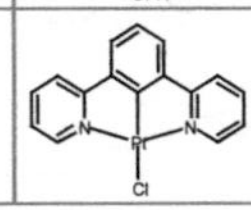	5.23	2.77	485	491	60	2.53 eV, 7.2 μs	(0.23, 0.57)	~16 %	[206]

(continued)

Table 6.5 (continued)

Material	Structure	HOMO (eV)	LUMO (eV)	Abs (nm)	λ_{max} (nm)	PL QE (%)	E_T(eV) τ (μs)	CIE (x, y)	Efficiency	Refs.
Pt(dpt)Cl	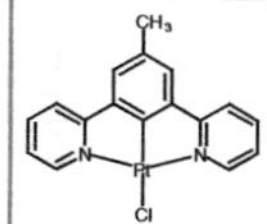	5.17	2.78	495	505	58	2.46 eV, 7.8 μs	(0.31, 0.61)	~10 %	[206]
Pt(dpppy)Cl	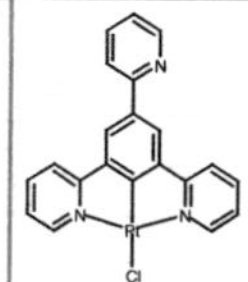	5.33	2.80	493	506	57	2.45 eV, 9.2 μs	(0.20, 0.76)	~9 %	[206]
Pt(dppmst)Cl	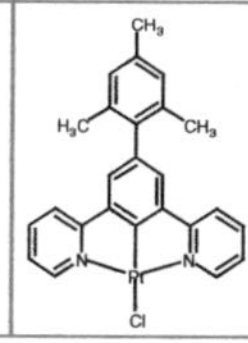	5.38	2.81	492	501	62	2.48 eV, 7.9 μs	(0.24, 0.63)	~4 %	[206]

Sm/Au as the semi-transparent cathode, and the device shows a maximum current efficiency of 69 cd/A with a turn-on voltage of 3.2 V. Recently, a new efficiency record of green phosphorescence OLED belongs to. Lu's group [195] and demonstrate a highly simplified green OLED on a chlorinated ITO transparent electrode with a work function of >6.1 eV which provides a direct match to the energy levels of the active light-emitting materials. EQE of the Cl-ITO device reaches 29.1 % (93 cd/A) at 100 cd/m^2, 29.2 % (94 cd/A) at 1000 cd/m^2, and 25.4 % (81 cd/A) at 10,000 cd/m^2. More importantly, the device shows significantly suppressed efficiency roll-off. Even at an ultrahigh luminance of 100,000 cd/m^2, the EQE of the Cl-ITO device is still as high as 14.3 % (46 cd/A). By applying a simple lens-based structure to extract light being trapped at the glass-air interface (the substrate mode), the EQE can be further increased up to 54 % at 1000 cd/m^2, and the maximum power efficiency is 230 lm/W.

Based on Ir(ppy)$_3$ and (ppy)$_2$Ir(acac), more green phosphorescent Ir complexes have been developed by directly grafting carbazole moiety onto ppy cyclometalation ligand, such as Ir(Czppy)$_3$ [196], Ir(FCzppy)$_3$ [197], Ir(CCzppy)$_3$ [197], Ir (CzppyF)$_3$ [197], Ir(CzppyC)$_3$ [197], Ir(CzppyF)$_2$Ir(acac) [197], and Ir(CzppyC)$_2$Ir (acac) [197]. Because of the hole transporting nature of the carbazole moiety, one of the most commonly used structures for the host material, HOMO energy level has been raised less than 5.0 eV. It is also appreciative of green color chromaticity of these Ir complexes and is attributed to the rigid and bulky carbazole moiety which prevents the close π–π interaction among molecules. The efficiency roll-off of their OLEDs is thus somewhat alleviated. Ir(mCP)$_3$ [198] and (mCP)$_2$Ir(bpp) [198] are designed and applied for OLEDs based on a similar principle.

By substituting various functional groups (X-), such as SiPh3, GePh3, NPh2, POPh2, OPh, SPh, and SO2Ph2 on the 4-position of ppy cyclometalation ligand, more green to yellow phosphorescence materials Ir(X-ppy)$_3$ [199] and (X-ppy)$_2$Ir (acac) [200] have been developed recently. Except for Ir(Oppy)$_3$ [199] and (Oppy)$_2$Ir(acac) [200], most of these phosphorescence materials have CIE x-coordinates between 0.32 and 0.37, and hence they are not truly green but yellow-green in color. The experimental results on OLEDs also show that most Ir(X-ppy)$_3$ are better than Ir(ppy)$_3$ because of the presence of electron transporting X-substituent. Jung et al. [201] reported a homoleptic Ir(III) complex of fac-tris[2-(3′-trimethylsilylphenyl)-5-trimethylsilylpyridinato]iridium [Ir(dsippy)$_3$] with narrow green emission (full-width at half-maximum: 50 nm) and higher efficiency than Ir(ppy)$_3$. The bulky silyl group on the ppy ring seems to play a key role in suppressing various intermolecular excited-state interactions, which illustrates a way to solve the problem of color purity for green phosphorescence.

Liu et al. [202] designed two iridium complexes, both containing carbazole-functionalized beta-diketonate, Ir(ppy)$_2$(CBDK) [bis(2-phenylpyridinato-*N*,C-2) iridium(1-(carbazol-9-yl)-5,5-dimethylhexane-2,4-diketonate)], Ir(dfppy)$_2$(CBDK) [bis(2-(2,4-difluorophenyl)pyridinato-*N*,C-2)iridium(1-(carbazol-9-yl)-5,5-dimethylhexane-2,4-diketonate)] for non-doped devices. The designed Ir(ppy)$_2$(CBDK) shows not only a good hole transporting ability, but also a good electron transporting ability. The improved performance of Ir(ppy)$_2$(CBDK) and Ir

$(dfppy)_2(CBDK)$ can be attributed to the introduced bulky carbazole-functionalized beta-diketonate, resulting in improved carrier transporting properties and suppressed TTA.

Another high-efficiency series of iridium complexes is based on benzoimidazole ligands complex such as in $Ir(bim)_2acac$ [203]. $Ir(bim)_2acac$ exhibits green phosphorescence with a maximum peak emission at 509 nm and a quantum yield of about 40 %.

Similar to red phosphorescent emitters, another class of green phosphorescent material is based on the Pt heavy atom. The flat square planar molecular shape is not conducive to solid state luminescence because of molecular π–π stacking which quenches the emission readily. One of the Pt-based green phosphorescent materials, $(dmbipy)Pt(CCPh)_2$ [204], was reported in the early stages of phosphorescent material development. However, the solution fabricated OLEDs (using PVK as the host material) show unsatisfactory performance. It is not until the changes of cyclometalate ligand to tridentate N^C^N that the performance of OLEDs is improved. Pt(dpt)(oph) [205], Pt(dpp)Cl [206], Pt(dpt)Cl [206], Pt(dpppy)Cl [206], and Pt(dppmst)Cl [206] are some typical examples. The highest EQE can be up to 16 % with no significant roll-off over a four-decade current intensity span roll-off. The emission color from the yellow to the green-bluish region can be simply tuned by changing the substituents at the central 5-position of the cyclometalating ligand. However, OLEDs fabricated with these Pt complexes are sensitive to the concentration of phosphorescent dopant materials. One of them, Pt(dpppy)Cl OLED, exhibits a very high CIE y-coordinate of 0.76, very close to the saturated green color. This can be attributed to its quite narrow emission peak with FWHM about 30 nm, which is probably because of the small and rigid molecular structure of Pt (dpppy)Cl. Chang et al. [208] synthesized new series of luminescent platinum(II) azolate complexes $Pt(N\text{^}N)_2$, in which N^N = mppz, bppz, bzpz, bmpz, bqpz, fppz, hppz, bptz, and hptz. Large EL red-shift in the solid state is observed caused by the aggregation of platinum(II) complexes.

6.4.3 Blue Phosphorescent Materials

Compared with other phosphorescent materials, blue phosphorescent materials have been less developed, less satisfactory, and generally recognized as the "weakest link" in the realization of a high-efficiency all phosphorescence OLED. One particular problem is finding a material with a high-enough triplet state to correspond to a blue emission wavelength. Furthermore, in addition to the necessary hole-transporting or blocking properties, sufficiently high triplet energy becomes an obstacle and challenge in the design of effective host material for blue phosphorescent dopants. Finding a host into which this material may be doped without quenching the emissive state is a major challenge because the triplet state of the host must be even higher in energy than the emitter triplet state. The high-energy triplet state of a blue phosphorescent emitter appears to be difficult to maintain without degradation, and therefore life times of phosphorescent blues are a problem. Hence, the host materials for blue phosphorescent OLEDs are more crucial than the red and green ones.

Recently much effort has been devoted to the selection of suitable ligands and central heavy atoms. Cyclometalated iridium complex is one of the best candidates for phosphorescent dyes with high quantum yields because of the short life time of triplet excited states. The most famous blue phosphorescent materials are heteroleptic iridium(III)bis[(4,6-di-fluorophenyl)-pyridinato-*N*,C2′-picolinate] (FIrpic) [209] and bis(4′,6′-difluorophenylpyridinato)tetrakis-1-pyrazolyl)borate (FIr6) [210] using fluoro-substituted phenylpyridine ligands and an anionic 2-picolinic acid or poly(pyrazolyl)borate as an auxiliary ligand, respectively. However, these emissions are either greenish-blue (0.17, 0.34) or sky-blue (0.16, 0.26), far from a saturated blue color as the National Television Standards Committee (NTSC) standard of (0.14, 0.08).

As the first blue phosphorescent material, FIrpic has two emission peaks at 470 and 494 nm with phosphorescence quantum yield of 50–60 %. The strong vibration peak which appears at a longer wavelength of the main emission peak makes the color actually cyan containing a lot of green emission. With improved blue color, FIr6 has a shorter emission wavelength (λ_{max} at 457 and 485 nm) and a higher phosphorescence quantum yield (96 %) than those of FIrpic, but its device stability was worse. Because the emission is mainly from the $\pi\pi^*$ state of cyclometalate ligands mixing with d-orbital of heavy transition metal via MLCT, such an emission state because of the mixing of d-orbital is often inadequate, particularly for the short wavelength blue phosphorescence. Consequently, such phosphorescence usually has a longer triplet life time closer to that of pure organic species (organic compound without containing heavy atom). Triplet state life time of μs range is required for a fast emissive decay process and a high phosphorescence quantum yield is desirable to achieve a theoretical maximum EQE over 20 %. It has been known that Ir-based organometallic compounds show a short excited state life time because of singlet-triplet mixing and easy control of emission spectra by managing the chemical structure of ligands. In general, Ir-based phosphorescent dopant materials have been developed to meet the requirements of deep blue phosphorescent dopant materials.

Phenylpyridine groups are widely used as the ligand structure of Ir-based phosphorescent dopant. The HOMO is mainly composed of the π-orbitals of the phenyl ring and the d-orbitals of the metal ion, whereas the pyridine moiety is the principal contributor to the LUMO. Electron-withdrawing groups attached to the phenyl group or changing the ancillary ligand to be more electron accepting decrease the HOMO level, whereas electrondonating groups substituted to the pyridine unit or replacing the pyridine ring to a N-heterocyclic ligand with a higher LUMO increase the LUMO level (Table 6.6).

FIrpic has two fluorines on the phenyl unit to shift the HOMO level downward for high triplet energy. The triplet energy was increased by the electron-withdrawing fluorine and a bulky picolinic acid ancillary ligand. Although the emission spectrum of FIrpic is blue-shifted, it exhibits only a sky blue color with y-coordinate over 0.30.

Table 6.6 Chemical structures and properties of blue phosphorescent materials

Material	Structure	HOMO (eV)	LUMO (eV)	Abs (nm)	λ_{max} (nm)	PL QE (%)	E_T(eV) τ (μs)	CIE (x, y)	Efficiency (%)	Refs.
FIrpic	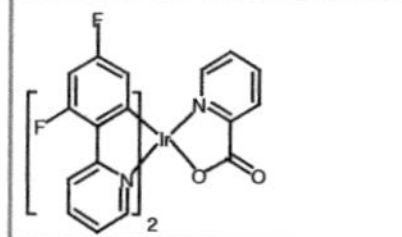	5.8	2.9	256, 379, 455	470, 494	79	2.62 eV, 0.5 μs	(0.16, 0.29)	5.7	[209, 211]
FIr_6	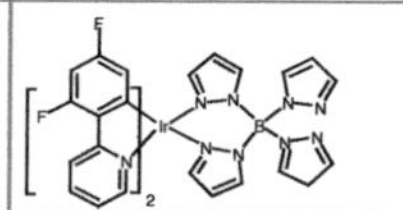	6.2	3.1	325	458	73	2.72 eV	(0.16, 0.26)	11.6	[210]
FIrtaz	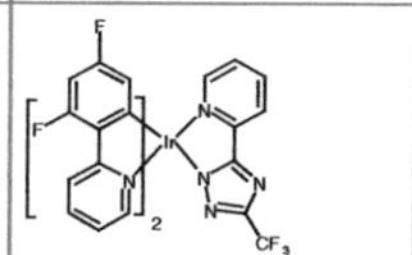				460, 489					[212]
$FIrN_4$	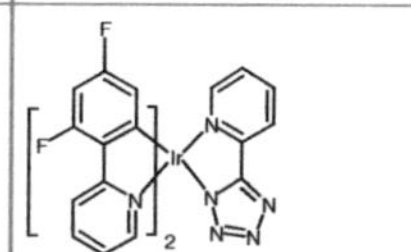			253, 367, 449	459, 489	13	2.76 eV, 0.15 μs	(0.15, 0.24)	9.4	[211, 212]
Ir $(fppz)_2(dfbdp)$	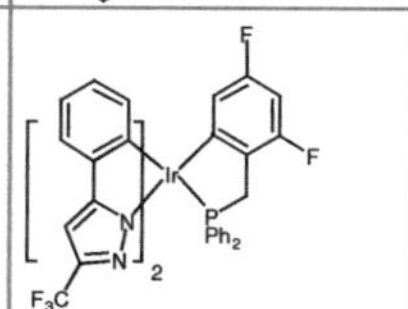			262, 300, 347	428, 455		2.88 eV, 7.42 μs	(0.15, 0.11)	11.9	[213]
$Ir(F_4ppy)_3$	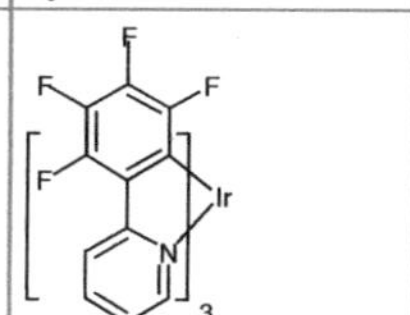	5.73	2.31		464, 493	3.1	0.20 μs		5.5	[214]

(continued)

Table 6.6 (continued)

Material	Structure	HOMO (eV)	LUMO (eV)	Abs (nm)	λ_{max} (nm)	PL QE (%)	E_T(eV) τ (μs)	CIE (x, y)	Efficiency (%)	Refs.
FCNIr	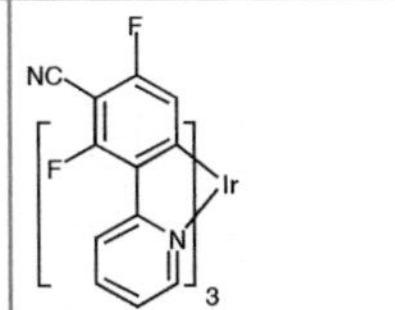	5.8	3.0		448, 476		2.8 eV,	(0.15, 0.19)	4.9	[215]
FCNIrpic	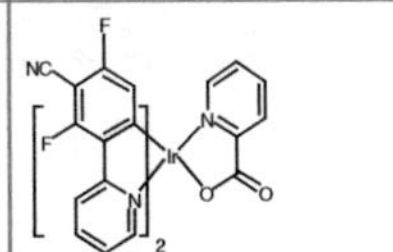	5.72	2.98				3.0 eV,	(0.14, 0.17)	25.1	[215, 216]
FCF_3Irpic	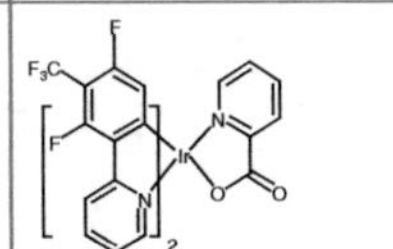	6.2	3.2		454, 482	27	1.48 μs	(0.15, 0.21)	23.3	[217]
$Ir(dfpypy)_3$	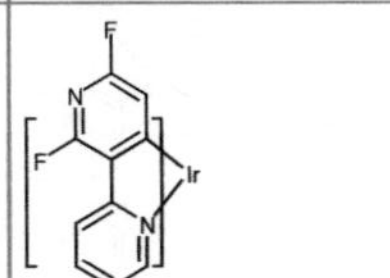	6.39	2.26		438, 463	77	2.97 eV	(0.14, 0.12)		[218]
$Ir(ppz)_3$	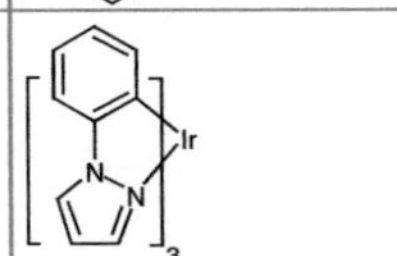	5.02	0.57	244, 261, 292, 321, 366	414		14 μs			[153]

(continued)

Table 6.6 (continued)

Material	Structure	HOMO (eV)	LUMO (eV)	Abs (nm)	λ_{max} (nm)	PL QE (%)	E_T(eV) τ (μs)	CIE (x, y)	Efficiency (%)	Refs.
Ir(mpptz)$_3$	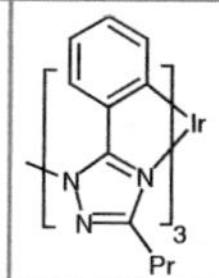	4.94	1.14		449, 479	66	2.79 eV, 1.08 μs	(0.16, 0.12)		[219]
Ir(pmb)$_3$	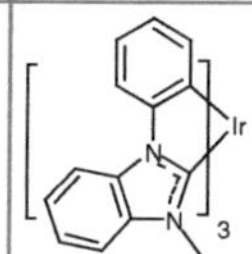	5.1	1.8			4	2.78 eV, 1 μs	(0.17, 0.06)	5.8	[220]
(dfbmb)$_2$Ir (tptz)	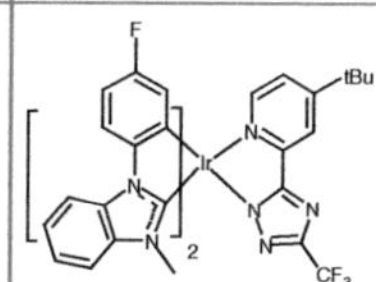			298, 318, 356	392, 461	0.05	0.83 ns			[221]
(fbmb)$_2$Ir (bptz)	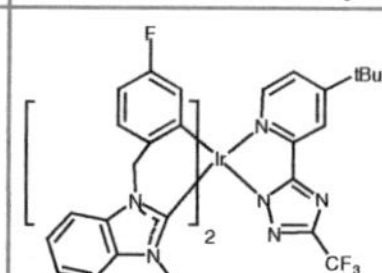			298, 322, 363	460	22	0.219 μs			[221]
(dfbmb)$_2$Ir (fptz)	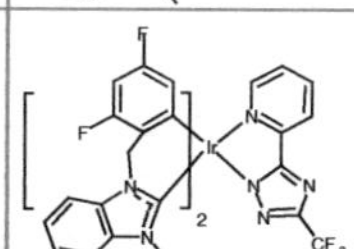			293, 316, 363	458	73	0.378 μs	(0.16, 0.13)	6.0	[221]

The color coordinate is further blue-shifted by changing the ancillary ligand from picolinate (pic) to bulky tetrakis(1-pyrazolyl)borate ligand (FIr6). The Fir6 has a bulky tetrakis(1-pyrazolyl)borate ligand to reduce the conjugation of the main ligand through steric hindrance. The peak wavelength of the FIr6 is 458 nm compared with the 470 nm of FIrpic. Similarly, triazole (FIrtaz) [212] and tetrazole (FIrN4) [212] derivatives were adopted as the ancillary ligands. The emission wavelengths of FIrtaz and FIrN4 were 459 and 460 nm, respectively. The color coordinates of the blue PHOLED are (0.15, 0.24) with a quantum efficiency of 5.8 %.

Chiu et al. [213] synthesized Ir(fppz)$_2$(dfbdp) by changing ppy ligand with a larger π–π^* energy gap of ppz together with weak conjugated and strong ligand-field strength benzyl phosphine (C^P) chelate, and obtained a deep blue light with CIE coordinates of (0.15, 0.11). A maximum EQE of 12 % was obtained from a device of ITO/NPB 30 nm/TCTA 20 nm/CzSi 3 nm/CzSi:6 % of Ir(fbppz)$_2$(dfbdp) 35 nm/UGH2:6 % Ir(fbppz)$_2$(dfbdp) 3 nm/UGH2 2 nm/BCP 50 nm/Cs_2CO_3 2 nm/Ag 150 nm, which is the highest true blue PHOLED reported.

As mentioned above, a deep HOMO level can also be obtained by adding an electron-withdrawing substituent on the phenyl ring of ppy ligand. Ragni et al. [214] prepared a blue emitting iridium(III) complex containing 4-fluorine-substituted phenylpyridine ligands, tris[2-(3′,4′,5′,6′-tetrafluorophenyl)pryidinato-*N*,C$^{2\prime}$] iridium(III) (Ir(F4ppy)$_3$). A blue PHOLED, with the structure of ITO/PEDOT: PSS/PVK:fac-Ir(F4ppy)$_3$:Br-PBD/TPBI/Ba/Al, emitted blue light with the emission peak at 471 nm and an EQE of 5.5 %. Phenylpyridine derivatives with cyano-(FCNIr or FCNIrpic) [215] or trifluoromethyl-(FCF$_3$Irpic) [222] groups along with two fluorine units further shift the color coordinate of blue PHOLEDs. A color coordinate of (0.15, 0.16) with a quantum efficiency of 9 % are obtained by doping FCNIr into the common mCP host [223]. The device performances of the FCNIr-doped blue PHOLED are further improved by carefully tuning the charge balance in PPO1, PPO2, and PPO21 host materials [224]. The highest EQE of deep blue PHOLED is achieved by heteroleptic FCNIrpic emitter with picolinic acid ancillary ligand, which is as high as 25.1 % with color coordinate of (0.14, 0.17), and the quantum efficiency remains at 22.3 % at 1000 cd/m^2 [216]. Seo et al. [217] employed FCF$_3$Irpic emitter in a double-emission layer structure, which was also evaluated as a deep blue emitting dopant using a double layer emitting structure; a maximum EQE of 23.3 % was obtained with color coordinates of (0.15, 0.21). However, the efficiency roll-off is significant and the efficiency at 20 mA/cm^2 was only 4.2 %.

Kang and co-workers [218] reported a fac-tris-(2′,4-difluoro-2,3-bipyridinato-*N*, C4′)iridium(III) (Ir(dfpypy)$_3$) containing fluorine-substituted bipyridine ligand. The use of a difluoropyridine unit instead of common difluorophenyl further shifts the emission wavelength to the blue region because of electron deficiency of the pyridine ring, and significantly improves the thermal stability and molecular rigidity because of intermolecular interactions in the solid state. It exhibits intensive pure blue PL (quantum yield of 0.77 in CH_2Cl_2) with the peaks blue-shifted to 438 and 463 nm and the color coordinate was (0.14, 0.12). This indicates that N-heterocyclic substitution of phenyl in the ppy ligand is the most effective way to decrease the

HOMO level of the complex to enlarge the energy gap of phosphorescent dye without much deteriorative effect on the PL quantum yield.

Although many arylpyridine-type dopant materials have been synthesized to shift the emission wavelength to short wavelength, the most effective way to get a deep blue emission is to modify the aryl group with strong electron-withdrawing groups such as F, CF_3, and CN. However, those strong electron-withdrawing groups have been known to degrade the long-term stability of deep blue PHOLEDs, dramatically shortening the life times of devices. Therefore, the arylpyridine type dopant materials modified with strong electron-withdrawing groups may not be used in commercial application and fluorine-free phosphorescent dopant materials are being developed.

To avoid using strong electron-withdrawing groups, one strategy is to increase the LUMO level through the addition of electron-donating groups to the pyridine ring or replacing the pyridine ring with an N-heterocyclic ligand with a higher LUMO. Tamayo et al. [153] replaced the pyridine ring by *N*-pyrazole with a higher LUMO and obtained tris(phenylpyrazolyl)-iridium(III) [$Ir(ppz)_3$] which exhibited PL at 414 nm for its facial isomer and 427 nm for its meridional isomer at 77 K, with no observable emission at RT. However when the pyridine ring was replaced by triazole, which has a higher LUMO energy, to obtain tris-(1-methyl-5-phenyl-3-propyl-[1, 2, 4] triazolyl)iridium (III) [$Ir(mpptz)_3$], PL at 449 and 479 nm with a high quantum yield of 0.66 was achieved [219].

The best blue chromaticity phosphorescence OLED is achieved by tris(phenyl-methyl-benzimidazolyl) iridium(III) [$Ir(pmb)_3$] [220] emission wavelength to 395 nm corresponding to CIE coordinates of (0.17, 0.06). However, its EL efficiency and stability were low and far from satisfactory. Nonetheless, $Ir(pmb)_3$ was the first non-fluorine-containing blue phosphorescent dopant material and the first iridium complex using carbene (=C:) as the coordinating atom. Carbene-containing ligand is stronger than azolate ligand in terms of ligand-field splitting and hence LUMO energy is higher and the band gap is larger. However, low phosphorescence quantum yield ($\sim$0.04) is the weakness of f-$Ir(pmb)_3$ and its maximum EQE is only 5.8 %. Moreover, $Ir(pmb)_3$ OLEDs exhibit serious efficiency roll-off: EQE decreases to 2.3 % at a current density of 10 mA/cm^2 and EQE is as low as 0.5 % at a current density of 100 mA/cm^2. Another carbene-containing blue phosphorescence material is $(dfbmb)_2Ir(tptz)$ [221] and it has a much better quantum yield of $\sim$0.73 but somewhat worse in color coordinates of (0.16, 0.13). EQE of $(dfbmb)_2Ir$ (tptz) OLED can reach 6.0 % but efficiency roll-off is intense. At display brightness (100 cd/m^2), EQE decreases to 2.7 %; EQE decreases further to 0.7 % at lighting condition (brightness of 1000 cd/m^2). Such EL efficiency is worse than many other fluorescence materials with similar blue chromaticity.

Various heteroleptic bis(carbene)-type Ir dopant materials have also been studied for the last several years. The carbene derivatives, [$(fbmb)_2Ir(bptz)$] (fbmb = 1-(4-fluorobenzyl)-3-methylbenzimidazolium, bptz = 4-*tert*-butyl-2-(5-(trifluoromethyl)-1, 2, 4-triazol-3-yl)pyridine) and [(dfbmb) 2Ir(fptz)] (dfbmb = 1-(2, 4-difluorob-enzyl)-3-methylben-zimidazolium, fptz = 2-(5-(trifluoromethyl)-1, 2,4-triazol-3-yl) pyridine) exhibit deep blue PL emission peak at 460 and 458 nm, respectively

[221].Their phosphorescence quantum yield are much higher than phenylimidazole-based [(fpmb)2Ir(bptz)] (fpmb = 1-(4-fluorophenyl)-3-methylbenzoimidazolium, bptz = 4-*tert*-butyl-2-(5-(trifluoromethyl)-1, 2, 4-triazol-3-yl)pyridine) because of distinctively low nonradiative decay constant of the benzylimidazole-based dopant materials. The best performing $(dfbmb)_2Ir(fptz)$-doped PHOLED shows a maximum quantum efficiency of 6 % with deep blue color coordinates of (0.16, 0.13). The use of CzSi and UGH2 host materials with mismatched energy levels with the dopant materials degraded the device performances of deep blue PHOLEDs doped with benzylimidazole-type carbene phosphorescent emitting materials.

Although platinum complexes [167, 225] have been reported to give blue phosphorescent emission at room temperature, the light purity and the efficiency of platinum complexes are not comparable to those of iridium complexes because of the longer life times of the triplet states. However in 2008 Yang et al. [226] reported platinum(II) [1,3-difluoro-4,6-di(2-pyridinyl) benzene] chloride (Pt-4), and the device based on this Pt complex exhibits a peak EQE of 16 %, a power efficiency of 20 lm/W, and CIE coordinates of (0.15, 0.26), which are comparable to FIr6 (0.16, 0.26). It suggests that the utilization of Pt complexes is a viable approach for the development of blue PHOLED. The high quantum yield and narrow emission spectra are mainly attributed to its strong mixing with ^{1}MLCT character to the lowest excited state.

6.5 Summary and Outlook

OLEDs have achieved tremendous progresses in the application of flat panel display and solid state lighting since their invention in 1987. Small and medium size OLED panels have been successfully applied in portable devices, such as mobile phones, digital cameras, mp3 players, and so on. Large size OLED TVs are also available from Samsung and LG. In the development of OLEDs, the electroluminescent materials play a critical role. The first generation of electroluminescent materials is fluorescent, which can provide satisfactory red, green, and blue colors and life times, but efficiency is limited because of the spin statistics. The second generation phosphorescent materials can utilize nearly 100 % excitons, and significantly boost the efficiencies. The red and green phosphorescent materials have already been commercialized because of their high efficiencies and fairly long life times. However, highly efficient deep blue phosphorescence light is still difficult to achieve because of the lack of highly efficient emitters and corresponding wide gap charge transporting materials and hosts. Furthermore, the shortages of iridium resources makes it difficult to lower the cost of phosphorescent materials. Considering the key role in OLED technology, high efficiency and low-cost electroluminescent materials need to be developed in the future.

References

1. Tang CW, VanSlyke SA (1987) Organic electroluminescent diodes. Appl Phys Lett 51 (12):913–915
2. Pope M, Kallmann H, Magnante P (1963) Electroluminescence in organic crystals. J Chem Phys 38(8):2042–2043
3. Vincett PS, Barlow WA, Hann RA, Roberts GG (1982) Electrical conduction and low voltage blue electroluminescence in vacuum-deposited organic films. Thin Solid Films 94 (2):171–183
4. Burroughes JH, Bradley DDC, Brown AR, Marks RN, Mackay K, Friend RH, Burns PL, Holmes AB (1990) Light-emitting diodes based on conjugated polymers. Nature 347 (6293):539–541
5. Baldo MA, O'Brien DF, You Y, Shoustikov A, Sibley S, Thompson ME, Forrest SR (1998) Highly efficient phosphorescent emission from organic electroluminescent devices. Nature 395(6698):151–154
6. Endo A, Ogasawara M, Takahashi A, Yokoyama D, Kato Y, Adachi C (2009) Thermally activated delayed fluorescence from Sn4+–porphyrin complexes and their application to organic light-emitting diodes—a novel mechanism for electroluminescence. Adv Mater 21 (47):4802–4806
7. Goushi K, Yoshida K, Sato K, Adachi C (2012) Organic light-emitting diodes employing efficient reverse intersystem crossing for triplet-to-singlet state conversion. Nat Photonics 6 (4):253–258
8. Kim JS, Granstrom M, Friend RH, Johansson N, Salaneck WR, Daik R, Feast WJ, Cacialli F (1998) Indium-tin oxide treatments for single- and double-layer polymeric light-emitting diodes: the relation between the anode physical, chemical, and morphological properties and the device performance. J Appl Phys 84(12):6859–6870
9. So SK, Choi WK, Cheng CH, Leung LM, Kwong CF (1999) Surface preparation and characterization of indium tin oxide substrates for organic electroluminescent devices. Appl Phys A-mater 68(4):447–450
10. Mason MG, Hung LS, Tang CW, Lee ST, Wong KW, Wang M (1999) Characterization of treated indium-tin-oxide surfaces used in electroluminescent devices. J Appl Phys 86 (3):1688–1692
11. Van Slyke S, Chen C, Tang C (1996) Organic electroluminescent devices with improved stability. Appl Phys Lett 69:2160–2162
12. Shirota Y, Okumoto K, Inada H (2000) Thermally stable organic light-emitting diodes using new families of hole-transporting amorphous molecular materials. Synth Met 111:387–391
13. Kim WH, Mäkinen AJ, Nikolov N, Shashidhar R, Kim H, Kafafi ZH (2002) Molecular organic light-emitting diodes using highly conducting polymers as anodes. Appl Phys Lett 80 (20):3844–3846
14. Jung S-H, Choi J-H, Yang S-M, Cho W-J, Ha C-S (2001) Syntheses and characterization of soluble phthalocyanine derivatives for organic electroluminescent devices. Mat Sci Eng B 85 (2):160–164
15. Son S-H, Jang J-G, Jeon S-Y, Yoon S-H, Lee J-C, Kim K-K (2004) Electroluminescent devices with low work function anode. WO Patent 2, 004, 054, 326, (2004)
16. Yang Y, Heeger A (1994) Polyaniline as a transparent electrode for polymer light-emitting diodes: Lower operating voltage and higher efficiency. Appl Phys Lett 64(10):1245–1247
17. Gao J, Heeger AJ, Lee JY, Kim CY (1996) Soluble polypyrrole as the transparent anode in polymer light-emitting diodes. Synth Met 82(3):221–223
18. Deng ZB, Ding XM, Lee ST, Gambling WA (1999) Enhanced brightness and efficiency in organic electroluminescent devices using SiO_2 buffer layers. Appl Phys Lett 74(15):2227–2229

19. Poon CO, Wong FL, Tong SW, Zhang RQ, Lee CS, Lee ST (2003) Improved performance and stability of organic light-emitting devices with silicon oxy-nitride buffer layer. Appl Phys Lett 83(5):1038–1040
20. Qiu Y, Gao YD, Wang LD, Zhang DQ (2002) Efficient light emitting diodes with Teflon buffer layer. Synth Met 130(3):235–237
21. Zhao JM, Zhang ST, Wang XJ, Zhan YQ, Wang XZ, Zhong GY, Wang ZJ, Ding XM, Huang W, Hou XY (2004) Dual role of LiF as a hole-injection buffer in organic light-emitting diodes. Appl Phys Lett 84(15):2913–2915
22. Zhang ZF, Deng ZB, Liang CJ, Zhang MX, Xu DH (2003) Organic light-emitting diodes with a nanostructured TiO_2 layer at the interface between ITO and NPB layers. Displays 24 (4–5):231–234
23. Romero DB, Schaer M, Zuppiroli L, Cesar B, Francois B (1995) Effects of doping in polymer light-emitting diodes. Appl Phys Lett 67(12):1659–1661
24. Huang F, MacDiarmid AG, Hsieh BR (1997) An iodine-doped polymer light-emitting diode. Appl Phys Lett 71(17):2415–2417
25. Blochwitz J, Pfeiffer M, Fritz T, Leo K (1998) Low voltage organic light emitting diodes featuring doped phthalocyanine as hole transport material. Appl Phys Lett 73(6):729–731
26. Yamamori A, Adachi C, Koyama T, Taniguchi Y (1998) Doped organic light emitting diodes having a 650-nm-thick hole transport layer. Appl Phys Lett 72(17):2147–2149
27. Tang C, VanSlyke S, Chen C (1989) Electroluminescence of doped organic thin films. J Appl Phys 65:3610–3612
28. Naka S, Tamekawa M, Terashita T, Okada H, Anada H, Onnagawa H (1997) Electrical properties of organic electroluminescent devices with aluminium alloy cathode. Synth Met 91(1):129–130
29. Wakimoto T, Fukuda Y, Nagayama K, Yokoi A, Nakada H, Tsuchida M (1997) Organic EL cells using alkaline metal compounds as electron injection materials. IEEE Trans Electron Dev 44(8):1245–1248
30. Ganzorig C, Suga K, Fujihira M (2001) Alkali metal acetates as effective electron injection layers for organic electroluminescent devices. Mat Sci Eng B 85(2–3):140–143
31. Ganzorig C, Fujihira M (2004) Evidence for alkali metal formation at a cathode interface of organic electroluminescent devices by thermal decomposition of alkali metal carboxylates during their vapor deposition. Appl Phys Lett 85(20):4774–4776
32. Stossel M, Staudigel J, Steuber F, Blassing J, Simmerer J, Winnacker A (2000) Space-charge-limited electron currents in 8-hydroxyquinoline aluminum. Appl Phys Lett 76 (1):115–117
33. Brown TM, Friend RH, Millard IS, Lacey DJ, Butler T, Burroughes JH, Cacialli F (2003) Electronic line-up in light-emitting diodes with alkali-halide/metal cathodes. J Appl Phys 93 (10):6159–6172
34. Mori T, Fujikawa H, Tokito S, Taga Y (1998) Electronic structure of 8-hydroxyquinoline aluminum/LiF/Al interface for organic electroluminescent device studied by ultraviolet photoelectron spectroscopy. Appl Phys Lett 73(19):2763–2765
35. Schlaf R, Parkinson BA, Lee PA, Nebesny KW, Jabbour G, Kippelen B, Peyghambarian N, Armstrong NR (1998) Photoemission spectroscopy of LiF coated Al and Pt electrodes. J Appl Phys 84(12):6729–6736
36. Heil H, Steiger J, Karg S, Gastel M, Ortner H, von Seggern H, Stossel M (2001) Mechanisms of injection enhancement in organic light-emitting diodes through an Al/LiF electrode. J Appl Phys 89(1):420–424
37. Mason MG, Tang CW, Hung LS, Raychaudhuri P, Madathil J, Giesen DJ, Yan L, Le QT, Gao Y, Lee ST, Liao LS, Cheng LF, Salaneck WR, dos Santos DA, Bredas JL (2001) Interfacial chemistry of Alq3 and LiF with reactive metals. J Appl Phys 89(5):2756–2765
38. Hung LS, Zhang RQ, He P, Mason G (2002) Contact formation of LiF/Al cathodes in Alq-based organic light-emitting diodes. J Phys D Appl Phys 35(2):103–107
39. Schmitz C, Schmidt HW, Thelakkat M (2000) Lithium-quinolate complexes as emitter and interface materials in organic light-emitting diodes. Chem Mater 12(10):3012–3019

40. Liu ZG, Salata OV, Male N (2002) Improved electron injection in organic LED with lithium quinolate/aluminium cathode. Synth Met 128(2):211–214
41. Fukase A, Kido J (2002) Organic electroluminescent devices having self-doped cathode interface layer. Jpn J Appl Phys 41(3B):L334–L336
42. Liang FS, Chen JS, Wang LX, Ma DG, Jing XB, Wang FS (2003) A hydroxyphenyloxadiazole lithium complex as a highly efficient blue emitter and interface material in organic light-emitting diodes. J Mater Chem 13(12):2922–2926
43. Zheng XY, Wu YZ, Sun RG, Zhu WQ, Jiang XY, Zhang ZL, Xu SH (2005) Efficiency improvement of organic light-emitting diodes using 8-hydroxy-quinolinato lithium as an electron injection layer. Thin Solid Films 478(1–2):252–255
44. Kuwabara Y, Ogawa H, Inada H, Noma N, Shirota Y (1994) Thermally stable multilared organic electroluminescent devices using novel starburst molecules, 4, 4′, 4″-Tri (N-carbazolyl) triphenylamine (TCTA) and 4, 4′, 4″-Tris (3-methylphenylphenylamino) triphenylamine (m-MTDATA), as hole-transport materials. Adv Mater 6(9):677–679
45. Shirota Y (2000) Organic materials for electronic and optoelectronic devices. J Mater Chem 10(1):1–25
46. Adachi C, Tsutsui T, Saito S (1989) Organic electroluminescent device having a hole conductor as an emitting layer. Appl Phys Lett 55(15):1489–1491
47. Shih HT, Lin CH, Shih HH, Cheng CH (2002) High-performance blue electroluminescent devices based on a biaryl. Adv Mater 14(19):1409–1412
48. Adachi C, Baldo MA, Forrest SR, Thompson ME (2000) High-efficiency organic electrophosphorescent devices with tris(2-phenylpyridine)iridium doped into electron-transporting materials. Appl Phys Lett 77(6):904–906
49. Jandke M, Strohriegl P, Berleb S, Werner E, Brutting W (1998) Phenylquinoxaline polymers and low molar mass glasses as electron-transport materials in organic light-emitting diodes. Macromolecules 31(19):6434–6443
50. O'Brien DF, Baldo MA, Thompson ME, Forrest SR (1999) Improved energy transfer in electrophosphorescent devices. Appl Phys Lett 74(3):442–444
51. Baldo MA, Lamansky S, Burrows PE, Thompson ME, Forrest SR (1999) Very high-efficiency green organic light-emitting devices based on electrophosphorescence. Appl Phys Lett 75(1):4–6
52. Adamovich VI, Cordero SR, Djurovich PI, Tamayo A, Thompson ME, D'Andrade BW, Forrest SR (2003) New charge-carrier blocking materials for high efficiency OLEDs. Org Electron 4(2–3):77–87
53. Naka S, Okada H, Onnagawa H, Tsutsui T (2000) High electron mobility in bathophenanthroline. Appl Phys Lett 76(2):197–199
54. D'Andrade BW, Forrest SR, Chwang AB (2003) Operational stability of electrophosphorescent devices containing p and n doped transport layers. Appl Phys Lett 83(19):3858–3860
55. Tamao K, Uchida M, Izumizawa T, Furukawa K, Yamaguchi S (1996) Silole derivatives as efficient electron transporting materials. J Am Chem Soc 118(47):11974–11975
56. Yamaguchi S, Tamao K (1998) Silole-containing sigma- and pi-conjugated compounds. J Chem Soc Dalton 22:3693–3702
57. Su S-J, Chiba T, Takeda T, Kido J (2008) Pyridine-containing triphenylbenzene derivatives with high electron mobility for highly efficient phosphorescent OLEDs. Adv Mater 20 (11):2125–2130
58. Su S-J, Takahashi Y, Chiba T, Takeda T, Kido J (2009) Structure-property relationship of pyridine-containing triphenyl benzene electron-transport materials for highly efficient blue phosphorescent OLEDs. Adv Funct Mater 19(8):1260–1267
59. Tollin G, Kearns DR, Calvin M (1960) Electrical properties of organic solids. I. Kinetics and mechanism of conductivity of metal-free phthalocyanine. J Chem Phys 32:1013
60. Andre J, Simon J, Even R, Boudjema B, Guillaud G, Maitrot M (1987) Molecular semiconductors and junction formation: phthalocyanine derivatives. Synth Met 18(1):683–688

61. Maitrot M, Guillaud G, Boudjema B, André J, Simon J (1986) Molecular material-based junctions: formation of a Schottky contact with metallophthalocyanine thin films doped by the cosublimation method. J Appl Phys 60(7):2396–2400
62. Pfeiffer M, Beyer A, Fritz T, Leo K (1998) Controlled doping of phthalocyanine layers by cosublimation with acceptor molecules: a systematic Seebeck and conductivity study. Appl Phys Lett 73(22):3202–3204
63. Gao WY, Kahn A (2001) Controlled p-doping of zinc phthalocyanine by coevaporation with tetrafluorotetracyanoquinodimethane: a direct and inverse photoemission study. Appl Phys Lett 79(24):4040–4042
64. Blochwitz J, Fritz T, Pfeiffer M, Leo K, Alloway D, Lee P, Armstrong N (2001) Interface electronic structure of organic semiconductors with controlled doping levels. Org Electron 2 (2):97–104
65. Gao ZQ, Mi BX, Xu GZ, Wan YQ, Gong ML, Cheah KW, Chen CH (2008) An organic p-type dopant with high thermal stability for an organic semiconductor. Chem Commun 1:117–119
66. Koech PK, Padmaperuma AB, Wang LA, Swensen JS, Polikarpov E, Darsell JT, Rainbolt JE, Gaspar DJ (2010) Synthesis and Application of 1, 3, 4, 5, 7, 8-Hexafluorotetracyanonaphthoquinodimethane (F6-TNAP): a conductivity dopant for organic light-emitting devices. Chem Mater 22(13):3926–3932
67. Kleemann H, Schuenemann C, Zakhidov AA, Riede M, Lussem B, Leo K (2012) Structural phase transition in pentacene caused by molecular doping and its effect on charge carrier mobility. Org Electron 13(1):58–65
68. Tietze ML, Burtone L, Riede M, Lussem B, Leo K (2012) Fermi level shift and doping efficiency in p-doped small molecule organic semiconductors: a photoelectron spectroscopy and theoretical study. Phys Rev B 86(3):035320
69. Lee JH, Lee J, Kim YH, Yun C, Lussem B, Leo K (2014) Effect of trap states on the electrical doping of organic semiconductors. Org Electron 15(1):16–21
70. Wellmann P, Hofmann M, Zeika O, Werner A, Birnstock J, Meerheim R, He G, Walzer K, Pfeiffer M, Leo K (2005) High-efficiency p-i-n organic light-emitting diodes with long lifetime. J Soc Inf Display 13(5):393–397
71. Blochwitz-Nimoth J, Langguth O, Murano S, He G, Romainczyk T, Birnstock J (2010) PIN-OLEDs for active-matrix-display use. J Soc Inf Display 18(8):596–605
72. Kido J, Matsumoto T (1998) Bright organic electroluminescent devices having a metal-doped electron-injecting layer. Appl Phys Lett 73(20):2866–2868
73. Parthasarathy G, Shen C, Kahn A, Forrest SR (2001) Lithium doping of semiconducting organic charge transport materials. J Appl Phys 89(9):4986–4992
74. Nollau A, Pfeiffer M, Fritz T, Leo K (2000) Controlled n-type doping of a molecular organic semiconductor: naphthalenetetracarboxylic dianhydride (NTCDA) doped with bis (ethylenedithio)-tetrathiafulvalene (BEDT-TTF). J Appl Phys 87(9):4340–4343
75. Tanaka S, Kanai K, Kawabe E, Iwahashi T, Nishi T, Ouchi Y, Seki K (2005) Doping effect of tetrathianaphthacene molecule in organic semiconductors on their interfacial electronic structures studied by UV photoemission spectroscopy. Jpn J Appl Phys 44(6A):3760–3763
76. Chan CK, Amy F, Zhang Q, Barlow S, Marder S, Kahn A (2006) N-type doping of an electron-transport material by controlled gas-phase incorporation of cobaltocene. Chem Phys Lett 431(1–3):67–71
77. Menke T, Wei P, Ray D, Kleemann H, Naab BD, Bao ZA, Leo K, Riede M (2012) A comparison of two air-stable molecular n-dopants for C-60. Org Electron 13(12):3319–3325
78. Lussem B, Tietze ML, Kleemann H, Hossbach C, Bartha JW, Zakhidov A, Leo K (2013) Doped organic transistors operating in the inversion and depletion regime. Nat Commun 4
79. Wang J, Liu J, Huang S, Wu X, Shi X, Chen C, Ye Z, Lu J, Su Y, He G (2013) High efficiency green phosphorescent organic light-emitting diodes with a low roll-off at high brightness. Org Electron 14(11):2854–2858

80. Birnstock J, Canzler T, Hofmann M, Lux A, Murano S, Wellmann P, Werner A (2008) PINOLEDs—improved structures and materials to enhance device lifetime. J Soc Inf Display 16(2):221–229
81. He G, Rothe C, Murano S, Werner A, Zeika O, Birnstock J (2009) White stacked OLED with 38 lm/W and 100, 000-hour lifetime at 1000 cd/m^2 for display and lighting applications. J Soc Inf Display 17(2):159–165
82. Huang JS, Pfeiffer M, Werner A, Blochwitz J, Leo K, Liu SY (2002) Low-voltage organic electroluminescent devices using pin structures. Appl Phys Lett 80(1):139–141
83. Walzer K, Maennig B, Pfeiffer M, Leo K (2007) Highly efficient organic devices based on electrically doped transport layers. Chem Rev 107(4):1233–1271
84. Pfeiffer M, Forrest SR, Leo K, Thompson ME (2002) Electrophosphorescent p-i-n organic light-emitting devices for very-high-efficiency flat-panel displays. Adv Mater 14(22):1633–1636
85. Meerheim R, Walzer K, Pfeiffer M, Leo K (2006) Ultrastable and efficient red organic light emitting diodes with doped transport layers. Appl Phys Lett 89(6):061111
86. Tang CW, VanSlyke SA (1989) Organic electroluminescent diodes. Electroluminescence. In: Proceedings of the Fourth International Workshop, pp 356–357
87. Shi J, Tang C (1997) Doped organic electroluminescent devices with improved stability. Appl Phys Lett 70(13):1665–1667
88. Chen CH, Tang CW, Shi J, Klubek KP (2000) Recent developments in the synthesis of red dopants for Alq(3) hosted electroluminescence. Thin Solid Films 363(1–2):327–331
89. Hamada Y, Kanno H, Tsujioka T, Takahashi H, Usuki T (1999) Red organic light-emitting diodes using an emitting assist dopant. Appl Phys Lett 75(12):1682–1684
90. Liu TH, Iou CY, Wen SW, Chen CH (2003) 4-(Dicyanomethylene)-2-t-butyl-6-(1, 1, 7, 7-tetramethyljulolidyl-9-enyl)-4H-pyran doped red emitters in organic light-emitting devices. Thin Solid Films 441(1–2):223–227
91. Liu TH, Iou CY, Chen CH (2005) Development of highly stable organic electroluminescent devices with a doped co-host emitter system. Curr Appl Phys 5(3):218–221
92. Yang LF, Guan M, Nie DB, Lou B, Liu ZW, Bian ZQ, Bian J, Huang C (2007) Efficient, saturated red electroluminescent devices with modified pyran-containing emitters. Opt Mater 29(12):1672–1679
93. Zhang ZL, Jiang XY, Zhu WQ, Zhang BX, Xu SH (2001) A white organic light emitting diode with improved stability. J Phys D Appl Phys 34(20):3083–3087
94. Yao YS, Zhou QX, Wang XS, Wang Y, Zhang BW (2007) A DCM-type red-fluorescent dopant for high-performance organic electroluminescent devices. Adv Funct Mater 17(1):93–100
95. Okumoto K, Kanno H, Hamada Y, Takahashi H, Shibata K (2006) Green fluorescent organic light-emitting device with external quantum efficiency of nearly 10 %. Appl Phys Lett 89 (16):063504
96. Chen CH, Tang CW, Shi J, Klubek KP (2000) Green organic electroluminescent devices. US Patent 6, 020, 078
97. Chen CH, Tang CW (2001) Efficient green organic light-emitting diodes with stericly hindered coumarin dopants. Appl Phys Lett 79(22):3711–3713
98. Lee MT, Yen CK, Yang WP, Chen HH, Liao CH, Tsai CH, Chen CH (2004) Efficient green coumarin dopants for organic light-emitting devices. Org Lett 6(8):1241–1244
99. Murayama R, Wakimoto T, Nakada H, Nomura M, Sato G (1993) Electroluminescent device. US Patent 5, 227, 252
100. Murata H, Merritt CD, Inada H, Shirota Y, Kafafi ZH (1999) Molecular organic light-emitting diodes with temperature-independent quantum efficiency and improved thermal durability. Appl Phys Lett 75(21):3252–3254
101. Tao YT, Balasubramaniam E, Danel A, Tomasik P (2000) Dipyrazolopyridine derivatives as bright blue electroluminescent materials. Appl Phys Lett 77(7):933–935
102. Shi J, Tang CW (1997) Organic electroluminescent devices with high operational stability. US Patent 5, 593, 788

103. Wakimoto T, Yonemoto Y, Funaki J, Tsuchida M, Murayama R, Nakada H, Matsumoto H, Yamamura S, Nomura M (1997) Stability characteristics of quinacridone and coumarin molecules as guest dopants in the organic LEDs. Synth Met 91(1–3):15–19
104. Wang LD, Gao YD, Wei P, Qiu Y (2004) Novel structure organic light-emitting diodes with high performance. In: SID symposium digest of technical papers 2004, vol 1, pp 703–705. Wiley Online Library
105. Ricks ML, Vargas JR, Klubek KP, Jarikov VV, Liao LS, Helber MJ, Begley WJ, Hatwar TK, Conley SR, Cosimbescu L (2007) Efficient, long-lifetime OLED host and dopant formulations for full-color displays. In: SID symposium digest of technical papers 2007, vol 1, pp 830–833. Wiley Online Library
106. Adachi C, Tsutsui T, Saito S (1990) Blue light-emitting organic electroluminescent devices. Appl Phys Lett 56(9):799–801
107. Shi JM, Tang CW (2002) Anthracene derivatives for stable blue-emitting organic electroluminescence devices. Appl Phys Lett 80(17):3201–3203
108. Shi J (2001) Method of using predoped materials for making an organic light-emitting device. EP Patent 1, 156, 536
109. Zhang P, Xia BH, Sun YH, Yang B, Tian WJ, Wang Y, Zhang G (2006) Electronic structures and optical properties of two anthracene derivatives. Chin Sci Bull 51(20):2444–2450
110. Kan Y, Wang LD, Gao YD, Duan L, Wu GS, Qiu Y (2004) Highly efficient blue electroluminescence based on a new anthracene derivative. Synth Met 141(3):245–249
111. Du PW, Eisenberg R (2010) Energy upconversion sensitized by a platinum(II) terpyridyl acetylide complex. Chem Sci 1(4):502–506
112. Tao SL, Xu SD, Zhang XH (2006) Efficient blue organic light-emitting devices based on novel anthracence derivatives with pronounced thermal stability and excellent film-forming property. Chem Phys Lett 429(4–6):622–627
113. Lee MT, Wu YS, Chen HH, Tsai CH, Liao CH, Chen CH (2004) Efficient blue organic electroluminescent devices based on a stable blue host material. In: SID symposium digest of technical papers 2004, vol 1, pp 710–713. Wiley Online Library
114. Li MT, Li WL, Su WM, Zang FX, Chu B, Xin Q, Bi DF, Li B, Yu TZ (2008) High efficiency and color saturated blue electroluminescence by using 4, 4'-bis N-(1-naphthyl)-N-phenylamino biphenyl as the thinner host and hole-transporter. Solid State Electron 52 (1):121–125
115. Lee MT, Chen HH, Liao CH, Tsai CH, Chen CH (2004) Stable styrylamine-doped blue organic electroluminescent device based on 2-methyl-9, 10-di(2-naphthyl)anthracene. Appl Phys Lett 85(15):3301–3303
116. Jeon S-O, Jeon Y-M, Kim J-W, Lee C-W, Gong M-S (2007) A blue organic emitting diode derived from new styrylamine type dopant materials. Synth Met 157(13):558–563
117. Ho MH, Wu YS, Wen SW, Lee MT, Chen TM, Chen CH, Kwok KC, So SK, Yeung KT, Cheng YK, Gao ZQ (2006) Highly efficient deep blue organic electroluminescent device based on 1-methyl-9, 10-di(1-naphthyl)anthracene. Appl Phys Lett 89(25):252903
118. Ho MH, Balaganesan B, Chen CH (2012) Blue fluorescence and bipolar transport materials based on anthracene and their application in OLEDs. Isr J Chem 52(6):484–495
119. Kim K-S, Lee HS, Jeon Y-M, Kim J-W, Lee C-W, Gong M-S (2009) Blue light-emitting diodes from 2-(10-naphthylanthracene)-spiro [fluorene-7, 9′-benzofluorene] host material. Dyes Pigm 81(3):174–179
120. Hosokawa C, Higashi H, Nakamura H, Kusumoto T (1995) Highly efficient blue electroluminescence from a distyrylarylene emitting layer with a new dopant. Appl Phys Lett 67(26):3853–3855
121. Li J, Xu Z, Zhang F, Zhao S, Song D, Zhu H, Song J, Wang Y, Xu X (2010) Electroplex emission of the blend film of PVK and DPVBi. Solid State Electron 54(4):349–352
122. Wen SW, Lee MT, Chen CH (2005) Recent development of blue fluorescent OLED materials and devices. J Display Technol 1(1):90–99
123. Zheng X, Zhu W, Wu Y, Jiang X, Sun R, Zhang Z, Xu S (2003) A white OLED based on DPVBi blue light emitting host and DCJTB red dopant. Displays 24(3):121–124

124. Yang S-H, Hong B-C, Huang S-F (2009) Luminescence enhancement and emission color adjustment of white organic light-emitting diodes with quantum-well-like structures. J Appl Phys 105(11):113105
125. Lee SJ, Park JS, Yoon KJ, Kim YI, Jin SH, Kang SK, Gal YS, Kang S, Lee JY, Kang JW, Lee SH, Park HD, Kim JJ (2008) High-efficiency deep-blue light-emitting diodes based on phenylquinoline/carbazole-based compounds. Adv Funct Mater 18(24):3922–3930
126. Wu YZ, Zheng XY, Zhu WQ, Sun RG, Jiang XY, Zhang ZL, Xu SH (2003) Highly efficient pure blue electroluminescence from 1, 4-bis[2-(3-N-ethylcarbazoryl)vinyl]benzene. Appl Phys Lett 83(24):5077–5079
127. Tyagi P, Srivastava R, Kumar A, Tuli S, Kamalasanan M (2013) Effect of doping of cesium carbonate on electron transport in Tris (8-hydroxyquinolinato) aluminum. Org Electron 14 (5):1391–1395
128. Ali TA, Jones GW, Howard WE (2004) Dual doped high Tg white organic light emitting devices on silicon. In: SID Symposium Digest of Technical Papers 2004, vol 1, pp 1012–1015. Wiley Online Library
129. Zhang M, Wang F, Wei N, Zhou P, Peng K, Yu J, Wang Z, Wei B (2013) High color rending index and high-efficiency white organic light-emitting diodes based on the control of red phosphorescent dye-doped hole transport layer. Opt Express 21(101):A173–A178
130. Cariati E, Dragonetti C, Lucenti E, Nisic F, Righetto S, Roberto D, Tordin E (2014) An acido-triggered reversible luminescent and nonlinear optical switch based on a substituted styrylpyridine: EFISH measurements as an unusual method to reveal a protonation–deprotonation NLO contrast. Chem Commun 50(13):1608–1610
131. Wee KR, Han WS, Kim JE, Kim AL, Kwon S, Kang SO (2011) Asymmetric anthracene-based blue host materials: synthesis and electroluminescence properties of 9-(2-naphthyl)-10-arylanthracenes. J Mater Chem 21(4):1115–1123
132. Hosokawa C, Sakamoto S, Kusumoto T Organic electroluminescence device. US Patent 5, 389, 444
133. Jiang XY, Zhang ZL, Zheng XY, Wu YZ, Xu SH (2001) A blue organic emitting diode from anthracene derivative. Thin Solid Films 401(1–2):251–254
134. Kondakov DY (2009) Role of triplet-triplet annihilation in highly efficient fluorescent devices. J Soc Inf Display 17(2):137–144
135. Kondakov DY, Pawlik TD, Hatwar TK, Spindler JP (2009) Triplet annihilation exceeding spin statistical limit in highly efficient fluorescent organic light-emitting diodes. J Appl Phys 106(12):124510
136. Endo A, Sato K, Yoshimura K, Kai T, Kawada A, Miyazaki H, Adachi C (2011) Efficient up-conversion of triplet excitons into a singlet state and its application for organic light emitting diodes. Appl Phys Lett 98(8):083302
137. Wu SH, Aonuma M, Zhang QS, Huang SP, Nakagawa T, Kuwabara K, Adachi C (2014) High-efficiency deep-blue organic light-emitting diodes based on a thermally activated delayed fluorescence emitter. J Mater Chem C 2(3):421–424
138. Tanaka H, Shizu K, Nakanotani H, Adachi C (2013) Twisted intramolecular charge transfer state for long-wavelength thermally activated delayed fluorescence. Chem Mater 25 (18):3766–3771
139. Serevicius T, Nakagawa T, Kuo MC, Cheng SH, Wong KT, Chang CH, Kwong RC, Xia S, Adachi C (2013) Enhanced electroluminescence based on thermally activated delayed fluorescence from a carbazole-triazine derivative. Phys Chem Chem Phys 15(38):15850–15855
140. Masui K, Nakanotani H, Adachi C (2013) Analysis of exciton annihilation in high-efficiency sky-blue organic light-emitting diodes with thermally activated delayed fluorescence. Org Electron 14(11):2721–2726
141. Li J, Nakagawa T, MacDonald J, Zhang QS, Nomura H, Miyazaki H, Adachi C (2013) Highly efficient organic light-emitting diode based on a hidden thermally activated delayed fluorescence channel in a heptazine derivative. Adv Mater 25(24):3319–3323

142. Lee J, Shizu K, Tanaka H, Nomura H, Yasuda T, Adachi C (2013) Oxadiazole- and triazole-based highly-efficient thermally activated delayed fluorescence emitters for organic light-emitting diodes. J Mater Chem C 1(30):4599–4604
143. Ishimatsu R, Matsunami S, Shizu K, Adachi C, Nakano K, Imato T (2013) Solvent effect on thermally activated delayed fluorescence by 1, 2, 3, 5-Tetrakis(carbazol-9-yl)-4, 6-dicyanobenzene. J Phys Chem A 117(27):5607–5612
144. Uoyama H, Goushi K, Shizu K, Nomura H, Adachi C (2012) Highly efficient organic light-emitting diodes from delayed fluorescence. Nature 492(7428):234–238
145. Komino T, Nomura H, Koyanagi T, Adachi C (2013) Suppression of efficiency roll-off characteristics in thermally activated delayed fluorescence based organic light-emitting diodes using randomly oriented host molecules. Chem Mater 25(15):3038–3047
146. Yao L, Yang B, Ma Y (2014) Progress in next-generation organic electroluminescent materials: material design beyond exciton statistics. Sci China Chem 57(3):335–345
147. Tanaka D, Agata Y, Takeda T, Watanabe S, Kido J (2007) High luminous efficiency blue organic light-emitting devices using high triplet excited energy materials. Jpn J Appl Phys 46 (4–7):L117–L119
148. Sasabe H, Gonmori E, Chiba T, Li Y-J, Tanaka D, Su S-J, Takeda T, Pu Y-J, Nakayama K-I, Kido J (2008) Wide-energy-gap electron-transport materials containing 3, 5-dipyridylphenyl moieties for an ultra high efficiency blue organic light-emitting device. Chem Mater 20 (19):5951–5953
149. Chopra N, Lee J, Zheng Y, Eom SH, Xue JG, So F (2008) High efficiency blue phosphorescent organic light-emitting device. Appl Phys Lett 93(14):143307
150. Kwong RC, Sibley S, Dubovoy T, Baldo M, Forrest SR, Thompson ME (1999) Efficient, saturated red organic light emitting devices based on phosphorescent platinum(II) porphyrins. Chem Mater 11(12):3709–3713
151. Adachi C, Baldo MA, Forrest SR, Lamansky S, Thompson ME, Kwong RC (2001) High-efficiency red electrophosphorescence devices. Appl Phys Lett 78(11):1622–1624
152. Hay PJ (2002) Theoretical studies of the ground and excited electronic states in cyclometalated phenylpyridine Ir(III) complexes using density functional theory. J Phys Chem A 106(8):1634–1641
153. Tamayo AB, Alleyne BD, Djurovich PI, Lamansky S, Tsyba I, Ho NN, Bau R, Thompson ME (2003) Synthesis and characterization of facial and meridional tris-cyclometalated iridium(III) complexes. J Am Chem Soc 125(24):7377–7387
154. Tsuboyama A, Iwawaki H, Furugori M, Mukaide T, Kamatani J, Igawa S, Moriyama T, Miura S, Takiguchi T, Okada S, Hoshino M, Ueno K (2003) Homoleptic cyclometalated iridium complexes with highly efficient red phosphorescence and application to organic light-emitting diode. J Am Chem Soc 125(42):12971–12979
155. Su YJ, Huang HL, Li CL, Chien CH, Tao YT, Chou PT, Datta S, Liu RS (2003) Highly efficient red electrophosphorescent devices based on iridium isoquinoline complexes: remarkable external quantum efficiency over a wide range of current. Adv Mater 15(11):884–888
156. Tsuji T, Kawami S, Miyaguchi S, Naijo T, Yuki T, Matsuo S, Miyazaki H (2005) Red-phosphorescent OLEDs employing bis(8-quinolinolato)phenolato-aluminum(III) complexes as emission-layer hosts. J Soc Inf Display 13(2):117–121
157. Jiang CY, Yang W, Peng JB, Xiao S, Cao Y (2004) High-efficiency, saturated red-phosphorescent polymer light-emitting diodes based on conjugated and non-conjugated polymers doped with an Ir complex. Adv Mater 16(6):537–541
158. Yang XH, Muller DC, Neher D, Meerholz K (2006) Highly efficient polymeric electrophosphorescent diodes. Adv Mater 18(7):948–954
159. Tao Y, Wang Q, Yang C, Wang Q, Zhang Z, Zou T, Qin J, Ma D (2008) A simple carbazole/oxadiazole hybrid molecule: an excellent bipolar host for green and red phosphorescent OLEDs. Angew Chem 120(42):8224–8227
160. Yang CH, Tai CC, Sun IW (2004) Synthesis of a high-efficiency red phosphorescent emitter for organic light-emitting diodes. J Mater Chem 14(6):947–950

161. Okada S, Okinaka K, Iwawaki H, Furugori M, Hashimoto M, Mukaide T, Kamatani J, Igawa S, Tsuboyama A, Takiguchi T, Ueno K (2005) Substituent effects of iridium complexes for highly efficient red OLEDs. Dalton Trans 9:1583–1590
162. Duan JP, Sun PP, Cheng CH (2003) New iridium complexes as highly efficient orange-red emitters in organic light-emitting diodes. Adv Mater 15(3):224–228
163. Li CX, Sun PP, Yan LJ, Pan Y, Cheng CH (2008) Synthesis and electroluminescent properties of Ir complexes with benzo[c] acridine or 5, 6-dihydro-benzo[c]acridine ligands. Thin Solid Films 516(18):6186–6190
164. Xu M, Wang G, Zhou R, An Z, Zhou Q, Li W (2007) Tuning iridium (III) complexes containing 2-benzo [b] thiophen-2-yl-pyridine based ligands in the red region. Inorg Chim Acta 360(10):3149–3154
165. Mäkinen A, Hill I, Kafafi Z (2002) Vacuum level alignment in organic guest-host systems. J Appl Phys 92(3):1598–1603
166. Chang S-C, He G, Chen F-C, Guo T-F, Yang Y (2001) Degradation mechanism of phosphorescent-dye-doped polymer light-emitting diodes. Appl Phys Lett 79(13):2088–2090
167. Brooks J, Babayan Y, Lamansky S, Djurovich PI, Tsyba I, Bau R, Thompson ME (2002) Synthesis and characterization of phosphorescent cyclometalated platinum complexes. Inorg Chem 41(12):3055–3066
168. Xu ML, Li MT, Hong ZR, Li WL, An ZW, Zhou Q (2006) Highly efficient red electrophosphorescent device based on a new iridium complex with trifluoromethyl-substituted 2-benzo[b]thiophen-2-yl-pyridine ligand. Opt Mater 28(8–9):1025–1028
169. Chin C-L, Chen WC, Cheng K-L Organic metal complexes. US Patent 7, 799, 918
170. Gao J, You H, Fang JF, Ma DG, Wang LX, Jing XB, Wang FS (2005) Pure red electrophosphorescent organic light-emitting diodes based on a new iridium complex. Synth Met 155(1):168–171
171. Ding JQ, Gao J, Fu Q, Cheng YX, Ma DG, Wang LX (2005) Highly efficient phosphorescent bis-cyclometalated iridium complexes based on quinoline ligands. Synth Met 155(3):539–548
172. Thomas KRJ, Velusamy M, Lin JT, Chien CH, Tao YT, Wen YS, Hu YH, Chou PT (2005) Efficient red-emitting cyclometalated iridium(III) complexes containing lepidine-based ligands. Inorg Chem 44(16):5677–5685
173. Niu YH, Tung YL, Chi Y, Shu CF, Kim JH, Chen BQ, Luo JD, Carty AJ, Jen AKY (2005) Highly efficient electrophosphorescent devices with saturated red emission from a neutral osmium complex. Chem Mater 17(13):3532–3536
174. Tung YL, Wu PC, Liu CS, Chi Y, Yu JK, Hu YH, Chou PT, Peng SM, Lee GH, Tao Y, Carty AJ, Shu CF, Wu FI (2004) Highly efficient red phosphorescent osmium(II) complexes for OLED applications. Organometallics 23(15):3745–3748
175. Wu FI, Shih PI, Tseng YH, Chen GY, Chien CH, Shu CF, Tung YL, Chi Y, Jen AKY (2005) Highly efficient red-electrophosphorescent devices based on polyfluorene copolymers containing charge-transporting pendant units. J Phys Chem B 109(29):14000–14005
176. Wu CH, Shih PI, Shu CF, Chi Y (2008) Highly efficient red organic light-emitting devices based on a fluorene-triphenylamine host doped with an Os(II) phosphor. Appl Phys Lett 92 (23):233303
177. Chien CH, Hsu FM, Shu CF, Chi Y (2009) Efficient red electrophosphorescence from a fluorene-based bipolar host material. Org Electron 10(5):871–876
178. Male NAH, Salata OV, Christou V (2002) Enhanced electroluminescent efficiency from spin-coated europium(III) organic light-emitting device. Synth Met 126(1):7–10
179. Tung YL, Lee SW, Chi Y, Chen LS, Shu CF, Wu FI, Carty AJ, Chou PT, Peng SM, Lee GM (2005) Organic light-emitting diodes based on charge-neutral Ru-II phosphorescent emitters. Adv Mater 17(8):1059–1064
180. Li F, Zhang M, Feng J, Cheng G, Wu ZJ, Ma YG, Liu SY, Sheng JC, Lee ST (2003) Red electrophosphorescence devices based on rhenium complexes. Appl Phys Lett 83(2):365–367

181. Zhou GJ, Wong WY, Yao B, Xie ZY, Wang LX (2007) Triphenylamine-dendronized pure red iridium phosphors with superior OLED efficiency/color purity trade-offs. Angew Chem Int Edit 46(7):1149–1151
182. Jiang XZ, Jen AKY, Carlson B, Dalton LR (2002) Red electrophosphorescence from osmium complexes. Appl Phys Lett 80(5):713–715
183. Zheng YX, Liang YJ, Zhang HJ, Lin Q, Chuan G, Wang SB (2002) Red electroluminescent device with europium 1, 1, 1-trifluoroacetylacetonate complex as emissive center. Mater Lett 53(1–2):52–56
184. Tung YL, Chen LS, Chi Y, Chou PT, Cheng YM, Li EY, Lee GH, Shu CF, Wu TI, Carty AJ (2006) Orange and red organic light-emitting devices employing neutral Ru(II) emitters: rational design and prospects for color tuning. Adv Funct Mater 16(12):1615–1626
185. Fukase A, Dao KLT, Kido J (2002) High-efficiency organic electroluminescent devices using iridium complex emitter and arylamine-containing polymer buffer layer. Polym Adv Technol 13(8):601–604
186. He G, Pfeiffer M, Leo K, Hofmann M, Birnstock J, Pudzich R, Salbeck J (2004) High-efficiency and low-voltage pin electrophosphorescent organic light-emitting diodes with double-emission layers. Appl Phys Lett 85(17):3911–3913
187. Mikami A, Koyanagi T (2009) High efficiency 200-lm/W green light emitting organic devices prepared on high-index of refraction substrate. In: SID symposium digest of technical papers 2009, vol 1, pp 907–910. Wiley Online Library
188. Ho MH, Balaganesan B, Chu TY, Chen TM, Chen CH (2008) A morphologically stable host material for efficient phosphorescent green and red organic light emitting devices. Thin Solid Films 517(2):943–947
189. Tsuzuki T, Tokito S (2009) Highly efficient and stable organic light-emitting diode using 4, 4 (')-bis(N-carbazolyl)-9, 9(')-spirobifluorene as a thermally stable host material. Appl Phys Lett 94(3):03302
190. Kang JW, Lee DS, Park HD, Kim JW, Jeong WI, Park YS, Lee SH, Go K, Lee JS, Kim JJ (2008) A host material containing tetraphenylsilane for phosphorescent OLEDs with high efficiency and operational stability. Org Electron 9(4):452–460
191. Tanaka D, Sasabe H, Li Y-J, Su S-J, Takeda T, Kido J (2007) Ultra high efficiency green organic light-emitting devices. Jpn J Appl Phys 46(1–3):L10–L12
192. Sasabe H, Chiba T, Su S-J, Pu Y-J, K-i Nakayama, Kido J (2008) 2-Phenylpyrimidine skeleton-based electron-transport materials for extremely efficient green organic light-emitting devices. Chem Commun 44:5821–5823
193. Su S-J, Tanaka D, Li Y-J, Sasabe H, Takeda T, Kido J (2008) Novel four-pyridylbenzene-armed biphenyls as electron-transport materials for phosphorescent OLEDs. Org Lett 10 (5):941–944
194. Su S-J, Gonmori E, Sasabe H, Kido J (2008) Highly efficient organic blue-and white-light-emitting devices having a carrier- and exciton-confining structure for reduced efficiency roll-off. Adv Mater 20(21):4189–4194
195. Helander MG, Wang ZB, Qiu J, Greiner MT, Puzzo DP, Liu ZW, Lu ZH (2011) Chlorinated indium tin oxide electrodes with high work function for organic device compatibility. Science 332(6032):944–947
196. Wong WY, Ho CL, Gao ZQ, Mi BX, Chen CH, Cheah KW, Lin Z (2006) Multifunctional iridium complexes based on carbazole modules as highly efficient electrophosphors. Angew Chem Int Ed 45(46):7800–7803
197. Ho CL, Wang Q, Lam CS, Wong WY, Ma DG, Wang LX, Gao ZQ, Chen CH, Cheah KW, Lin ZY (2009) Phosphorescence color tuning by ligand, and substituent effects of multifunctional iridium(III) cyclometalates with 9-arylcarbazole moieties. Chemistry-An Asian J 4(1):89–103
198. Iguchi N, Pu Y-J, Nakayama K-I, Yokoyama M, Kido J (2009) Synthesis, photoluminescence and electroluminescence properties of iridium complexes with bulky carbazole dendrons. Org Electron 10(3):465–472

199. Zhou GJ, Wang Q, Ho CL, Wong WY, Ma DG, Wang LX, Lin ZY (2008) Robust tris-cyclometalated iridium(III) phosphors with ligands for effective charge carrier injection/transport: synthesis, redox, photophysical, and electrophosphorescent behavior. Chemistry-an Asian J 3(10):1830–1841
200. Zhou GJ, Ho CL, Wong WY, Wang Q, Ma DG, Wang LX, Lin ZY, Marder TB, Beeby A (2008) Manipulating charge-transfer character with electron-withdrawing main-group moieties for the color tuning of iridium electrophosphors. Adv Funct Mater 18(3):499–511
201. Jung SO, Zhao Q, Park JW, Kim SO, Kim YH, Oh HY, Kim J, Kwon SK, Kang Y (2009) A green emitting iridium(III) complex with narrow emission band and its application to phosphorescence organic light-emitting diodes (OLEDs). Org Electron 10(6):1066–1073
202. Liu ZW, Bian ZQ, Ming L, Ding F, Shen HY, Nie DB, Huang CH (2008) Green and blue-green phosphorescent heteroleptic iridium complexes containing carbazole-functionalized beta-diketonate for non-doped organic light-emitting diodes. Org Electron 9(2):171–182
203. Ma B, Knowles DB, Brown CS, Murphy D, Thompson ME Organic light emitting materials and devices. US Patent 6, 687, 266
204. Chan SC, Chan MC, Wang Y, Che CM, Cheung KK, Zhu N (2001) Organic light-emitting materials based on bis (arylacetylide) platinum (II) complexes bearing substituted bipyridine and phenanthroline ligands: photo-and electroluminescence from 3MLCT excited states. Chem-eur J 7(19):4180–4190
205. Sotoyama W, Satoh T, Sawatari N, Inoue H (2005) Efficient organic light-emitting diodes with phosphorescent platinum complexes containing (NCN)-C-boolean AND-N-boolean AND 4-coordinating tridentate ligand. Appl Phys Lett 86(15):153505
206. Cocchi M, Virgili D, Fattori V, Rochester DL, Williams JAG (2007) N boolean AND C boolean AND N-coordinated platinum(II) complexes as phosphorescent emitters in high-performance organic light-emitting devices. Adv Funct Mater 17(2):285–289
207. Zhao WQ, Ran GZ, Liu ZW, Bian ZQ, Sun K, Xu WJ, Huang CH, Qin GG (2008) Combination of passivated Si anode with phosphor doped organic to realize highly efficient Si-based electroluminescence. Opt Express 16(7):5158–5163
208. Chang SY, Kavitha J, Li SW, Hsu CS, Chi Y, Yeh YS, Chou PT, Lee GH, Carty AJ, Tao YT, Chien CH (2006) Platinum(II) complexes with pyridyl azolate-based chelates: synthesis, structural characterization, and tuning of photo- and electrophosphorescence. Inorg Chem 45 (1):137–146
209. Adachi C, Kwong RC, Djurovich P, Adamovich V, Baldo MA, Thompson ME, Forrest SR (2001) Endothermic energy transfer: a mechanism for generating very efficient high-energy phosphorescent emission in organic materials. Appl Phys Lett 79(13):2082–2084
210. Holmes RJ, D'Andrade BW, Forrest SR, Ren X, Li J, Thompson ME (2003) Efficient, deep-blue organic electrophosphorescence by guest charge trapping. Appl Phys Lett 83(18):3818–3820
211. Tsuboi T, Murayama H, Yeh S-J, Wu M-F, Chen C-T (2008) Photoluminescence characteristics of blue phosphorescent Ir3+-compounds FIrpic and FIrN4 doped in mCP and SimCP. Opt Mater 31(2):366–371
212. Yeh SJ, Wu MF, Chen CT, Song YH, Chi Y, Ho MH, Hsu SF, Chen CH (2005) New dopant and host materials for blue-light-emitting phosphorescent organic electroluminescent devices. Adv Mater 17(3):285–289
213. Chiu YC, Hung JY, Chi Y, Chen CC, Chang CH, Wu CC, Cheng YM, Yu YC, Lee GH, Chou PT (2009) En route to high external quantum efficiency (similar to 12 %), organic true-blue-light-emitting diodes employing novel design of iridium (III) phosphors. Adv Mater 21 (21):2221–2225
214. Ragni R, Plummer EA, Brunner K, Hofstraat JW, Babudri F, Farinola GM, Naso F, De Cola L (2006) Blue emitting iridium complexes: synthesis, photophysics and phosphorescent devices. J Mater Chem 16(12):1161–1170
215. Kim SH, Jang J, Lee SJ, Lee JY (2008) Deep blue phosphorescent organic light-emitting diodes using a Si based wide bandgap host and an Ir dopant with electron withdrawing substituents. Thin Solid Films 517(2):722–726

216. Jeon SO, Jang SE, Son HS, Lee JY (2011) External quantum efficiency above 20 % in deep blue phosphorescent organic light-emitting diodes. Adv Mater 23(12):1436–1441
217. Seo HJ, Yoo KM, Song M, Park JS, Jin SH, Kim YI, Kim JJ (2010) Deep-blue phosphorescent iridium complexes with picolinic acid N-oxide as the ancillary ligand for high efficiency organic light-emitting diodes. Org Electron 11(4):564–572
218. Lee SJ, Park KM, Yang K, Kang Y (2009) Blue phosphorescent Ir(III) complex with high color purity: fac-Tris(2', 6'-difluoro-2, 3'-bipyridinato-N, C-4')iridium(III). Inorg Chem 48 (3):1030–1037
219. Lo SC, Shipley CP, Bera RN, Harding RE, Cowley AR, Burn PL, Samuel IDW (2006) Blue phosphorescence from iridium(III) complexes at room temperature. Chem Mater 18 (21):5119–5129
220. Holmes RJ, Forrest SR, Sajoto T, Tamayo A, Djurovich PI, Thompson ME, Brooks J, Tung YJ, D'Andrade BW, Weaver MS, Kwong RC, Brown JJ (2005) Saturated deep blue organic electrophosphorescence using a fluorine-free emitter. Appl Phys Lett 87(24):243507
221. Chang CF, Cheng YM, Chi Y, Chiu YC, Lin CC, Lee GH, Chou PT, Chen CC, Chang CH, Wu CC (2008) Highly efficient blue-emitting iridium(III) carbene complexes and phosphorescent OLEDs. Angew Chem Int Edit 47(24):4542–4545
222. Takizawa S, Echizen H, Nishida J, Tsuzuki T, Tokito S, Yamashita Y (2006) Finely-tuned blue-phosphorescent iridium complexes based on 2-phenylpyridine derivatives and application to polymer organic light-emitting device. Chem Lett 35(7):748–749
223. Yook KS, Jeon SO, Joo CW, Lee JY (2009) High efficiency deep blue phosphorescent organic light-emitting diodes. Org Electron 10(1):170–173
224. Chopra N, Lee J, Zheng Y, Eom SH, Xue JE, So F (2009) Effect of the charge balance on high-efficiency blue-phosphorescent organic light-emitting diodes. ACS Appl Mater Inter 1 (6):1169–1172
225. D'Andrade BW, Brooks J, Adamovich V, Thompson ME, Forrest SR (2002) White light emission using triplet excimers in electrophosphorescent organic light-emitting devices. Adv Mater 14(15):1032–1036
226. Yang XH, Wang ZX, Madakuni S, Li J, Jabbour GE (2008) Efficient blue- and white-emitting electrophosphorescent devices based on platinum(II) [1, 3-difluoro-4, 6-di(2-pyridinyl)benzene] chloride. Adv Mater 20(12):2405–2409

Chapter 7
Conjugated Polymer Electroluminescent Materials

Xing Guan, Shenjian Liu and Fei Huang

Abstract Polymer light-emitting diodes (PLEDs) are promising devices for use in large-area, flat-panel displays and next-generation solid-state lighting because of the advantages of ease of fabrication and low production cost for the large-size devices. This chapter provides an overview of the recent development of polymer electroluminescent materials as the active layer in PLEDs. These polymer electroluminescent materials are reviewed according to the classification of traditional electroluminescent polymers, luminescent polymers based on dopant/host systems, hyperbranched polymers, and supramolecular luminescent polymers. Emphasis is placed on the relationships between molecular structure and device performance. Finally, some scientific problems and developing trends on PLEDs are discussed.

Keywords Conjugated polymers · Electroluminescence · Polymer light-emitting diodes

7.1 Introduction

Organic light-emitting diodes (OLEDs) have drawn intense attention in both scientific and industrial communities in the past decade because of their potential applications in large-area, flat-panel displays and next-generation solid-state lighting [1–3]. Considerable progress has been made in this area, and vacuum deposited small molecule OLEDs have been successfully commercialized [4, 5]. Compared to small-molecular-based OLEDs, polymer light-emitting diodes (PLEDs), which can be processed by spin-coating or ink-jet printing techniques via solution processing [6, 7], have also attracted great attention because of the advantages of ease of

X. Guan · S. Liu · F. Huang (✉)
State Key Laboratory of Luminescent Materials and Devices, Institute of Polymer Optoelectronic Materials and Devices, South China University of Technology, Guangzhou 510640, People's Republic of China
e-mail: msfhuang@scut.edu.cn

Y. Li (ed.), *Organic Optoelectronic Materials*,
Lecture Notes in Chemistry 91, DOI 10.1007/978-3-319-16862-3_7

fabrication and low production cost for the large-size devices [8, 9]. For high performance PLEDs, highly efficient conjugated polymer electroluminescent materials with good color purity are critical. As a consequence, various strategies toward high performance polymer electroluminescent materials have been developed [9, 10]. In this chapter, we first present a brief description of the basic principles of electroluminescence (EL) and PLEDs and then introduce the progress of luminescent polymers in the recent 5 years.

7.1.1 Electroluminescence and PLEDs

EL is the result of radiative recombination of electrons and holes injected into a semiconductor in the presence of an external circuit. This electron-hole recombination leads to the formation of singlet and triplet excitons in a ratio of 1:3 [11]. The relaxation of the singlet excitons results in emission of light (fluorescence), whereas triplet excitons do not produce fluorescence in most cases, other than by indirect processes such as triplet-triplet annihilation, or by phosphorescence [12]. The color of the emission is determined by the value of the highest occupied molecular orbital (HOMO)–lowest unoccupied molecular orbital (LUMO) energy gap of a semiconductor, which for visible light (380–780 nm) corresponds to 1.5–3.2 eV.

PLED is an EL device that uses a conjugated polymer as the active layer. In a simple, single layer PLED, a thin film of an emissive polymer is sandwiched between two electrodes. Generally, indium tin oxide (ITO) on a glass or polymer substrate is chosen as the (transparent) anode, and the cathode consists of a vacuum-deposited metal layer. To get a better performance, most PLEDs need a multilayer device structure, where a hole-transporting/injection layer (HTL) and an electron-transporting/injection layer (ETL) are used to facilitate hole and electron injection/transport from anode and cathode, respectively, to maximize the device performance, resulting in a generally used multilayer device structure: ITO anode/HTL/emissive layer (EML)/ETL/metallic electrode cathode (Fig. 7.1) [13, 14]. During device operation, a voltage is applied across the PLED such that the anode is positive with respect to the cathode. The electrons are injected into the LUMO of the EML from the cathode and the holes are injected into the HOMO of the EML from the anode. The electrons and holes recombine in the EML and lead to light emission. Therefore, balancing the charge injection and transporting plays an important role in achieving high EL efficiency PLEDs.

With regards to materials design, many factors should be considered in designing a new light-emitting polymer for high performance PLEDs. For examples, proper HOMO and LUMO energy levels which match the desired electrode materials give better charge carrier transporting/injection properties, and lead to an improvement in the charge balance of electrons and holes in PLEDs; a decent solubility affords facilities for fabricating PLED devices through spin-coating; and a strong intermolecular interaction brings on an enhancement for transporting of

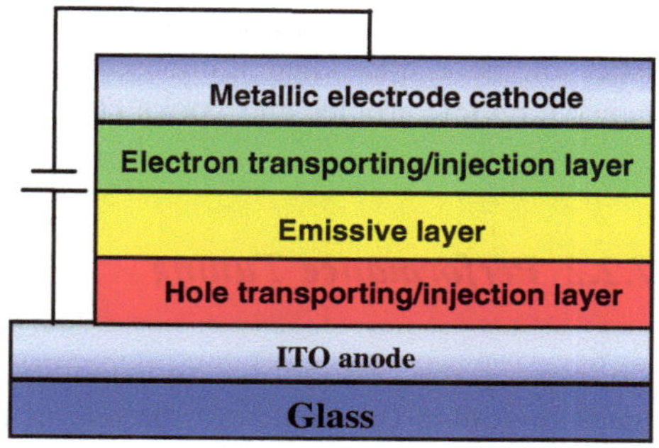

Fig. 7.1 Schematic drawing of a typical multilayer PLED device

charges and/or emission quenching which could influence the performance of PLEDs. The recent progress on newly developed conjugated EL polymers and their application in PLEDs is discussed in the following.

7.2 Conjugated Electroluminescent Polymers and Performance Tuning

7.2.1 Early Efforts

Conjugated polymers are organic semiconductors with delocalized π-molecular orbitals along the polymeric chain. In the past decade a variety of conjugated polymers such as poly(*p*-phenylene vinylene) (PPV), poly(*p*-phenylene) (PPP), polyfluorene (PF), polycarbazole (PCz), and polythiophene (PT) have been widely used as electroluminescent polymers in PLEDs. PPV and its soluble derivatives, which have relatively small optical band gaps and emit green to orange fluorescence, are among the most widely studied luminescent polymers [8]. PPP derivatives are of particular interest as blue-emitting polymers because their ring twisting caused by steric interactions effectively limits the intrinsic conjugation length and hence endow their relative large band gaps [15]. PFs are promising blue-light emission conjugated polymers because of their wide band gaps and high photoluminescence (PL) quantum efficiency as well as good thermal and chemical stability [16]. However, PLEDs based on PF homopolymers usually exhibit poor EL efficiency and insufficient color stability because of their strong interchain interactions, aggregations of polymer chains, and oxidative degradation during device operation [17–21]. The rigid biphenyl structure, good hole-transporting ability, and short π-conjugation length of 3,6-PCzs make them potential candidates as blue-light emitters or hosts for phosphorescence materials in PLEDs [22, 23]. 2,7-PCzs are also very attractive for the development of blue light PLEDs [24]. PTs are relatively stable light-emitting materials and their properties can be easily varied by changing the substituents on the starting monomer. However, the luminescence efficiency (LE) of PTs in the solid state is relatively low because of a tendency towards strong interchain interactions [25]. More details of the chemical and physical characteristics of these traditional luminescent polymers can be found in some comprehensive review articles and books [10, 26, 27].

7.2.2 Performance Tuning

To improve the performance and stability of PLEDs, considerable effort has been made on tuning the chemical structure of these traditional EL polymers and many useful strategies have been developed. For example, it was found that the introduction of substituents into the EL polymers' skeleton or side chains could allow fine tuning of their electronic properties (e.g., emission color, band gap, electron affinity, and ionization potential). The charge (hole/electron) carrier units could also be introduced to the polymer's main chain or side chains to achieve more balanced hole and electron transporting in EL polymers,and thus yield the maximum exciton formation.

7.2.2.1 Main-Chain Tuning

Tuning Conjugation Length

The optoelectronic properties of conjugated polymers are highly dependent on their conjugation length. One approach to tune the conjugation length is to develop polymers composed of short conjugated fluorescent segments interconnected via nonconjugated linkers [28]. Wide band-gap polymers with pure blue emission and hosts for phosphorescent dyes with high triplet energy levels (E_T) could be acquired in this way.

A series of polymers (P1–P5, Fig. 7.2) had been prepared by incorporating a nonconjugated polyurethane (PU) segment with 9-butyl-3,6-bis(4-hydroxyphenyl) carbazole and 2,5-bis(4-hydroxyphenyl)-1,3,4-oxadiazole segments through condensation polymerization [29]. PU moieties endowed polymers with large band gaps and high E_T, and the carbazole and oxadiazole units endowed polymers with balanced hole and electron transport/injection properties. Therefore, these polymers could be used as hosts for phosphorescent dyes. Red EL emission was obtained when Ir(btp)$_2$(acac) or Ir(2-phq)$_2$(acac) was used as the phosphorescent dyes in P2–P5. PLEDs based on P2 with a structure of ITO/poly(3,4-ethylenedioxythiophene): poly(styrenesulfonate) (PEDOT:PSS)/Ir(2-phq)$_2$(acac) (8 %) in polymer/(2,9-dimethyl-4,7-diphenyl-1,10-phenanthroline) (BCP)/tris(8-hydroxyquinolinolato) aluminum (Alq$_3$)/LiF/Al showed the maximum luminescence (L_{max}) and a LE of 394 cd/m^2 and 1 cd/A, respectively.

δ-Si structure could also be introduced into conjugated polymers to interrupt the extended π-conjugation and achieve high values of E_T. Cho et al. [30] studied two novel types of PF copolymers containing siloxane linkages or distilbene moieties on

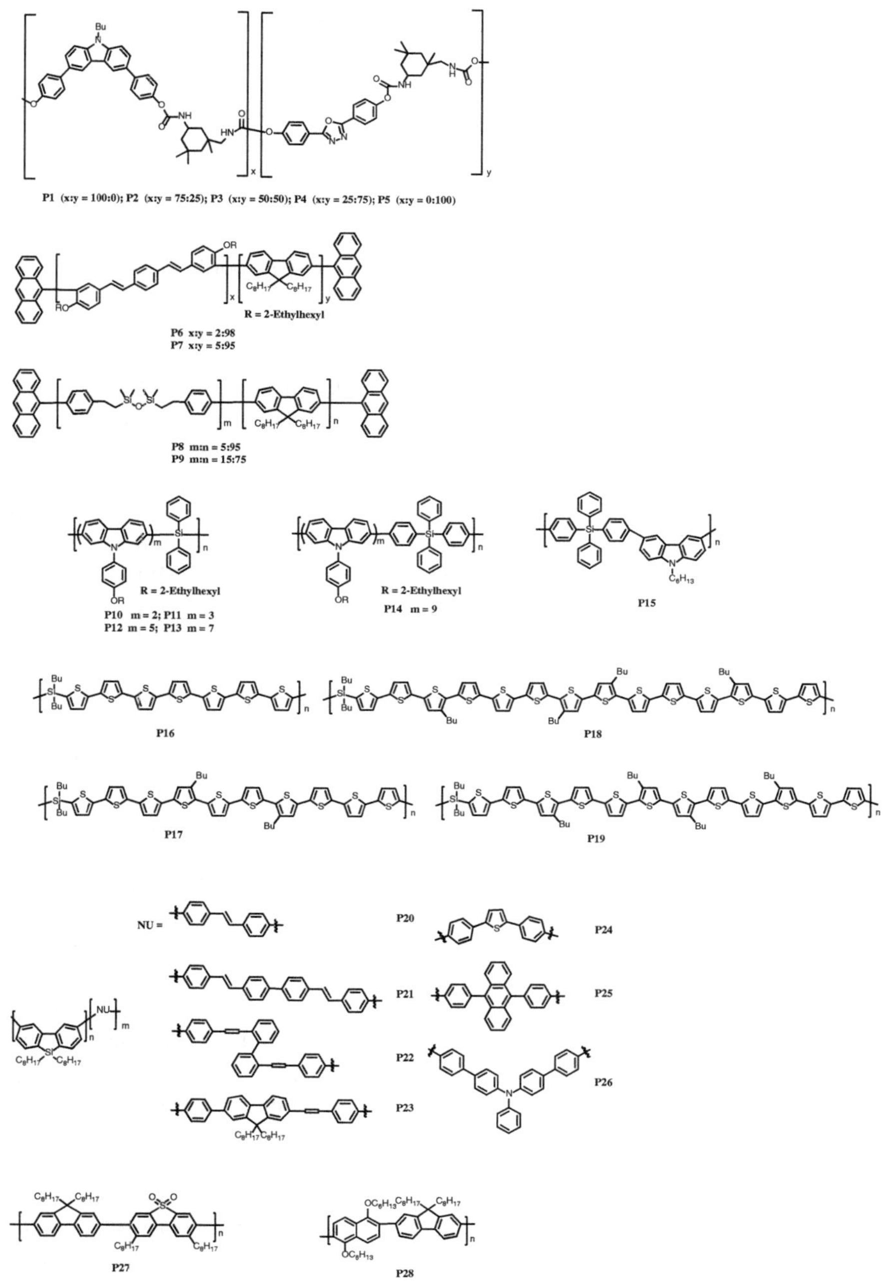

Fig. 7.2 Chemical structures of P1–P28

their main chains (P6–P9, Fig. 7.2). More obvious blue shift of the peak in UV-visible absorption profiles was observed for the polymers with a higher molar percentage of siloxane linkages or distilbene moieties The blue shift of the maximum in the UV-visible absorption was greater in polymers with a higher molar percentage of siloxane linkages or distilbene moieties than in homo poly(dihexylfluorene) (PDHF). However, the PL spectra of the polymers were similar to those of PDHF in terms of the onsets and patterns. The maximum EL emission wavelengths of the polymers were at around 425–450 nm, corresponding to a pure blue light. The Commission Internationale de L' Eclairage (CIE) coordinates of the PFs containing siloxane linkages or distilbene moieties ranged from (0.21, 0.21) to (0.17, 0.10), indicating a deeper blue light than that of PDHF. PLEDs based on P8 with a structure of ITO/PEDOT:PSS/polymers/Ca/Al showed the maximum external quantum efficiency (EQE_{max}) of 0.11 %. Koguchi et al. [31] developed a series of poly[oligo(2,7-carbazolylene)-*alt*-diphenylsilylene]s by Suzuki couplings (P10–P14, Fig. 7.2). The polymers consist of the conjugated units of dimer, trimer, pentamer, heptamer, and nonamer of the *N*-[*p*-(2-ethylhexyloxy)phenyl]carbazole-2,7-diyl and the connecting diphenylsilylene portion between the oligomer units. The full width at half-maximum in PL spectra of these polymers in the solid state became narrower as the conjugation length got longer. The HOMO and LUMO energy levels of all of the polymers were located at about −5.6 and −2.6 eV, respectively. PLEDs based on P12 and P13 with a structure of ITO/PEDOT:PSS/copolymer/CsF/Al exhibited better performances with L_{max} (>8000 cd/m^2) and LE (>0.5 cd/A) than PLEDs based on other poly(oligomer)s. A similar wide band-gap polymer (P15, Fig. 7.2), derived from 3,6-carbazole and tetraphenylsilane, was used as the host for green and blue emission phosphorescent materials [32]. The conjugation length of P15 was effectively confined because of the δ-Si interrupted polymer backbone. The polymer exhibited a violet emission with a peak at 392 nm in solution, and an optical band gap of 3.26 eV. The green and blue phosphorescent PLEDs based on P15 with a structure of ITO/PEDOT:PSS/polymer:dopants/(2,2′,2″-(1,3,5-benzenetriyl)-tris-(1-phenyl-1*H*- benzimidazole)) (TPBI)/LiF/Al exhibited the LE of 27.6 and 3.4 cd/A, respectively. Jiang et al. [33] reported a series of monosilylene-oligothienylene copolymers (P16–P19, Fig. 7.2). All the resulting copolymers exhibited obvious red shifted absorption and PL spectra with increasing the number of thienylene rings in the repeat units of their main chains. When used in single-layer PLEDs, the turn-on voltages (V_{on}) of the resulting devices decreased with increase in the number of thienylene rings, which was caused by the increase of charge carrier mobility and decrease of the bandgap as the π-conjugation length increased.

Another strategy for conjugation length tuning is changing the connection ways of the repeat units on the conjugated main chains. Fluorene, carbazole, silafluorene, and dibenzothiophene-*S,S*-dioxide (SO) are popular building blocks for light-emitting conjugated polymers. There are always two connection ways for these units on the skeleton, the 2,7- or 3,6- linkage, except for SO units, which can be

2,8- or 3,7-linkage. Generally, because of a reduction of conjugation in the 3,6- (or 2,8- for SO)—vs a 2,7- (or 3,7- for SO)—linkage in these units, the 3,6- (or 2,8- for SO) based polymers exhibited a larger band gap and a higher E_T. Both 2,7-fluorene-based and 3,6-fluorene-based polymers are widely use as electroluminescent materials. As the 3,6-fluorene-based polymers exhibited a larger band gap and a higher E_T, they are a good candidate for pure blue light-emitting devices and a good host material for green and blue phosphorescent complexes [34–36]. Similar to 2,7-fluorene, 2,7-carbazole-based polymers have longer effective conjugation lengths, and the 3,6-carbazole-based polymers exhibit larger band gaps and higher E_T than the 2,7-carbazole-based polymers. Therefore 3,6-carbazole-based polymers are suitable hosts for green iridium complexes [37, 38]. Polysilafluorene is a typical wide band-gap conjugated polymer with low-lying LUMO. Silafluorene has a chemical structure similar to that of fluorene, but because the silicon cannot be oxidized to a ketone, the emission of polysilafluorene is much more stable than that of PF, and no green emission band appeared even after prolonged heating in air. Therefore, 2,7-silafluorene-based polymers could be used as blue or deep blue light-emitting materials. As the silicon bridges could interrupt the through conjugation, 3,6-silafluorene-based polymers exhibit ultraviolet light emission. Incorporating 3,6-silafluorene with another unit, such as fluorene, could give a stable deep blue emission [39–41]. Recently, Mo et al. [42] reported novel 3,6-silafluorene-based copolymers (P20–P26, Fig. 7.2). They incorporated the monomer containing vinylene, anthracene, and tri-arylamine moieties into the poly(3,6-silafluorene) backbone and formed efficient deep-blue emitting copolymers with a EL efficiency of 1.1–1.9 %. Those silicon-containing copolymers with tri-arylamine derivatives had higher electroluminescent efficiencies than other 3,6-fluorene-based copolymers. As the electron-withdrawing SO unit could effectively suppress the formation of keto-defects, the SO unit become a novel building block for stable blue light-emitting polymers. Recently, Liu et al. [43] reported a 3,7-SO-based polymer (P27, Fig. 7.2). Blue PLEDs (ITO/PEDOT:PSS/polymer/Ba/Al) based on it showed a V_{on} of 4.8 V, a L_{max} of 200 cd/m^2, and a LE of 0.2 cd/A.

Having remarkable inherent properties such as intense fluorescence, strong π-stacking, and good chemical stabilities, naphthalene-containing polymers have also attracted attention as luminescent materials. Park et al. [44] developed a new blue light-emitting polymer that alternated between fluorene and alkoxynaphthalene structure (P28, Fig. 7.2). Because of steric interaction between the hydrogen at the 3-position of fluorene and the hexyloxy group at the 3-position of naphthalene, rigid fluorene and naphthalene units were distorted with each other, and thus the main chain would be distorted, resulting in a stable blue emission. The film PL spectrum (peaking at 405 nm) of P28 was consistent with that of solution and the polymer did not show any emission in the long wavelength region. The double-layered device with an ITO/PEDOT/polymer/LiF/Al structure had a V_{on} of about 5.4 V, a L_{max} of 110 cd/m^2, and a LE of 0.09 cd/A. The PLED generated pure blue EL emission (λ_{max} = 405 nm) with good CIE coordinates (0.15, 0.10).

Tuning Charge Transporting Properties

It is well known that balanced transporting/injection of electrons and holes is critical to achieving high performance PLEDs, and thus it is highly desirable to obtain EL materials possessing excellent hole and electron transporting/injection capabilities simultaneously. Electron-rich moieties with high-lying HOMO energy levels could give a reduced hole injection barrier, and electron-deficient moieties with low-lying LUMO energy levels could give a reduced electron injection barrier. Consequently, these electron-rich and electron-deficient moieties are often introduced into the EL polymers' main chains to improve the resulting polymers' hole and electron transporting/injection capabilities [45, 46].

Pyridine and oxadiazole are two widely used electron-deficient moieties which can improve the electron affinity of the resulting materials. Qi et al. [47] reported a series of highly soluble copolymers *p*-/*n*-poly[(2,5-divinyl-3,4-dialkylthiophene)-*alt*-2,6-pyridine] (P29, P30, Fig. 7.3) and poly[(2,5-divinyl-3,4-dialkylthiophene)-*alt*-(2,5-diphenyl-1,3,4-oxadiazole)] (P31, P32, Fig. 7.3). The optical band gap energy of these polymers was similar, ranging from 2.68 to 2.80 eV in solid films. The electron affinities of these polymers range from 2.79 to 3.09 eV, which were propitious to electron injection from the cathode. These novel copolymers presented some EL performance in their single layer PLED with configuration of ITO/polymer/Al, which showed V_{on} between 4.0 and 5.8 V and emitted bright green-yellow (538 nm) and yellow (545–552 nm) EL light. Gong et al. [48] developed an ambipolar charge-transport semiconducting polymer containing both a hole-transporting moiety triphenylaniline, and an electron-transporting moiety oxadiazole, in the main chain (P33, Fig. 7.3). One hundred times enhanced LE was observed from single layer PLEDs (ITO/PEDOT:PSS/polymers/Al) made by the polymer with oxadiazole units compared with the PLEDs made by the polymer without oxadiazole units. This could effect an improvement in the charge balance of electrons and holes inside this ambipolar polymer. The existence of 5,8-quinolinenevinylene units in the polymer main chain could not only improve the electron-injection and transport in the polymer, but also tune the color of the copolymer. Therefore, Liu et al. [49] developed such a nitrogen-containing electroluminescent copolymer (P34, Fig. 7.3) by Wittig–Horner polymerization. The absorption peaks of the copolymer in solution and a thin film are at 490 and 516 nm and the PL emission peaks in solution and thin film are at 571 and 629 nm, respectively. The HOMO and LUMO of the polymer were −5.30 and −3.30 eV, respectively. PLEDs (configuration: ITO/PEDOT:PSS/polymer/Ca/Al) based on P34 showed a very pure red light emission with maximum peaks around 618 nm, and exhibited a L_{max} of 188 cd/m^2 with a LE of 0.01 cd/A. Zhu et al. [50] reported a series of blue light-emitting copolymers based on 9,9′-dioctylfluorene and 2,2′-(1,4-phenylene)-bis (benzimidazole) moieties (P35, P36, Fig. 7.3). The benzimidazole moiety in P35, P36 could improve the electron-transport property of the copolymer. P35 emitted blue light efficiently, with the quantum yield up to 99 % in chloroform. PLEDs

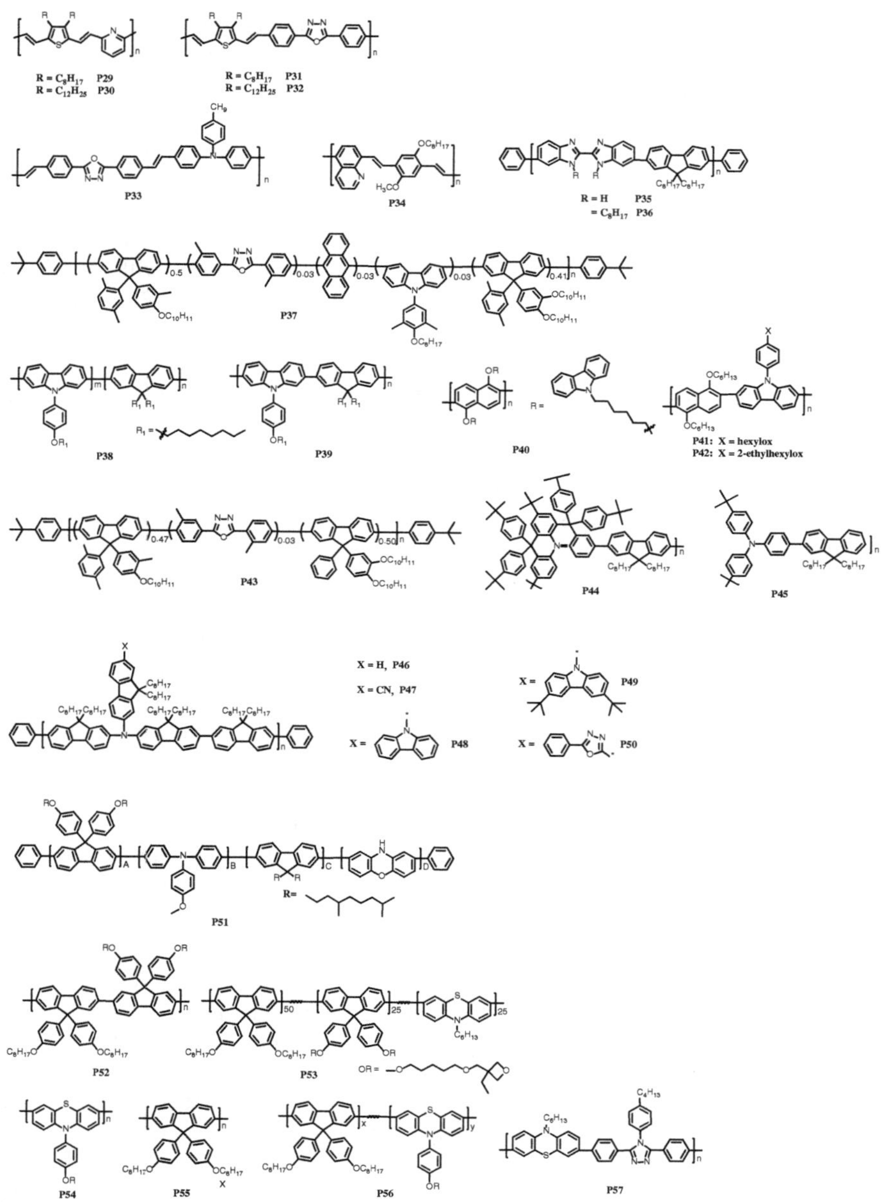

Fig. 7.3 Chemical structures of P29–P57

based on P36 with the configuration of ITO/PEDOT:PSS/polymer/LiF/Al emitted blue light with the maximum at 448 nm, and exhibited a L_{max} of 1534 cd/m^2 with the LE and power efficiency (PE) of 0.67 cd/A and 0.20 lm/W, respectively.

Carbazole and triphenylamine (TPA) are the most widely used hole transporting moieties for EL polymers. Chen et al. [51] presented a stable blue-light-emitting copolyfluorene (P37, Fig. 7.3) consisting of carbazole, oxadiazole, and charge trapping anthracene groups. It was found that the hole-transporting carbazole and electron-transporting oxadiazole units improved charge injection and transporting properties of the resulting polymers, whereas the anthracene was the ultimate emitting chromophore. In the film state, P37 showed a blue emission at 451 nm attributed to the anthracene chromophore. PLEDs using P37 as the emitting layer (ITO/PEDOT:PSS/polymer/Ca/Al) exhibited good performance with a LE of 5.1 cd/A) and a CIE coordinate of (0.16, 0.11). A carbazole and fluorene-based random (P38, Fig. 7.3) and an alternating copolymer (P39, Fig. 7.3) were developed for high-performance blue light-emitting polymers [52]. These copolymers absorb light energy at about λ_{max} 400 nm in the film state, and emit light at about λ_{max} 430 nm in the thin film state. Energy gaps between their HOMO and LUMO are about 2.9 eV. PLEDs (configuration: ITO/PEDOT:PSS/polymer/CsF/Al) based on P38 showed notably higher performance with the L_{max} of 31,200 cd/m^2, and the maximum LE (LE_{max}) of 1.68 cd/A. Mori et al. [53] reported a series of carbazole-containing 1,5-disubstituted poly(2,6-naphthalene) derivatives, including a 2,6-naphthalene homopolymer which had a carbazolyl side chain at 1,5-positions (P40, Fig. 7.3), a random copolymer, and an alternating copolymer that consisted of 1,5-dialkoxynaphthalene-2,6-diyl and *N*-phenylcarbazole-2,7-diyl (P41, P42, Fig. 7.3). These polymers exhibited blue PL in the film states. PLEDs (configuration: ITO/PEDOT:PSS/polymer/CsF/Al) based on these copolymers emitted blue-green to green emissions with an EL λ_{max} at around 490 nm. PLEDs fabricated with P41 exhibited the best performance, showing a L_{max} of 8370 cd/m^2 at 13 V and a LE_{max} of 2.16 cd/A at 7 V. By introducing nonsymmetric and bulky aromatic groups at C-9 position, a PF derivative (P43, Fig. 7.3) was developed as an efficient deep blue-emitting materials [54]. With a configuration of ITO/PEDOT:PSS/poly(fluorene-*co*-triphenylamine) (PFO-TPA)/polymer/Cs_2CO_3/Al, PLEDs with P43 as the EML, PFO-TPA derivative as the hole-injection/transporting layer exhibited a deep blue emission centered at 430–450 nm with the CIE chromaticity coordinate of (0.15, 0.14), a L_{max} of 35,054.2 cd/m^2 and a LE of 14.0 cd/A (at 2975.0 cd/m^2). Jiang et al. [55] reported a series of novel aryl-bridged TPA alternating copolymers, 7-*tert*-butyl-5,5,9,9-tetraaryl-13*b*-aza-naphtho[3,2,1-*de*]anthracene/dihexylfluorene P44 and TPA/dihexylfluorene P45 (Fig. 7.3). The HOMO energy levels of the two polymers were very close (−5.15 eV for P44 and −5.13 eV for P45). The maximum absorption peak of P44 was 398 nm in film, which was red shifted by 21 nm with respect to P45. Because of the well-matching HOMO energy level of the polymers with ITO and its good hole-transporting properties which were endowed by the incorporation with the TPA unit, the device configuration could be fabricated without using PEDOT:PSS as the hole injection layer. PLED with a simple configuration of ITO/P44/tetranaphthalen-2-yl-silane/Alq_3/Al emitted a blue light with emission peak at 436 nm, and exhibited a LE_{max} of 1.89 cd/A and a L_{max} of 4183 cd/m^2, which was superior to the device with P45 as EML under the identical condition. P46–P50 (Fig. 7.3) were a series of blue-light-emitting conjugated polymers containing TFA as a building block [56]. The emission color could be effectively tuned in the region of deep-blue and light-blue by introducing various substituents onto the TFA unit as the pendants. Because of the high-lying HOMO, energy levels of the trifluoren-2-yl-amine (TFA) unit could give a reduced hole injection barrier, PLEDs based on P48 (structure: ITO/PEDOT:PSS/polymers/TPBI/CsF/Al) showed a LE_{max} of 2.44 cd/A, corresponding to an EQE_{max} of 3.00 %, with a CIE coordinate of (0.16, 0.12). Harkema et al. [57] carried out a study of PLEDs based on blue-emitting fluorene polymers containing various hole-transporting units. Ten polymers containing a systematically varied amount of two different (benzidine-based and phenoxazine-based) aromatic amine comonomers (P51, Fig. 7.3) were developed. It was found that, with decreasing hole mobility, EQE_{max} increased and the peak voltage decreased. They explained this voltage dependence of the efficiency from a drift–diffusion device model [58].

N-Hexylphenotiazine is another type of electron-rich moiety which could enhance the hole transporting/injection of the resulting EL polymers. Park et al. [59] reported an alternating copolymer composed of bis-(4-octyloxyphenyl)fluorene and bis((3-hexyloxy-3-ethyl)oxetane) fluorene P52, and a polymer composed of the above two monomers with *N*-hexylphenotiazine P53 (Fig. 7.3). These polymers would become insoluble after UV irradiation in the presence of a proper photo-acid generator. PLEDs (configuration: ITO/PEDOT:PSS/polymer/LiF/Al) using the photo-cross-linked polymer films also showed lower operating voltages than the devices using the corresponding polymer films without cross-link. Moreover, PLEDs based on P52 with photo-cross-link showed the best device performances with a L_{max} of 4750 cd/m^2 and a LE of 0.68 cd/A, respectively. Furthermore, Kim et al. [60] developed poly[10-(4′-octyloxyphenyl)-phenothiazine-3,7-diyl] P54, poly[9,9-bis(4′-octyloxyphenyl) fluorene-2,7-diyl] P55, and their random copolymers P56 (Fig. 7.3). PLEDs (configuration: ITO/PEDOT:PSS/polymer/Ca/Al) constructed with homopolymers P54 or P55 exhibited poor device performance, whereas the device based on the random copolymer P56 exhibited much higher PE and brightness. This enhanced efficiency of the copolymer devices resulted from the improved hole injection and much better charge carrier balance. To improve the electron injection/transporting ability, electronegative 1,2,4-triazole group was introduced into an *N*-hexylphenotiazine-based polymer (P57, Fig. 7.3) [61]. The PL maximum wavelength, the band gap energy, and the HOMO energy level of P57 film were 509 nm, 2.67, and −5.06 eV, respectively. The LE_{max} and L_{max} of the PLEDs based on P57 were 0.247 cd/A and 771 cd/m^2, respectively.

7.2.2.2 Side-Chain Tuning

Tuning Solubility and Intermolecular Interaction

Good solubility of conjugated polymers is a crucial premise for their solution-processable application. To increase the solubility of luminescent conjugated polymers, solubilizing side chains (such as long alkyl chains or bulky substituent groups) are usually introduced to their rigid backbones. Besides the improved solubility, it was found that the introduced side chains have multiple effects on the resulting polymers, including improving their thermal properties, impeding intermolecular aggregation, depressing PL quenching, and enhancing PL quantum efficiency, etc. [62].

Saikia et al. [63] synthesized a series of PPP derivatives (P58, Fig. 7.4) with high molecular weights. It was found that the introduction of alkoxy side chains not only

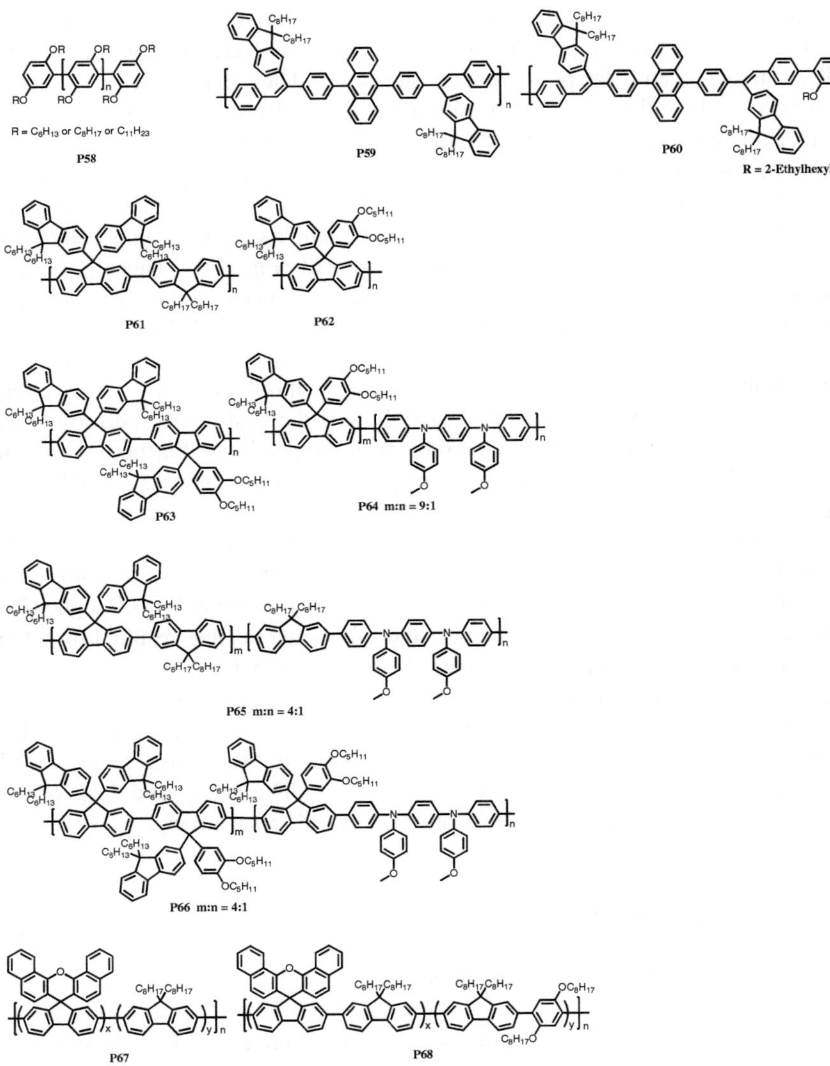

Fig. 7.4 Chemical structures of P58–P68

improved the solubility of polymers in organic solvents but also affected their optical properties. Because of crystallization, simple, single-layer PLEDs made with neat polymers exhibited high threshold voltages from 8×10^5 to 2.5×10^6 V/cm and enhanced emission at the second peak around 480 nm. Kim et al. [64] studied a new series of conjugated polymers having diphenylanthracene vinylene biphenylene and diphenylanthracene vinylene terphenylene in the main chain and fluorene as the pendant group (P59, P60, Fig. 7.4). The introduced bulky fluorene pendant groups could not only enhance solubility of the resulting polymer, but also shorten the effective conjugation length of the backbone because of steric hindrance. These polymers showed blue emission (λ_{max} 461 nm for P59 and λ_{max} 455 nm for P60) when excited by UV light. PLEDs with the structure ITO/PEDOT:PSS/P60/LiF/Al exhibited a V_{on} of about 5.8 V, a L_{max} of 152 cd/m^2, and an electroluminescent efficiency of 0.143 lm/W.

The incorporation of bulky side functional groups into PFs, in particular the attachment of bulky aromatic substituents at the 9-position of fluorene, has been reported to be an effective way to improve the PFs derivatives' device performance [65]. Recently, Jin et al. [66] reported a series of highly efficient and pure blue-emitting PF-type polymers (P61–P66, Fig. 7.4) which contained 9,9-di(9,9-dihexylfluoren-2-yl)fluorene and 9-(9,9-dihexylfluoren-2-yl)-9-(3,4-di(2-methyl)butyloxyphenyl)fluorene units. It was found that these bulky aromatic side chains could effectively suppress interchain interactions of PFs. As a result, these polymers exhibited good PL characteristics as a blue-light-emitting polymer and showed highly enhanced color integrity and color stability against oxidative conditions. When N^4,$N^{4'}$-bis(4-methoxyphenyl)-N^4,$N^{4'}$-diphenylbiphenyl-4,4′-diamine was incorporated into the polymer backbone, the polymer exhibited a good EL performance with a LE of 4.12 cd/A and a CIE coordinate of (0.15, 0.15) in the device with the structure of ITO/PEDOT:PSS/polymer/LiF/Al.

Constructing conjugated polymers with larger spiro-units which stretch out to prevent further inter-chain interaction could enhance the spectral stability in PLEDs [67]. Wang et al. [68] prepared conjugated polymers (P67–P68, Fig. 7.4) based on naphthalene-containing spirofluorene units and 9,9-dioctylfluorene, 2,5-dioctylbezene. These two polymers showed good blue emission both in solution and thin film. The white PLEDs by using P67 as the host material, containing two typical phosphorescent Ir complex dopants, green emitter Ir(mppy)$_3$ and red emitter Ir (piq)$_2$, were fabricated. The LE$_{max}$ and L_{max} of these devices were determined to be 2.8 cd/A and 13,500 cd/m^2 at 11.2 V, respectively. The color coordinate CIE stayed nearly constant, changing from (0.32, 0.31) to (0.28, 0.31) when the current density varied from 2 to 20 mA/cm^2.

Tuning Charge Transporting/Injection Properties

The charge transporting/injection properties of the conjugated polymers can also be tuned through side chain modification. A typical example is so-called water/alcohol soluble conjugated polymers (WSCPs), which have been widely used as ETL in

PLEDs. It was found that the pendant highly polar side chains among WSCPs endow them with not only good solubility in highly polar solvents but also a unique electron-injection ability from high work-function metal cathodes, because of the dipole interaction at the interface [69]. Hence, these WSCPs can be used as active layer in PLEDs with stable high work-function metals (such as Al, Ag, Au) as cathode [70].

Recently, Shi et al. [71] reported a series of novel cationic conjugated polyelectrolytes with alternating TPA and fluorene as backbones and attached by ammonium (amino)alkyl side chains (P69–P72, Fig. 7.5). Both the hole and the electron injection capabilities of these polymers were improved because of the simultaneous introduction of the TPA segment and the aminoalkyl (ammonium) group. PLEDs based on these polymers (ITO/PEDOT (or poly(9-vinylcarbazole) (PVK))/polymer/Al (or Ba/Al)) showed a L_{max} of 1194 cd/m^2 and a EQE_{max} of 0.61 %, respectively. Liu et al. [72] synthesized a series of blue, green, and red-emitting aminoalkyl functionalized PF derivatives (P73–P84, Fig. 7.5) containing SO, benzothiadiazole (BT), and thiophene-BT-thiophene (DTBT) as chromophores, respectively. It was found that the variation of molar ratio of aminoalkyl functional groups did not significantly influence thermal stability, UV-vis absorption, PL, or electrochemical properties of copolymers. The application of the resulting aminoalkyl functionalized copolymers in PLEDs exhibited dual-function including efficient light-emission and electron injection from Al cathode. The increase of molar ratio of aminoalkyl side groups leads to enhanced device performances for both green- and red-emitting copolymers. Yu et al. [73] developed a WSCP containing pendant crown ether moieties (P85, Fig. 7.5). In the film state, its PL spectrum (peaking at 430 and 452 nm) showed a noticeable red shift relative to that of poly (9,9-dihexylfluorene) (peaking at 423 and 448 nm). The HOMO and LUMO levels of P85 were estimated to be 5.68 and 2.65 eV, respectively. PLEDs based on P85 (ITO/PEDOT:PSS/P85/Ca/Al) exhibited higher L_{max} (7910 cd/m^2) and LE_{max} (2.3 cd/A) than those of PF devices (860 cd/m^2, 0.29 cd/A). Moreover, inserting a P85 layer between the PF emitting layer and the calcium cathode led to a reduced V_{on} and an enhanced device performance.

Similar to the strategy used in main chain tuning, the electron-rich hole-transporting and electron-deficient electron-transporting moieties can be introduced to the conjugated polymers' side chains to enhance their charge transporting capability, where the resulting bulky side chains can also suppress the interchain aggregation of the resulting polymers [74]. Recently, Lin et al. [75] developed a novel blue light-emitting PF-based copolymer (P86, Fig. 7.5) containing electron-rich TPA and electron-deficient phenylquinoline side chains in the C-9 position of the fluorene unit. PLEDs based on this polymer (configuration: ITO/PEDOT:PSS/polymer (70 %) + 2-(4-*tert*-butylphenyl)-5-(4-biphenylyl)-1,3,4-oxadiazole (PBD) (30 %)/ 1,3,5-tris(*N*-phenylbenzimidizol-2-yl)benzene (TPBI)/LiF/Al) exhibited a main EL emission peak at 428 nm with two shoulder peaks at 452 and 484 nm, and showed a L_{max} of 2367 cd/m^2 and a LE_{max} of 1.02 cd/A. Lin et al. [76] also prepared a similar bipolar conjugated PF copolymer with TPA and cyanophenylfluorene as side chains (P87, Fig. 7.5). The HOMO and LUMO energy levels of the polymer were

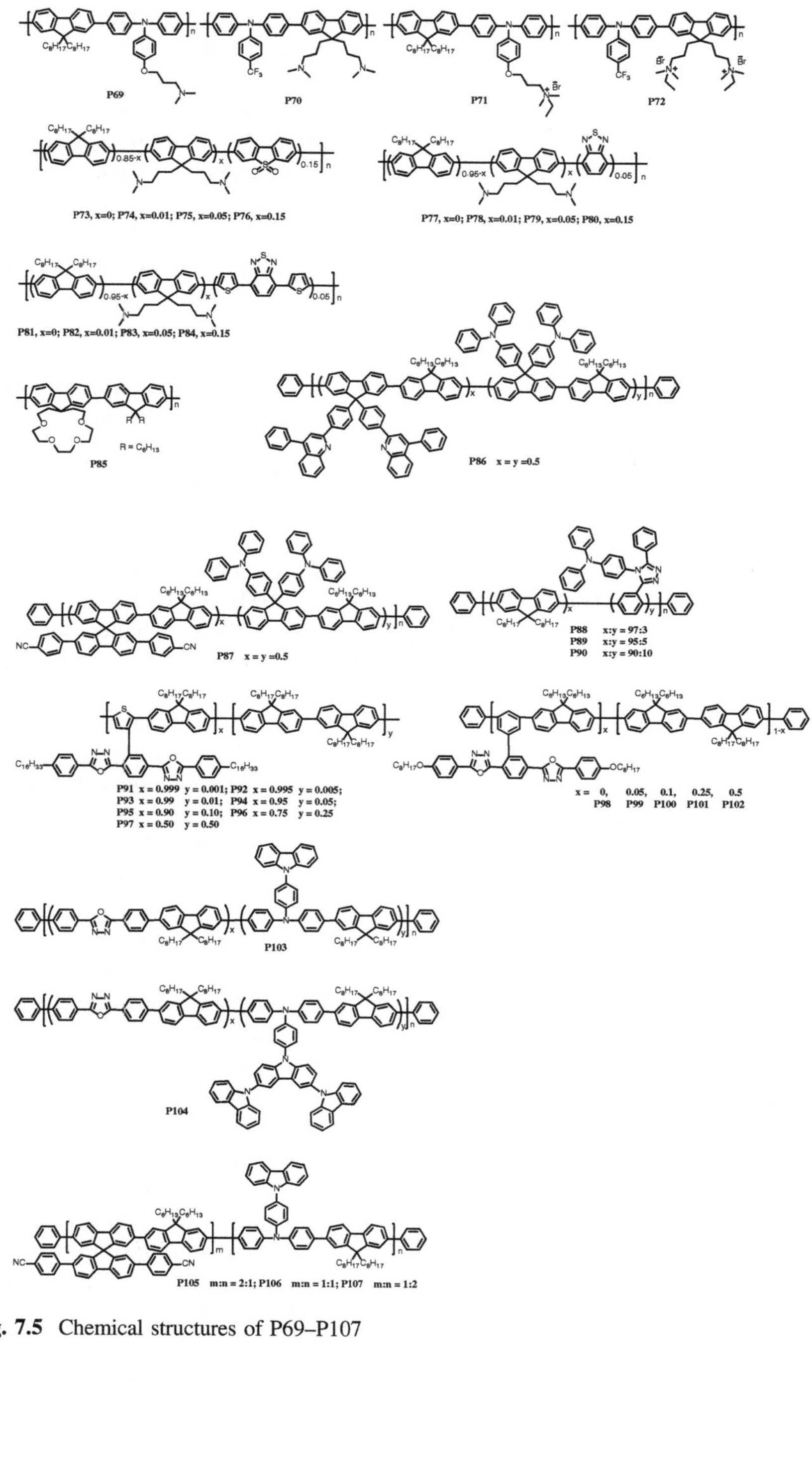

Fig. 7.5 Chemical structures of P69–P107

electrochemically estimated as −5.68 and −2.80 eV, respectively. PLEDs based on this polymer (configuration: ITO/PEDOT:PSS/polymer/CsF/Ca/Al) exhibited a deep-blue emission, and showed a V_{on} of 4.2 V, a L_{max} of 1189 cd m^2, and a LE of 0.72 cd/A. Furthermore, a bipolar unit consisting of electron-transporting aromatic 1,2,4-triazole directly linked with hole-transporting TPA was introduced into the polymer (P88–P90, Fig. 7.5) as pedant groups [77], and the resulting polymers were blended with MEH-PPV as the emission layer in PLEDs. The resulting device (configuration: ITO/PEDOT:PSS/polymer + MEH-PPV/Ca/Al) based on P90 exhibited the best performance with a L_{max} of 11,090 cd/m^2 and a LE of 0.56 cd/A (ca. 0.4 wt% of bipolar residue).

When side chains are attached laterally to the gravity centers of the main chain without spacers, the main chains of polymers are forced to extend and to conform rigidly because of the high population of both bulky and rigid side groups around the backbone, and the "jacket" is formed [78]. Therefore, jacketed derivatives with electron-deficient side-chain building blocks could suppress the long wavelength emission and improve the luminescent efficiencies of PF. Following this theory, Wang et al. [79] reported a series of new copolymers (P91–P97, Fig. 7.5) comprised of 9,9-dioctylfluorene and jacketed units 2,5-bis[(5-octyloxy-phenyl)-1,3,4-oxadiazole]-1-(3,5-dibromophenyl)-benzene. PLEDs (configuration: ITO/PEDOT:PSS/polymer/Ca/Ag, ITO/PEDOT:PSS/polymer/TPBI/Mg:Ag(10:1, wt)/Ag, and ITO/PEDOT:PSS/PVK/polymer/TPBI/Ca/Ag) based on these polymers exhibited pure blue EL emissions. P96- and P97-based devices showed a L_{max} of 5097.8 and 3122.8 cd/m^2, a LE of 0.484 and 0.416 cd/A, respectively. Yang et al. [80] developed a series of similar copolymers (P98–P102, Fig. 7.5) based on 3-{2,5-bis[(4-hexadecyloxy-phenyl)-1,3,4-oxadiazole]phenyl}-2,5-dibromothiophene. The HOMO and LUMO energy levels of the resulting polymers were both lower than those of PF, which resulted in better electron injection and transport but still good blue-green emission. PLEDs based on P99 (configuration: ITO/PEDOT-PSS/polymer/Ca/Al) exhibited the best device performance, with a L_{max} of 5558 cd/m^2 and a LE of 0.39 cd/A.

Electron-rich carbazole side chains are also widely used to enhance hole transporting ability of light-emitting polymers in PLEDs. Lin et al. [81] reported two new blue light-emitting polymers (Fig. 7.5), poly{[2,5-bis(4-phenylene)-1,3,4-oxadiazole]-[9,9-dihexylfluorene-2,7-diyl]-[*N*-(4-(9*H*-carbazol-9-yl)phenyl)-*N*,*N*-bis(*p*-phenylene)aniline]} (P103) and poly{[2,5-bis(4-phenylene)-1,3,4-oxadiazole]-[9,9-dihexylfluorene-2,7-diyl]-[4-(3,6-(di-9*H*-carbazol-9-yl)-9*H*-carbazol-9-yl)-*N*,*N*-bis(*p*-phenylene)-aniline]} (P104). PLEDs (configuration: ITO/PEDOT:PSS/polymer/TPBi/LiF/Al) based on P104 displayed a stable blue emission having color coordinate of (0.15, 0.20), a L_{max} of 4762 cd/m^2, and a LE_{max} of 1.79 cd/A. By using this polymer as the host material doped with 1 wt% 4,4′-bis[2-(4-(*N*,*N*-diphenylamino)phenyl)vinyl]biphenyl, the achieved L_{max}, LE_{max} and maximum PE of the resulting device were 13,613 cd/m^2, 3.38 cd/A, and 1.84 lm/W, respectively. Lin et al. [82] reported a series of novel blue light-emitting copolymers (P105–P107, Fig. 7.5), composed of different ratios of electron-withdrawing segments (spirobifluorene substituted with cyanophenyl groups) and electron-donating segments (carbazole-

TPAs). Incorporation of the rigid spirobifluorene units substituted with cyanophenyl groups into the polymer backbone improved not only the thermal stabilities but also the PL efficiencies of the resulting polymers. PLEDs based on these polymers (configuration: ITO/PEDOT:PSS/polymers:PBD/CsF/Ca/Al) showed the best performance when P106 was used as the active layer, with the lowest V_{on} of 3.1 V, the highest L_{max} of 6369 cd/m^2, and the highest LE of 1.97 cd/A.

By introducing the carbazole pendant groups, new polyfluorenevinylenes, P108–P110 (Fig. 7.6) with cyano-substituted vinylene units had been synthesized by Song et al. [83] The EL emission maxima of the polymers appeared at around 496–504 nm. PLEDs (configuration: ITO/PEDOT:PSS/polymer/Ca/Al) based on P109 exhibited a L_{max} of 1724 cd/m^2 and a LE$_{max}$ of 0.18 cd/A. Peng et al. [84] reported a series of new PFs with dendritic functional carbazole and oxazole side chains (P111–P113, Fig. 7.6). With the steric hindrance of dendritic functional carbazole and oxazole units, the PL and EL emission color quality was improved because of the lower aggregate of the main chains. Huang et al. [85] designed a series of highly efficient electroluminescent polymers (P114–P121, Fig. 7.6) by introducing multiple charge transport moieties into spiro-PF for efficient charge injection. The TPA and carbazole units were integrated in the same side chain of spiro-PF for more efficient hole injection and the electron transport moiety with an electron-accepting unit triazole was introduced to enhance the electron injection. Deep blue PLEDs (configuration: ITO/Al/polymer/CsF/Al) based on these polymers exhibited a EQE$_{max}$ of 7.28 %. A series of new 9,10-diphenylanthracene-based, 2,6-linked blue-light-emitting copolymers (P122–P127, Fig. 7.6) bearing hole- or electron-transporters as well as bulky substituents were reported by Chen et al. [86]. Among all the copolymers, P125 exhibited a balanced hole/electron injection/transporting capability because of the substituted electron-transporting oxadiazole units. As a result, the PLED based on P125 showed a very mild efficiency roll-off: only 0.13 cd/A LE drops from current densities of 10–100 mA/cm^2, corresponding to EL brightness of 169–1558 cd/m^2.

Dong et al. [87] synthesized a series of 9,9-dioctylfluorene-*alt*-benzothiadiazole (FOBT)-based copolymers (P128–P131, Fig. 7.6) with carbazole substituted side chains. The ratios of fluorene and BT units in polymers were adjusted to balance the electron and hole transport, and it was found that P130 exhibited 13 times higher LE$_{max}$ (10.8 cd/A) than PFOBT (0.8 cd/A) in a double-layered device (configuration: ITO/PEDOT: PSS/polymer/TPBI/CsF/Al). Moreover, the carbazole unit was found to be a highly electroactive group with a relatively low oxidation potential. By using the novel multifunctional conjugated polymer precursor (P132, Fig. 7.6) as electrochemical deposition precursor, Gu et al. [88] showed for the first time that cross-linked polymer electrochemical deposition films exhibit high quality and good device performance. This is a breakthrough for cross-linking conjugated polymers from uncontrollable to controllable preparation. PLEDs (configuration: ITO/electrochemical deposition film/TPBi/LiF)/Al) based on it exhibited a LE of 3.8 cd/A.

The cyano group with high electron affinity could modulate the electron affinity of the π-conjugated system and balance the charge transport. The introduction of a

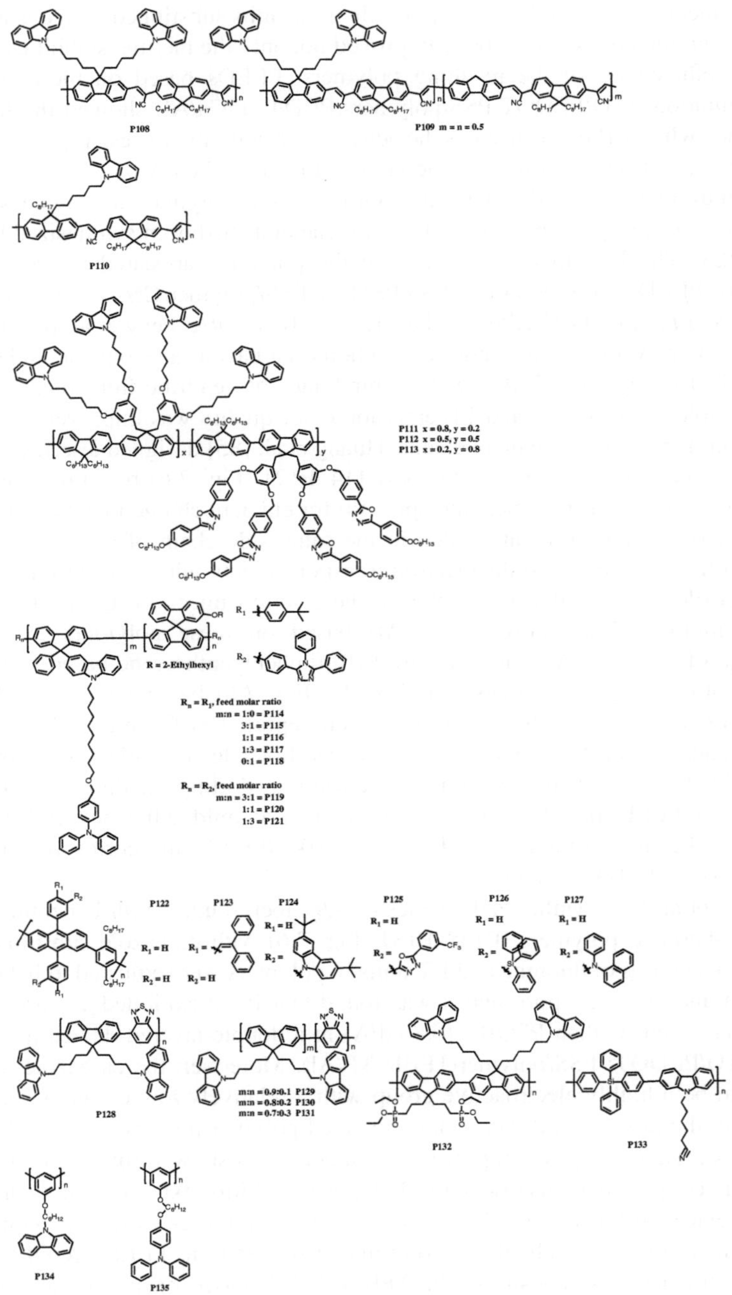

Fig. 7.6 Chemical structures of P108–P135

peripheral CN group could enhance the electron injection property of the polymer, and does not affect the energy gap. Following this strategy, Hu et al. [89] reported a novel blue phosphor host (P133, Fig. 7.6) using 3,6-linked carbazole with a δ-π tetraphenylsilane segment as the main chain modified by a peripheral cyanohexyl group, where the backbone provided the polymer with wide bandgap, high E_T, and good hole-transporting ability. PLEDs (configuration: ITO/PEDOT:PSS/polymer: iridium(III) bis(4,6-(difluorophenyl)pyridinato-N,C_2)picolinate (FIrpic) (8 wt%)/TPBi/LiF/Al), using P133 as the host for blue iridium complex FIrpic, exhibited a LE_{max} of 15 cd/A and a EQE_{max} of 6.7 %, respectively.

To achieve a high E_T and good hole transport/injection host for phosphorescent materials, Liu et al. [90] developed a novel conjugated polymer (P134, Fig. 7.6) by attaching carbazole unit to poly(*m*-phenylene) backbone. An E_T as high as 2.64 eV was achieved and this was the first conjugated polymer with E_T higher than that of state-of-the-art blue phosphorescent dopant, FIrpic. Because of the high E_T and good miscibility, blends of P134 with FIrpic showed no triplet energy back-transfer and exhibited emission exclusively from FIrpic, even at FIrpic concentrations as low as 1 wt%. Single-layer blue phosphorescent PLEDs (structure: ITO/PEDOT:PSS/polymer:dopants/CsF/Al) based on the blend exhibited a LE of 4.69 cd/A. Furthermore, Liu et al. [91] reported a poly(*m*-phenylene) derivative (P135, Fig. 7.6) tethering TPA unit. The poly(*m*-phenylene) backbone endowed P135 with an E_T as high as 2.65 eV, which was sufficiently high to prevent triplet energy back transfer. As a result, the blue phosphorescent PLEDs (structure: ITO/PEDOT:PSS/polymer:FIrpic/2,2′-(1,3-phenylene)bis[5-(4-*tert*-butylphenyl)-1,3,4-oxadiazole] (OXD-7)/CsF/Al) based on P135 showed a LE of 17.9 cd/A and an EQE_{max} of 9.3 %. The white phosphorescent PLEDs based on P135 showed a LE of 22.1 cd/A and an EQE_{max} of 10.6 %.

7.3 Luminescent Polymers Based on Dopant/Host System

In most cases, blue-emitting homopolymers have a large band gap and emit pure blue emission in solution and thin films when excited by UV light. However, when used in PLEDs, they always exhibited poor color stability and low EL efficiency because of aggregate/excimer or defect formation [18, 21]. Despite their poor device performance, these large band gap blue emitters can be developed into high efficiency EML materials by using dopant-host strategy. It was found that by incorporating small amounts of narrow band gap units (dopants) into the blue emitting homopolymers' (hosts) main chain or side chain, the resulting materials' device performance can be greatly improved because of charge trapping and energy transfer mechanisms. Moreover, in these systems, the dopant is covalently connected with the hosts to realize molecular dispersion of the dopant and avoid phase separation. This results in good color stability, which is an obvious advantage compared to the homopolymers. In this way, high-efficiency green-, red-, and blue-emitting polymers have been achieved [92–94]. Later on, phosphorescent complexes were also introduced to polymers' main chain or side chain as dopants to enhance further the resulting materials' EL efficiencies [95]. Another obvious advantage of dopant-host copolymer design is that the emission color and device performance can easily be tuned by choosing different dopants or controlling their ratios, without complicated synthesis procedures [96]. Such fine-tuning strategy could also be used in developing high performance white emission polymers for potential applications in solid-state lighting [97].

7.3.1 Electrofluorescent Polymers

7.3.1.1 Polymers with Dopants in the Main Chains

Red pigment diketopyrrolopyrrole (DPP) derivatives have attracted much attention because of their excellent photostability, high quantum yield of fluorescence, and potential applications for PLEDs. Qiao et al. [98] reported high efficiency red polymers (P136–P140, Fig. 7.7) via introducing DPP into host polymer main chain as acceptor. Only very few DPP units (0.2 %) were needed to quench completely the emission from fluorene segments in the copolymers because of the efficient energy transfer from fluorene to exciton trapping on the narrow band gap DPP sites. Emission colors of PLEDs changed from orange to red, and the EL peaks were gradually red-shifted from 582 to 600 nm with increase of DPP content in polymer chains. The best EL performance (EQE_{max} of 0.25 % and L_{max} of 259 cd/m^2) was achieved when P140 was used as EML in PLED with an ITO/PEDOT:PSS/PVK/polymer/Ba/Al structure. Because BT and DTBT moieties have smaller gaps than PF, the excitation energy on the fluorene segments can be transferred to BT and DTBT units efficiently. By incorporating BT, DTBT moieties into cross-linkable polymer main chains, the polymers (P141–P143, Fig. 7.7) emitted blue, green, and red light, respectively [99]. A double-layer device with an ITO/PEDOT/polymer/Ca/Al structure by using P141–P143 as the EML showed good performance. Moreover, a white-light-emitting device with a CIE coordinate of (0.34, 0.33) was achieved by blending P142–P143 into a host material P142 as EML. Liao et al. [100] and Su et al. [101] synthesized copolymers (P144–P153, Fig. 7.7) slightly doped with a small amount of green chromophore 2,5-dihexyloxy-1,4-bis(2-phenyl-2-cyanovinyl)benzene or 2,5-bis(2-phenyl-2-cyanovinyl)thiophene. The intensity of green emission increases significantly with increasing doped chromophore content, indicating the efficient energy transfer from the fluorene segments to the green chromophores. The performance of the PLEDs with an ITO/PEDOT:PSS/polymer/Ca/Al structure was improved with increased chromophore content in emitting copolyfluorenes, i.e., the P148-based PLED showed the best performance with a L_{max} of 6790 cd/m^2 and a LE_{max} of 1.69 cd/A. Moreover, PLED was successfully fabricated by blending P146 with 0.1 wt% of red phosphor [Ir(piq)$_2$(acac)], with the L_{max} and a CIE coordinate being 4120 cd/m^2 and (0.31, 0.28), respectively.

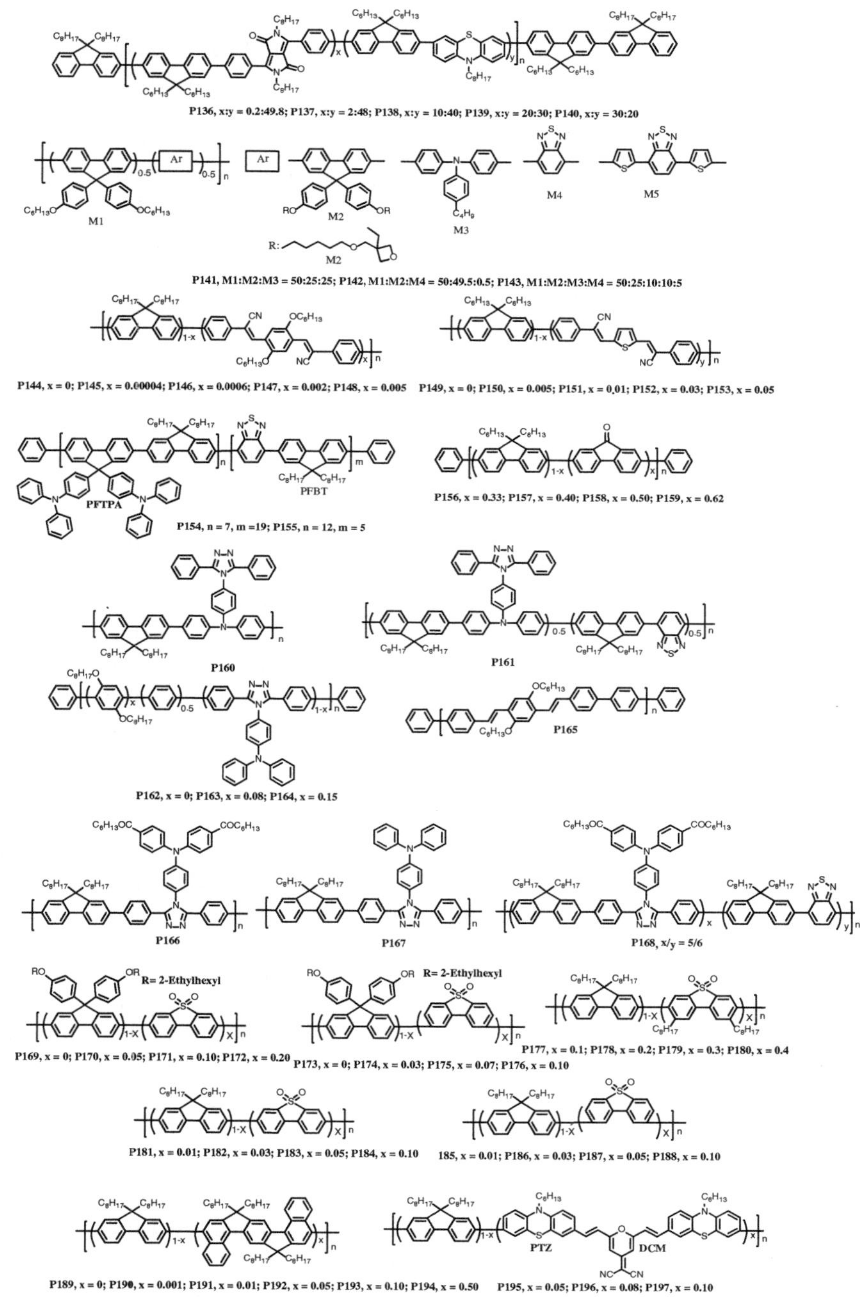

Fig. 7.7 Chemical structures of P136–P197

Diblock copolymers (P154–P155, Fig. 7.7) [102], where the emission of PFO-TPA block overlaps very well with the absorption of PFOBT block, allowing efficient both intra- and inter-chain energy transfer processes, were developed to improve the stability and the efficiency of PLEDs. It was found that the two blocks showed different morphological and structural properties. PFO-TPA showed a stable amorphous phase, whereas the PFOBT phase exhibited a noticeable degree of order. After thermal annealing, the P155 based PLED device exhibited good device performance with an EQE_{max} of 5.5 %, a L_{max} above 50,000 cd/m^2, a LE_{max} of 22.5 cd/A, and a CIE coordinate of (0.36; 0.59) in the structure of ITO/PEDOT: PSS/PVK/polymer/Ba/Al. By introducing electron-withdrawing fluorenone units, both LUMO and HOMO levels of poly(fluorene-*co*-fluorenone)s (P156–P159, Fig. 7.7) were slightly lowered with the increase in fluorenone contents [103]. Their PL spectra in film were mainly originated from fluorenone chromophores because of the efficient energy transfer from the fluorene to fluorenone segments. The performances of the PLEDs with an ITO/PEDOT:PSS/polymer/Ca/Al structure were significantly enhanced depending on the contents of fluorenone. The L_{max} and LE_{max} were 4400 cd/m^2 and 1.52 cd/A for P159-based devices.

Fluorene-based conjugated copolymers (P160–P161, Fig. 7.7) contained both the electron-rich TPA groups and electron-deficient 1,2,4-triazole groups [104]. A small number of BT units were introduced as the emission centre and the energy transfer efficiency from fluorene segment to low bandgap BT unit increased significantly with the increase of BT content. The P161-based PLED with a structure of ITO/PEDOT:PSS/polymer/Ca/Al showed emission peak at 542 nm with a CIE coordinate of (0.345, 0.625), a low V_{on} of 5 V, a L_{max} of 696 cd/m^2, and a LE_{max} of 2.02 cd/A. By introducing a directly linked hole transporting TPA and electron transporting aromatic 1,2,4-triazole segment into a PPV derivative's (P165) skeleton, Wu et al. [105] synthesized two bipolar copolymers (P163–P164, Fig. 7.7). By blending P163, P164 with P165 as an EML, the PLEDs' (structure: ITO/ PEDOT:PSS/polymer blend/Ca/Al) emission efficiency was effectively improved. The L_{max} and LE_{max} were significantly enhanced from 310 cd/m^2 and 0.03 cd/A (EML: P165) to 1450 cd/m^2 and 0.20 cd/A (EML: P165/P163), respectively. Wang et al. [106] designed three conjugated polymers (P166–P168, Fig. 7.7) with electron-deficient main chains and electron-donating pendant groups. The incorporation of BT into P168 reduced the coplanarity of the conjugated polymer, which prohibited the polymer from aggregation. Because of a good charge balance, PLEDs with an ITO/PEDOT:PSS/P168/Ca/Ag structure exhibited a LE_{max} as high as 2.60 cd/A.

In most cases, those blue-emitting homopolymers (such as PFs, PCZs. etc.) have a large band gap and emit pure blue light in solution and thin films when excited by UV light, whereas they always exhibit poor color stability and low EL efficiency in PLED devices because of the aggregate/excimer or defect formation [18, 21]. Dopant-host strategy has also been successfully employed to solve these problems and achieve high efficiency blue-emitting polymers.

Liu et al. [107] reported spectrally stable blue-light-emitting PFs with high efficiencies via introducing SO isomer (3,7-diyl or 2,8-diyl) into P169–P176's (Fig. 7.7) main chain because of their excellent electron-transporting properties, good electron affinity, and high fluorescence efficiency. By introducing SO isomer into PF backbone, the spectral stability and efficiency of the blue-emitting PFs are significantly improved. Furthermore, green emission, which was usually associated with excimer/aggregation or defects, was absent in PL and EL spectra. PLED with an ITO/PEDOT:PSS/PVK/P173/Ba/Al structure exhibited the best performance with a LE_{max} of 6.0 cd/A, an EQE_{max} of 5.5 %, and a CIE coordinate of (0.16, 0.19). To improve further the solubility of the SO-based blue-light-emitting polymers, the 2,8-dioctyldibenzothiophene-*S,S*-dioxide (DOSO) was introduced into the P177–P180's (Fig. 7.7) backbone [43]. The optical band gap increased from 2.95 to 3.20 eV as the content of DOSO unit increased. The EL spectra of the resulting polymers showed a CIE coordinate around (0.16, 0.07), independent of the ratio of DOSO units in the polymers, because of the efficent internal charge transfer. A high efficiency PLED (configuration: ITO/PEDOT:PSS/P178/Ba/Al) with a LE_{max} of 3.1 cd/A and an EQE_{max} of 3.9 % was obtained.

Li et al. [108] also synthesized a series of SO-containing PFs (P181–P188, Fig. 7.7). PLEDs with an ITO/PEDOT:PSS/polymer/Ba/Al structure based on P181 and P188 showed a EQE_{max} of 3.6 %, a LE_{max} of 3.7 cd/A with a CIE coordinate of (0.16, 0.07), and a EQE_{max} of 3.8 %, a LE_{max} of 4.6 cd/A with a CIE coordinate of (0.15, 0.12), respectively. In other research work, King et al. [109] found that when the SO unit contents increased to 30 %, these copolymers display broad emission, observed as greenish-white light, which arises from dual fluorescence, viz. both local excited states and charge transfer states.

By introducing dinaphtho-s-indacene into a PF host, Guo et al. [110] developed a series of spectrally stable blue-light-emitting copolymers (P189–P194, Fig. 7.7) as the guest for energy transfer. High efficiency, good color stability, and purity of results were achieved because of the formation of low energy segments in the main chain. The best device performance with a LE_{max} of 3.43 cd/A, a L_{max} of 6539 cd/m^2, and a CIE coordinate of (0.152, 0.164) was achieved for P191-based PLED with an ITO/PEDOT:PSS/polymer/poly(9,9-bis(diethoxylphosphorylhexyl)fluorene) (PF-EP)/Al structure.

Lee et al. [111] developed a series of novel fluorene-based copolymers (P195–P197, Fig. 7.7) containing ambipolar moieties, such as 10-*n*-hexylphenothiazine or 2-(2,6-bis-2-(5-(cyanomethyl)thiophen-2-yl)vinyl)-4*H*-pyran-4-ylidene)malononitrile (DCM). These ambipolar moieties could prompt the formation of the β-phase along the polymer chain. PLEDs based on the β-phase PF copolymers P195 and P196 exhibited a deep blue emission which originated from an energy transfer from the amorphous matrix to the β-phase and a weaker long-wavelength tail. As an increasing fraction of the DCM units in P197, an additional peak which appeared at 606 nm was observed. The PLED with an ITO/PEDOT:PSS/polymer/Ca/Al structure based on P195 exhibited a greater LE_{max} of 0.74 cd/A and an improved L_{max} of 1900 cd/m^2 relative to the devices containing PF.

7.3.1.2 Polymers with Dopants in the Side Chains

An alternative approach to design high efficiency EL polymers of dopant/host system with molecular dispersion features is to attach covalently small amounts of dopant units to the side chains of the polymer host. Zhou et al. [112] synthesized a series of red EL polymers (P198–P200, Fig. 7.8) by using PF as blue host and BT derivatives with different emission wavelengths as red dopants on the side chains. By introduction of an ethanol soluble polymer PF-EP as the ETL, PLEDs with an ITO/PEDOT:PSS/polymer/PF-EP/LiF/Al structure showed a pure red emission at 624 nm with a LE_{max} of 5.50 cd/A and a CIE coordinate of (0.62, 0.35) for P198, a saturated red emission at 636 nm with a LE_{max} of 3.10 cd/A, and a CIE coordinate of (0.63, 0.33) for P199, respectively [113]. By attaching small amounts of highly efficient red emission D-A-D-type 2,1,3-benzoselenadiazole and 2,1,3-naphthothiadiazole derivatives to the side chains of PF hosts (P201–P204, Fig. 7.8), complete energy transfer from the PF host to the red dopants occurred [114]. The EL spectra

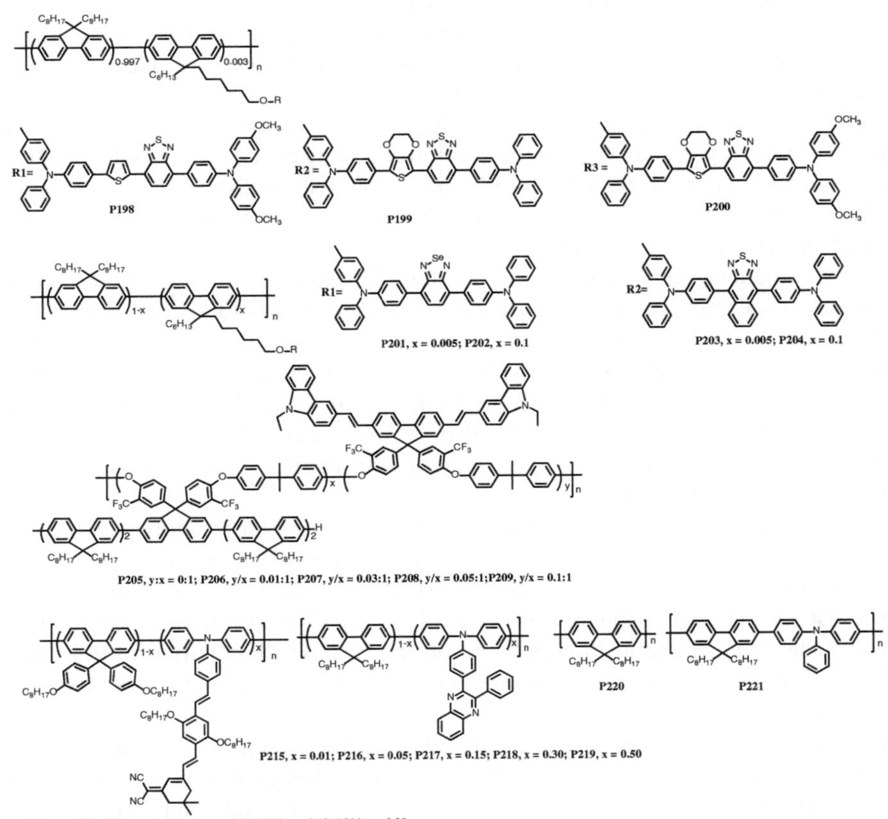

Fig. 7.8 Chemical structures of P198–P221

of the resulting polymers showed predominantly a red emission, attributed to the dopants. Among all the polymers, P204 exhibited the best device performance in the PLED with the structure of ITO/PEDOT:PSS/polymer/Ca/Al and showed a pure red emission with a peak at 632 nm, a LE_{max} of 3.04 cd/A, and a CIE coordinate of (0.63, 0.35).

The emission spectra of novel arylene ether polymers (P205–P209, Fig. 7.8) containing both pentafluorene (5F) and distyrylarylene derivative (BCzVF) units in the side chains indicated that color tuning could be achieved through efficient Förster energy transfer from the deep-blue emssion 5F host to the pure-blue emission BCzVF dopant [115]. Single-layer PLEDs with an ITO/PEDOT:PSS/polymer/Ca/Al structure based on P207 exhibited a voltage-independent and stable pure blue emission with a CIE coordinate of (0.15, 0.15), a L_{max} of 3576 cd/m^2, and a LE_{max} of 2.15 cd/A.

Park et al. [116] designed a series of new fluorene-based copolymers (P210–P214, Fig. 7.8) with varying molar ratios of the low band gap comonomer 2{3-vinyl)-2,5-bisoctyloxyphenyl]vinyl}-5,5-dimethyl-cyclohex-2-enylidene)malononitrile (BTBM). All the coplolymers exhibited both red PL and EL emissions produced by the efficient energy transfer between the blue-light-emitting fluorene segments and the orange-light-emitting BTBM units. The P212-based PLED with an ITO/PEDOT:PSS/polymer/bis(2-methyl-8-quinolinolato-*N*1,*O*8)-(1,1′-biphenyl-4-olato)aluminum/LiF/Al structure showed the best performance with a L_{max} of 510 cd/m^2 and a LE_{max} of 0.57 cd/A.

Conjugated PF derivatives (P215–P219, Fig. 7.8) were comprised of electron-donating TPA based backbone and electron-accepting quinoxaline based pendant groups [117]. These copolymers' PL spectra displayed a significant red-shift relative to that of P220 and P221, and peaked in the range of 475–510 nm. The disappearance of the emission bands at ~423 nm indicated that the efficient energy-transfer happened between the fluorene segment and D-A moiety. PLEDs with an ITO/PEDOT:PSS/polymer/Ba/Al structure exhibited superior performance compared to that of corresponding P220- and P221-based PLEDs.

7.3.2 Electrophosphorescent Polymers

Electrophosphorescence materials, especially heavy-metal complexes, have attracted much attention recently because they can make full use of both singlet and triplet excitons due to strong spin-orbital coupling of heavy-metal ions in phosphorescent complexes. As a result, electrophosphorescence can theoretically approach 100 % internal quantum efficiency [118, 119]. Although the high-efficiency PLEDs with phosphorescent dye doped into a small molecule or polymer host have been successfully realized, there are some problems that need to be overcome, such as complexity in fabrication procedure and fast decay of efficiency with increase in current density, which might become more serious during a long-term operation because of the intrinsic instability of such blend systems [120]. A possible solution to these problems is to introduce a phosphorescent dye into the polymer main chain or side chain, which is a simple and effective strategy to develop high-efficiency light emitting electrophosphorescent polymers.

Ma et al. [121] reported a series of red-light emitting electrophosphorescent PFs (P222–P227, Fig. 7.9) with various contents of a red dye quinoline-based iridium complex $(PPQ)_2Ir(acac)$ (bis(2,4-diphenylquinolyl-N,C2′)iridium(acetylacetonate)) in the side chains. Even at a low Ir content of 1 mol%, the resulting polymer's EL

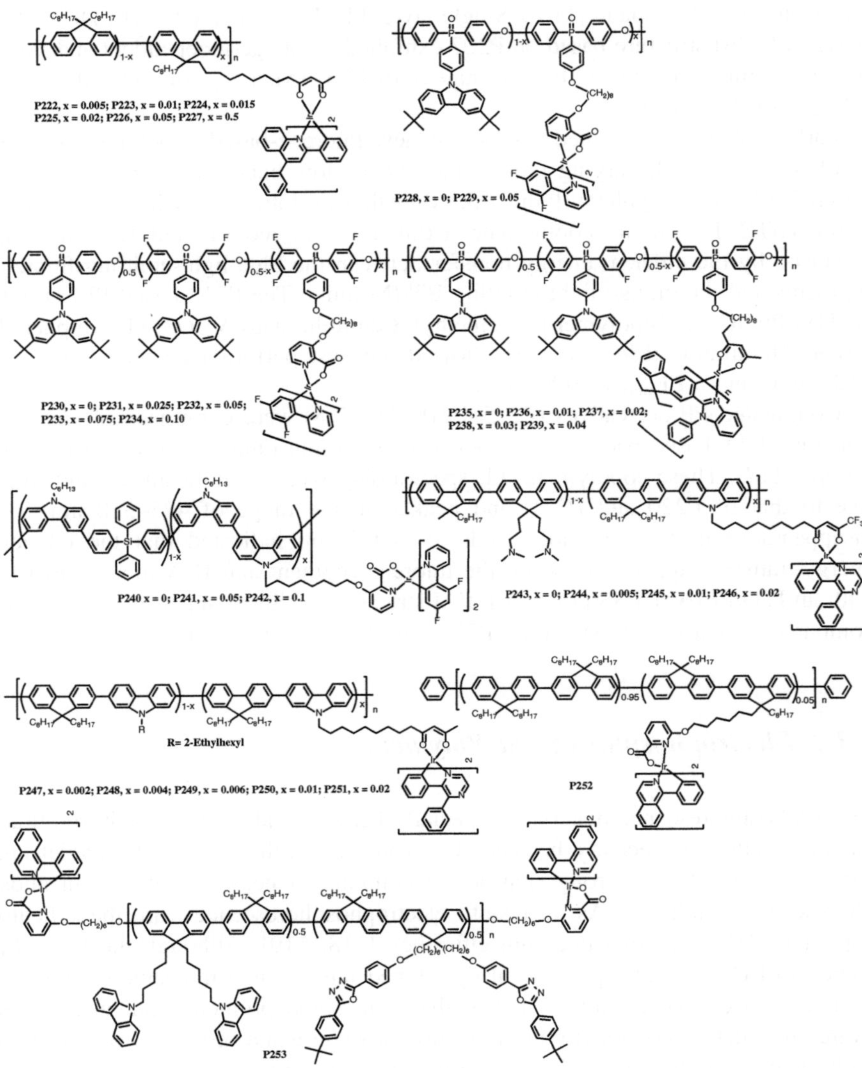

Fig. 7.9 Chemical structures of P222–P253

emission from fluorene unit was completely quenched because of the efficient energy transfer and direct charge trapping in the polymer. A single-layer device based on P223 showed a gentle efficiency roll-off at high current density, and a LE_{max} of 5.0 cd/A. In poly(arylene ether phosphine oxide)-based bipolar hosts (P228, P229, Fig. 7.9), oxygen atoms could interrupt the conjugation length along the main chain. In comparison to PVK, P228's HOMO and LUMO levels were tuned to −5.7 and −2.3 eV, respectively, resulting in a good charge injection property although the triplet energy remained as high as 2.96 eV. With P229 as the host and the blue phosphor FIrpic as the dopant, a promising LE_{max} of 23.3 cd/A had been attained [122]. Promoted by this state-of-the-art performance, FIrpic was introduced to the side chain of fluorinated polymers (P230–P234, Fig. 7.9) via covalent bonding. The resulting polymers emitted blue phosphorescence from FIrpic, and the corresponding devices gave a LE_{max} of 19.4 cd/A [123]. Moreover, P230 is also a suitable scaffold for efficient yellow phosphorescent complexes. A series of novel yellow-emitting electrophosphorescent polymers (P235–P239) had been developed by grafting a yellow dye (bis[2-(9,9-diethyl-9*H*-fluoren-2-yl)-1-phenyl-1*H*-benzimidazolate-N,C] (acetylacetonato)-iridium(III) ($(fbi)_2Ir(acac)$) on the side chains [50]. Because of the efficient intermolecular energy transfer from host to $(fbi)_2Ir(acac)$ and charge trapping on $(fbi)_2Ir(acac)$, the EL from host is almost completely quenched, even when the Ir complex content incorporated into the polymer was as low as 2 mol%. The resulting device exhibited good performance with a LE_{max} of 10.4 cd/A. Fei et al. reported a series of blue-light emitting electrophosphorescent polymers (P240–P242, Fig. 7.9) with 3,6-carbazole-*alt*-tetraphenylsilane copolymers as the host and blue emissive iridium complex FIrpic on the side chains as the dopant, in which the content of the complex can be controlled by adjusting the feed ratio of monomers [124]. In thin film, the polymer films mainly show blue emission from FIrpic, and the emission from the host was completely quenched because of the efficient energy transfer from host to covalently bonded FIrpic dopant. The resulting PLEDs with the structure of ITO/PEDOT:PSS/polymer/TPBI/LiF/Al exhibited good performance with a LE_{max} of 2.3 cd/A, and the efficiency roll-off at high current densities was suppressed. Amino-alkyl-containing light emitting conjugated polymers can significantly enhance the high work-function metal cathode devices' performance because of the interfacial dipole formation between the amino groups and the cathode, which can effectively reduce the work function of metal cathode and are beneficial for electron injection. Ying et al. reported a series of amino-alkyl-containing polymers (P243–P246, Fig. 7.9) with Ir complexes on the side chains [125]. All the polymers exhibited good performance in the device with a high work function metal (Al or Au) as cathode, and the EQE_{max} reached 3.7 and 1.6 % for Al and Au cathode devices, respectively.

Ying et al. [126] developed novel poly(fluorene-*alt*-carbazole)-based copolymers (P247–P251, Fig. 7.9) with 3,6-carbazole-*N*-alkyl grafted iridium complex using 2,3-diphenylpyrazine as ligand (IrBpz). The emission of host, poly(fluorene-*alt*-carbazole), was completely quenched when the copolymer contained 1 mol% of iridium complex. An orange-red emission with a CIE coordinate of (0.56, 0.42) was observed from PLEDs with an ITO/PEDOT:PSS/polymer + PBD/Ba/Al structure with a LE_{max} of 5.58 cd/A and a L_{max} of 8625 cd/m^2. By tuning the content of iridium complex, high efficiency white light with a CIE coordinate of (0.33, 0.27) was observed from PLEDs with a LE_{max} of 2.30 cd/A and a L_{max} of 2068 cd/m^2.

Tan et al. [127] reported two D-A-based phosphorescent PF derivatives (P252, P253, Fig. 7.9). These polymers contain the carrier-transporting units of carbazole and oxadiazole and the end-capped phosphorescent unit of red-emitting iridium(III) bi(phenylisoquilonato) (picolinato) [Ir(Piq)$_2$(pic)]. All carrier-transporting and phosphorescent units were appended by unconjugated linkage as substitutes in the C-9 position of fluorene. In the PLEDs with an ITO/PEDOT/polymer/LiF/Al structure, the P276 showed the best EL properties with a LE_{max} of 0.72 cd/A and a L_{max} of 1398 cd/m^2.

Besides introducing a phosphorescent dye into the polymer side chain, introducing a phosphorescent dye into the polymer main chain has also attracted much attention. By introducing extended π-conjugated iridium complex ligands ((ptb)$_2$Irdbm) and long alkyl chains which contribute to control the HOMO energy level and the solubility, Park et al. [128] synthesized a series of copolymers (P254–P258, Fig. 7.10). The P256-based PLED showed the best performance with a L_{max} of 2260 cd/m^2 and a LE_{max} of 1.1 cd/A at 7.5 V because of the balanced electron and hole injection in the device.

Huang et al. [129] developed a series of 2,8-disubstituted fluorene-dibenzothiophene (D)- and 2,8-disubstituted fluorene-dibenzothiophene-*S,S*-dioxide (DO)-based copolymers (P259–P267, Fig. 7.10) with 3 and 10 mol% of covalently-bonded iridium segments in the backbones. With increasing contents of D or DO segments in the copolymers, the HOMO/LUMO levels and E_T values were enhanced. PLEDs with an ITO/PEDOT:PSS/polymer/TPBI/LiF/Al structure showed high performance with a LE_{max} of 6.20 cd/A and a CIE coordinate of (0.47, 0.43).

Huang et al. [130] synthesized a series of novel electroluminescent copolymers (P268–P277, Fig. 7.10) containing fluorene-1,4-bis(9-octyl-9*H*carbarzol-3yl)-2,5-dioctyloxy-benzene (BCB) segments and phosphorescent benzimidazole-based iridium (Ir) complexes in the backbones. As the content of BCB units in the copolymers increased, the triplet energy of copolymers grew; thus the energy transfers induced from the polymer backbones to the iridium units were much more efficient, and energy back transfers induced from the iridium units to the polymer backbones were much less efficient. When the Ir-containing copolymer P275 was used as EML in PLED with an ITO/PEDOT:PSS/polymer/TPBI/LiF/Al structure, a high performance white-light emission with a LE_{max} of 1.88 cd/A and a L_{max} of 1960 cd/m^2 was achieved.

Zhang et al. [131] introducd large steric hindrances oxadiazole derivatives into phosphorescent polymers (P278–P281, Fig. 7.10) effectively to suppress the aggregation of the polymer backbones. With the excellent electron-transporting and hole-blocking property of the oxadiazole units, the LUMO level of the polymers reduced. Among simple devices with an ITO/PEDOT:PSS/polymer/Ca/Al structure, the device with P281 showed the best performance with a L_{max} of 846 cd/m^2 and a LE_{max} of 0.61 cd/A because of the charge balance and high PL efficiency.

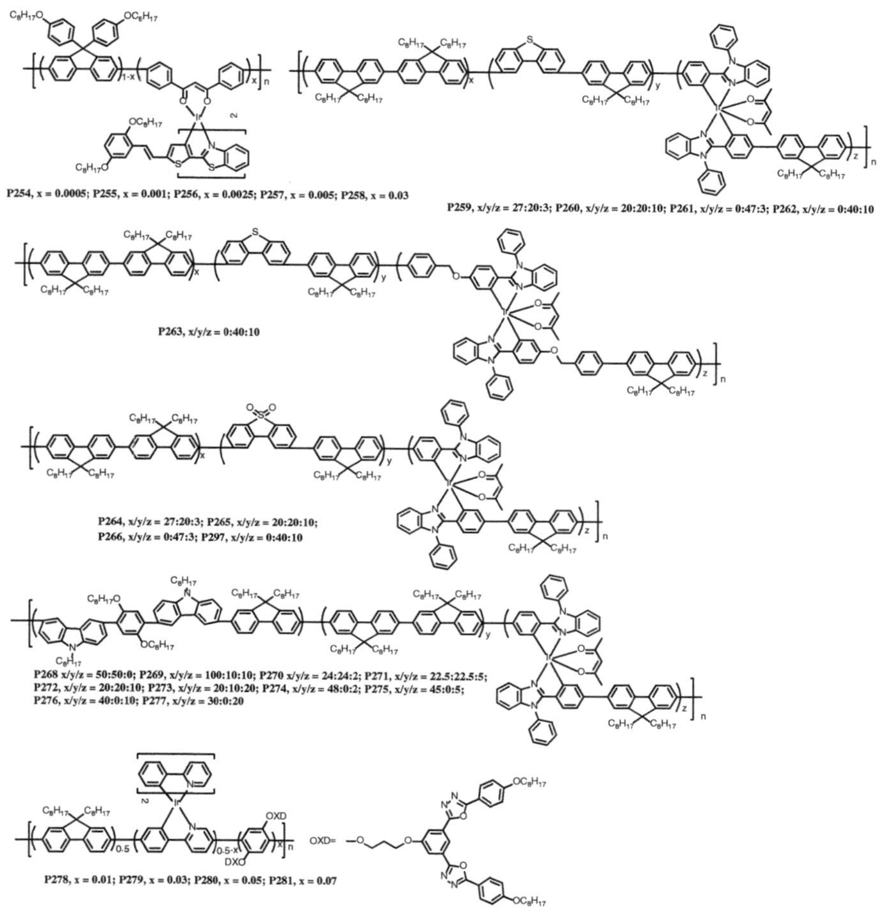

Fig. 7.10 Chemical structures of P254–P281

Another way to develop electrophosphorescent polymers is to introduce a nonconjugated covalent linkage between phosphorescent dopants and nonconjugated polymer hosts. In such systems, the polymer's backbone does not take part in charge transport or emission because of its extremely high bandgap of more than 4 eV. A series of different iridium complex dopants were attached to the polystyrene backbone as the emission units. Moreover, *N,N*-di-*p*-tolyl-aniline and 2-(4-biphenyl)-5-(4-*tert*-butylphenyl)-1,3,4-oxadiazole (*tert*-BuPBD) were also attached to the polystyrene backbone to improve the resulting polymers' charge injection and transport ability [132]. PLEDs with an ITO/PEDOT:PSS/polymer/CsF/Ca/Ag structure based on these multiple functionalized polymers (P282–P287, Fig. 7.11) exhibited high efficiencies of 28 cd/A at 6 V (green), 4.9 cd/A at 5 V (red), and 4.3 cd/A at 6 V (blue).

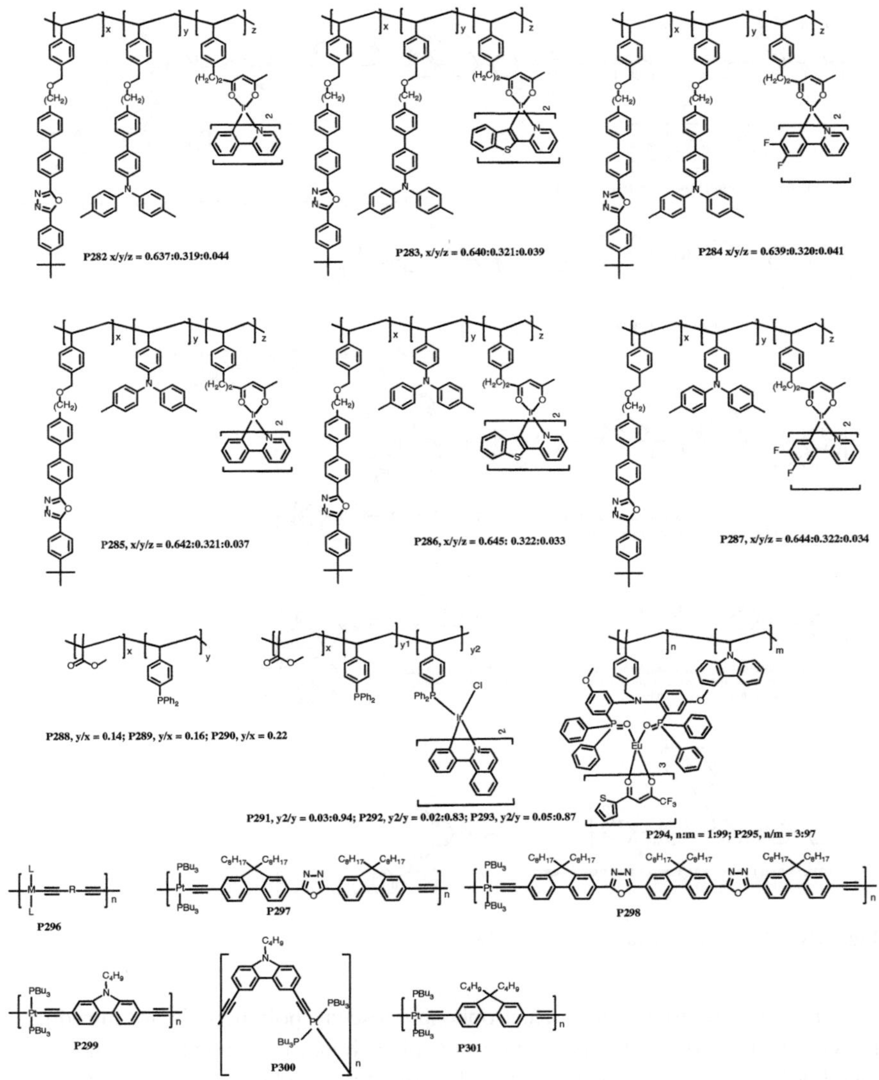

Fig. 7.11 Chemical structures of P282–P301

Koga et al. [133] developed phosphorescent cyclometalated iridium-containing metallopolymers (P288–P293, Fig. 7.11), in which near-red luminescent iridium complexes bearing phosphine-containing copolymers are used as polymer ligands. PLEDs fabricated from the P288–P293 and its nonpolymer analog, [IrCl(piq)$_2$PPh$_3$], exhibited quite similar luminescence behavior, except for the emission from the free-phosphine-units in the polymer side chain and their energy-transferring properties from host to guest materials. PLED with a structure of ITO/PEDOT:PSS/P293

(20 wt%):PFO-TPA/Ba/Al exhibited the best result with a L_{max} of 73.4 cd/m^2 and a LE_{max} of 0.13 cd/A, respectively.

By using a novel polymerizable Eu^{3+} complex as the ligands, Xu et al. [134] developed two metallopolymers (P294, P295, Fig. 7.11). The Eu^{3+}-complexed moieties served as the double-carrier traps (Eu trap) in the copolymers. The single-layer spin-coated devices with an ITO/PEDOT:PSS/polymer:PBD/Ba/Al structure realized the pure red emissions from Eu^{3+} ions with the L_{max} of 149.1 cd/m^2, which is one of the highest EL among Eu^{3+}-containing copolymers.

A prototype polymetallayne (P296, Fig. 7.11) is a type of polymer in which the polymer had a linear backbone comprised of the metal center M, the spacer group R, and the auxiliary ligand L on the metal center [135]. These polymers were based on the group 10 platinum metal with the formulation *trans*-[-Pt$(PBu_3)_2$C≡C(p-C_6H_4)C≡C-]$_n$. Recently, a series of soluble and thermally stable group 10 platinum (II) polyyne polymers (P297, P298, Fig. 7.11) were synthesized by Pd catalyzed sp^2-sp alkynylation reactions [136]. Geometry optimizations predict totally planar molecules for these metalated polymers, allowing better π-conjugation across the main chain. The ligands were strongly fluorescent but also became phosphorescent when the Pt atom was introduced in the backbone of the conjugated organometallic polymers. PLEDs using P298 as a phosphorescent dopant with a structure of ITO/PEDOT-PSS/polymer + PVK + PBD/BCP/Alq_3/LiF/Al had been prepared with good morphological properties, and exhibited a EQE_{max} of 0.15 % and a LE_{max} of 0.58 cd/A.

Ho et al. [137] developed a series of platinum(II)-acetylide polymers by using 2,7-carbazole (P299, Fig. 7.11), 3,6-carbazole (P300, Fig. 7.11), and 2,7-fluorene (P301, Fig. 7.11) as building blocks. P299 and P301 had similar photophysical properties, but there was a significant different between the photophysical properties of P299 and P300. Multilayer PLEDs with an ITO/PEDOT:PSS/polymer + PVK + PBD/TPBI/LiF/Al structure fabricated with P301 as the EML gave a strong green-yellow electrophosphorescence and exhibited a LE_{max} of 4.7 cd/A and a EQE_{max} of 1.5 %.

7.3.3 Single White Emitting Polymers

Single white emitting polymers (SWEPs) are capable of white emission from simultaneous blue, green, and red emissions or complementary blue, orange emissions [97]. The basic strategy to obtain such a polymer is covalently binding chromophores with RGB or complementary colors in the main or side chain of polymer hosts. By decreasing the doping content of the incorporated chromophores at a certain low level, the white emission could be achieved from the simultaneous emission of the host and the chromophores because of incomplete energy transfers. Lee et al. [138] reported a series of SWEPs (P302–P309, Fig. 7.12) containing dioctylfluorene, 2,3-bis(4-methyloxyphenyl)quinoxaline (moQ) and 5,8-bis(*N,N*-diphenylamino)-2,3-bis(4-methyloxyphenyl)quinoxaline (DPAmoQ) units. The

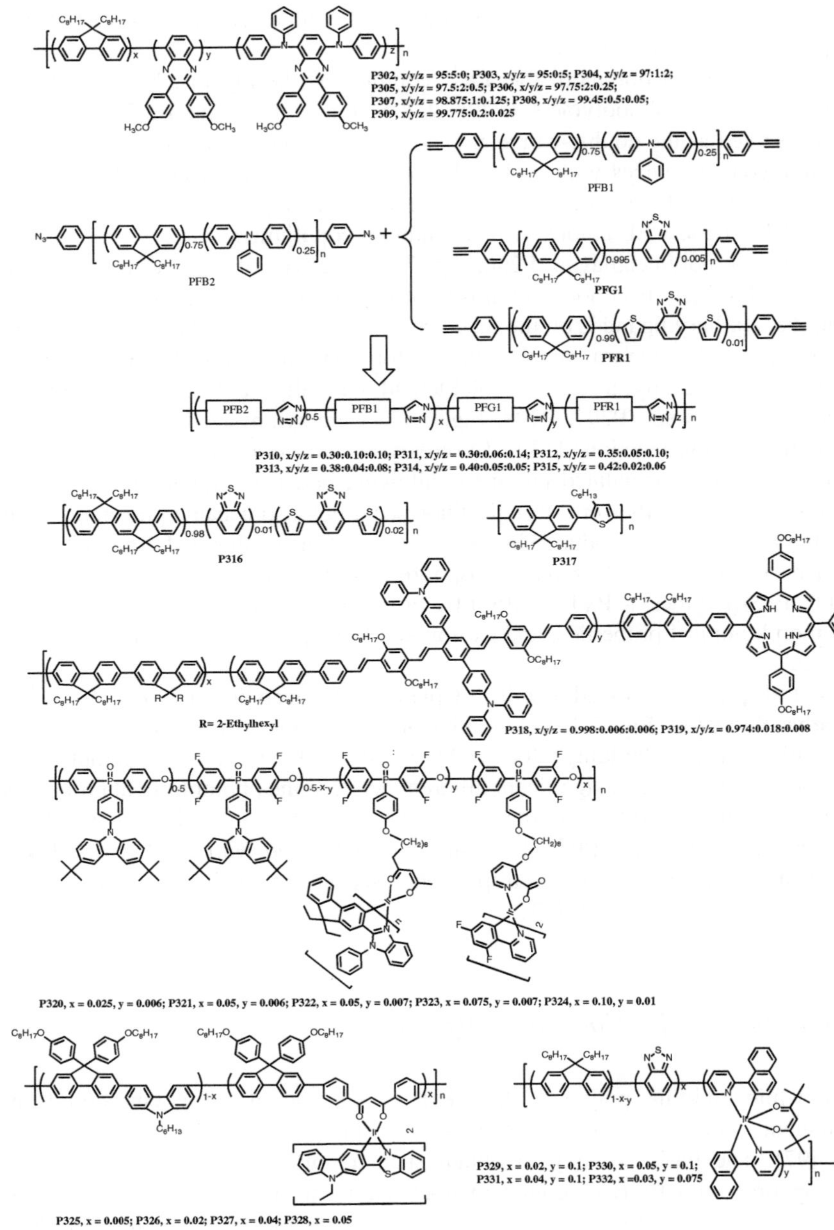

Fig. 7.12 Chemical structures of P302–P332

DPAmoQ units exhibited a longer wavelength emission because of the intramolecular charge transfer, where the biphenylamine group acts as an electron donor, and the quinoxaline unit acts as an electron acceptor. By adjusting the molar ratios of the monomers, white emission could be observed from the devices of P308 and P309. The devices (structure: ITO/PEDOT:PSS/polymer/Ca/Ag) of P308 and P309 exhibited a L_{max} of 12,300 and 7256 cd/m^2 as well as a LE_{max} of 2.02 and 1.87 cd/A with CIE coordinates of (0.33, 0.40) and (0.30, 0.36), respectively.

Lo et al. [139] reported a series of SWEPs synthesized by using azide-alkyne click reaction between RGB polymer precursors (PFB1, PFB2, PFG1, and PFR1, Fig. 7.12) with specific end-capping groups. The EL spectra of the resulting SWEPs (P310–P315, Fig. 7.12) showed three distinguishable emission peaks in the blue, green, and red regions produced by the incomplete energy transfer from the blue emissive segments PFB1 and PFB2 to the green emissive segment PFG1 or red emissive segment PFR1. The relative intensities of RGB emissions vary in accordance with the relative composition proportion of the different segments. By using P314 as emitting material, PLEDs with an ITO/PEDOT:PSS/TFB/polymer/Cs/Al structure exhibited the best EL performance with a L_{max} of 7551 cd/m^2, a LE_{max} of 6.21 cd/A, and a CIE coordinate of (0.30, 0.33).

A number of PFs and their derivatives have been studied as blue-emitting host polymers because of their large band gap, high PL and EL efficiency. However, the intrinsic emission of PF is located around 420 nm, where the human eye is not very sensitive. In comparison, polyindenofluorenes have emission maxima in regions two or three times as sensitive, relative to PFs. Jeong et al. [140] developed an indeno[1,2-*b*]fluorene-based copolymer (P316, Fig. 7.12) derived from tetraoctylindenofluorene, BT and DTBT derivatives as blue-, green-, and red-emitting moieties, respectively. Stable and pure white emission with simultaneous balanced RGB emission was achieved. The resulted PLEDs with an ITO/PEDOT:PSS/polymer/Ca/Al structure showed a L_{max} of 4088 cd/m^2 and a LE_{max} of 0.36 cd/A with a CIE coordinate of (0.34, 0.32).

Su et al. [141] demonstrated a two-color-combined SWEP poly(3-hexylthiophene-*alt*-fluorene) (P317, Fig. 7.12), comprising blue-green light-emitting fluorene and red-orange light-emitting 3-hexylthiophene. To enhance the optoelectronic performance of polymers, graphene is added into P317, and the electric conductivity increases with an increase in the amount of graphene. The resulting PLEDs with an ITO/PEDOT:PSS/(P317 1 % graphene)/Ca/Al structure showed two-color white EL with a L_{max} of 1273 cd/m^2, a LE_{max} of 0.83 cd/A, and a CIE coordinate of (0.28, 0.34).

Li et al. [142] developed two multicomponent copolymers (P318, P319, Fig. 7.12) containing PF, oligo(phenylenevinylene) (OPV), and porphyrin (Por) derivatives. PF acted as the host and blue-emitting unit because of its large band gap, and small amounts of OPV and Por derivatives acted as the guests for the green- and red-emission, respectively. By carefully controlling the concentrations of the guest species in the resulting copolymers, white EL, with contributions from all the three primary red, green, and blue colors, was achieved. The single layer PLEDs with an ITO/PEDOT:PSS/polymer/Ca/Al structure based on P319 emitted white light with a CIE coordinate of (0.29, 0.30), and a L_{max} of 443 cd/m^2.

Besides all-fluorescent SWEPs, all-phosphorescent SWEPs have also attracted much attention because this class of SWEPs can potentially make full use of both singlet and triplet excitons, leading theoretically to 100 % internal quantum efficiency. However, highly efficient all-phosphorescent SWEPs remains as a challenge because of the lack of suitable polymer hosts with high triplet energies above 2.75 eV as well as suitable HOMO/LUMO levels matched with the Fermi levels of the electrodes. Shao et al. [143] reported a series of all-phosphorescent SWEPs (P320–P324, Fig. 7.12) based on a fluorinated poly(arylene ether phosphine oxide), which had high triplet energy (2.96 eV) and matched HOMO/LUMO levels, simultaneously grafted with blue FIrpic and yellow $(fbi)_2Ir(acac)$ phosphors. By tuning the incorporated contents of FIrpic and $(fbi)_2Ir(acac)$, individual blue and yellow emissions were generated to give standard white EL emssions and the resulting devices exhibited a prominent efficiency as high as 18.4 cd/A with a CIE coordinate of (0.31, 0.43).

Park et al. [144] developed SWEPs (P325–P328, Fig. 7.12) containing a novel iridium complex with a β-diketonate unit, bis(2-benzothiazol-2-yl-*N*-ethylcarbazole)iridium-1,3-bis(*p*-bromophenyl)1,3-propanedione ((bec)2IrdbmBr), as the red emission unit. The resulted polymers exhibited two strong emission bands in both the blue and red spectral regions. By controlling monomer ratios and energy transfer, white light emissions were realized. PLEDs based on P326 with an ITO/PEDOT:PSS/poly(TPD)/polymers/TPBI/LiF/Al structure exhibited a white light emission with a CIE coordinate of (0.31, 0.32) and a LE_{max} of 0.05 cd/A. Moreover, the EL spectrum of P326 was stable with respect to applied voltage, and a CIE coordinate was almost not changed at various driving voltages.

Chen et al. [145] reported a series of SWEPs (P329–P332, Fig. 7.12), where the fluorescent chromophore BT and phosphorescence chromophore iridium(III)bis(2-(1-naphthalene)pyridine-$C^{2'}$,N)-2,2,6,6- tetramethyl-3,5-heptanedion $[(1\text{-}npy)_2Ir(tmd)]$ units were incorporated into PF's backbone as green and red emission units, respectively. By adjusting the monomer feed ratios, the white emission from three individual emission species in a single polymer was achieved. The device with an ITO/PEDOT:PSS/PVK/polymer/CsF/Al structure from P356 showed a LE_{max} of 5.3 cd/A and a L_{max} of 9900 cd/m^2 at a current density of 453 mA cm^{-2} and a CIE coordinate of (0.32, 0.34). Moreover, the EL efficiencies decline slightly with increase of current density.

7.4 Hyperbranched Polymers

As an alternative to linear polymers, hyperbranched polymers (HBPs) have received considerable attention because of their unusual molecular structures and properties. The hyperbranched structure helps to reduce the intramolecular and intermolecular interactions of the polymers, which in turn gives them much better solubility and highly fluorescent quantum yields in the solid state [146–148].

Chen et al. [149] developed star-like single-polymer systems (P333, P334, Fig. 7.13) through incorporating six blue PF arms onto star-shaped D-A type orange cores. The six blue PF chains acted as the branching arms and were expected to prevent the orange cores from aggregation and suppress their concentration quenching effect more effectively than previous linear and star-shaped single-polymers, which resulted in a higher PLEDs efficiency. As a result, their single layer devices with an ITO/PEDOT:PSS/polymer/Ca/Al structure achieved a high LE_{max} of 18.01 cd/A and an EQE_{max} of 6.36 % with a CIE coordinate of (0.33, 0.35), which was one of the best single-layer white PLEDs based on single-polymer fluorescence systems. To improve further the solubility of P334, three methyl groups were introduced, and a novel star-shaped orange core was designed [150]. The single-polymer (P335, Fig. 7.13), which was obtained by incorporating three PF arms into the TPB3 orange core, also showed a high EL efficiency. A typical single-layer device with an ITO/PEDOT:PSS/polymer/Ca/Al structure showed pure white emission with a LE_{max} of 16.62 cd/A, an EQE_{max} of 6.28 %, and a CIE coordinate of (0.33, 0.36) for P335 containing 0.02 mol% orange core.

Lin et al. [151] reported a series of new star-shape white electroluminescent polymers (P336–P342, Fig. 7.13) containing an orange core made of maleimide and three blue arms made of either PF or PCZ. These materials exhibited blue emission from PF and yellow to red emission from the maleimide as a result of partial energy transfer between the fluorene and maleimide segments. The EL spectra of devices can be adjusted by changing the molar ratio of maleimide to fluorene. A typical device based on the star-shape polymer P341 containing 0.01 mol% of maleimide exhibited a LE_{max} of 7.2 cd/A and an EQE_{max} of 3.2 %. The device based on P338 with a methyl substituent on the indole group can be improved to reach a L_{max} of 11,450 cd/m^2, whereas the device based on P341 with arms comprising equal amount of PF and PCZ can be boosted to a maximum PE of 4.8 lm W^{-1}.

For the purpose of achieving high efficiency HBPs-based red, green, blue, and white emission materials (Fig. 7.14), three HBPs—HB-terfluorene (P343), HB-4,7-bis(9,9′-dioctylfluoren-2-yl)-2,1,3-benzothiodiazole (P344), and HB-4,7-bis[(9,9′-dioctylfluoren-2-yl)-thien-2-yl]-2,1,3-benzothiodiazole (P345)—were developed by Shih et al. [152]. The L_{max} of the double-layer devices with an ITO/PEDOT:PSS/polymer/CsF/Al structure of P343, P344, and P345 were 48, 42, and 29 cd/m^2, respectively, and the LE_{max} of the devices were 0.01, 0.02, and 0.01 cd/A, respectively.

TPA derivatives with excellent hole mobility can be viewed as 3D systems, in which several electron-donating ethynylene units (carbazole, fluorene, or dialkoxybenzenes) have been introduced in the branched sides with TPA as the core [153]. Moreover, ethynylene units in the HBPs (P346–P349, Fig. 7.14) could provide an enhanced π-electron delocalization because of the coplanar molecular structure, which resulted in better charge-balancing. These HBPs were found to be highly fluorescent with PL quantum yields around 33–42 %. The EL maxima of these HBPs (P346–P349) were found to be in the range 507–558 nm. A L_{max} of 316–490 cd/m^2 could be achieved for the PLEDs with an ITO/PEDOT:PSS/polymer/LiF/Al structure.

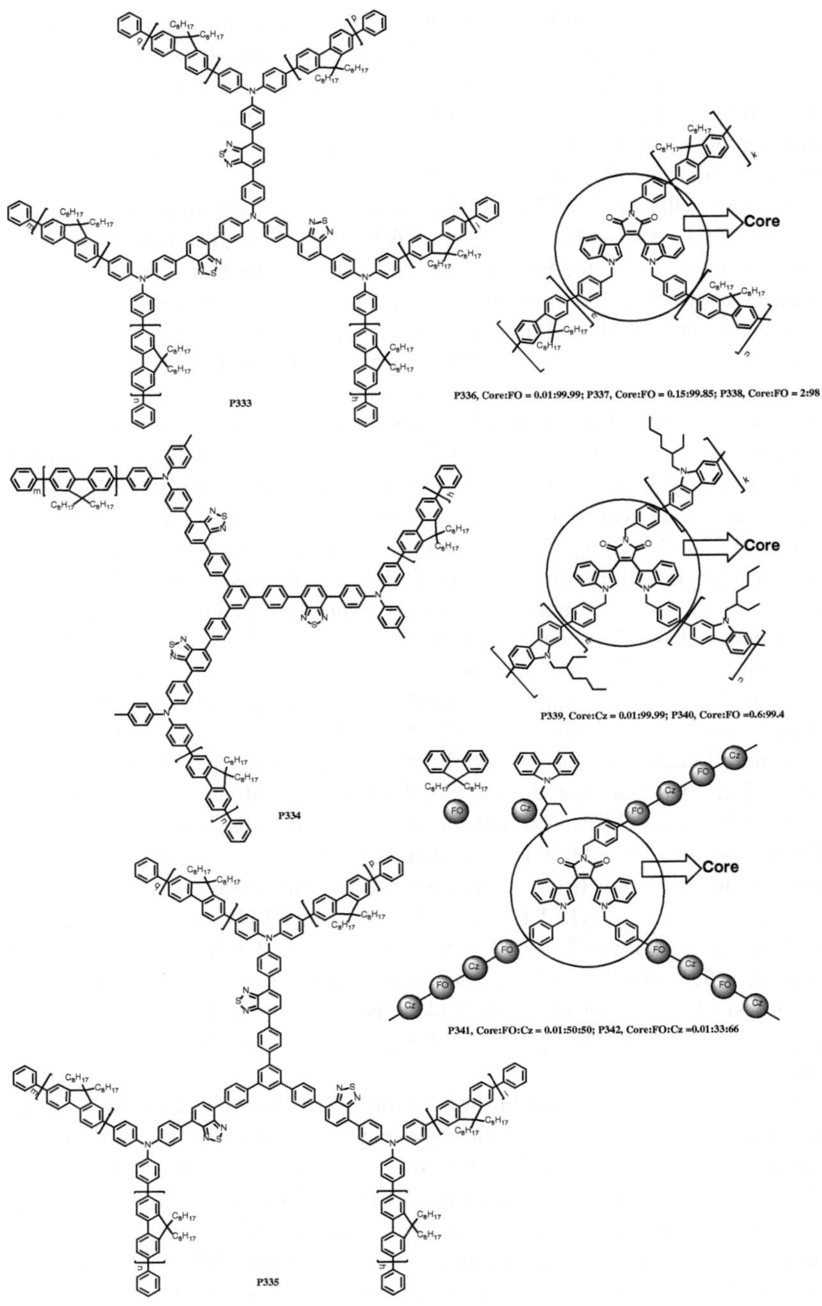

Fig. 7.13 Chemical structures of P333–P342

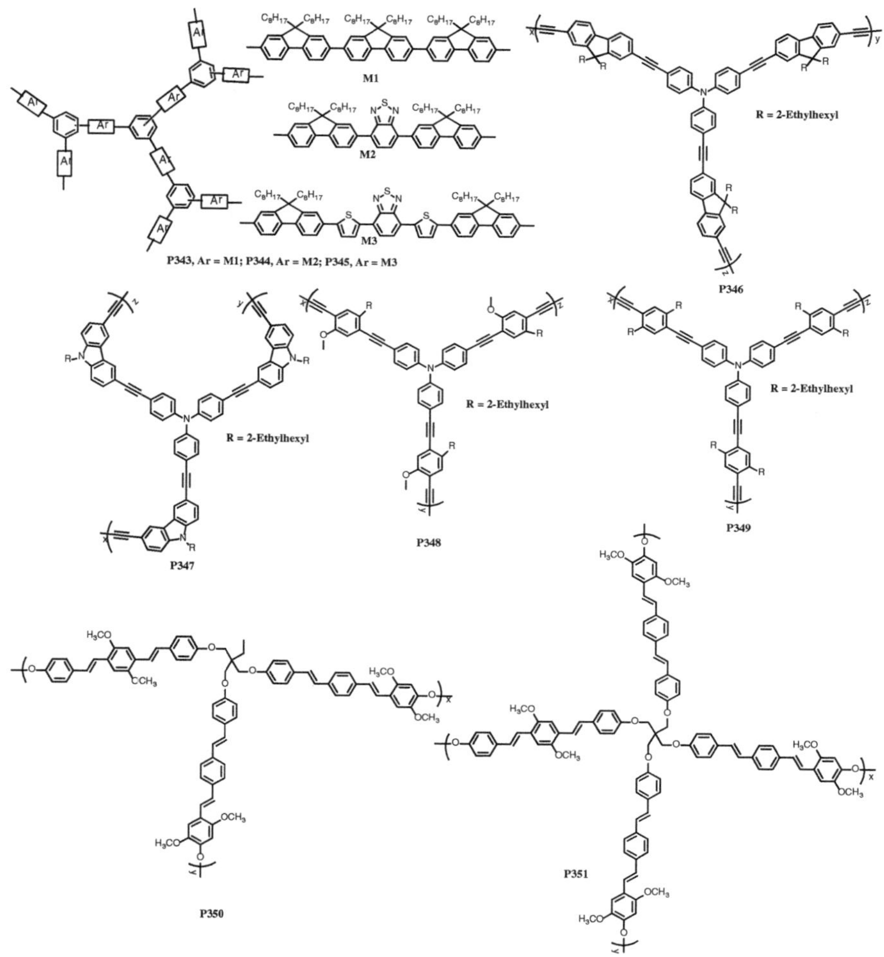

Fig. 7.14 Chemical structures of P343–P351

The AB2 type HBPs, where AB2 monomers polymerize intermolecularly to build up hyperbranched structures, have a mass of conjugated defects which may influence luminescent performance and result in increased probability of self-quenching and decreased luminescent efficiency. To overcome these problems, Lu et al. [154] synthesized a partially conjugated HBP (2,5-dimethoxy-substituted hyperbranched PPV, MOHPV) (P350, Fig. 7.14) based on rigid fluorescent conjugated segments, 2,5-dimethoxy-substituted distyrylbenzene (a derivative of oligo-PPV), and flexible non-conjugated spacers, trioxymethylpropane via a A2 + B3 approach. The three-dimensional and hindered structure of the HBPs endow them with a high fluorescence quantum yield because of their lower tendency to self-aggregate. A L_{max} of 1500 cd/m^2 and a LE_{max} of 1.38 cd/A were achieved when

P350 was used as EML in double-layer PLEDs with an ITO/PEDOT:PSS/P374 + PBD/Alq_3/Ca/Al structure. Moreover, Lu et al. [155] synthesized a novel partially conjugated HBP (P351, Fig. 7.14) via A2 + B4 approach, in which A2 is 1,4-distyrylol-2,5-butoxybenzene and B4 is pentaerythritol tetra(methylbenzene sulfonate). A single-layer device with an ITO/P351/Ca/Al structure reached an optimistic L_{max} of 190 cd/m^2 at 8.2 V.

Tsai et al. [156] prepared fluorescent hyperbranched copolymers (P352, Fig. 7.15) with inherent tetraphenylthiophene, TPA, and quinoline (Qu) moieties. In P352, TPA-Qu was used as the branching point of the HBPs, the electron-donating TPA units being separated by the electron-accepting Qu unit. With the wholly aromatic-heterocyclic structures, the resultant P352 were rigid materials with high

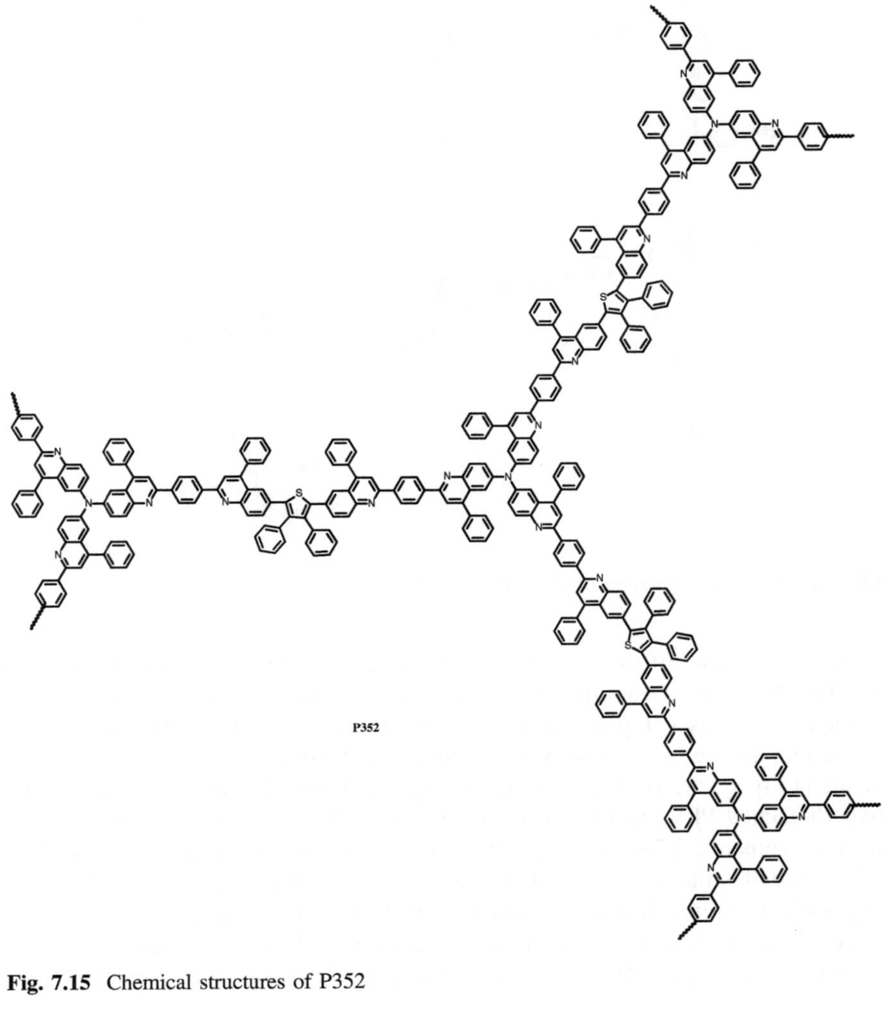

P352

Fig. 7.15 Chemical structures of P352

thermal transitions and stabilities, which enhanced their corresponding thermal and spectral stability during annealing at temperatures below glass transition temperature (*T*g). With a high *T*g of 315 °C, the EL spectrum of the P352-based device after heating at 300 °C for 4 h remained essentially the same as the pristine one before heating.

P353–P356 (Fig. 7.16), in which fluorene or TPA were used as donor chromophores and DCM was used as acceptor chromophore, were obtained through A2 + B3, A2 + B3 + B2, and A2 + B2 approaches by Vanjinathan et al. [157]. The symmetrical combination of DCM with fluorene or TPA assures better conjugated length and coplanarity. DCM is a strong electron-accepting group, which can

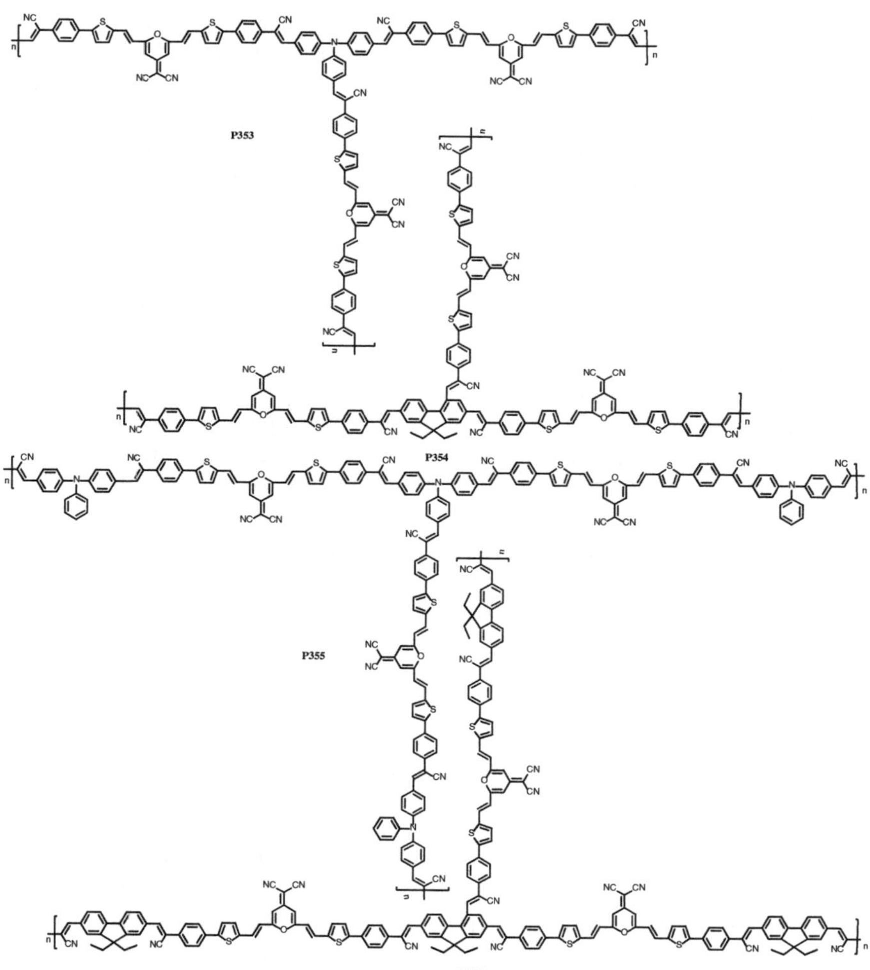

Fig. 7.16 Chemical structures of P353–P356

increase electron affinity and reduce the band gap of the conjugated system, and TPA is a well-known hole-transporting unit. As a result, the PLEDs with an ITO/PEDOT:PSS/polymer P355/BCP/Alq_3/LiF/Al structure showed high efficiency red-light emission with a L_{max} of 4104 cd/m^2 and a LE_{max} of 0.55 cd/A.

A series of HBPs (P357–P360, Fig. 7.17) with electron-deficient group triazine and hole-conducting group TPA as core and end-cap groups was developed to clarify systematically the influences of the type of core and end-cap group on the PL and EL properties [158]. Among all the HBPs-based PLEDs with an ITO/PEDOT:PSS/polymer/Ca/Al structure, the device based on P359 containing the triazine core and tetraphenylsilane end-cap group exhibited the best performance with a L_{max} of 1702 cd/m^2 and a LE_{max} of 0.72 cd/A.

A novel series of hyperbranched interrupted π-conjugated polymers (P361–P363, Fig. 7.17) based on complicated 9,9-diarylfluorenes branching core and end-capped with high carrier-mobility pyrene moieties were synthesized via the "A2 + A2 + B3" approach [159]. The TPA linked fluorene unit via the C9 position in the polymers break the conjugation between fluorene and TPA groups. Thus, the resulting HBPs had a better confined conjugation length and improved spectrum stability. The P363-based device with an ITO/PEDOT:PSS/PVK/P363/Ba/Al structure showed stable blue emission with the peaks at 422 and 447 nm and a CIE coordinate of (0.18, 0.16). The brightness of the device reached 1051 cd/m^2 at 15.7 V.

Li et al. [160] reported a new "A3 + B2"-type blue-emitting HBPs (P364–P367, Fig. 7.17) using 1,2,4-tribromohexaphenyl-benzene as the core and 1,3,4-oxadiazole with high PL quantum yields and high electron-transport mobility as branch units. The deep blue light emission of their films was very pure and stable, and no long wavelength excimer-like emissions at 500–600 nm were observed, even after annealing at 150 °C for 0.5 h because of the effective suppression of the formation of aggregation/excimer and keto defects. A two-layer PLED with an ITO/PEDOT:PSS/polymer/TPBI/Ca/Ag structure based on P366 showed the best performance with a LE_{max} of 0.72 cd/A and a L_{max} of 549 cd/m^2.

To improve further the efficiency of HBPs-based PLEDs, the aggregation-induced emission (AIE) active tetraphenylethene (TPE) units (B4) were used to construct all-conjugated polymer HBPs (P368–P370, Fig. 7.17) with carbazole, fluorene, and benzene moieties (A2), through an "A2 + B4" approach [161]. These polymers exhibited interesting aggregation-induced enhanced emission (AIEE) behavior. The advantages of the 3D topological structure of HBPs were reflected clearly in the fabricated PLEDs. All the performances (such as V_{on}, LE, L, etc.) of these HBPs-based PLEDs were much better than those of their analog linear copolymers. The PLED devices based on P368 with an ITO/PEDOT:PSS/Poly-TPD/polymer/TPBI/Cs_2CO_3/Ag structure showed a L_{max} of 948 cd/m^2, and a LE_{max} of 1.15 cd/A, which was higher than most of conjugated HBPs. Moreover, the same group [162] developed AIE active TPE units (A2) contained in HBP (P371, Fig. 7.17) with carbazole moieties (B4), a good hole-transporting and electroluminescent group, through an "A2 + B4" approach. The PLED with an ITO/PEDOT:PSS/poly-TPD/P371/TPBI/Cs_2CO_3/Ag structure exhibited a remarkably enhanced LE_{max} of 2.13 cd/A and a L_{max} of 5914 cd/m^2.

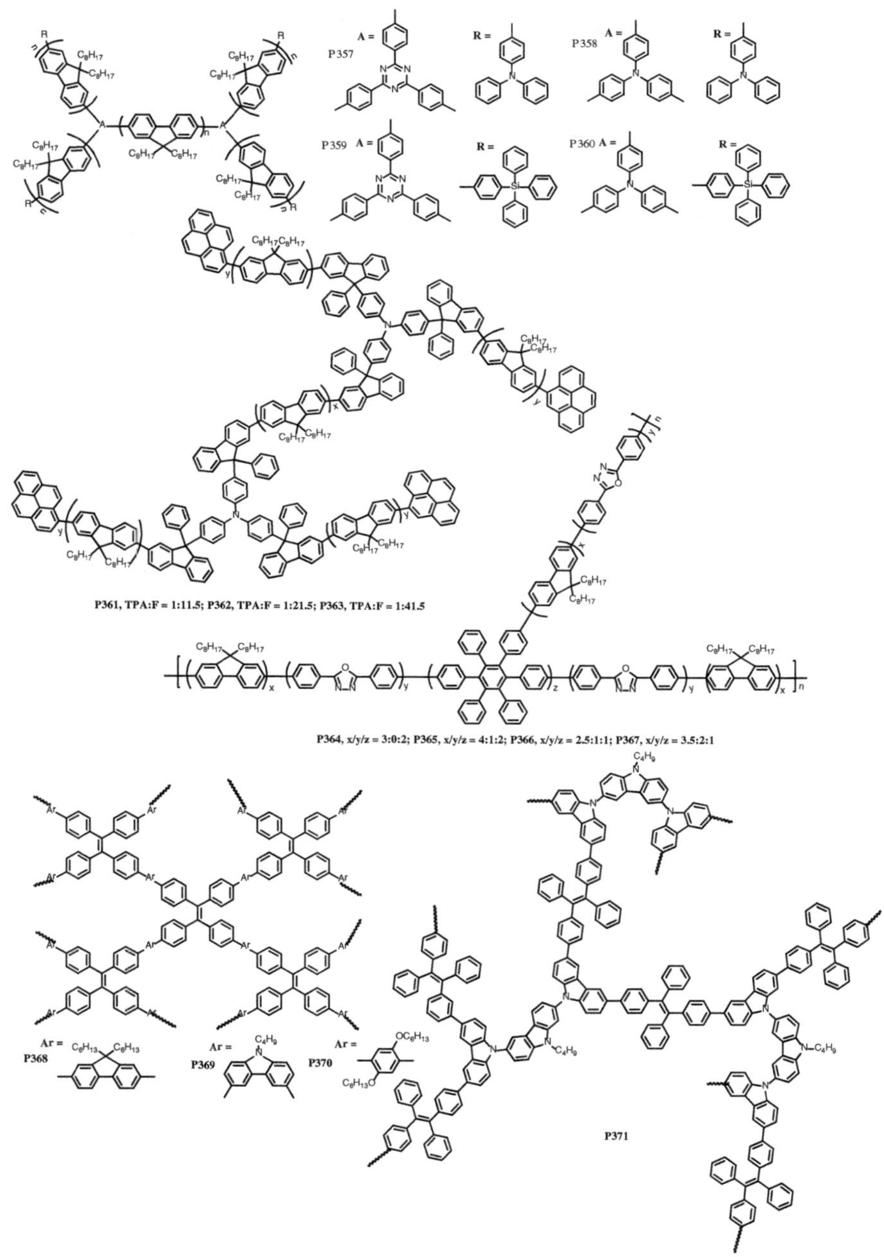

Fig. 7.17 Chemical structures of P357–P371

Guan et al. [163] designed a series of phosphorescent HBPs (P372–P376, Fig. 7.18), in which a charge transport balanced 3,6-carbazole, 2,6-pyridine-based copolymer with high triplet energy level acted as branch, and green light-emitting iridium, tris[2-(2-pyridinyl-κN)phenyl-κC] ($Ir(ppy)_3$) complex acted as core. Such a highly branched framework provided a novel molecular design for highly efficient phosphorescent green light-emitting polymers. The HOMO and the LUMO levels of copolymers reduced gradually on increasing the content of 2,6-pyridine units from 10 to 30 mol%. The incorporation of the pyridine moiety into the PCZ backbone resulted in significantly enhanced device efficiencies. The PLEDs with an ITO/PEDOT/P374 + PBD/Ba/Al structure showed high performance with an EQE_{max} of 13.3 % and a LE of 30.1 cd/A at 5.6 V. To improve further the electron

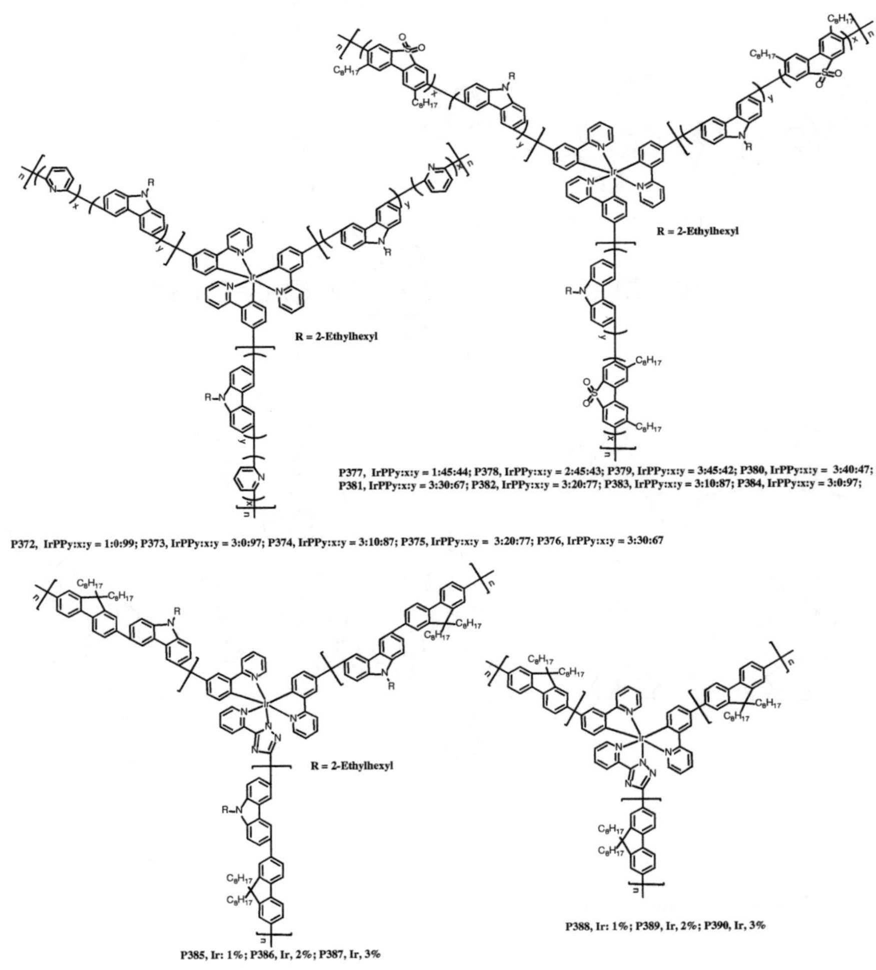

Fig. 7.18 Chemical structures of P372–P390

injection/transportation and charge transport balance, the 2,6-pyridine units were replaced by SO which has high fluorescence efficiency, high electron affinity energy and excellent electron transport properties [164]. The P377–P384's (Fig. 7.18) branches have a triplet energy level above 2.5 eV and can effectively prevent energy feedback from the phosphorescent core to carbazole-*co*-DOSO. A P380-based device with an ITO/PEDOT:PSS/polymer/CsF/Al structure exhibited the best performance, with a LE_{max} of 50.5 cd/A and a EQE_{max} of 15.3 %. When the core Ir $(ppy)_3$ complex was replaced by $(1\text{-phenylisoquinoline})_2Ir$(3-(pyridin-2-yl)-1*H*-1,2,4-triazole) ($[(Piq)_2Ir(Pytz)]$), a series of red-emitting HBPs were achieved via the "A2 + A2′ + B3" approach by utilizing poly(fluorene-*alt*-carbazole) (P385–P387, Fig. 7.18) and PF (P388–P390, Fig. 7.18) as the branches [165]. The HBPs P388–P390 with fluorene branches exhibited a better device performance. A LE_{max} of 6.54 cd/A, an EQE_{max} of 4.88 % with a CIE coordinate of (0.65, 0.34) were obtained from the P388-based PLED with an ITO/PEDOT/PVK/polymer/CsF/Al structure. Moreover, the device efficiencies from these materials show a reduced roll-off upon the increase of the current density.

7.5 Supramolecular Luminescent Polymers

Although significant progress has been made in PLEDs, there are still many challenges in developing high performance and long lifetime PLEDs. For example, most conjugated polymers are synthesized by the transition-metal-catalyzed cross-coupling reactions. It has also been shown that the remaining traces of metal catalysts have a detrimental effect on the resulting thin film device performance. In addition, the molecular weights as well as polydispersity of conjugated polymers that also play an important role on the performance of the resulting devices are hard to be defined during the polymerization process, which usually generate the batch to batch variation [166, 167].

To overcome these challenges, a possible solution is to replace the traditional conjugated polymers with supramolecular light-emitting polymers (SLEPs) generally formed from the monomeric units by directional and reversible secondary interactions such as hydrogen bonding, π-stacking, host-guest interactions, hydrophobic interactions, etc. [166, 167].

Abbel et al. [168] reported the pioneering work of using hydrogen bonded supramolecular copolymers (M1–M3, Fig. 7.19) for PLEDs, although the resulting devices exhibited poor performances with LE less than 0.1 cd/A. It is challenging to develop high performance supramolecular copolymers for optoelectronic device applications. SLEPs based on host-guest interactions were developed by Zhang et al. [169] in which the dibenzo-24-crown-8 functionalized blue-emitting conjugated oligomer (M4, Fig. 7.19) and green-emitting conjugated oligomer (M6, Fig. 7.19) were used as the host materials, and the dibenzylammonium salt functionalized blue-emitting conjugated oligomer (M5, Fig. 7.19) was used as the guest material. It was found that the SLEPs' emission colors can be well tuned from blue

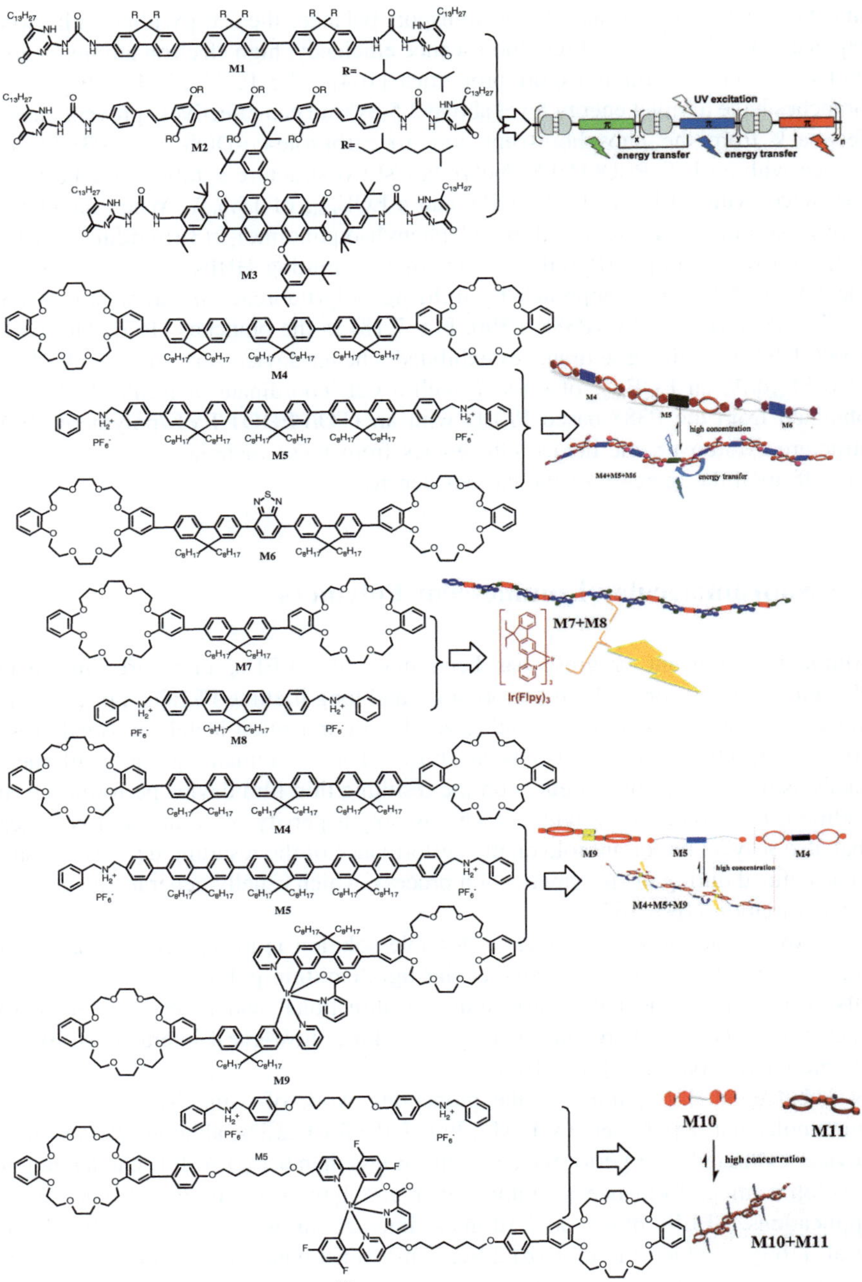

Fig. 7.19 Chemical structures of supramolecular luminescent polymers

to green with significantly enhanced photoluminescent efficiencies by using M6 as the dopant, which is because of the efficient energy transfer caused by the exciton trapping on narrow band gap oligomer M6 in the SLEPs. As a result, the devices with an ITO/PEDOT:PSS/SLEPs/TPBI/CsF/Al structure of SLEPs based on M4, M5, and M6 showed a LE_{max} more than 3 cd/A, which is much higher than that of devices of SLEPs based on M4 and M5. A large band gap supramolecular polymer based on host-guest interactions, was developed as the host material for phosphorescent PLEDs [170]. Because the host-guest interactions cannot enhance the conjugated length of the resulting SLEPs (M7 + M8) (Fig. 7.19), the SLEPs (M7 + M8) retains a high E_T similar to those of M7 and M8 monomers. Moreover, the SLEPs can be formed by the precisely defined monomers without using any metal catalysts, and have good film formation capability. As a result, the yellow-light emitting LED with an ITO/PEDOT:PSS/PBD:SLEP:Ir(Flpy)$_3$/TPBI/CsF/Al structure showed good performance with a LE_{max} of 18.2 cd/A.

Besides the supramolecular fluorescence polymer, supramolecular phosphorescent polymers (SPPs) were also developed and used as EML in PLEDs by Liang et al. [171]. The SPPs (M4 + M5 + M9) (Fig. 7.19) were formed by utilizing the efficient nonbonding self-assembly of luminescent iridium monomer M4 and "terfluorenyl"-based monomers M5 and M9, tethered with either a crown ether or dibenzylammonium unit. The SPPs (M4 + M5 + M9) exhibited an obvious glass transition with a *Tg* of 72.5–81.5 °C, which is absent in the monomers. The PLED with an ITO/PEDOT:PSS/SPPs + PBD/CsF/Al structure gave an LE of 14.6 cd/A at a *L* of 450 cd/m^2. Moreover, novel supramolecular sky-blue phosphorescent iridium complexes through efficient non-bonding assembly of dibenzo-24-crown-8 and dibenzylammonium-functionalized monomer units (M10, M11, Fig. 7.19) were also reported [172]. Of great importance is that the sky-blue SPPs exhibited attractive PL and EL efficiency.

7.6 Conclusion

This chapter provides an overview of recent development of polymer electroluminescent materials as the active layer in PLEDs. A variety of electroluminescent polymers have been developed in the past few years because of the aspiration for the commercialization potential of PLEDs in the future. In addition to the new chemical structure design, the strategies which have been adopted in small molecule OLEDs (such as dopant/host system, electrophosphorescent materials, etc.), have also been successfully applied in polymer electroluminescent materials, resulting in significantly improved performance of PLEDs. Encouraging results have been reported by Cambridge Display Technology, which announced a blue-emitting PLED with a LE_{max} of 10 cd/A and a lifetime of 13,000 h, a green-emitting PLED with a LE_{max} of 73 cd/A and a lifetime of about 100,000 h, and a red-emitting PLED with a LE_{max} of 21.8 cd/A and a lifetime above 350,000 h [173]. The new fabrication techniques for PLEDs also developed rapidly and 1.5-inch PLED displays have been fabricated without any dead pixels or dead lines through all-solution processing [7], indicating the unique advantage of PLEDs for low-cost large-area devices. All these results show us the promising commercialization future for PLEDs. However, to realize this, the performance of polymer electroluminescent materials, especially blue-emitting polymers, need to be further improved. The recent significant progress on new generation small-molecule light-emitting materials [174, 175] may provide new direction to improve further the performance of polymer electroluminescent materials and hence offer the opportunities to exploit fully the potential of low-cost fabrication of polymer optoelectronic devices with a large area size.

References

1. Tang CW, VanSlyke SA (1987) Organic electroluminescent diodes. Appl Phys Lett 51 (12):913–915. doi:10.1063/1.98799
2. Gong X, Wang S, Moses D, Bazan GC, Heeger AJ (2005) Multilayer polymer light-emitting diodes: white-light emission with high efficiency. Adv Mater 17(17):2053–2058. doi:10.1002/adma.200500727
3. Sasabe H, Kido J (2013) Development of high performance OLEDs for general lighting. J Mater Chem C 1(9):1699–1707. doi:10.1039/C2TC00584K
4. SAMSUNG (2013) http://www.samsung.com/uk/oled/. Accessed 1 Feb 2014
5. LG (2013) http://www.lg.com/global/oledtv/index.html. Accessed 1 Feb 2014
6. Zeng WJ, Wu HB, Zhang C, Huang F, Peng JB, Yang W, Cao Y (2007) Polymer light-emitting diodes with cathodes printed from conducting Ag paste. Adv Mater 19(6): 810–814. doi:10.1002/adma.200602567
7. Zheng H, Zheng Y, Liu N, Ai N, Wang Q, Wu S, Zhou J, Hu D, Yu S, Han S, Xu W, Luo C, Meng Y, Jiang Z, Chen Y, Li D, Huang F, Wang J, Peng J, Cao Y (2013) All-solution processed polymer light-emitting diode displays. Nat Commun 4:1971–1974. doi:10.1038/ncomms2971
8. Burroughes JH, Bradley DDC, Brown AR, Marks RN, Mackay K, Friend RH, Burns PL, Holmes AB (1990) Light-emitting diodes based on conjugated polymers. Nature 347 (6293):539–541. doi:10.1038/347539a0
9. Zhong CM, Duan CH, Huang F, Wu HB, Cao Y (2011) Highly efficient electron injection from Indium Tin Oxide/cross-linkable amino-functionalized polyfluorene interface in inverted organic light emitting devices. Chem Mater 23(21):4870–4876. doi:10.1021/cm2025685
10. Grimsdale AC, Leok Chan K, Martin RE, Jokisz PG, Holmes AB (2009) Synthesis of light-emitting conjugated polymers for applications in electroluminescent devices. Chem Rev 109 (3):897–1091. doi:10.1021/cr000013v
11. Baldo MA, O'Brien DF, Thompson ME, Forrest SR (1999) Excitonic singlet-triplet ratio in a semiconducting organic thin film. Phys Rev B 60(20):14422–14428. doi:10.1103/PhysRevB.60.14422
12. Friend RH, Gymer RW, Holmes AB, Burroughes JH, Marks RN, Taliani C, Bradley DDC, Dos Santos DA, BreÂ das JL, LoÈ gdlund M, Salaneck WR (1999) Electroluminescence in conjugated polymers. Nature 397(6715):121–128. doi:10.1038/16393
13. Thelakkat M (2002) Star-shaped, dendrimeric and polymeric triarylamines as photoconductors and hole transport materials for electro-optical applications. Macromol Mater Eng 287(7):442–461. doi:10.1002/1439-2054(20020701)

14. Kulkarni AP, Tonzola CJ, Babel A, Jenekhe SA (2004) Electron transport materials for organic light-emitting diodes. Chem Mater 16(23):4556–4573. doi:10.1021/cm049473l
15. Jing W-X, Kraft A, Moratti SC, Grüner J, Cacialli F, Hamer PJ, Holmes AB, Friend RH (1994) Synthesis of a polyphenylene light-emitting polymer. Synth Met 67(1-3):161–163. doi:10.1016/0379-6779(94)90032-9
16. Fukuda M, Sawada K, Yoshino K (1993) Synthesis of fusible and soluble conducting polyfluorene derivatives and their characteristics. J Polym Sci Pol Chem 31(10):2465–2471. doi:10.1002/pola.1993.080311006
17. Scherf U, List EJW (2002) Semiconducting polyfluorenes—towards reliable structure–property relationships. Adv Mater 14(7):477–487. doi:10.1002/1521-4095(20020404)14:7<477::AID-ADMA477>3.0.CO;2-9
18. Lu H-H, Liu C-Y, Jen T-H, Liao J-L, Tseng H-E, Huang C-W, Hung M-C, Chen S-A (2005) Excimer formation by electric field induction and side chain motion assistance in polyfluorenes. Macromolecules 38(26):10829–10835. doi:10.1021/ma051594z
19. Montilla F, Mallavia R (2007) On the origin of green emission bands in fluorene-based conjugated polymers. Adv Funct Mater 17(1):71–78. doi:10.1002/adfm.200600141
20. Rozanski LJ, Cone CW, Ostrowski DP, Bout DAV (2007) Effect of film morphology on the energy transfer to emissive green defects in dialkyl polyfluorenes. Macromolecules 40 (13):4524–4529. doi:10.1021/ma0706069
21. List EJW, Guentner R, de Freitas PS, Scherf U (2002) The effect of Keto defect sites on the emission properties of polyfluorene-type materials. Adv Mater 14(5):374–378. doi:10.1002/1521-4095(20020304)14:5<374::AID-ADMA374>3.0.CO;2-U
22. Romero DB, Schaer M, Leclerc M, Ade`s D, Siove A, Zuppiroli L (1996) The role of carbazole in organic light-emitting devices. Synth Met 80(3):271–277. doi:10.1016/0379-6779(96)80213-X
23. Lamansky S, Djurovich P, Murphy D, Abdel-Razzaq F, Lee HE, Adachi C, Burrows PE, Forrest SR, Thompson ME (2001) Highly phosphorescent bis-cyclometalated iridium complexes: synthesis, photophysical characterization, and use in organic light emitting diodes. J Am Chem Soc 123(18):4304–4312. doi:10.1021/ja003693s
24. Morin JF, Leclerc M (2001) Syntheses of conjugated polymers derived from N-Alkyl-2,7-carbazoles. Macromolecules 34(14):4680–4682. doi:10.1021/ma010152u
25. Chen F, Mehta PG, Takiff L, McCullough RD (1996) Improved electroluminescence performance of poly (3-alkylthiophenes) having a high head-to-tail (HT) ratio. J Mater Chem 6(11):1763–1766. doi:10.1039/JM9960601763
26. Gong SL, Yang CL, Qin JG (2012) Efficient phosphorescent polymer light-emitting diodes by suppressing triplet energy back transfer. Chem Soc Rev 41(14):4797–4807. doi:10.1039/C2CS35056D
27. Tang C, Liu XD, Liu F, Wang XL, Xu H, Huang W (2013) Recent progress in polymer white light-emitting materials and devices. Macromol Chem Phys 214(3):314–342. doi:10.1002/macp.201200305
28. Hay M, Klavetter FL (1995) Aliphatic phenylene vinylene copolymers: tuning the color of luminescence through co-monomer feed ratios. J Am Chem Soc 117(27):7112–7118. doi:10.1021/ja00132a011
29. Ku CH, Kuo CH, Leung MK, Hsieh KH (2009) Carbazole–oxadiazole containing polyurethanes as phosphorescent host for organic light emitting diodes. Eur Polym J 45 (5):1545–1553. doi:10.1016/j.eurpolymj.2009.01.024
30. Cho HJ, Park MJ, Hwang DH, Ahn T, Hong JW, Lee J, Cho NS, Shim HK (2009) Synthesis and characterization of fluorene-based copolymers containing siloxane or distilbene moieties on their main chain. J Polym Sci Pol Chem 47(6):1595–1608. doi:10.1002/pola.23259
31. Koguchi R, Kobayashi N, Kijima M (2009) Appropriately conjugated poly[oligo(N-phenyl-2,7-carbazolylene)-alt-diphenylsilylene]s. Macromolecules 42(16):5946–5952. doi:10.1021/ma900908b

32. Fei T, Cheng G, Hu DH, Lu P, Ma YG (2009) A wide band gap polymer derived from 3,6-carbazole and tetraphenylsilane as host for green and blue phosphorescent complexes. J Polym Sci Pol Chem 47(18):4784–4792. doi:10.1002/pola.23532
33. Jiang XQ, Zheng ZM, Harima Y, Ohshita J, Sun PP (2011) Optical properties of a series of monosilylene–oligothienylene copolymers and the application to light-emitting diodes. J Mater Chem 21:1902–1906. doi:10.1039/C0JM02761H
34. Mo YQ, Jiang X, Cao DR (2007) Synthesis and electroluminescent properties of soluble poly (3,6-fluorene) and its copolymer. Org Lett 9(21):4371–4373. doi:10.1021/ol7019907
35. Wu ZL, Xiong Y, Zou JH, Wang L, Liu JC, Chen QL, Yang W, Peng JB, Cao Y (2008) High-triplet-energy poly(9,9′-bis(2-ethylhexyl)-3,6-fluorene) as host for blue and green phosphorescent complexes. Adv Mater 20(12):2359–2364. doi:10.1002/adma.200800213
36. Yeh HC, Chien CH, Shih PI, Yuan MC, Shu CF (2008) Polymers derived from 3,6-fluorene and tetraphenylsilane derivatives: solution-processable host materials for green phosphorescent OLEDs. Macromolecules 41(11):3801–3807. doi:10.1021/ma800391e
37. van Dijken A, Bastiaansen JJAM, Kiggen NMM, Langeveld BMW, Rothe C, Monkman A, Bach I, Sto¨ssel P, Brunner K (2004) Carbazole compounds as host materials for triplet emitters in organic light-emitting diodes: polymer hosts for high-efficiency light-emitting diodes. J Am Chem Soc 126(24):7718–7727. doi:10.1021/ja049771j
38. Chen Y-C, Huang G-S, Hsiao C-C, Chen S-A (2006) High triplet energy polymer as host for electrophosphorescence with high efficiency. J Am Chem Soc 128(26):8549–8558. doi:10.1021/ja060936t
39. Mo YQ, Tian RY, Shi W, Cao Y (2005) Ultraviolet-emitting conjugated polymer poly(9,9′-alkyl-3,6-silafluorene) with a wide band gap of 4.0 eV. Chem Commun 39:4925–4926. doi:10.1039/B507518A
40. Chan KL, Watkins SE, Mak CSK, McKiernan MJ, Towns CR, Pascu SI, Holmes AB (2005) Poly(9,9-dialkyl-3,6-dibenzosilole)—a high energy gap host for phosphorescent light emitting devices. Chem Commun 46:5766–5768. doi:10.1039/B511208G
41. Wang EG, Li C, Peng JB, Cao Y (2007) High-efficiency blue light-emitting polymers based on 3,6-silafluorene and 2,7-silafluorene. J Polym Sci Pol Chem 45(21):4941–4949. doi:10.1002/pola.22243
42. Mo YQ, Deng XY, Jiang X, Cui QH (2009) Blue electroluminescence from 3,6-silafluorene-based copolymers. J Polym Sci Pol Chem 47(13):3286–3295. doi:10.1002/pola.23393
43. Liu J, Hu SJ, Zhao W, Zou QH, Luo W, Yang W, Peng JBA, Cao Y (2010) Novel spectrally stable saturated blue-light-emitting poly[(fluorene)-co-(dioctyldibenzothiophene-S,S-dioxide)]s. Macromol Rapid Commun 31(5):496–501. doi:10.1002/marc.200900547
44. Park JW, Park SJ, Kim YH, Shin DC, You H, Kwon SK (2009) Pure color and stable blue-light emission-alternating copolymer based on fluorene and dialkoxynaphthalene. Polymer 50(1):102–106. doi:10.1016/j.polymer.2008.10.056
45. Gong X, Ostrowski JC, Moses D, Bazan GC, Heeger AJ (2003) Electrophosphorescence from a polymer guest–host system with an Iridium complex as guest: förster energy transfer and charge trapping. Adv Funct Mater 13(6):439–444. doi:10.1002/adfm.200304334
46. Lamansky S, Djurovich P, Abdel-Razzaq F, Garon S, Murphy D, Thompson ME (2002) Cyclometalated Ir complexes in polymer organic light-emitting devices. J Appl Phys 92 (3):1570–1575. doi:10.1063/1.1491587
47. Qi ZJ, Wang XM, Wei B, Kang F, Tang LL, Hong MX, Sun YM (2011) Optical and electronic properties of 3,4-dialkylthiophene-based p-/n-alternating copolymers. J Appl Polym Sci 120(5):2678–2684. doi:10.1002/app.33407
48. Gong X, Yang YL, Xiao S (2009) Ambipolar charge transport in polymer light-emitting diodes. J Phys Chem C 113(17):7398–7404. doi:10.1021/jp811396j
49. Liu ZP, Zhang JJ, Qiu YR, Qin L, Zhang P (2010) Synthesis and characterization of a red-emitting copolymer containing 5,8-quinoline units. Macromol Chem Phys 211(18):1960–1968. doi:10.1002/macp.201000236
50. Zhu HF, Tong H, Gong YY, Shao SY, Deng CM, Yuan WZ, Zhang YM (2012) Fluorene- and benzimidazole-based blue light-emitting copolymers: Synthesis, photophysical

properties, and PLED applications. J Polym Sci Pol Chem 50(11):2172–2181. doi:10.1002/pola.25984
51. Chen RT, Su WF, Chen Y (2011) Highly efficient and stable blue-light-emitting copolyfluorene consisting of carbazole, oxadiazole, and charge-trapping anthracene groups. J Polym Sci Pol Chem 49(1):184–191. doi:10.1002/pola.24439
52. Horii T, Shinnai T, Tsuchiya K, Mori T, Kijima M (2012) Synthesis and properties of conjugated copolycondensates consisting of carbazole-2,7-diyl and fluorene-2,7-diyl J Polym Sci Pol Chem 50(21):4557–4562. doi:10.1002/pola.26268
53. Mori T, Kijima M (2009) Synthesis and electroluminescence properties of carbazole-containing 2,6-naphthalene-based conjugated polymers. Eur Polym J 45(4):1149–1157. doi:10.1016/j.eurpolymj.2008.12.042
54. Lin MW, Chen RT, Yeh CH, Wen TC, Guo TF (2012) Bright, efficient, deep blue-emissive polymer light-emitting diodes of suitable hole-transport layer and cathode design. Org Electron 13(12):3067–3073. doi:10.1016/j.orgel.2012.09.009
55. Jiang ZQ, Zhang WJ, Yao HQ, Yang CL, Cao Y, Qin JG, Yu G, Liu YQ (2009) Copolyfluorenes containing bridged triphenylamine or triphenylamine: Synthesis, characterization, and optoelectronic properties. J Polym Sci Pol Chem 47(14):3651–3661. doi:10.1002/pola.23442
56. Ye H, Zhao BF, Liu M, Zhou X, Li YH, Li DY, Su SJ, Yang W, Cao Y (2011) Dual-functional conjugated polymers based on trifluoren-2-yl-amine for RGB organic light-emitting diodes. J Mater Chem 21(43):17454–17461. doi:10.1039/C1JM13533C
57. Harkema S, Kicken R, Langeveld-Voss BMW, van Mensfoort SLM, de Kok MM, Coehoorn R (2010) Tuning the voltage dependence of the efficiency of blue organic light-emitting diodes based on fluorene–amine copolymers. Org Electron 11(5):755–766. doi:10.1016/j.orgel.2010.01.015
58. Coehoorn R, Mensfoort SLMV (2009) Effects of disorder on the current density and recombination profile in organic light-emitting diodes. Phys Rev B: Condens Matter 80(8):085302. doi:10.1103/PhysRevB.80.085302
59. Park MJ, Lee JI, Chu HY, Kim SH, Zyung T, Eom JH, Shim HK, Hwang DH (2009) Light-emitting properties of photo-curable polyfluorene derivatives. Synth Met 159(14):1393–1397. doi:10.1016/j.synthmet.2009.03.027
60. Kim SK, Eom JH, Mi D, Jung CH, Lee JH, Kang IN, Kim JH, Hwang DH (2009) Synthesis and light-emitting properties of copolymers composed of fluorene and n-alkoxyphenyl phenothiazine. Synth Met 159(15-16):1672–1676. doi:10.1016/j.synthmet.2009.05.005
61. Choi J, Lee B, Kim JH (2009) Synthesis and electroluminescent properties of π-conjugated copolymer based on 10-hexylphenothiazine and aromatic 1,2,4-triazole. Synth Met 159(19–20):1922–1927. doi:10.1016/j.synthmet.2009.03.029
62. Ahn T, Jang MS, Shim HK, Hwang DH, Zyung T (1999) Blue electroluminescent polymers: control of conjugation length by kink linkages and substituents in the poly(p-phenylenevinylene)-related copolymers. Macromolecules 32(10):3279–3285. doi:10.1021/ma981864w
63. Saikia G, Singh R, Sarmah PJ, Akhtar MW, Sinha J, Katiyar M, Iyer PK (2009) Synthesis and characterization of soluble poly(p-phenylene) derivatives for PLED applications. Macromol Chem Phys 210(24):2153–2159. doi:10.1002/macp.200900387
64. Kim SO, Jung HC, Lee MJ, Jun C, Kim YH, Kwon SK (2009) Synthesis and characterization of 9,10-diphenylanthracene-based blue light emitting materials. J Polym Sci Pol Chem 47(21):5908–5916. doi:10.1002/pola.23636
65. Muller CD, Falcou A, Reckefuss N, Rojahn M, Wiederhirn V, Rudati P, Frohne H, Nuyken O, Becker H, Meerholz K (2003) Multi-colour organic light-emitting displays by solution processing. Nature 421:829–833. doi:10.1038/nature01390
66. Jin JK, Kwon SK, Kim YH, Shin DC, You H, Jung HT (2009) Synthesis and device performance of a highly efficient fluorene-based blue emission polymer containing bulky 9,9-dialkylfluorene substituents. Macromolecules 42(17):6339–6347. doi:10.1021/ma901071d

67. Zhu R, Wen GA, Feng JC, Chen RF, Zhao L, Yao HP, Fan QL, Wei W, Peng B, Huang W (2005) Di-channel polyfluorene containing spiro-bridged oxadiazole branches. Macromol Rapid Commun 26(21):1729–1735. doi:10.1002/marc.200500412
68. Wang HY, Qian Q, Lin KH, Peng B, Huang W, Liu F, Wei W (2012) Stable and good color purity white light-emitting devices based on random fluorene/spirofluorene copolymers doped with iridium complex. J Polym Sci Pol Phys 50(3):180–188. doi:10.1002/polb.22391
69. Huang F, Wu HB, Cao Y (2010) Water/alcohol soluble conjugated polymers as highly efficient electron transporting/injection layer in optoelectronic devices. Chem Soc Rev 39:2500–2521. doi:10.1039/B907991M
70. Huang F, Hou LT, Wu HB, Wang XH, Shen HL, Cao W, Yang W, Cao Y (2004) High-efficiency, environment-friendly electroluminescent polymers with stable high work function metal as a cathode: green- and yellow-emitting conjugated polyfluorene polyelectrolytes and their neutral precursors. J Am Chem Soc 126(31): 845–9853. doi:10.1021/ja0476765
71. Shi W, Jiang X, Zen WJ, Huang F, Yang W, Liu RS, Cao Y (2009) Triphenylamine and fluorene based cationic conjugated polyelectrolytes: synthesis and characterization. Macromol Chem Phys 210(2):150–160. doi:10.1002/macp.200800436
72. Liu SJ, Zhong CM, Dong S, Zhang J, Huang XL, Zhou C, Lu JM, Ying L, Wang L, Huang F, Cao Y (2014) Novel aminoalkyl-functionalized blue-, green- and red-emitting polyfluorenes.Org Electron 15(4):850–857. doi:10.1016/1477j.orgel.2014.1001.1016
73. Yu JM, Chen Y (2009) Synthesis, characterization, and electroluminescent performance of a novel copolyfluorene containing pendant crown ether groups. J Polym Sci Pol Chem 47 (12):2985–2995. doi:10.1002/pola.23380
74. Shu CF, Dodda R, Wu FI, Liu MS, Jen AKY (2003) Highly efficient blue-light-emitting diodes from polyfluorene containing bipolar pendant groups. Macromolecules 36(18):6698–6703. doi:10.1021/ma030123e
75. Lin Y, Chen Y, Chen ZK, Ma DG, Zhang B, Ye TL, Dai YF (2010) Triphenylamine and quinoline-containing polyfluorene for blue light-emitting diodes. Eur Polym J 46(5):997–1003. doi:10.1016/j.eurpolymj.2010.02.013
76. Lin Y, Ye TL, Chen Y, Ma DG, Chen ZK, Dai YF, Li YX (2010) Blue-light-emitting polyfluorene functionalized with triphenylamine and cyanophenylfluorene bipolar side chains. J Polym Sci Pol Chem 48(24):5930–5937. doi:10.1002/pola.24406
77. Wu CS, Chen Y (2010) Copolyfluorenes containing pendant bipolar groups: synthesis, optoelectronic properties and applications. J Mater Chem 20(36):7700–7709. doi:10.1039/C0JM00707B
78. Zhou QF, Li HM, Feng XD (1987) Synthesis of liquid-crystalline polyacrylates with laterally substituted mesogens. Macromolecules 20(2):233–234. doi:10.1021/ma00167a042
79. Wang P, Jin H, Yang Q, Liu WL, Shen ZH, Chen XF, Fan XH, Zou DC, Zhou QF (2009) Synthesis, characterization, and electroluminescence of novel copolyfluorenes and their applications in white light emission. J Polym Sci Pol Chem 47(18):4555–4565. doi:10.1002/pola.23508
80. Yang Q, Jin H, Xu YD, Wang P, Liang XC, Shen ZH, Chen XF, Zou DC, Fan XH, Zhou QF (2009) Synthesis, photophysics, and electroluminescence of mesogen-jacketed 2d conjugated copolymers based on fluorene−thiophene−oxadiazole derivative. Macromolecules 42 (4):1037–1046. doi:10.1021/ma802414s
81. Lin Y, Chen Y, Ye TL, Ma DG, Li YX (2012) Carbazole-modified blue light-emitting copolymers with the backbones integrated by diphenyloxadiazole, fluorene, and triphenylamine. Eur Polym J 48(2):416–424. doi:10.1016/j.eurpolymj.2011.12.004
82. Lin Y, Chen ZK, Ye TL, Dai YF, Ma DG, Ma Z, Liu QD, Chen Y (2010) Novel fluorene-based light-emitting copolymers containing cyanophenyl pendants and carbazole-triphenylamines: synthesis, characterization and their pled application. Polymer 51 (6):1270–1278. doi:10.1016/j.polymer.2010.01.024
83. Song S, Jin Y, Kim SH, Shim JY, Son S, Kim I, Lee K, Suh H (2009) Synthesis and characterization of polyfluorenevinylene with cyano group and carbazole unit. J Polym Sci Pol Chem 47(23):6540–6551. doi:10.1002/pola.23697

84. Peng Q, Xu J, Li MJ, Zheng WX (2009) Blue emitting polyfluorenes containing dendronized carbazole and oxadiazole pendants: synthesis, optical properties, and electroluminescent properties.Macromolecules 42(15):5478–5485. doi:10.1021/ma9008737
85. Huang CW, Tsai CL, Liu CY, Jen TH, Yang NJ, Chen SA (2012) Design of deep blue electroluminescent spiro-polyfluorenes with high efficiency by facilitating the injection of charge carriers through incorporation of multiple charge transport moieties. Macromolecules 45(3):1281–1287. doi:10.1021/ma202413g
86. Chen HY, Chen CT, Chen CT (2010) Synthesis and characterization of a new series of blue fluorescent 2,6-linked 9,10-diphenylanthrylenephenylene copolymers and their application for polymer light-emitting diodes. Macromolecules 43(8):3613–3623. doi:10.1021/ma100195m
87. Dong WY, Xue SF, Lu P, Deng J, Zhao DL, Gu C, Ma YG (2011) Functionality of peripheral side chain for enhanced performance of conjugated polymer—f8bt as an example. J Polym Sci Pol Chem 49(21):4549–4555. doi:10.1002/pola.24898
88. Gu C, Dong WY, Yao L, Lv Y, Zhang ZB, Lu D, Ma YG (2012) Cross-linked multifunctional conjugated polymers prepared by in situ electrochemical deposition for a highly-efficient blue-emitting and electron-transport layer. Adv Mater 24(18):2413–2417. doi:10.1002/adma.201200559
89. Hu DH, Cheng G, Lu P, Liu H, Shen FZ, Li FH, Lv Y, Dong WY, Ma YG (2011) Peripheral cyanohexyl substituent in wide bandgap polymer: increase the electron injection property for blue phosphorescence light emitting device. Macromol Rapid Commun 32(18):1467–1471. doi:10.1002/marc.201100179
90. Liu J, Pei QB (2010) Poly(m-phenylene): conjugated polymer host with high triplet energy for efficient blue electrophosphorescence. Macromolecules 43(23):9608–9612. doi:10.1021/ma102091g
91. Liu J, Li L, Pei QB (2011) Conjugated polymer as host for high efficiency blue and white electrophosphorescence. Macromolecules 44(8):2451–2456. doi:10.1021/ma200282x
92. Yang RQ, Tian RY, Hou Q, Yang W, Cao Y (2003) Synthesis and optical and electroluminescent properties of novel conjugated copolymers derived from fluorene and benzoselenadiazole . Macromolecules 36(20):7453–7460. doi:10.1021/ma034134j
93. Yang J, Jiang CY, Zhang Y, Yang RQ, Yang W, Hou Q, Cao Y (2004) High-efficiency saturated red emitting polymers derived from fluorene and naphthoselenadiazole. macromolecules 37(4):1211–1218. doi:10.1021/ma035743u
94. Hou Q, Zhou QM, Zhang Y, Yang W, Yang RQ, Cao Y (2004) Synthesis and electroluminescent properties of high-efficiency saturated red emitter based on copolymers from fluorene and 4,7-di(4-hexylthien-2-yl)-2,1,3-benzothiadiazole. Macromolecules 37 (17):6299–6305. doi:10.1021/ma049204g
95. Jiang JX, Jiang CY, Yang W, Zhen HG, Huang F, Cao Y (2005) High-efficiency electrophosphorescent fluorene-alt-carbazole copolymers N-Grafted with cyclometalated Ir Complexes. Macromolecules 38(10):4072–4080. doi:10.1021/ma0474473
96. Liu J, Zhou QG, Cheng YX, Geng YH, Wang LX, Ma DG, Jing XB, Wang FS (2005) The first single polymer with simultaneous blue, green, and red emission for white electroluminescence. Adv Mater 17(24):2974–2978. doi:10.1002/adma.200501850
97. Wu HB, Ying L, Yang W, Cao Y (2009) Progress and perspective of polymer white light-emitting devices and materials . Chem Soc Rev 38:3391–3400. doi:10.1039/b816352a
98. Qiao Z, Peng JB, Jin Y, Liu QL, Weng JEN, He ZC, Han SH, Cao DR (2010) Synthesis and electroluminescence properties of fluorene-co-diketopyrrolopyrrole-co-phenothiazine polymers . Polymer 51(5):1016–1023. doi:10.1016/j.polymer.2009.12.044
99. Wang PH, Ho MS, Yang SH, Chen KB, Hsu CS (2010) Synthesis of thermal-stable and photo-crosslinkable polyfluorenes for the applications of polymer light-emitting diodes. J Polym Sci Pol Chem 48(3):516–524. doi:10.1002/pola.23712
100. Liao CF, Hsieh BY, Chen Y (2009) Synthesis, characterization, and application of light-emitting copolyfluorenes slightly doped with distyrylbenzene derivatives. J Polym Sci Pol Chem 47(1):149–160. doi:10.1002/pola.23133

101. Su WF, Chen TT, Chen Y (2010) Synthesis and optoelectronic properties of luminescent copolyfluorenes slightly doped with thiophene chromophore. Polymer 51(7):1555–1562. doi:10.1016/j.polymer.2010.02.012
102. Giovanella U, Betti P, Botta C, Destri S, Moreau J, Pasini M, Porzio W, Vercelli B, Bolognesi A (2010) All-conjugated diblock copolymer approach to improve single layer green electroluminescent devices. Chem Mater 23(3):810–816. doi:10.1021/cm102154q
103. Wu JR, Chen Y, Wu TY (2011) Synthesis of copolyfluorenes with high fluorenone contents and its application in electroluminescent device by simple blending. J Appl Polym Sci 119 (5):2576–2583. doi:10.1002/app.32705
104. Huang ST, Liaw DJ, Hsieh LG, Chang CC, Leung MK, Wang KL, Chen WT, Lee KR, Lai JY, Chan LH, Chen CT (2009) Synthesis and electroluminescent properties of polyfluorene-based conjugated polymers containing bipolar groups. J Polym Sci Pol Chem 47(22):6231–6245. doi:10.1002/pola.23667
105. Wu CS, Chen Y (2010) Copoly(p-phenylene)s containing bipolar triphenylamine and 1,2,4-triazole groups: Synthesis, optoelectronic properties, and applications. J Polym Sci Pol Chem 48:5727–5736. doi:10.1002/pola.24374
106. Wang KL, Leung MK, Hsieh LG, Chang CC, Lee KR, Wu CL, Jiang JC, Tseng CY, Wang HT (2011) Conjugated polymers containing electron-deficient main chains and electron-rich pendant groups: Synthesis and application to electroluminescence. Org Electron 12(6):1048–1062. doi:10.1016/j.orgel.2011.03.020
107. Liu J, Zou J, Yang W, Wu H, Li C, Zhang B, Peng J, Cao Y (2008) Highly efficient and spectrally stable blue-light-emitting polyfluorenes containing a dibenzothiophene-S,S-dioxide unit. Chem Mater 20(13):4499–4506. doi:10.1021/cm800129h
108. Li YY, Wu HB, Zou JH, Ying L, Yang W, Cao Y (2009) Enhancement of spectral stability and efficiency on blue light-emitters via introducing dibenzothiophene-S,S-dioxide isomers into polyfluorene backbone. Org Electron 10(5):901–909. doi:10.1016/j.orgel.2009.04.021
109. King SM, Perepichka II, Perepichka IF, Dias FB, Bryce MR, Monkman AP (2009) Exploiting a dual-fluorescence process in fluorene–dibenzothiophene-S,S-dioxideco-polymers to give efficient single polymer LEDs with broadened emission. Adv Funct Mater 19(4):586–591. doi:10.1002/adfm.200801237
110. Guo X, Cheng YX, Xie ZY, Geng YH, Wang LX, Jing XB, Wang FS (2009) Fluorene-based copolymers containing dinaphtho-s-indacene as new building blocks for high-efficiency and color-stable blue LEDs. Macromol Rapid Commun 30(9–10):816–825. doi:10.1002/marc.200800765
111. Lee SK, Ahn T, Park JH, Jung YK, Chung DS, Park CE, Shim HK (2009) β-Phase formation in poly(9,9-di-n-octylfluorene) by incorporating an ambipolar unit containing phenothiazine and 4-(dicyanomethylene)-2-methyl-6-[p-(dimethylamino)styryl]-4H-pyran. J Mater Chem 19(38):7062–7069. doi:10.1039/B909300A
112. Zhou G, Qian G, Ma L, Cheng Y, Xie Z, Wang L, Jing X, Wang F (2005) Polyfluorenes with phosphonate groups in the side chains as chemosensors and electroluminescent materials. Macromolecules 38(13):5416–5424. doi:10.1021/ma050807h
113. Chen L, Zhang BH, Cheng YX, Xie ZY, Wang LX, Jing XB, Wang FS (2010) Pure and saturated red electroluminescent polyfluorenes with dopant/host system and PLED efficiency/color purity trade-offs. Adv Funct Mater 20(18):3143–3153. doi:10.1002/adfm.201000840
114. Chen L, Tong H, Xie ZY, Wang LX, Jing XB, Wang FS (2011) Red electroluminescent polyfluorenes containing highly efficient 2,1,3-benzoselenadiazole- and 2,1,3-naphthothiadiazole-based red dopants in the side chain. J Mater Chem 21(39):15773–15779. doi:10.1039/C1JM12549D
115. Bian CL, Jiang GX, Tong H, Cheng YX, Xie ZY, Wang LX, Jing XB, Wang FS (2011) Pure blue electroluminescent poly(aryl ether)s with dopant–host systems. J Polym Sci Pol Chem 49(18):3911–3919. doi:10.1002/pola.24828
116. Park MJ, Lee J, Jung IH, Park JH, Kong H, Oh JY, Hwang DH, Shim HK (2010) Synthesis, characterization, and electroluminescence of polyfluorene copolymers containing T-shaped isophorone derivatives. J Polym Sci Pol Chem 48(1):82–90. doi:10.1002/pola.23761

117. Shi W, Wang L, Zhen HY, Zhu DX, Awut T, Mi HY, Nurulla I (2009) Novel luminescent polymers containing backbone triphenylamine groups and pendant quinoxaline groups. Dyes Pigm 83(1):102–110. doi:10.1016/j.dyepig.2009.03.016
118. Ma Y, Zhang H, Shen J, Che C (1998) Electroluminescence from triplet metal—ligand charge-transfer excited state of transition metal complexes. Synth Met 94(3):245–248. doi:10.1016/S0379-6779(97)04166-0
119. Baldo MA, O'Brien DF, You Y, Shoustikov A, Sibley S, Thompson ME, Forrest SR (1998) Highly efficient phosphorescent emission from organic electroluminescent devices. Nature 395(6998):151–154. doi:10.1038/25954
120. Jiang JX, Xu YH, Yang W, Guan R, Liu ZQ, Zhen HY, Cao Y (2006) High-efficiency white-light-emitting devices from a single polymer by mixing singlet and triplet emission. Adv Mater 18(13):1769–1773. doi:10.1002/adma.200502740
121. Ma ZH, Ding JQ, Zhang BH, Mei CY, Cheng YX, Xie ZY, Wang LX, Jing XB, Wang FS (2010) Red-emitting polyfluorenes grafted with quinoline-based Iridium complex: "simple polymeric chain, unexpected high efficiency". Adv Funct Mater 20(1):138–146. doi:10.1002/adfm.200901595
122. Shao S, Ding J, Ye T, Xie Z, Wang L, Jing X, Wang F (2011) A novel, bipolar polymeric host for highly efficient blue electrophosphorescence: a non-conjugated poly(aryl ether) containing triphenylphosphine oxide units in the electron-transporting main chain and carbazole units in hole-transporting side chains. Adv Mater 23(31):3570–3574. doi:10.1002/adma.201101074
123. Shao S, Ding J, Wang L, Jing X, Wang F (2012) Highly efficient blue electrophosphorescent polymers with fluorinated poly(arylene ether phosphine oxide) as backbone. J Am Chem Soc 134(37):15189–15192. doi:10.1021/ja305634j
124. Fei T, Cheng G, Hu DH, Dong WY, Lu P, Ma YG (2010) Iridium complex grafted to 3,6-carbazole-alt-tetraphenylsilane copolymers for blue electrophosphorescence. J Polym Sci Pol Chem 48(9):1859–1865. doi:10.1002/pola.23934
125. Ying L, Xu YH, Yang W, Wang L, Wu HB, Cao Y (2009) Efficient red-light-emitting diodes based on novel amino-alkyl containing electrophosphorescent polyfluorenes with Al or Au as cathode. Org Electron 10(1):42–47. doi:10.1016/j.orgel.2008.10.001
126. Ying L, Zou JH, Zhang AQ, Chen B, Yang W, Cao Y (2009) Novel orange-red light-emitting polymers with cyclometaled Iridium complex grafted in alkyl chain. J Organomet Chem 694(17):2727–2734. doi:10.1016/j.jorganchem.2009.05.007
127. Tan H, Yu JT, Chen JH, Nie KX, Chen Q, Huang Y, Zhang ZY, Wang YF, Liu Y, Lei GT, Zhu WG (2012) Polyfluorene derivatives pending iridium complexes: improved optoelectronic properties by introducing D-A units and altering pendent mode. J Polym Sci Pol Chem 50(10):1900–1905. doi:10.1002/pola.25959
128. Park MJ, Lee J, Kwak J, Jung IH, Park JH, Kong H, Lee C, Hwang DH, Shim HK (2009) Synthesis and electroluminescence of new polyfluorene copolymers containing Iridium complex coordinated on the main chain. Macromolecules 42(15): 5551–5557. doi:10.1021/ma9007175
129. Huang WS, Wu YH, Hsu YC, Lin HC, Lin JT (2009) Synthesis, characterization, and photophysics of electroluminescent fluorene/dibenzothiophene- and fluorene/dibenzothiophene-S,S-dioxide-based main-chain copolymers bearing benzimidazole-based Iridium complexes as backbones or dopants Polymer 50(25):5945–5958. doi:10.1016/j.polymer.2009.10.011
130. Huang WS, Wu YH, Lin HC, Lin JT (2010) Electroluminescent main-chain copolymers containing phosphorescentbenzimidazole-based Iridium complexes as copolymerization backbone units or dopants. Polym Chem 1:494–505. doi:10.1039/B9PY00276F
131. Zhang W, Jin H, Zhou F, Shen ZH, Zou DC, Fan XH (2012) Synthesis and characterization of electrophosphorescent jacketed conjugated polymers. J Polym Sci Pol Chem 50(18):3895–3903. doi:10.1002/pola.26189

132. Thesen MW, Krueger H, Janietz S, Wedel A, Graf M (2010) Investigation of spacer influences in phosphorescent-emitting nonconjugated PLED systems. J Polym Sci Pol Chem 48(2):389–402. doi:10.1002/pola.23796
133. Koga Y, Yoshida N, Matsubara K (2009) Phosphorescent organic light-emitting diodes using an iridium complex polymer as the solution-processible host material. J Polym Sci Pol Chem 47(17):4366–4378. doi:10.1002/pola.23489
134. Xu H, Zhu R, Zhao P, Huang W (2011) Monochromic Red-emitting nonconjugated copolymers containing double-carrier-trapping phosphine oxide Eu3+ segments: toward bright and efficient electroluminescence. J Phys Chem C 115(31):15627–15638. doi:10.1021/jp2029714
135. Wong W-Y, Ho C-L (2006) Di-, oligo- and polymetallaynes: syntheses, photophysics, structures and applications. Coord Chem Rev 250(19–20):2627–2690. doi:10.1016/j.ccr.2006.04.014
136. Goudreault T, He Z, Guo Y, Ho CL, Zhan HM, Wang QW, Ho KYF, Wong KL, Fortin D, Yao B, Xie ZY, Wang LX, Kwok WM, Harvey PD, Wong WY (2010) Synthesis, light-lmitting, and two-photon absorption properties of platinum-containing poly(arylene-ethynylene)s linked by 1,3,4-oxadiazole units. Macromolecules 43(19):7936–7949. doi:10.1021/ma1009319
137. Ho CL, Chui CH, Wong WY, Aly SM, Fortin D, Harvey PD, Yao B, Xie ZY, Wang LX (2009) Efficient electrophosphorescence from a platinum metallopolyyne featuring a 2,7-carbazole chromophore. Macromol Chem Phys 210(21):1786–1798. doi:10.1002/macp.200900351
138. Lee PI, Hsu SLC, Lin PY (2010) White-light-emitting diodes from single polymer systems based on polyfluorene copolymers with quinoxaline derivatives. Macromolecules 43(19):8051–8057. doi:10.1021/ma101673a
139. Lo CN, Hsu CS (2011) Synthesis and electroluminescence properties of white-light single polyfluorenes with high-molecular weight by click reaction. J Polym Sci Pol Chem 49(15):3355–3365. doi:10.1002/pola.24772
140. Jeong E, Kim SH, Jung IH, Xia Y, Lee K, Suh H, Shim HK, Woo HY (2009) Synthesis and characterization of indeno[1,2-b]fluorene-based white light-emitting copolymer. J Polym Sci Pol Chem 47(14):3467–3479. doi:10.1002/pola.23422
141. Su RYT, Chiu LK, Weng SW, Chou YS, Tsiang RCC (2012) White light electroluminescence from graphene-enhanced single polymer comprising two color emitters of equal molar ratios. J Polym Sci Pol Phys 50(4):280–288. doi:10.1002/polb.23008
142. Li H, Xiang N, Wang L, Zhao B, Shen P, Lu JJ, Tan ST (2009) Synthesis and white electroluminescent properties of multicomponent copolymers containing polyfluorene, oligo(phenylenevinylene), and porphyrin derivatives. J Polym Sci Pol Chem 47(20):5291–5303. doi:10.1002/pola.23578
143. Shao SY, Ding JQ, Wang LX, Jing XB, Wang FS (2012) White electroluminescence from all-phosphorescent single polymers on a fluorinated poly(arylene ether phosphine oxide) backbone simultaneously grafted with blue and yellow phosphors. J Am Chem Soc 134(50):20290–20293. doi:10.1021/ja310158j
144. Park MJ, Kwak J, Lee J, Jung IH, Kong H, Lee C, Hwang DH, Shim HK (2010) Single chain white-light-emitting polyfluorene copolymers containing iridium complex coordinated on the main chain. Macromolecules 43(3):1379–1386. doi:10.1021/ma902318t
145. Chen Q, Liu N, Ying L, Yang W, Wu H, Xu W, Cao Y (2009) Novel white-light-emitting polyfluorenes with benzothiadiazole and Ir complex on the backbone. Polymer 50(6):1430–1437. doi:10.1016/j.polymer.2009.01.017
146. Chen J, Peng H, Law CCW, Dong Y, Lam JWY, Williams ID, Tang BZ (2003) Hyperbranched poly(phenylenesilolene)s: synthesis, thermal stability, electronic conjugation, optical power limiting, and cooling-enhanced light emission. Macromolecules 36(12):4319–4327. doi:10.1021/ma034012r

147. Tao X-T, Zhang Y-D, Wada T, Sasabe H, Suzuki H, Watanabe T, Miyata S (1998) Hyperbranched polymers for electroluminescence applications. Adv Mater 10(3):226–230. doi:10.1002/(SICI)1521-4095(199802)10:3<226::AID-ADMA226>3.0.CO;2-E
148. Li J, Bo ZS (2004) "AB2 + AB" approach to hyperbranched polymers used as polymer blue light emitting materials. Macromolecules 37(6):2013–2015. doi:10.1021/ma0357422
149. Chen L, Li PC, Cheng YX, Xie ZY, Wang LX, Jing XB, Wang FS (2011) White electroluminescence from star-like single polymer systems: 2,1,3-benzothiadiazole derivatives dopant as orange cores and polyfluorene host as six blue arms. Adv Mater 23 (26):2986–2990. doi:10.1002/adma.201100297
150. Chen L, Li PC, Tong H, Xie ZY, Wang LX, Jing XB, Wang FS (2012) White electroluminescent single-polymer achieved by incorporating three polyfluorene blue arms into a star-shaped orange core. J Polym Sci Pol Chem 50(14):2854–2862. doi:10.1002/pola.26061
151. Lin ZH, Lin YD, Wu CY, Chow PT, Sun CH, Chow TJ (2010) White light-emitting devices based on star-shape polymers with a bisindolylmaleimide core. Macromolecules 43 (14):5925-5931. doi:10.1021/ma101007x
152. Shih HM, Wu RC, Shih PI, Wang CL, Hsu CS (2012) Synthesis of fluorene-based hyperbranched polymers for solution-processable blue, green, red, and white light-emitting devices. J Polym Sci Pol Chem 50(4):696–710. doi:10.1002/pola.25080
153. Palai AK, Mishra SP, Kumar A, Srivastava R, Kamalasanan MN, Patri M (2011) Hyperbranched poly (arylene ethynylene) s with triphenylamine core for polymer light-emitting diodes. J Polym Sci Pol Chem 49(4):832–841. doi:10.1002/pola.24459
154. Wang HQ, Qiu T, Song N, Li XY (2010) Hyperbranched polymer for light-emitting applications. Polym Int 59(10):1384–1389. doi:10.1002/pi.2879
155. Lu CL, Wang HQ, Wang XC, Li YF, Oiu T, He LF, Li XY (2010) Synthesis and properties of partially conjugated hyperbranched light-emitting polymers. J Appl Polym Sci 117 (1):517–523. doi:10.1002/app.30359
156. Tsai YT, Lai CT, Chien RH, Hong JL, Yeh AC (2012) Thermal and spectral stability of electroluminescent hyperbranched copolymers containing tetraphenylthiophene-quinoline-triphenylamine moieties. J Polym Sci Pol Chem 50(2):237–249. doi:10.1002/pola.25020
157. Vanjinathan M, Lin HC, Nasar AS (2012) Design, synthesis, photophysical, and electrochemical properties of DCM-based conjugated polymers for light-emitting devices. J Polym Sci Pol Chem 50(18):3806–3818. doi:10.1002/pola.26169
158. Lee RH, Chen WS, Wang YY (2009) Synthesis and electroluminescence properties of a series of hyper-branched light-emitting polymers. Thin Solid Films 517(19):5747–5756. doi:10.1016/j.tsf.2009.03.216
159. Liu F, Liu JQ, Liu RR, Hou XY, Xie LH, Wu HB, Tang C, Wei W, Cao Y, Huang W (2009) Hyperbranched framework of interrupted π-conjugated polymers end-capped with high carrier-mobility moieties for stable light-emitting materials with low driving voltage. J Polym Sci Pol Chem 47(23):6451–6462. doi:10.1002/pola.23685
160. Li ZA, Ye SH, Liu YQ, Yu G, Wu WB, Qin JG, Li Z (2010) New hyperbranched conjugated polymers containing hexaphenylbenzene and oxadiazole units: convenient synthesis and efficient deep blue emitters for PLEDs application. J Phys Chem B 114(28):9101–9108. doi:10.1021/jp1014077
161. Wu WB, Ye SH, Yu G, Liu YQ, Qin JG, Li Z (2012) Novel functional conjugative hyperbranched polymers with aggregation-induced emission: synthesis through one-pot "A2 + B4" polymerization and application as explosive chemsensors and PLEDs. Macromol Rapid Commun 33(2):164–171. doi:10.1002/marc.201100503
162. Wu WB, Ye SH, Huang LJ, Xiao L, Fu YJ, Huang Q, Yu G, Liu YQ, Qin JG, Li QQ, Li Z (2012) A conjugated hyperbranched polymer constructed from carbazole and tetraphenylethylene moieties: convenient synthesis through one-pot "A 2+ B 4" Suzuki polymerization, aggregation-induced enhanced emission, and application as explosive chemosensors and PLEDs. J Mater Chem 22(13):6374–6382. doi:10.1039/C2JM16514G

163. Guan R, Xu YH, Ying L, Yang W, Wu HB, Chen QL, Cao Y (2009) Novel green-light-emitting hyperbranched polymers with iridium complex as core and 3, 6-carbazole-co-2, 6-pyridine unit as branch. J Mater Chem 19(4):531–537. doi:10.1039/B813927J
164. Liu J, Yu L, Zhong C, He R, Yang W, Wu H, Cao Y (2012) Highly efficient green-emitting electrophosphorescent hyperbranched polymers using a bipolar carbazole-3, 6-diyl-co-2, 8-octyldibenzothiophene-S, S-dioxide-3, 7-diyl unit as the branch. RSC Adv 2(2):689–696. doi:10.1039/C1RA00610J
165. Guo T, Yu L, Zhao BF, Li YH, Tao Y, Yang W, Hou Q, Wu HB, Cao Y (2012) Highly efficient, red-emitting hyperbranched polymers utilizing a phenyl-isoquinoline iridium complex as the core. Macromol Chem Phys 213(8):820–828. doi:10.1002/macp.201100676
166. Hoeben FJM, Jonkheijm P, Meijer EW, Schenning APHJ (2005) About supramolecular assemblies of π-conjugated systems. Chem Rev 105(4):1491–1546. doi:10.1021/cr030070z
167. Brunsveld L, Folmer BJB, Meijer EW, Sijbesma RP (2001) Supramolecular polymers. Chem Rev 101(12):4071–4098. doi:10.1021/cr990125q
168. Abbel R, Grenier C, Pouderoijen MJ, Stouwdam JW, Leclère PELG, Sijbesma RP, Meijer EW, Schenning APHJ (2008) White-light emitting hydrogen-bonded supramolecular copolymers based on π-conjugated oligomers. J Am Chem Soc 131(2):833–843. doi:10.1021/ja807996y
169. Zhang J, Zhang K, Huang XL, Cai WZ, Zhou C, Liu SJ, Huang F, Cao Y (2012) Supramolecular light-emitting polymers for solution-processed optoelectronic devices. J Mater Chem 22(25):12759–12766. doi:10.1039/C2JM31773G
170. Zhang J, Zhang K, Liu S, Liang A, Huang X, Huang F, Peng J, Cao Y (2013) A supramolecular large band gap host for phosphorescent organic light-emitting diodes. RSC Adv 3(12):3829–3835. doi:10.1039/C3RA40249E
171. Liang A-H, Zhang K, Zhang J, Huang F, Zhu X-H, Cao Y (2013) Supramolecular phosphorescent polymer Iridium complexes for high-efficiency organic light-emitting diodes. Chem Mater 25(6):1013–1019. doi:10.1021/cm400333c
172. Liang AH, Dong S, Zhang K, Xiao X, Huang F, Zhu XH, Cao Y (2013) Supramolecular sky-blue phosphorescent polymer Iridium complexes for single-emissive-layer organic light-emitting diodes. Macromol Rapid Commun 34(16):1301–1305. doi:10.1002/marc.201300254
173. Cambridge Display Technology (2012) Progress in polymer OLED efficiency. http://www.cdtltd.co.uk/technology/technical-papers/. Accessed 1 Feb 2014
174. Uoyama H, Goushi K, Shizu K, Nomura H, Adachi C (2012) Highly efficient organic light-emitting diodes from delayed fluorescence. Nature 492(13):234–238. doi:10.1038/nature11687
175. Li W, Pan Y, Xiao R, Peng Q, Zhang S, Ma D, Li F, Shen F, Wang Y, Yang B, Ma Y (2013) Employing ~100% excitons in OLEDs by utilizing a fluorescent molecule with hybridized local and charge-transfer excited state. Adv Funct Mater 24(11):1609–1614. doi:10.1002/adfm.201301750

Chapter 8
Transparent Conducting Polymers

Yijie Xia and Jianyong Ouyang

Abstract A conjugated polymer in neutral state is semiconductive. It becomes conductive after it is doped by oxidation or reduction. The optical properties of neutral conjugated polymers are predominantly determined by the electron transition from the highest occupied molecular orbital (HOMO) to the lowest unoccupied molecular orbital (LUMO), whereas the optical properties of conducting polymers are related to new energy levels, polaron and bipolaron levels, which are generated during oxidation or reduction. The appearance of the new energy levels significantly changes the optical properties. Some conjugated polymers with low energy band gap can have high transparency in the visible range after they are oxidized and become conductive. Poly(3,4-ethylenedioxythiophene) (PEDOT), its derivatives and analogues are the most popular transparent conducting polymers. This chapter reviews the preparation, structure, properties, and application of transparent conducting polymers, particularly the PEDOTs. PEDOTs can be prepared by solution chemical polymerization, vapor-phase polymerization, or electrochemical polymerization of its monomer. Poly(3,4-ethylenedioxythiophene):poly(styrenesulfonate) (PEDOT:PSS) prepared by chemical polymerization in solution is particularly interesting because it can be dispersed in water and some polar organic solvents. High-quality PEDOT:PSS films can be readily prepared through solution processing techniques. Less conductive PEDOT:PSS has been used as the buffer layer in optoelectronic devices, such as organic and polymer light-emitting diodes (OLEDs and PLEDs) and organic and polymer solar cells (OSCs and PSCs). However, as-prepared PEDOT:PSS films from aqueous solution cannot be directly used as the transparent electrode of optoelectronic devices because its conductivity is below 1 S cm^{-1}. This conductivity is lower than the conductivity of indium tin oxide (ITO), the conventional transparent electrode material, by three to four orders of magnitude. A couple of methods have been developed to improve significantly

Y. Xia
Institute of Materials Research and Engineering, Singapore, Singapore

J. Ouyang (✉)
Department of Materials Science and Engineering, National University of Singapore, Singapore 117576, Singapore
e-mail: mseoj@nus.edu.sg

Y. Li (ed.), *Organic Optoelectronic Materials*,
Lecture Notes in Chemistry 91, DOI 10.1007/978-3-319-16862-3_8

the conductivity of PEDOT:PSS. Conductivity of more than 3000 S cm^{-1} was recently observed with PEDOT:PSS films after treatment with acids. This conductivity is higher than that of ITO on plastic and comparable to ITO on glass. It is anticipated that the conductivity of PEDOT:PSS can be further increased if the molecular weight of PEDOT can be increased. Thus, transparent conducting polymers are very promising candidates to replace ITO as the next-generation of transparent electrode materials.

Keywords PEDOT:PSS · Conductivity · Optoelectronics · Transparency

8.1 Electronic Structure and Optical Properties of Conducting Polymers

Since the report on conducting polyacetylene by MacDiarmid, Shirakawa, and Heeger in the 1970s, conducting polymers have been attracting considerable attention [1]. The high conductivity of conducting polymers makes them quite different from conventional polymers such as polyethylene and polystyrene. Those conventional polymers are insulators. The different electrical behavior of conducting polymers from conventional polymers arises from their different chemical structures. Conducting polymers have conjugated π-bonds along the main chain. The 2s and 2p atomic orbitals of carbon atoms hybridize in an sp^2 manner. The p_z orbital which does not participate in the hybridization is perpendicular to the plane formed by the three sp^2 orbitals. The three sp^2 orbitals form three σ-bonds whereas the p_z orbital form a π-bond with neighboring atoms. As a result, there are alternative σ- and π-bonds along the backbone. The π-electrons are delocalized along the whole polymer chains, leading to the formation of valence and conduction bands. However, the bandwidth of the conduction band and valence band is so narrow for neutral conjugated polymers so that to use the lowest unoccupied molecular orbital (LUMO) and highest occupied molecular orbital (HOMO) is more popular for describing their electronic structure.

The optical properties of conjugated polymers in the neutral state are mainly determined by the electron transition from HOMO to LUMO. The oxidation or reduction doping of conjugated polymers turns them highly conductive. The dramatic increase in the conductivity after doping is related to the appearance of new energy levels (Fig. 8.1). For most conjugated polymers, two polaron levels are generated when the conjugated polymers are lightly doped. After the conjugated polymers are further doped, the polarons interact and the coupling turns two polarons into a bipolaron. Correspondingly, the two polaron levels become two bipolaron levels. There is one electron on the lower polaron level, whereas there is no electron on the bipolaron levels. When the conjugated polymers are heavily doped, the bipolarons interact, causing the bipolaron levels to become bipolaron bands.

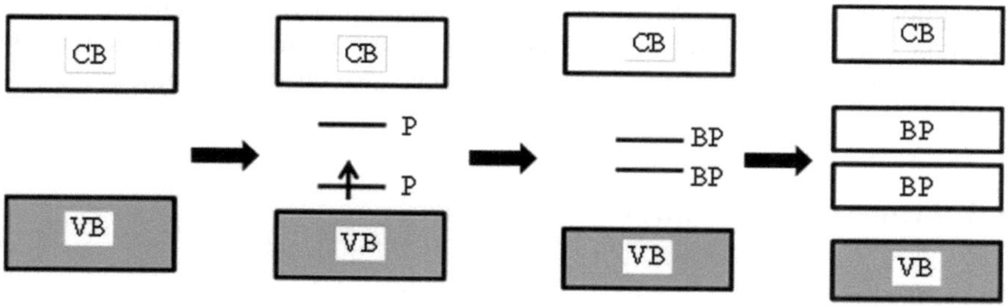

Fig. 8.1 Change of the electronic structure of conducting polymers during oxidation doping. *VB* valence band, *CB* conduction band, *P* polaron, *BP* bipolaron

Finally, the bipolaron bands overlap and merge as one band. As a result, the energy gap disappears, and the doped conjugated polymers behave as metals.

The change in the electronic structure gives rise to significant changes in the optical properties when conjugated polymers change from neutral to a doped state. There are remarkable color changes for conjugated polymers during this process. In fact, the in situ optical absorption spectra of conducting polymers during oxidation are important evidence for the occurrence of the polarons and bipolaron levels. Figure 8.2 shows the change in the absorption of polypyrrole during oxidation [2]. There is no absorption in the infrared range for neutral polypyrrole. The absorption in the infrared range increases after polypyrrole is further doped. The strong light

Fig. 8.2 Optical absorption spectra of polypyrrole doped with perchlorate counter anions (oxidation doping) with the doping degree increasing from the *bottom curve* (almost neutral) to the *top curve* (doping degree is 0.33). Reproduced from [2]

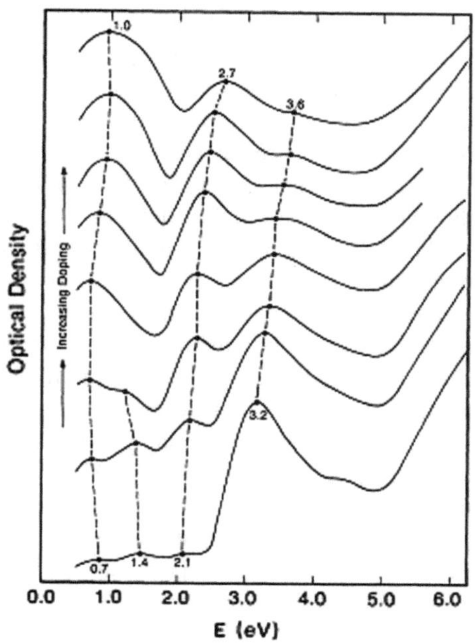

absorption in the infrared range for heavily doped polypyrrole is attributed to the plasmon oscillation of the metallic polymer. Apart from the absorption in the infrared range, the absorption in the visible range also changes significantly.

8.2 Transparent Conducting Polymers

The optical properties of conjugated polymers are mainly determined by the band gap between HOMO and LUMO. The band gap of conjugated polymers can be altered by modifying the chemical structure. For example, the band gap of poly(3-hexylthiophene) is about 2.0 eV. It decreases to 1.5 eV for PEDOT. The chemical structure of poly(3,4-ethylenedioxythiophene):poly(styrenesulfonate) (PEDOT:PSS) is shown in Scheme 8.1. PEDOT is the most important transparent conducting polymer. For undoped PEDOT, the maximum absorption appears at about 2.0 eV [3]. The absorption maximum shifts to 0.5 eV after it is doped, so that the conductive PEDOT films are highly transparent in the visible range.

Conductive polymers with high transmittance in the visible range can have important applications for optoelectronic devices, including liquid crystal displays (LCDs), light-emitting diodes (LEDs), solar cells, touch panel displays, lasers, and detectors. At least one electrode of an optoelectronic device should be transparent to emit or harvest light. Although the most popular material for the transparent electrode is indium tin oxide (ITO), ITO has several severe problems. Indium is a scarce element on Earth. The indium price has been skyrocketing, and the long-term availability of indium is a big concern. In addition, ITO is a brittle material so it is unsuitable for flexible electronic devices which are regarded as the next-generation electronic devices [4, 5]. For instance, Samsung and LG recently demonstrated flexible mobile phones. Thus, new transparent conductive materials are in urgent need to substitute ITO as the transparent electrode. Among the candidates, including conducting polymers [6–19], carbon nanotubes [19–24], graphene [25–27], and metal nanowires [28, 29], conducting polymers are very promising because they are cheap and highly flexible.

Apart from application as the transparent electrode of optoelectronic devices, transparent conducting polymers can also be used as the active materials for electrochromic displays. Transparent conducting polymers have advantages over other conducting polymers for use in smart windows. Materials for smart windows should be highly transparent when in the bleached state, whereas they should block most of the visible light when in the color state.

The transparent conducting polymers discovered hitherto are mainly PEDOT, its derivatives and analogues. PEDOT is the most important transparent conducting polymer because of its merits in processability, optical properties, and thermal properties. Figure 8.3a presents the absorption spectrum from 400 to 900 nm for PEDOT in neutral and doped states [30]. Neutral PEDOT has a blue color with an absorption maximum at about 600 nm, and doped PEDOT is highly transparent in the visible range. Although alkylation of PEDOT can change the electronic

Scheme 8.1 Chemical structure of PEDOT:PSS

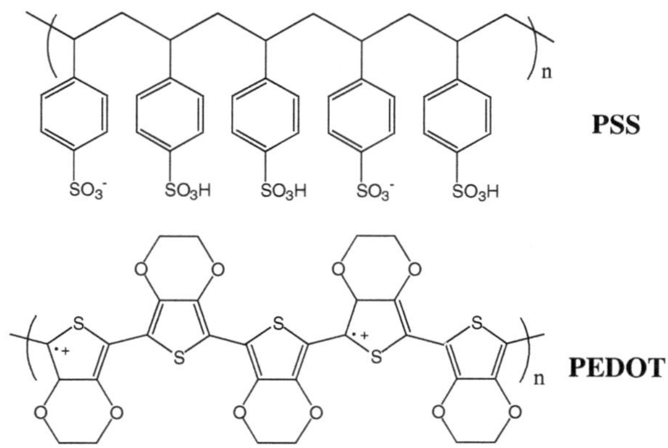

Fig. 8.3 UV-vis absorption spectra of **a** PEDOT and **b** $C_{14}H_{29}$-PEDOT in neutral (*solid curve*) and doped (*dashed curve*) states. Reproduced from [30]

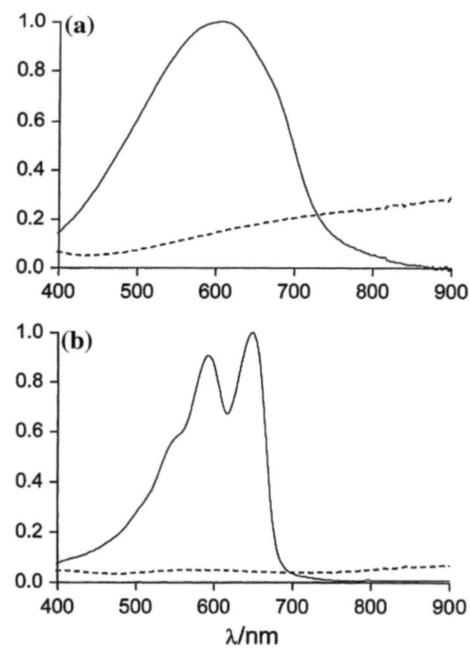

structure and optical properties, those derivatives are usually highly transparent in the visible range as well. As shown in Fig. 8.3b, the absorption spectrum of neutral $C_{14}H_{29}$-PEDOT, which has a $C_{14}H_{29}$ side group connecting to the ethylene group, is different from that of neutral PEDOT. However, both of them are highly transparent in the visible range when in the doped state.

Apart from changes in the electronic structure and optical properties, the conformation and configuration of PEDOT change as well when PEDOT changes from the neutral to conductive state. Lenz et al. [31] calculated the geometry of PEDOT in neutral and doped states. They found that the thiophene ring has the benzoid structure for neutral PEDOT and it turns to the quinoid structure after being doped. Because the doping changes the charges on the conducting polymers, the dielectric constant of PEDOT increases after it is doped [32].

8.3 Preparation of PEDOTs by Electrochemical Polymerization

PEDOTs are prepared through the polymerization of its monomer, ethylenedioxythiophene (EDOT), through the cation radical polymerization. EDOT becomes a positively charged radical after it loses an electron. The radical structure makes the monomer chemically active, and the monomer molecules can connect into a polymer chain. The monomer can be oxidized by applying an electrochemical potential or using an oxidizing agent.

Pei et al. [3] were the first to report the electrochemical polymerization of EDOT. The oxidation of EDOT in an acetonitrile solution of 0.1 M Bu_4NClO_4 starts at 1.04 V. Oxidation at this potential leads to the growth of a PEDOT on the anode. They observed a conductivity of about 200 S cm^{-1} for the PEDOT doped with ClO_4^-.

The electropolymerization of EDOT can also proceed in aqueous solution. Surfactant is used for the dispersion of the monomer in water [33]. For example, Bhandari et al. [33] prepared PEDOT by electropolymerization of EDOT in an aqueous micellar solution comprising camphorsulfonic acid, lithium trifluoromethanesulfonate ($LiCF_3SO_3$), and the monomer. Camphorsulfonic acid is a surfactant and forms micelles in water. The solvent, surfactant, and counter anion can affect the morphology of PEDOT films [34]. By carefully controlling the experimental conditions during the electropolymerization, nanostructured PEDOTs were demonstrated [35, 36].

When the substrate surface was passivated, PEDOT films could be electrochemically deposited on active metals [37]. Sakmeche et al. electrochemically deposited PEDOT on iron and mild steel substrates by using sodium phthalate as a passivating agent in a 0.1 M sodium dodecylsulfate (SDS) micellar aqueous solution. Homogeneous, regular, adhesive and thick ($\sim$15 μm) PEDOT films were electrochemically deposited on iron.

8.4 Preparation of PEDOTs by Chemical Synthesis

Scheme 8.2 presents the chemical polymerization of EDOT in solution. Oxidizing agents, such as persulfate salts, iron(III) salts, and bromine, were used for the polymerization [38].

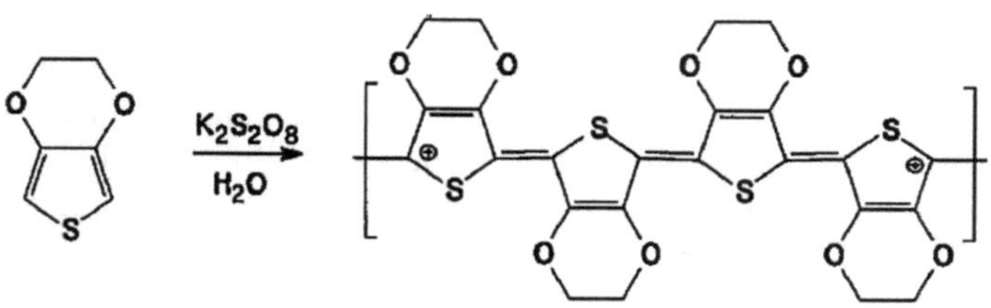

Scheme 8.2 Oxidative polymerization of EDOT with potassium persulfate. Reproduced from [39]

Seo and Chung [39] studied the polymerization mechanism of EDOT in aqueous solution for different concentrations of potassium persulfate. The polymerization and doping take place in one step, which yields conductive PEDOT. The electrical conductivity and yield of PEDOT increased as the molar ratio of potassium persulfate to EDOT increased until reaching the molar ratio of unity and then decreased. The reaction rate (R_p) was found to be governed by the concentration of the reactants: $R_p \propto [\text{potassium persulfate}]^{0.64} \times [\text{EDOT}]^{0.16}$. The reaction rate is also affected by the acidity of the solution. It reaches the maximum at a pH of 2.7.

PEDOT doped with tosylate (PEDOT:TsO) exhibited the highest conductivity. Ha et al. [40] carried out a comprehensive investigation on the chemical polymerization of EDOT with iron tosylate as the oxidizing agent. By optimizing experimental conditions, they attained a conductivity of 750 S cm^{-1} for PEDOT:TsO and a conductivity of 900 S cm^{-1} when methanol-substituted EDOT was used. PEDOTs doped with small counter anions are insoluble in any solvent. Conductive PEDOT thin films can be prepared by spin coating a solution consisting of EDOT and an oxidizing agent prior to the polymerization. To get good quality for the conductive PEDOT films, the experimental conditions should be precisely controlled.

When polystyrenesulfonate (PSS^-) is used as the counter anion of PEDOT, the polymer can be dispersed with good stability in water and some polar organic solvents [41]. Positively charged PEDOT chains are stabilized by negatively charged PSS^- chains in solvent. PEDOT is in fact an oligomer with molecular weight less than 1000–2500 Da (about 6–18 repeating units). Oligomeric PEDOT chains are tightly attached to the PSS chains which have a much higher molecular weight. PEDOT:PSS forms gel particles consisting of roughly 90–95 % water. The PEDOT:PSS aqueous solution can be readily processed into polymer films on various substrates.

The conductivity of PEDOT:PSS depends on the ratio of PEDOT to PSS. Increasing the PSS content reduces the electrical conductivity of PEDOT:PSS. It is about 1 S cm^{-1} when the molar ratio of PEDOT to PSS is 1:2.5, and it decreases to 10^{-3} S cm^{-1} when the molar ratio changes to 1:6. The conductivity of the PEDOT:PSS films is also affected by the polymer gel particle size [42]. The smaller the gel particles, the lower the conductivity of the PEDOT:PSS films.

Doped PEDOT can be reduced to some extent with hydrazine, hydroxylamine, etc. [43]. However, a truly neutral state of PEDOT has never been observed. Residual charged moieties could not be removed completely. Electrochemical methods can lead to a more pronounced reduction of PEDOT [44, 45]. However, even in this case a complete reduction is impossible. The residual conductivity of the reduced PEDOT layers by electrochemical dedoping is about 5×10^{-4} S cm^{-1}. Electron paramagnetic resonance investigations demonstrate the presence of the radicals in the electrochemically reduced PEDOT, even after the most extensive dedoping [46–48]. Thus, the intense blue colour of reduced PEDOT can be caused by the polythiophene radical cations (polarons).

8.5 Vapor-Phase Polymerization of EDOT

A variation for the chemical polymerization of EDOT in solution is the vapor-phase polymerization (VPP) [49–54]. An oxidizing agent, typically iron(III) tosylate, is coated onto a substrate. It forms a solid thin film on substrate after drying. Exposure of this substrate to EDOT vapor produces a PEDOT:TsO film on substrate. The conductivity of the PEDOT:TsO films can be as high as 1500 S cm^{-1} by this method. Recently, Evans et al. modified the VPP method by blending the oxidizing agent with a tri-block polymer, poly(ethylene glycol–propylene glycol–ethylene glycol) (PEG–PPG–PEG) [55]. They observed a conductivity of about 2500 S cm^{-1}.

The VPP method was combined with the electrospinning technique to fabricate PEDOT nanofibers [56]. The fibers had average diameters of around 350 nm. The nanofibers are highly ordered at the molecular level, and the non-woven nanofiber mats had a high conductivity of ~60 S cm^{-1}.

The conductivities of PEDOTs prepared by these methods are summarized in Table 8.1. PEDOT:PSS prepared by chemical synthesis in aqueous solution has a conductivity of less than 1 S cm^{-1}, but it can be dispersed in water and some polar organic solvents. This advantage in solution processability renders PEDOT:PSS important for application in many areas. Moreover, many methods have been developed to enhance significantly the conductivity of PEDOT:PSS.

Table 8.1 Conductivities (σ) of PEDOTs prepared by different methods

Synthesis method	Doping anions	σ (S cm^{-1})	References
Chemical synthesis	Small anions	Up to 750	[40]
	PSS (1:2.5 ratio)	<1	[41, 42]
VPP	TsO^-	Up to 1500	[49–54]
Modified VPP	TsO^-	Up to 2500	[55]
Electrochemical polymerization	Small and polymer anions	~200	[40]

8.6 Development of Highly Conductive PEDOT:PSS

Among the PEDOTs, solution-processable PEDOT:PSS has gained the most attention. High-quality PEDOT:PSS films can be readily prepared on various substrates through the conventional solution processing techniques such as coating and printing [57, 58]. However, an as-prepared PEDOT:PSS film from its aqueous solution usually has a problem of low conductivity, which is usually below 1 S cm^{-1} and remarkably lower than ITO. The conductivities of ITO on glass and plastic are 3000–6000 and $\sim$2000 S cm^{-1}, respectively. Thus, PEDOT:PSS is usually used as a buffer layer rather than the transparent electrode in electronic devices.

The conductivity of PEDOT:PSS should be greatly improved to be used as the transparent electrode of optoelectronic devices. As a substitute of ITO as the transparent electrode of optoelectronic devices, the conductivity of PEDOT:PSS should be higher than 2000 S cm^{-1}. This is particularly important for the development of flexible electronic devices, the next-generation electronic devices. The first work on the conductivity enhancement of PEDOT:PSS was reported by Kim et al. in 2002 [59]. The addition of dimethyl sulfoxide (DMSO) or dimethylformamide (DMF) into PEDOT:PSS aqueous solution can improve the conductivity of PEDOT:PSS films by one to two orders of magnitude. Since then, many ways have been developed for the conductivity enhancement of PEDOT:PSS [60–82]. Recent work demonstrated that a treatment of PEDOT:PSS films with inorganic or organic acids can enhance the conductivity of PEDOT:PSS films to be more than 3000 S cm^{-1} [83, 84].

8.6.1 Structure of PEDOT:PSS

There are two components, the positively charged PEDOT and negatively charged PSS, in PEDOT:PSS. Its conductivity is affected by the doping degree, the microstructure, and amount of PSS. PEDOT is hydrophobic and cannot be directly dispersed in water. Excess PSS is adopted for the dispersion PEDOT in water and polar organic solvents [85, 86]. PSS serves for two roles, as the counter anions for the charge neutralization and as a surfactant to stabilize PEDOT in solvents. Nevertheless, the presence of insulating PSS can suppress the charge transport and thus decrease the conductivity of PEDOT:PSS.

PEDOT:PSS aqueous solutions are commercially available from both Heraeus and Agfa. There are several grades, such as Clevios P and Clevios PH 1000 by Heraeus, on the market in terms of their conductivity and application. In these aqueous solutions, the molecular weight of PEDOT is about 1000–2500 g mol^{-1}, which is much lower than the molecular weight of PSS ($\sim$400,000 g mol^{-1}) [87]. The PEDOT chains are thus much shorter than the PSS chains. The Coulombic attraction among PEDOT and PSS leads to the attachment of short PEDOT chains

Scheme 8.3 Schematic structure of PEDOT:PSS in water. The *thin and thick curves* represent the PSS and PEDOT chains, respectively. Reproduced from [72]

to the long PSS chains. The short PEDOT chains are hydrophobic whereas the long PSS chains are hydrophilic. Hence, PEDOT:PSS has a necklace structure in water. Because of the different Coulombic interactions in the polymer chains and different interactions between the segments of the polymers and water medium, the PSS segments mainly have two kinds of conformations. For the PSS segments attached with PEDOT, the Coulombic attraction between the positively charged PEDOT and negatively charged PSS is stronger than the repulsions among the PSS anions. They thus have a coil conformation (Scheme 8.3). These PSS segments form blobs to keep PEDOT from contacting water. There is a core/shell structure for the blobs. The core is rich in the hydrophobic PEDOT, whereas the shell is rich in the hydrophilic PSS. However, the PSSH segments that do not attach to PEDOT have a different conformation. These PSSH segments are in excess because their role is not concerned with charge balance. They dissociate into PSS anions and protons in water. Owing to the Coulombic repulsions among the PSS anions, they adopt a linear conformation so that the PSS anions have the greatest separation. The linear PSS segments behave as strings between the blobs. In addition, the blobs are separated as far as possible in order to lower the repulsion arising from the PSS anions in the shell of each blob. This necklace structure in water is conserved in the as-prepared PEDOT:PSS films. The shell that is rich in the insulating PSS constructs an energy barrier which inhibits the charge transport across the conductive PEDOT chains. The PEDOT conformation also affects the charge transport. A linear conformation is favorable for the charge transport. However, the PEDOT chains in the shell have to follow the coil conformation of PSS. The coil conformation causes the localization of the charge carriers. Both the excess PSS and coil PEDOT conformation are seen as the reasons for the low conductivity of as-prepared PEDOT:PSS films.

8.6.2 *Conductivity Enhancement by Adding Compounds to PEDOT:PSS Aqueous Solution*

In 2002, Kim et al. [59] found that the addition of a polar organic solvent, such as DMSO, DMF, or tetrahydrofuran (THF), to Clevios P PEDOT:PSS aqueous solution could lead to the conductivity enhancement of the PEDOT:PSS films. The conductivity was enhanced by about two orders of magnitude at room temperature by adding DMSO. The conductivity enhancement is less by adding DMF or THF. However, the charge transport has the same mechanism of charge hopping for the pristine and treated PEDPOT:PSS films as revealed by the temperature dependences of the resistance. In 2002, Ouyang et al. [88] also added polyoxyethylene(12) tridecyl ether, a nonionic surfactant, into PEDOT:PSS aqueous solution. However, the surfactant was used to improve the wettability of PEDOT:PSS aqueous solution on plastic substrate. They did not report whether the addition of the surfactant affected the conductivity of PEDOT:PSS. Later, it was found that the nonionic surfactant could also enhance the conductivity of PEDOT:PSS by a factor of about 20 [64].

Since then, many methods have been developed for the conductivity enhancement of PEDOT:PSS. Apart from DMSO, DMF, and THF, many other organic solvents or organic solids were also used to enhance the conductivity of PEDOT: PSS. It was reported that polar organic solvents with high boiling point, including ethylene glycol (EG), nitromethanol, glycerol, and other organic solvents with multiple hydroxyl groups, can significantly enhance the conductivity to about 200 S cm^{-1} for PEDOT:PSS prepared from Clevios P aqueous solution. The addition of organic solids such as D-Sorbitol can enhance the conductivity of PEDOT:PSS to a similar degree as the organic solvents with high boiling point. The conductivity of PEDOT:PSS prepared from the Clevios PH1000 aqueous solution can reach 600–700 S cm^{-1} by adding EG or DMSO [75]. Untreated Clevios P and Clevios PH1000 are two grades of PEDOT:PSS aqueous solutions supplied by Heraeus. The as-prepared PEDOT:PSS films from these two solutions have almost the same conductivity of about 0.3 S cm^{-1}. However, after the same treatment, the conductivity of the PEDOT:PSS films prepared from Clevios PH1000 is usually higher than that from Clevios P by about one order of magnitude. The different conductivities for the treated PEDOT:PSS films are ascribed to the different PEDOT molecular weights in the two grades of PEDOT:PSS aqueous solution [78]. The PEDOT chains in Clevios PH1000 have a higher molecular weight than in Clevios P.

Room-temperature ionic liquids are nonvolatile solvents. They were also investigated for the conductivity enhancement. It was observed that the structure and concentration of the ionic liquids affected the conductivity enhancement of PEDOT:PSS [64]. When 1-butyl-3-methylimidazolium tetrafluoroborate tetrafluoroborate [(BMIm)BF_4] or 1-butyl-3-methylimidazolium bromide [(BMIm)Br] was added into Clevios P aqueous solution, the maximum conductivity of PEDOT:PSS was higher than 100 S cm^{-1}. The optimal concentration of the ionic liquids is 40–70 % by weight. Badre et al. [89] recently studied the conductivity enhancement through the addition of ionic liquids into Clevios PH1000 aqueous solution. They observed a conductivity as high as 2084 S cm^{-1} when 1-ethyl-3-methylimidazolium tetracyanoborate (EMIM-TCB) was used. They observed a transmittance of 96 % for the highly conductive PEDOT:PSS films with a thickness of about 96 nm.

Surfactants have both hydrophobic and hydrophilic units. They were also added into PEDOT:PSS aqueous solution. The addition of polyoxyethylene(12) tridecyl ether, a nonionic surfactant, into PEDOT:PSS aqueous solution can enhance the

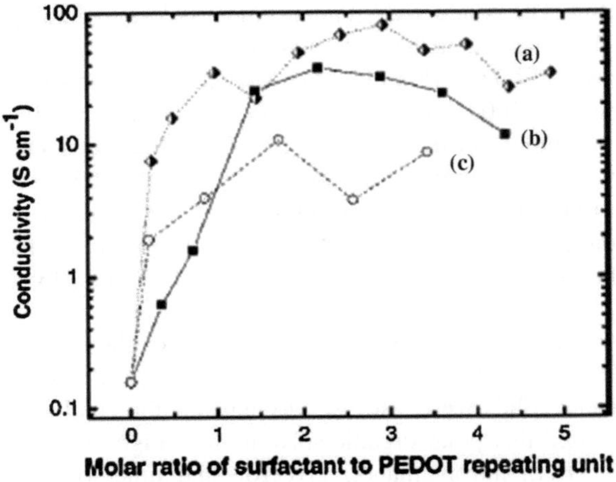

Fig. 8.4 Variations of the conductivity of PEDOT:PSS films with the molar ratio of the anionic surfactant to the PEDOT repeating unit in PEDOT:PSS aqueous solution. The additives are **a** sodium dodecyl sulfate (SDS), **b** sodium tosylate, and **c** dodecylbenzenesulfonic acid sodium salt. Reproduced from [65]

conductivity of PEDOT:PSS by a factor of 20 [65, 88]. Besides nonionic surfactants, ionic and cationic surfactants were also studied for the conductivity enhancement of PEDOT:PSS. When an anionic surfactant is added, the conductivity enhancement is more than two orders of magnitude (Fig. 8.4). However, the conductivity enhancement is negligible when a cationic surfactant is added. Anionic surfactants, particularly fluoro-containing surfactants, were also studied for the conductivity enhancement of Clevios PH1000 [90–92]. Zonyl was added into the Clevios PH1000 aqueous solution [90, 91]. Not only can it enhance the conductivity of the PEDOT:PSS films—it also improves the coating of the Clevios PH1000 aqueous solution on elastomers.

8.6.3 Conductivity Enhancement of PEDOT:PSS Through a Post-coating Treatment

As well as by the addition of compounds into the PEDOT:PSS aqueous solution, the conductivity of the PEDOT:PSS films can be increased through a post-coating treatment. Ouyang et al. [61] reported the conductivity enhancement through the treatment with EG or DMSO. The conductivity enhancement is similar for the two methods when organic compounds such as EG and DMSO are used. However, the conductivity enhancement is quite different when some other compounds, for example salts, cosolvents, and acids, were exploited. Although a post-coating treatment with these compounds can significantly enhance the conductivity of

Table 8.2 Conductivities of PEDOT:PSS films prepared from Clevios™ PH1000 after treatment with geminal diols

Name	Melting point (°C)	Boiling point (°C)	Conductivity (S cm^{-1})[a]
Cyclohexanehexone octahydrate	99	345	349 (0.1 M)
HFA trihydrate	−129	−28	1164
Formaldehyde	−92	−19	862 (122 M)
Acetaldehyde	−124	20	83 (18.2 M)
Acetone	−95	56	1
Perfluorobenzophenone	90–95	359	0.3 (1 M)

Reproduced from [77]
[a]Conductivity of PEDOT:PSS after treatment with a neat compound or aqueous solution of a compound with the highest concentration. The concentrations of aqueous solutions used for the treatments are presented in parentheses

PEDOT:PSS films, the conductivity enhancement is negligible if they are added into the PEDOT:PSS aqueous solution [69–73]. The post-coating treatment also has the advantage of easy removal of the compounds from the final PEDOT:PSS films after the treatment.

Organic solvents with multiple hydroxyl groups, such as EG or glycerol, have been extensively investigated for conductivity enhancement of PEDOT:PSS. Similar conductivity enhancement can be achieved when geminal diols that have two hydroxyl groups on one carbon atom are used to treat PEDOT:PSS. Hexafluoroacetone (HFA), which could hydrolyze into a geminal diol with water, was adopted to enhance the conductivity of PEDOT:PSS films prepared from Clevios™ PH1000 [75]. The conductivity of the PEDOT:PSS films was enhanced from 0.3 to 1164 S cm^{-1} after the treatment with HFA·$3H_2O$. It even reached higher than 1300 S cm^{-1} by treating the PEDOT:PSS films with HFA·$3H_2O$ three times.

Apart from HFA·$3H_2O$, other geminal diols were also studied to treat PEDOT: PSS films prepared from Clevios PH1000. The conductivity enhancement is related to the formation of the germinal diols. The geminal diol formation constants of HFA, formaldehyde, and acetone are 10^6, 10^3, and 10^{-3}, respectively. The conductivity is enhanced more for the compound with a higher germinal diol formation constant (Table 8.2).

Alcohols with one hydroxyl group were also investigated for conductivity enhancement. Chu et al. observed that a post-coating treatment with alcohols could significantly improve the conductivity of PEDOT:PSS films prepared from Clevios PH1000 [79]. The conductivity enhancement depends on the chemical structure of alcohols. It is 1015 S cm^{-1} when methanol is used for the treatment, whereas it is only 286 S cm^{-1} if the polymer films are treated with isobutanol. They also observed that the alcohol treatment gives rise to a better stability in conductivity than EG-treated and pristine PEDOT:PSS films.

Vapor of EG was also studied for conductivity enhancement of PEDOT:PSS film [93]. Na et al. treated PEDOT:PSS films with EG vapor at 150 °C (Fig. 8.5).

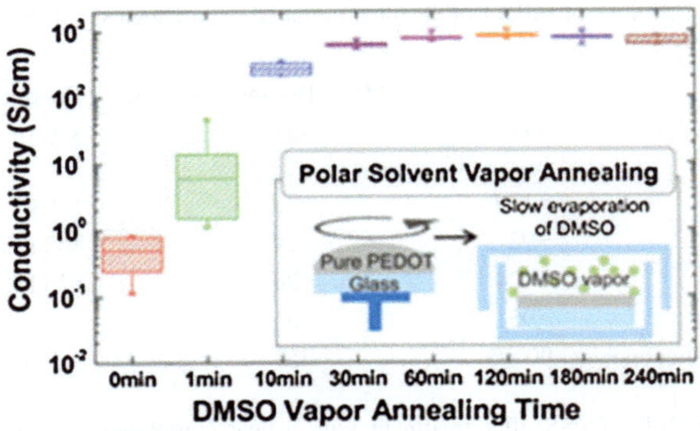

Fig. 8.5 Variation of the conductivity of PEDOT:PSS films treated with polar solvent vapor with treating time. The *inset* schematically illustrates the treatment process. Reproduced from [93]

Longer vapor treating time leads to higher conductivity of PEDOT:PSS films. The conductivity was higher than 1000 S cm^{-1} after the PEDOT:PSS films were treated for 120 min.

Kim et al. [74] carried out post-coating treatment on PEDOT:PSS films prepared from Clevios PH1000 with adding EG (Fig. 8.6). The conductivity of PEDOT:PSS prepared from Clevios PH1000 is more than 600 S cm^{-1}. It is enhanced to more than 1400 S cm^{-1} after the PEDOT:PSS films are further treated with EG vapor.

Salts were studied for conductivity enhancement of PEDOT:PSS films as well [70, 73]. A treatment of PEDOT:PSS films with aqueous solutions of $CuCl_2$ or $InCl_3$ improves the conductivity of PEDOT:PSS by more than two orders of magnitude. The softness parameter of the cation of the salts is a key factor for conductivity enhancement (Table 8.3). When a salt, whose cation has a positive softness parameter, such as InI_3, $InBr_3$, $InCl_3$, $CuCl_2$, $Cu(ClO_4)_2$, or $CuBr_2$, is used, the conductivity enhancement is significant. In contrast, when a salt, whose cation has a negative softness parameter, such as NaCl, $MgCl_2$, or $NaiCl_2$ is used, there is no remarkable conductivity enhancement. In terms of the definition of the softness parameter, the softness parameter of an ion implies the binding energy of this ion to other species [94]. A cation with a negative (or positive) softness parameter is a hard (or soft) Lewis acid. The softness parameters of cations are thus related to the binding strength between the cations and PSS. The experimental results indicate that the conductivity enhancement of the PEDOT:PSS films by the salt treatment is related to interaction between the metal ions of the salt and the PSS anions of PEDOT:PSS.

Apart from the cation, the anion of the salts also affect conductivity enhancement. An example is the treatment with Cu^{2+} salts. Treatment with $CuSO_4$ enhances the conductivity of PEDOT:PSS by about one order of magnitude, whereas treatment with $Cu(CH_3COO)_2$ hardly affects the conductivity of PEDOT:PSS. Further investigation suggests that the conductivity of PEDOT:PSS is not very consistent with the softness parameter of the anions but it is really consistent with the

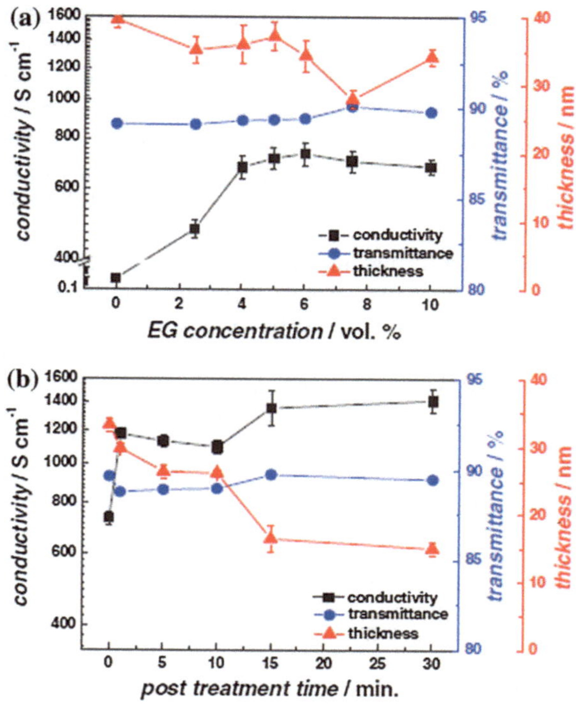

Fig. 8.6 Variations of the conductivity, transmittance and film thickness of PEDOT:PSS films treated with EG with respect to **a** EG concentration and **b** solvent post-treatment time (PEDOT:PSS doped with 6 vol.% EG). The *error bars* represent the standard deviation from several measurements. Reproduced from [74]

Table 8.3 Conductivities of PEDOT:PSS films after treatment with 0.1 M solution of a salt

Salt	Softness parameter of metal ion[a]	Softness parameter of anion[a]	Conductivity ($S\ cm^{-1}$)
$AgNO_3$	+0.18	+0.03	7.4
$CuCl_2$	+0.38	−0.09	29.0
$InCl_3$	+0.48	−0.09	95.5
LiCl	−1.02	−0.09	0.6
NaCl	−0.75	−0.09	1.5
$MgCl_2$	−0.41	−0.09	0.2
$NiCl_2$	−0.11	−0.09	0.3

Reproduced from [45]
[a]Soft parameters are obtained from [94]

association constant of the anions with proton. As the association constant can be expressed with the acid association constant (K_a) in the reversal process, the conductivity enhancement of PEDOT:PSS by the salt treatment is ascribed to the association of the cations with PSS anions and the association of the cations with PEDOT cations. The associations of the cations and anions with PEDOT:PSS can screen the Coulombic attraction between the PEDOT and PSS chains.

Small amount of ions can remain in PEDOT:PSS films after a salt treatment. These ions can diffuse under electric field into other parts of electronic devices and cause deterioration of the device performance. To solve this problem, zwitterions which have both positive and negative charges in the same molecules are used to enhance the conductivity of PEDOT:PSS. Three zwitterions were used by Xia et al. [71] to treat PEDOT:PSS films prepared from Clevios P. Similar to the treatment with $CuCl_2$ solution, the conductivity of PEDOT:PSS can be significantly enhanced by zwitterions.

PEDOT:PSS has the two components, PEDOT and PSS. They have different hydrophilicities. These give them different interactions with solvents of different polarities. A cosolvent can be formed by solvents of different polarities. Cosolvents of water with several organic solvents, including methanol, ethanol, isopropyl alcohol (IPA), acetonitrile (ACN), acetone, and THF, were investigated to treat PEDOT:PSS prepared from Clevios P (Fig. 8.7) [72]. The conductivity of PEDOT:

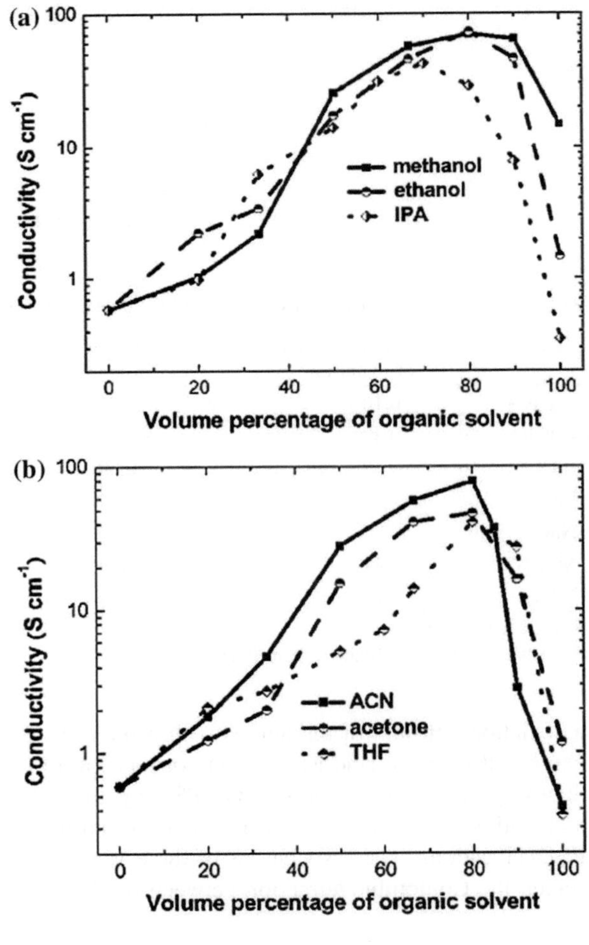

Fig. 8.7 Variation of the conductivities of PEDOT:PSS films treated with cosolvents of water and organic solvents with the volume percentage of the organic solvents: **a** methanol, ethanol, IPA; **b** ACN, acetone, and THF. Reproduced from [72]

PSS reached almost 100 S cm^{-1} at the optimal composition with about 80 vol.% organic solvent. Preferential solvations of PEDOT and PSS with the organic solvent and water are proposed for the conductivity enhancement.

A cosolvent can also be formed by a polar organic solvent with a nonpolar organic solvent. This type of cosolvent was also studied for conductivity enhancement of PEDOT:PSS films. When PEDOT:PSS films prepared from Clevios P were treated with a cosolvent of methanol and 1,2-dichlorobenzene, their conductivity hardly changed after the treatment. Nevertheless, a similar treatment for PEDOT:PSS films prepared from Clevios P VP Al 4083 can significantly enhance the conductivity [76]. At the optimal ratio of methanol to 1,2-dichlorobenzene, the conductivity increased from 10^{-3} S cm^{-1} to almost 10 S cm^{-1}. These results suggest that the preferential solvations of PEDOT and PSS are affected by the compositions of PEDOT:PSS.

The mechanism for the conductivity enhancement of PEDOT:PSS suggests that treatment can cause phase segregation between PEDOT and PSS. The phase segregation can take place when the Coulombic attraction between PEDOT and PSS is lowered. As discussed above, various compounds can screen the Coulombic attraction, so they can significantly enhance the conductivity of PEDOT:PSS. Phase segregation can also occur through the protonation of PSS, $PSS^- + H^+ \rightarrow PSSH$. There is no Coulombic attraction between the positively charged PEDOT chains and neutral PSSH chains, giving rise to the phase segregation between PEDOT and PSS. Hence, treatment of PEDOT:PSS with acids can also enhance the conductivity. Various organic and inorganic acids are used to treat PEDOT:PSS films prepared from Clevios P, and the conductivities are presented in Fig. 8.8 [69]. When oxalic acid or sulfurous acid is used to treat PEDOT:PSS, the conductivity can be higher than 100 or 200 S cm^{-1}. PEDOT prepared by vapor deposition was treated with acids by Howden et al. [95] and conductivity enhancement was also observed.

The conductivity enhancement is presumably caused by the protonation of PSS. The conductivity enhancement thus depends on the pK_a value of the acids. A strong acid with a low pK_a value in principle enhances the conductivity more significantly. Nevertheless, similar conductivity enhancement was observed on the PEDOT:PSS films treated with hydrochloric acid and weak organic or inorganic acids (Fig. 8.8). This is because HCl readily vaporizes during the treatment.

The treatment with sulfuric acid gives rise to very significant conductivity enhancement. PEDOT:PSS films prepared from Clevios PH1000 were treated with sulfuric acid (Fig. 8.9). The conductivity increases to about 2400 S cm^{-1} after a treatment with 1.5 M H_2SO_4. When the PEDOT:PSS films are treated with 1 M H_2SO_4 solution at 160 °C three times, the conductivity is further enhanced to 3065 S cm^{-1}. This conductivity is impressive, because it is higher than that (~2000 S cm^{-1}) of ITO on plastic and comparable to that (3000–6000 S cm^{-1}) of ITO on glass.

Moreover, the conduction mechanism of the PEDOT:PSS films significantly changes after the H_2SO_4 treatment. Figure 8.10 presents the temperature dependence of the resistances of PEDOT:PSS films before and after treatment. The

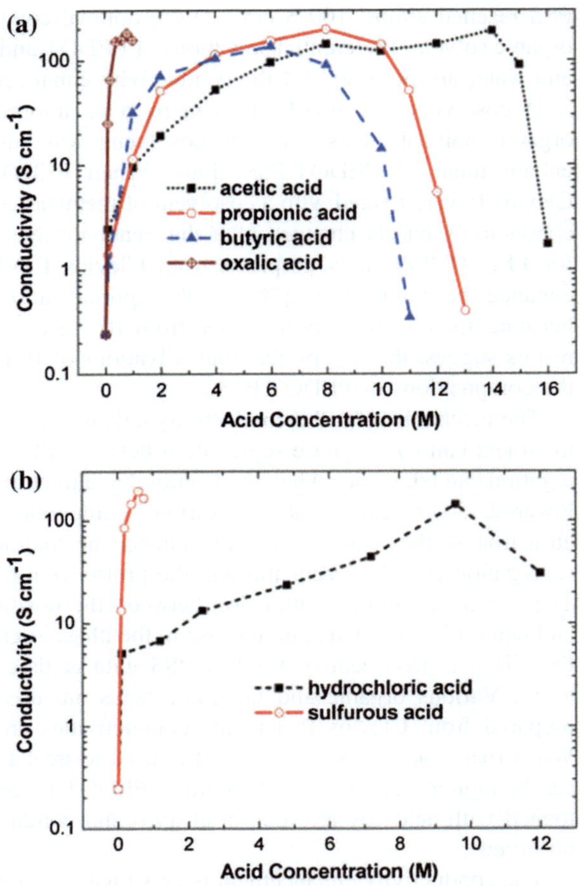

Fig. 8.8 Variation of the conductivities of PEDOT:PSS films treated with solutions of **a** organic and **b** inorganic acids with the acid concentration. The acids are acetic acid, propionic acid, butyric acid, oxalic acid, sulfurous acid, and hydrochloric acid. Reproduced from [69]

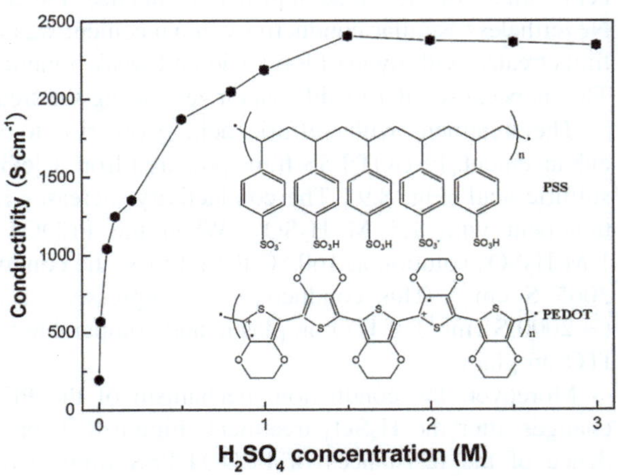

Fig. 8.9 Variation of the conductivities of H_2SO_4-treated PEDOT:PSS films with acid concentration at 160 °C. Reproduced from [83]

Fig. 8.10 Variation of the normalized resistance with temperature for untreated and H_2SO_4-treated PEDOT:PSS films. **a** Normalized resistances versus temperature. **b** Analysis of the temperature dependence of the resistance of the H_2SO_4-treated PEDOT:PSS film with the one-dimensional VRH model. The PEDOT:PSS film was treated with 1 M H_2SO_4 three times. Reproduced from [83]

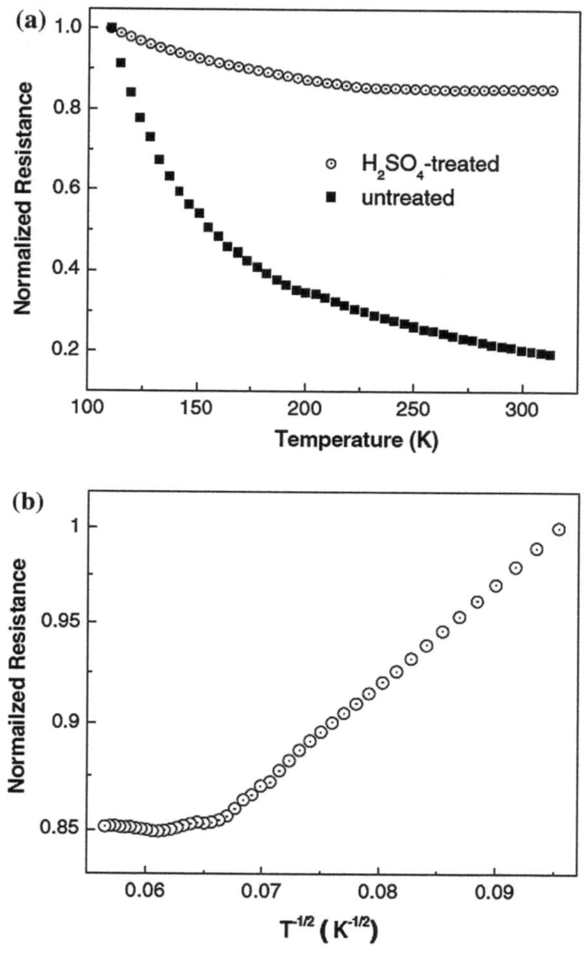

pristine PEDOT:PSS film exhibits a resistance increase when the temperature is lowered. Its temperature-dependence agrees well with the one-dimensional variable range hopping (VRH) model [96],

$$R(T) = R_0 \exp\left[\left(\frac{T_0}{T}\right)^{1/2}\right].$$

After the H_2SO_4 treatment, the temperature dependence of the resistance becomes quite different. The resistance is almost constant from 320 K down to 240 K. It then increases with lowering temperature. The temperature dependence of the H_2SO_4-treated PEDOT:PSS films is consistent with the hopping model only at temperature below 240 K. The H_2SO_4-treatment PEDOT:PSS is metallic or semi-metallic at

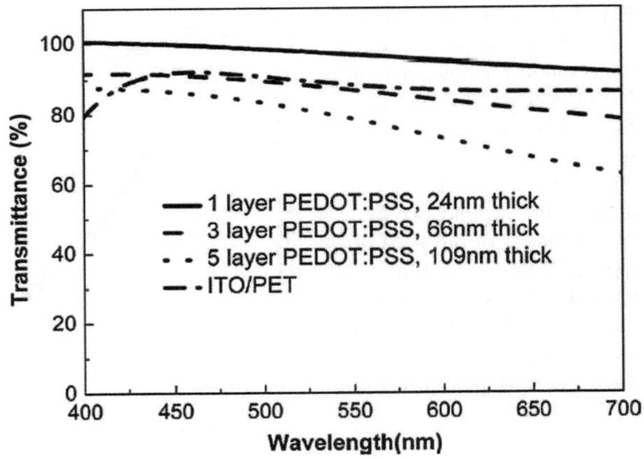

Fig. 8.11 Transmittance spectra of H_2SO_4-treated PEDOT:PSS films. The transmittance spectrum of ITO/PET is included for comparison. The sheet resistance of the ITO/PET is 50 $\Omega\,\square^{-1}$. Reproduced from [83]

room temperature. It is rare to observe metallic behavior on solution-processed conducting polymers.

Although the acid treatment significantly enhances the conductivity of PEDOT:PSS, it does not affect transmittance in the visible range. As shown in Fig. 8.11, the three H_2SO_4-treated PEDOT:PSS films with different thicknesses have high transmittance. For the 66 nm-thick PEDOT:PSS film, which has a sheet resistance of 67 $\Omega\,\square^{-1}$, its transmittance is 87 % at 550 nm. At wavelengths shorter than 500 nm, the transmittance is higher than 90 %. Although the transmittance decreases a little at longer wavelength, it is still higher than 80 % at wavelengths longer than 550 nm. Even for the 109 nm-thick PEDOT:PSS film, which has a sheet resistance of 39 $\Omega\,\square^{-1}$, its transmittance is more than 80 % at 550 nm. The last PEDOT:PSS film has a thickness close to that (100 nm) of the ITO layer of ITO/PET.

The PEDOT:PSS films treated with sulfuric acid can have high conductivity and high transparency in the visible range. However, sulfuric acid can bring some problems because it is a strong and corrosive acid. Other acids were investigated for the conductivity enhancement of PEDOT:PSS films prepared from Clevios PH1000. It was recently observed that mild organic acids can give rise to similar conductivity enhancement to sulfuric acid. The conductivity of PEDOT:PSS films becomes higher than 3300 S cm^{-1} after treatment with methanesulfonic acid, which is an organic acid and not corrosive [84]. Methanesulfonic acid treatment has a similar mechanism for conductivity enhancement to of sulfuric acid treatment. The acidity of methanesulfonic acid is almost the same as PSSH. However, it can protonate PSS^- into PSSH because PSSH and PEDOT have different hydrophilicities. The different hydrophilicities give rise to phase segregation of PSSH from PEDOT. The driving force for the protonation of PSS^- by methanesulfonic acid is the phase segregation.

The methods used to treat PEDOT:PSS and the conductivities of PEDOT:PSS are summarized in Tables 8.4 (Clevios P) and 8.5 (Clevios PH1000). The PEDOT:PSS grade is the dominant factor for conductivity enhancement. The conductivity

Table 8.4 Conductivities of PEDOT:PSS (Clevios P) treated with various methods

Method	Compound	σ (S cm^{-1})	References
Addition into solution	DMSO EG DMF D-Sorbitol	30–250	[59–82]
	Ionic liquids	Up to 100	[64]
	Anionic surfactants	Up to 100	[65]
Post treatment	DMSO EG	Up to 250	[61]
	Salts	Up to 230	[70, 73]
	Zwitterion	Up to 100	[71]
	Water/organic cosolvents	Up to 100	[72]
	Acids	Up to 230	[70, 95]

Table 8.5 Conductivities of PEDOT:PSS (Clevios PH1000) treated with various methods

Method	Compound	σ (S cm^{-1})	References
Addition into solution	DMSO EG	600–700	[75]
	Ionic liquids	Up to 2084	[89]
	Fluorosurfactants	Hundreds	[90, 91]
Post treatment	DMSO EG	600–700	[75]
	Geminal diols	Up to 1300	[75]
	Alcohols	Up to 1000	[79]
	DMSO vapor	~1000	[93]
	Acids	>3000	[83, 84]
Addition + post treatment	EG	~1460	[74]

enhancements for PEDOT:PSS from Clevios PH1000 are much more salient than from Clevios P.

8.6.4 Mechanisms for the Conductivity Enhancements of PEDOT:PSS

It took a long time to understand the mechanisms for the conductivity enhancement of PEDOT:PSS because of the difficulty in observing remarkable changes in its structure. The absorption of PEDOT:PSS aqueous solution and films almost do not change in the visible and infrared range. Differential scanning calorimetry (DSC) and thermal gravimetric analysis (TGA) also do not reveal any structural change induced by the treatments. Among the different mechanisms proposed for

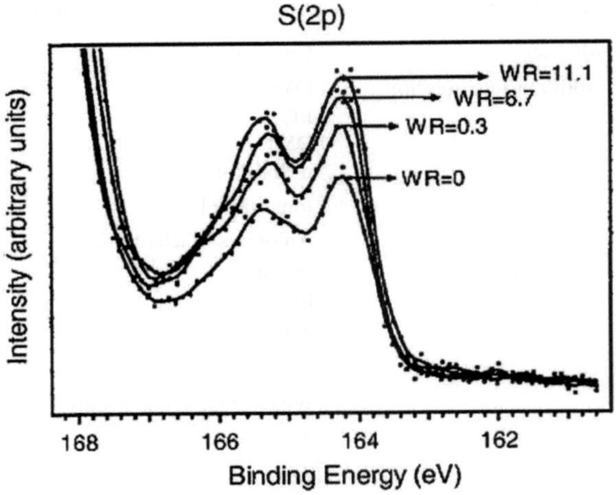

Fig. 8.12 S 2p XPS bands of PEDOT with the EG-to-PEDOT/PSS weight ratio. The intensities were normalized to the S 2p XPS band of PSS. Reproduced from [97]

conductivity enhancement, the treatment-induced phase segregation between PEDOT and PSS is believed to the major reason. It can cause the removal of some PSS chains from the PEDOT:PSS films and the change in the conformation of the PEDOT chains.

Jönsson et al. [67] employed X-ray photoelectron spectroscopy (XPS) to study the effect of sorbitol and 1-methyl-2-pyrrolidinone (NMP) on the composition of PEDOT:PSS films. The change in the S 2p signals suggested the wash-away of some PSS chains from the surface region of the PEDOT:PSS films after the treatment. Crispin et al. [97] also used XPS to investigate the mechanism for the conductivity enhancement by adding EG into PEDOT:PSS aqueous solution. Their results confirmed the removal of PSS from PEDOT:PSS. Figure 8.12 presents the XPS spectra of the PEDOT:PSS films with and without EG in the PEDOT:PSS aqueous solutions. The S 2p XPS bands in the range 163–167 eV originate from the S atoms on PEDOT. With respect to the S 2p XPS bands of PSS, the EG treatment induces the increase in the intensity of the S 2p bands of PEDOT. In another word, the loading of PSS in the PEDOT:PSS film decreases after the EG treatment. This is attributed to the wash-away of some PSS from PEDOT:PSS.

The UV-visible absorption spectroscopy confirms the treatment-induced removal of some PSS chains from the PEDOT:PSS films. Figure 8.13 presents the UV-visible absorption spectra of PEDOT:PSS films with and without $CuCl_2$ treatment [70]. The absorption in the infrared and visible range hardly changes after the treatment. In contrast, the treatment induces a remarkable decrease in the intensities of the two absorption bands in the UV range, which originate from the benzene rings of PSS. The drop in the two absorption bands confirms that the $CuCl_2$ treatment induces the removal of some PSS from PEDOT:PSS.

In addition, the treatment induces the change in the conformation of the PEDOT chains. However, it is hard to obtain any solid evidence for the conformational change by both DSC and UV-visible absorption spectroscopy. In early works, only

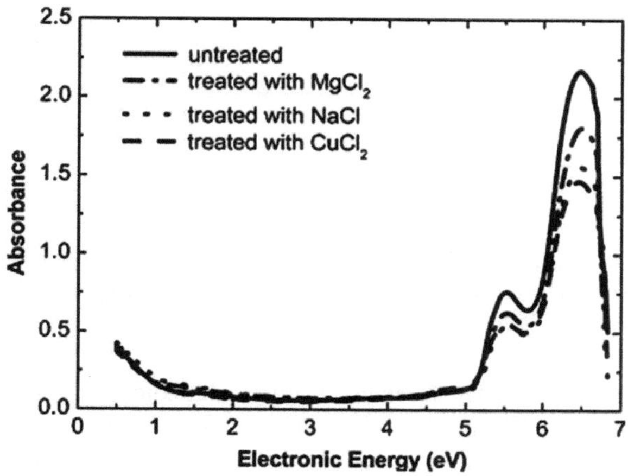

Fig. 8.13 UV-visible-NIR absorbance spectra of PEDOT:PSS films before (*solid curve*) and after the treatment with 0.1 M $MgCl_2$ (*dashed dotted curve*), 0.1 M NaCl (*dotted curve*), and 0.1 M $CuCl_2$ (*dashed curve*) solution. Reproduced from [70]

amorphous XRD patterns were observed on pristine and treated PEDOT:PSS films. Recently, Takano et al. [98] observed nanocrystal structure by wide-angle X-ray scattering on PEDOT:PSS after the EG treatment. Pristine PEDOT:PSS films exhibited only broad bands, whereas a relatively sharp band was observed at q = 18.5 nm after the EG treatment.

The Raman spectroscopy also provided solid evidence for the treatment-induced conformational change of PEDOT chains [61]. As shown in Fig. 8.14, the Raman band between 1400 and 1500 cm^{-1} is assigned to the stretching vibration of C_α-C_β on the five-member ring of PEDOT. After the EG treatment, this Raman band shifts to red and becomes narrower. Thus, the EG treatment induces the change of the

Fig. 8.14 Raman spectra of a PEDOT:PSS film **a** before and **b** after the EG treatment. The wavelength of the excitation laser is 632.8 nm. Reproduced from [61]

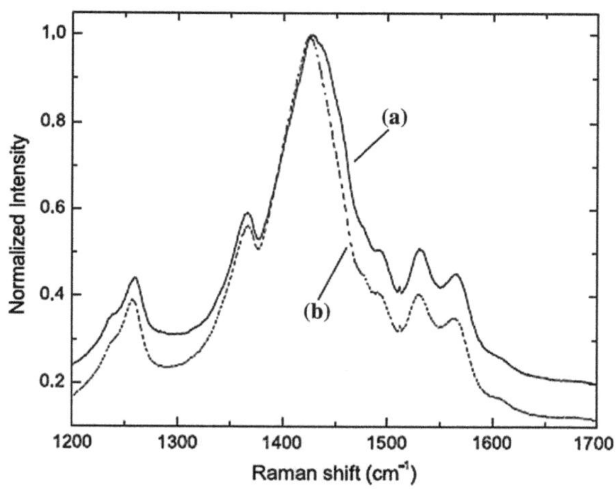

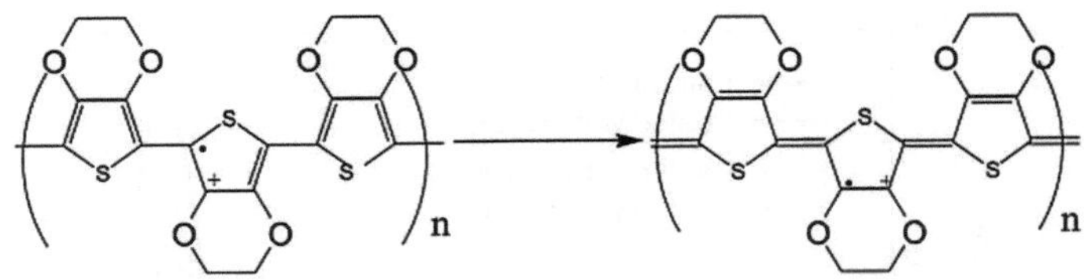

Scheme 8.4 Transformation of the PEDOT chain from the benzoid to quinoid structure. The '·' and '+' signs represents an unpaired electron and positive charge on the PEDOT chain, respectively. Reproduced from [61]

resonant structure of PEDOT chain from a benzoid to a quinoid structure (Scheme 8.4). For a coil conformation, the benzoid structure can be the favorite structure, whereas the favorite structure for a linear or expanded-coil conformation may be the quinoid structure. The Raman spectra thus imply the treatment-induced change of some PEDOT:PSS chains from the coil conformation into linear or expanded-coil conformation.

The treatment-induced conformational change is further evidenced by the electron spin resonance (ESR). The g-value does not change. However, the integrated ESR intensity of PEDOT:PSS decreases to about 50 % after the EG treatment. The ESR signal is produced by the polarons on PEDOT. Because a polaron has a spin of 1/2 whereas a bipolaron is spinless [99], roughly half of the polarons in the PEDOT:PSS film turn to bipolarons after the EG treatment. Because bipolarons are more delocalized on PEDOT than polarons, the change of polarons into bipolarons suggests the conformational change from the coil to linear or expanded-coil structure. The charges on the PEDOT chains become more delocalized after the EG treatment (Fig. 8.15).

Fig. 8.15 ESR signal of pristine (*solid line*) and EG-treated (*dash line*) PEDOT:PSS films at 300 K. Reproduced from [61]

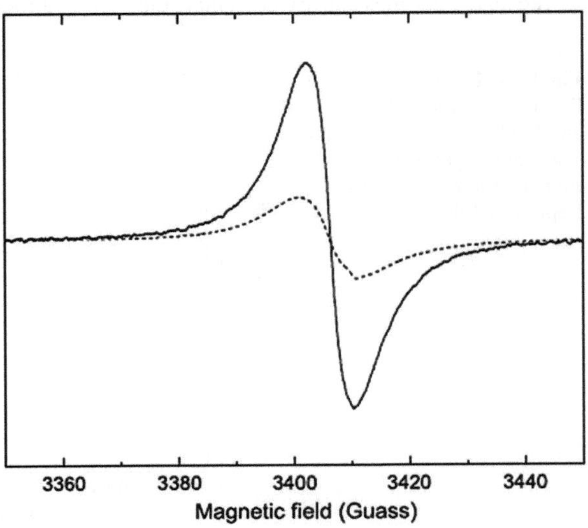

8.7 Application of PEDOT:PSS for Optoelectronic Devices

The application of PEDOT:PSS in electronic devices is related to its conductivity. Though PEDOT:PSS has strong potential as the transparent electrode, less conductive PEDOT:PSS is only used as the buffer layer in organic light-emitting diodes (OLEDs) and organic solar cells (OSCs), because PEDOT:PSS can be used to make the rough surface of ITO smooth. It can also help the hole injection for OLEDs and hole collection for OSCs, because it has a work function of ca. 5.0 eV. The molar ratio of PEDOT to PSS is 1:6 for PEDOT:PSS used for the buffer layer. The highly excess PSS leads to a low conductivity of 10^{-3}–10^{-4} S cm^{-1}. The less conductive PEDOT:PSS was treated with organic solvents and also exploited as the buffer layer for PSCs [76, 82, 100]. The treatment can improve the photovoltaic efficiency of PSCs.

It is certainly more important to use PEDOT:PSS directly to substitute ITO as the transparent electrode of optoelectronic devices. The treated PEDOT:PSS can have high conductivity, whereas the treatment hardly affects its transparency in the visible range and its work function (ca. 5.0 eV). A slight increase in the work function of PEDOT:PSS to 5.35 eV was observed by Na et al. after treatment with polar solvent vapor [93]. That is attributed to the treatment-induced change on the vertical structure of PEDOT:PSS films. Highly conductive PEDOT:PSS films have been reported as a replacement for ITO in the transparent electrode of PLEDs and PSCs [6–18, 101–105].

The EG-treated PEDOT:PSS films prepared from Clevios P aqueous solution was adopted as the transparent anode of PLEDs [16]. The current-voltage curves of a PLED with poly[2-methoxy-5-(2-ethylhexyloxy)-1,4-phenylenevinylene] (MEH-PPV) as the active materials are shown in Fig. 8.16. The current density for the device with EG-treated PEDOT:PSS film as the anode exhibits is quite close to that with ITO/PEDOT:PSS. In addition, this current is almost three orders of magnitude higher than the control devices with untreated PEDOT:PSS as the anode.

The PLEDs with an EG-treated PEDOT:PSS film as the transparent electrode emit light uniformly under external electric field. Figure 8.17 presents the luminance-voltage curves of such a PLED. An EG-treated PEDOT:PSS 300 nm thick was used as the transparent electrode. The luminance of the PLED is almost half as that of the control device with ITO/PEDOT. The low luminance is attributed to the low transmittance of the 300 nm-thick PEDOT:PSS film. The thickness of the PEDOT:PSS layer can be lowered by using treated PEDOT:PSS films prepared from Clevios PH1000. Cai et al. [105] recently reported that OLEDs with EG-treated PEDOT:PSS films as the transparent electrode could even exhibit a luminous efficiency and power efficiency higher than those of the control devices with ITO. They attributed the improvement to the advantageous optical properties of PEDOT:PSS films over ITO.

Treated PEDOT:PSS films with high conductivity were also studied as the transparent electrode of PSCs [74, 79, 83, 102–104]. Scheme 8.5 presents the architecture of a PSC with an H_2SO_4-treated PEDOT:PSS film as the transparent

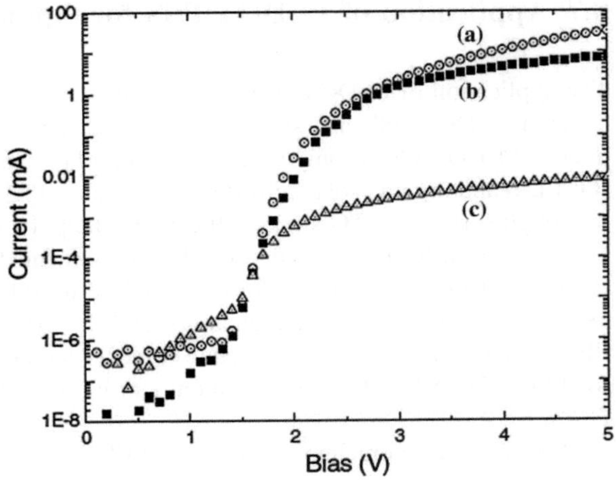

Fig. 8.16 Current–voltage curves of PLEDs, glass/anode/MEH-PPV/Ca/Al, with different anodes: **a** ITO/untreated PEDOT:PSS (40 nm thick); **b** EG-treated PEDOT:PSS (300 nm thick); **c** untreated PEDOT:PSS (300 nm thick). Reproduced from [16]

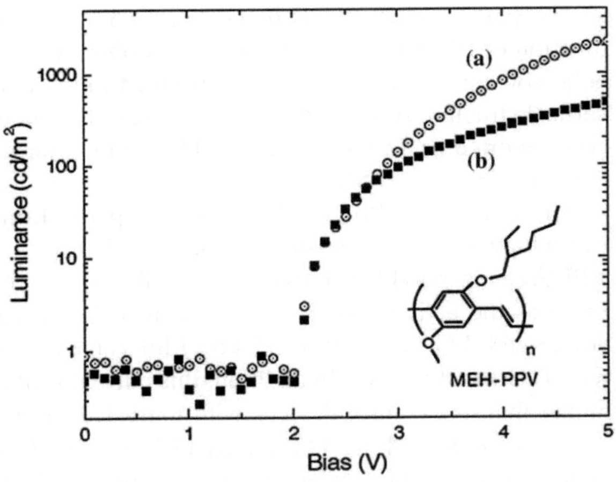

Fig. 8.17 Luminance–voltage curves of the PLEDs, glass/anode/MEH-PPV/Ca/Al, with different anodes: **a** ITO/untreated PEDOT:PSS (40 nm thick); **b** EG-treatment PEDOT:PSS (300 nm thick). The *inset* is the chemical structure of MEH-PPV. Reproduced from [16]

electrode. Because of the high conductivity of the H_2SO_4-treated PEDOT:PSS film, it only has a thickness of 70 nm. Figure 8.18 illustrates the current density–voltage —(*J*)–(*V*)—curves of the PSC. The device exhibited high photovoltaic performance with the short-circuit current (J_{sc}) of 9.29 mA cm^{-2}, open-circuit voltage (V_{oc}) of 0.59 V, fill factor (FF) of 0.65, and power conversion efficiency (PCE) of 3.56 %. The photovoltaic performance is only slightly lower than that of the control devices using ITO anodes. This is ascribed to the lower transmittance of the acid-treated PEDOT:PSS film than that of ITO in the visible and infrared ranges.

Treated PEDOT:PSS films with high conductivity were also studied as the top electrode of inverted PSCs. There is a wettability problem, because the active film of PSCs is hydrophobic whereas PEDOT:PSS aqueous solution is hydrophilic [106, 107]. Two methods were developed to coat PEDOT:PSS on the active organic film

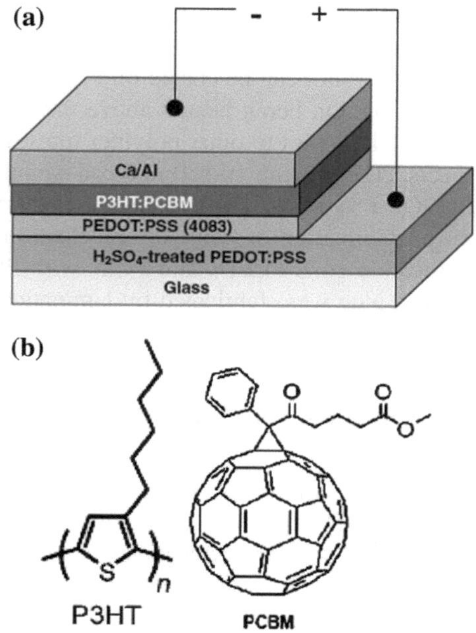

Scheme 8.5 **a** Architecture of a PSC with a H_2SO_4-treated PEDOT:PSS film to replace ITO as the transparent electrode. **b** Chemical structures of P3HT and PCBM

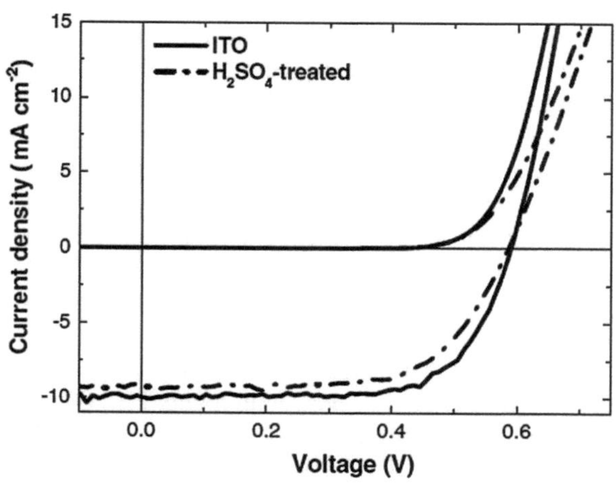

Fig. 8.18 *J–V* characteristics of PSCs with ITO and an H_2SO_4-treated PEDOT:PSS film prepared from Clevios PH1000 as the anodes under AM 1.5G illumination (100 mW cm^{-2}). Reproduced from [83]

of PSCs [102, 108, 109]. One is the addition of surfactant to PEDOT:PSS aqueous solution, and another is to spray coat PEDOT:PSS aqueous solution. Colsmann et al. prepared Clevios PH1000 aqueous solution by spray coating on P3HT:PCBM films for inverted PSCs [109]. Because both electrodes are transparent, the inverted PSCs are semitransparent and show a photovoltaic effect when they are exposed to light from either side of the devices. They observed a power conversion efficiency of about 2 % for the devices illuminated from the front side.

Another interesting application for PEDOT:PSS with D-sorbitol is the transparent electric glue. D-Sorbitol, a solid at room temperature, is adhesive in melting [17]. D-Sorbitol can be coated on PEDOT:PSS films by solution coating or thermal deposition. On being heated above the melting point of D-sorbitol, it can glue the PEDOT:PSS film to other polymer films, including conjugated polymers or plastic sheets. The melting D-sorbitol can simultaneously enhance the conductivity of PEDOT:PSS films. The transparent electric glue can play a crucial part in the roll-to-roll fabrication of polymer optoelectronic devices, which has advantage of low fabrication cost. PLEDs and PCSs with PEDOT:PSS/D-sorbitol as the transparent electric glue were fabricated by lamination [17, 104].

8.8 Outlook for Transparent Conducting Polymers

Transparent conducting polymers can have important applications for electronic devices. PEDOT:PSS is still the most important transparent conducting polymer. It can have important advantages as the transparent electrode of optoelectronic devices. The key issue for this application is to enhance significantly the conductivity of PEDOT:PSS. The addition of polar organic solvents with high boiling point into the PEDOT:PSS aqueous solutions can enhance the conductivity of PEDOT:PSS to higher than 200 S cm^{-1} for Clevios P and 600–700 S cm^{-1} for Clevios PH1000. The addition of ionic liquids into the Clevios PH1000 aqueous solution can further enhance the conductivity to higher than 2000 S cm^{-1}. Alternatively, the conductivity enhancement can be achieved by a post-coating treatment. Some compounds such as salts, zwitterions, cosolvents, and acids can significantly enhance the conductivity by post-coating treatment, but the addition of these compounds into a PEDOT:PSS aqueous solution hardly affects the conductivity. Recent results demonstrated that treatment with H_2SO_4 or methanesulfonic acid can enhance the conductivity of PEDOT:PSS films prepared from Clevios PH1000 to more than 3000 S cm^{-1}. This conductivity is even higher than that of ITO on plastic and comparable to that of ITO on glass.

We believe that the conductivity of PEDOT:PSS can be even higher than that of ITO on glass if PEDOT:PSS with higher PEDOT molecular weight is developed. In terms of the comparative study on Clevios P and Clevios PH1000, the conductivity of the treated PEDOT:PSS films depends on the PEDOT molecular weight. Moreover, the conductivity of the acid-treated PEDOT:PSS films is significantly higher even than the highest conductivity of PEDOT polymers doped with small anions. This implies the potential to develop PEDOT:PSS with even higher conductivity. Furthermore, the knowledge of enhancing the conductivity of PEDOT:PSS can be very useful for the development of other transparent conducting polymers. Many low-bandgap conjugated polymers have been developed as the donor materials for PSCs [103, 104]. However, they have rarely been studied as the transparent conducting polymers. They can have high transparency in the visible range after they are oxidized.

References

1. Skotheim TA (1986) Handbook of conducting polymers. M. Dekker, New York
2. Brédas JL, Scott JC, Yakushi K, Street GB (1984) Polarons and bipolarons in polypyrrole: evolution of the band structure and optical spectrum upon doing. Phys Rev B 30:1023
3. Pei Q, Zuccarello G, Ahlskog M, Inganäs O (1994) Electrochromic and highly stable poly (3,4-ethylenedioxythiophene) switches between opaque blue-black and transparent sky blue. Polymer 35:1347
4. Inganäs O (2011) Organic photovoltaics: avoiding indium. Nat Photon 5:201
5. Chipman A (2007) A commodity no more. Nature 449:131
6. Yoo JE, Yoo JE, Lee KS, Garcia A, Tarver J, Gomez ED, Baldwin K, Sun Y, Meng H, Nguyen TQ, Loo YL (2010) Directly patternable, highly conducting polymers for broad applications in organic electronics. Proc Nat Acad Sci 107:5712
7. Huang JH, Kekuda D, Chu CW, Ho KC (2009) Electrochemical characterization of the solvent-enhanced conductivity of poly(3,4-ethylenedioxythiophene) and its application in polymer solar cells. J Mater Chem 19:3704
8. Na SI, Wang G, Kim SS, Kim TW, Oh SH, Yu BK, Lee T, Kim DY (2009) Evolution of nanomorphology and anisotropic conductivity in solvent-modified PEDOT:PSS films for polymeric anodes of polymer solar cells. J Mater Chem 19:9045
9. Hsiao YS, Whang WT, Chen CP, Chen YC (2008) High-conductivity poly(3,4-ethylenedioxythiophene):poly(styrene sulfonate) film for use in ITO-free polymer solar cells. J Mater Chem 18:5948
10. Chang YM, Wang L, Su WF (2008) Polymer solar cells with poly(3,4-ethylenedioxythiophene) as transparent anode. Org Electron 9:968
11. Lee MW, Lee MY, Choi JC, Park JS, Song CK (2010) Fine patterning of glycerol-doped PEDOT:PSS on hydrophobic PVP dielectric with ink jet for source and drain electrode of OTFTs. Org Electron 11:854
12. Na SI, Kim SS, Jo J, Kim DK (2008) Efficient and flexible ITO-free organic solar cells using highly conductive polymer anodes. Adv Mater 20:4061
13. Kim WH, Karafi ZH (2002) Molecular organic light-emitting diodes using highly conducting polymers as anodes. Appl Phys Lett 80:3844
14. Zhang F, Johansson M, Andersson MR, Hummelen JC, Inganäs O (2002) Polymer photovoltaic cells with conducting polymer anodes. Adv Mater 14:662
15. Argun AA, Cirpan A, Reynolds JR (2003) The first truly all-polymer electrochromic devices. Adv Mater 15:1338
16. Ouyang J, Chu CW, Chen FC, Xu Q, Yang Y (2005) Highly conductive PEDOT:PSS film and its applications in optoelectronic devices. Adv Funct Mater 15:203
17. Ouyang J, Yang Y (2006) Conducting polymer as transparent electric glue. Adv Mater 18:2141
18. Lee CS, Kim JY, Lee DE, Koo YK, Joo J, Han S, Beag YW, Koh SK (2003) Organic based flexible speaker through enhanced conductivity of PEDOT/PSS with various solvents. Synth Met 135:13
19. Po R, Carbonera C, Bernardi A, Tinti F, Camaioni N (2012) Polymer- and carbon-based electrodes for polymer solar cells: toward low-cost, continuous fabrication over large area. Sol Energy Mater Sol Cells 100:97
20. Wu ZC, Chen ZH, Du X, Logan JM, Sippel J, Nikolou M, Kamaras K, Reynolds JR, Tanner DB, Hebard AF, Rinzler AG (2004) Transparent, conductive carbon nanotube films. Science 305:1273
21. Gruner G (2006) Carbon nanotube films for transparent and plastic electronics. J Mater Chem 16:3533
22. Zhang M, Fang S, Zakhidov AA, Lee SB, Aliev AE, Williams CD, Atkinson KR, Baughman RH (2005) Strong, transparent, multifunctional carbon nanotube sheets. Science 309:1215

23. Joshi P, Zhang L, Chen Q, Galipeau D, Fong H, Qiao Q (2010) Electrospun carbon nanofibers as low-cost counter electrode for dye-sensitized solar cells. ACS Appl Mater Interf 2:3572
24. Mei X, Ouyang J (2011) Highly conductive and transparent single-walled carbon nanotube thin films fabricated by gel coating. J Mater Chem 21:17842
25. Kim KS, Kim KS, Zhao Y, Jang H, Lee SY, Kim JM, Kim KS, Ahn JH, Kim P, Choi JY, Hong BH (2009) Large-scale pattern growth of graphene films for stretchable transparent electrodes. Nature 457:706
26. Becerril HA, Mao J, Liu Z, Stoltenberg RM, Bao Z, Chen Y (2008) Evaluation of solution-processed reduced graphene oxide films as transparent conductors. ACS Nano 2:463
27. Tung VC, Chen LM, Allen MJ, Wassei JK, Nelson K, Kaner RB, Yang Y (2009) Low-temperature solution processing of graphene–carbon nanotube hybrid materials for high-performance transparent conductors. Nano Lett 9:1949
28. Lee JY, Connor ST, Cui Y, Peumans P (2008) Solution-processed metal nanowire mesh transparent electrodes. Nano Lett 8:689
29. Yu Z, Zhang Q, Li L, Chen Q, Niu X, Liu J, Pei Q (2011) Highly flexible silver nanowire electrodes for shape-memory polymer light-emitting diodes. Adv Mater 23:664
30. Groenendaal L, Zotti G, Jonas F (2001) Optical, conductive and magnetic properties of electrochemically prepared alkylated poly(3,4-ethylenedioxythiophene)s. Synth Met 118:105
31. Lenz A, Kariis H, Pohl A, Persson P, Ojamäe L (2011) The electronic structure and reflectivity of PEDOT:PSS from density functional theory. Chem Phys 384:44
32. Li C, Imae T (2004) Electrochemical and optical properties of the poly(3,4-ethylenedioxythiophene) film electropolymerized in an aqueous sodium dodecyl sulfate and lithium tetrafluoroborate medium. Macromolecules 37:2411
33. Bhandari S, Deepa M, Singh S, Gupta G, Kant R (2008) Redox behavior and optical response of nanostructured poly(3,4-ethylenedioxythiophene) films grown in a camphorsulfonic acid based micellar solution. Electrochim Acta 53:3189
34. Han DH, Kim JW, Park SM (2006) Electrochemistry of conductive polymers 38. Electrodeposited poly(3,4-ethylenedioxy-thiophene) studied by current sensing atomic force microscopy. J Phys Chem B 110:14874
35. Musumeci C, Hutchison JA, Samorì P (2013) Controlling the morphology of conductive PEDOT by in situ electropolymerization: from thin films to nanowires with variable electrical properties. Nanoscale 5:7756
36. Gao Y, Zhao L, Li C, Shi G (2006) Electrosynthesis of poly(3,4-ethylenedioxythiophene) microcups in the aqueous solution of $LiClO_4$ and tri(ethylene glycol). Polymer 47:4953
37. Sakmeche N, Aaron JJ, Aeiyach S, Lacaze PC (2000) Usefulness of aqueous anionic micellar media for electrodeposition of poly-(3,4-ethylenedioxythiophene) films on iron, mild steel and aluminium. Electrochim Acta 45:1921
38. Chelawat H, Vaddiraju S, Gleason K (2010) Conformal, conducting poly(3,4-ethylenedioxythiophene) thin films deposited using bromine as the oxidant in a completely dry oxidative chemical vapor deposition process. Chem Mater 22:2864
39. Seo KI, Chung IJ (2000) Reaction analysis of 3,4-ethylenedioxythiophene with potassium persulfate in aqueous solution by using a calorimeter. Polymer 41:4491
40. Ha YH, Nikolov N, Pollack SK, Mastrangelo J, Martin BD, Shashidhar R (2004) Conformal coverage of poly(3,4-ethylenedioxythiophene) films with tunable nanoporosity via oxidative chemical vapor deposition. Adv Funct Mater 14:615
41. Lefebvre M, Qi Z, Rana D, Pickup PG (1999) Chemical synthesis, characterization, and electrochemical studies of poly(3,4-ethylenedioxythiophene)/poly(styrene-4-sulfonate) Composites. Chem Mater 11:262
42. Kirchmeyer S, Reuter K (2005) Scientific importance, properties and growing applications of poly(3,4-ethylenedioxythiophene). J Mater Chem 15:2077
43. Gustafsson JC, Liedberg B, Inganas O (1994) In situ spectroscopic investigations of electrochromism and ion transport in a poly(3,4-ethylenedioxythiophene) electrode in a solid state electrochemical cell. Solid State Ionics 69:145

44. Garreau S, Louarn G, Lefrant S, Buisson JP, Froyer G (1999) Optical study and vibrational analysis of the poly(3,4-ethylenedioxythiophene) (PEDT). Synth Met 101:312
45. Johansson T, Petterson LAA, Inganas O (2002) Conductivity of de-doped poly(3,4-ethylenedioxythiophene). Synth Met 129:269
46. Zykwinska A, Domagala W, Czardybon A, Pilawa B, Lapkowski M (2003) In situ EPR spectroelectrochemical studies of paramagnetic centres in poly(3,4-ethylenedioxythiophene) (PEDOT) and poly(3,4-butylenedioxythiophene) (PBuDOT) films. Chem Phys 292:31
47. Zykwinska A, Domagala W, Lapkowski M (2003) ESR spectroelectrochemistry of poly(3,4-ethylenedioxythiophene) (PEDOT). Electrochem Commun 5:603
48. Kim J, Kim E, Won Y, Lee H, Suh K (2003) The preparation and characteristics of conductive poly(3,4-ethylenedioxythiophene) thin film by vapor-phase polymerization. Synth Met 139:485
49. Padmalekha KG, Admassie S (2009) Electrochromic, magnetotransport and AC transport properties of vapor phase polymerized PEDOT (VPP PEDOT). Synth Met 159:1885
50. Fabretto M, Muller M, Zuber K, Murphy P (2009) Influence of PEG-ran-PPG surfactant on vapour phase polymerised PEDOT thin films. Macromol Rapid Commun 30:1846
51. Winther-Jensen B, West K (2004) Vapor-phase polymerization of 3,4-ethylenedioxythiophene: a route to highly conducting polymer surface layers. Macromolecules 37:4538
52. Winther-Jensen B, Breiby DW, West K (2005) Base inhibited oxidative polymerization of 3,4-ethylenedixoythiophene with iron(III)tosylate. Synth Met 152:1
53. Fabretto M, Jariego-Moncunill C, Autere JP, Michelmore A, Short R, Murphy P (2011) High conductivity PEDOT resulting from glycol/oxidant complex and glycol/polymer intercalation during vacuum vapour phase polymerization. Polymer 52:1725
54. Zuber K, Fabretto M, Hall C, Murphy P (2008) Improved PEDOT conductivity via suppression of crystallite formation in Fe(III) tosylate during vapor phase polymerization. Macromol Rapid Commun 29:1503
55. Evans D, Fabretto M, Mueller M, Zuber K, Short R, Murphy P (2012) Structure-directed growth of high conductivity PEDOT from liquid-like oxidant layers during vacuum vapor phase polymerization. J Mater Chem 22:14889
56. Laforgue A, Robitaille L (2010) Production of conductive pedot nanofibers by the combination of electrospinning and vapor-phase polymerization. Macromolecules 43:4194
57. Groenendaal L, Jonas F, Freitag D, Peilartzik H, Reynolds JR (2000) Poly(3,4-ethylenedioxythiophene) and its derivatives: past, present, and future. Adv Mater 12:481
58. Cao Y, Yu G, Menon R, Heeger AJ (1997) Polymer light-emitting diodes with polyethylene dioxythiophene–polystyrene sulfonate as the transparent anode. Synth Met 87:171
59. Kim JY, Jung JH, Lee DE, Joo J (2002) Enhancement of electrical conductivity of poly(3,4-ethylenedioxythiophene)/poly(4-styrenesulfonate) by a change of solvents. Synth Met 126:311
60. Ouyang J (2013) "Secondary doping" methods to significantly enhance the conductivity of PEDOT:PSS for its application as transparent electrode of optoelectronic devices. Displays 34:423
61. Ouyang J, Xu Q, Chu CW, Yang Y, Li G, Shinar J (2004) On the mechanism of conductivity enhancement in poly(3,4-ethylenedioxythiophene):poly(styrene sulfonate) film through solvent treatment. Polymer 45:8443
62. Crispin X, Jakobsson FLE, Crispin A, Grim PCM, Andersson P, Volodin A, van Haesendonck C, van der Auweraer M, Salaneck WR, Berggren M (2006) The origin of the high conductivity of poly(3,4-ethylenedioxythiophene)-poly(styrenesulfonate) (pedot-pss) plastic electrodes. Chem Mater 18:4354
63. Nardes AM, Janssen AJR, Kemerink MA (2008) A morphological model for the solvent-enhanced conductivity of PEDOT:PSS thin films. Adv Funct Mater 18:865
64. Döbbelin M, Marcilla R, Salsamendi M, Pozo-Gonzalo C, Carrasco PM, Pomposo JA, Mecerreyes D (2007) Influence of ionic liquids on the electrical conductivity and morphology of PEDOT:PSS films. Chem Mater 19:2147

65. Fan BH, Mei XG, Ouyang J (2008) Significant conductivity enhancement of conductive poly (3,4-ethylenedioxythiophene):poly(styrene sulfonate) (PEDOT:PSS) films by adding anionic surfactants into polymer solution. Macromolecules 41:5971
66. Pettersson LAA, Ghosh S, Inganäs O (2002) Optical anisotropy in thin films of poly(3,4-ethylenedioxythiophene)–poly(4-styrenesulfonate). Org Electron 3:143
67. Jönsson SKM, Birgersonb J, Crispin X, Greczynski G, Osikowicz W, van der Gon AWD, Salaneck JR, Fahlman M (2003) The effects of solvents on the morphology and sheet resistance in poly(3,4-ethylenedioxythiophene)–polystyrenesulfonic acid (PEDOT–PSS) films. Synth Met 139:1
68. Reyes-Reyes M, Cruz-Cruz I, Lopez-Sandoval R (2010) Enhancement of the electrical conductivity in PEDOT:PSS films by the addition of dimethyl sulfate. J Phys Chem C 114:20220
69. Xia Y, Ouyang J (2010) Significant conductivity enhancement of conductive poly(3,4-ethylenedioxythiophene):poly(styrenesulfonate) films through a treatment with organic carboxylic acids and inorganic acids. ACS Appl Mater Interf 2:474
70. Xia Y, Ouyang J (2009) Salt-induced conductivity enhancement of conducting poly(3,4-ethylenedioxythiophene):poly(styrenesulfonate) films. Macromolecules 42:4141
71. Xia Y, Zhang HM, Ouyang J (2010) Highly conductive PEDOT:PSS films prepared through a treatment with zwitterions and their application in polymer photovoltaic cells. J Mater Chem 20:9740
72. Xia Y, Ouyang J (2011) PEDOT:PSS films with significantly enhanced conductivities induced by preferential solvation with cosolvents and their application in polymer photovoltaic cells. J Mater Chem 21:4927
73. Xia Y, Ouyang J (2010) Anion effect on salt-induced conductivity enhancement of poly(3,4-ethylenedioxythiophene):poly(styrenesulfonate) films. Org Electron 11:1129
74. Kim YH, Sachse C, Machala ML, May C, Müller-Meskamp L, Leo K (2011) Highly conductive PEDOT:PSS electrode with optimized solvent and thermal post-treatment for ITO-free organic solar cells. Adv Funct Mater 21:1076
75. Xia Y, Sun K, Ouyang J (2012) Highly conductive poly(3,4-ethylenedioxythiophene):poly (styrene sulfonate) films treated with an amphiphilic fluoro compound as the transparent electrode of polymer solar cells. Energy Environ Sci 5:5325
76. Sun K, Xia Y, Ouyang J (2012) Improvement in the photovoltaic efficiency of polymer solar cells by treating the poly(3,4-ethylenedioxythiophene):poly(styrenesulfonate) buffer layer with co-solvents of hydrophilic organic solvents and hydrophobic 1,2-dichlorobenzene. Sol Energy Mater Sol Cells 97:89
77. Xia Y, Ouyang J (2012) Highly conductive PEDOT:PSS films prepared through a treatment with geminal diols or amphiphilic fluoro compounds. Org Electron 13:1785
78. Xia Y, Ouyang J (2012) Significant different conductivities of the two grades of poly(3,4-ethylenedioxythiophene):poly(styrenesulfonate), Clevios P and Clevios PH1000, arising from different molecular weights. ACS Appl Mater Interf 4:4131
79. Alemu D, Wei HY, Ho KC, Chu CW (2012) Highly conductive PEDOT:PSS electrode by simple film treatment with methanol for ITO-free polymer solar cells. Energy Environ Sci 5:9662
80. Cruz-Cruz I, Reyes-Reyes M, Aguilar-Frutis MA, Rodriguez AG, López-Sandoval R (2010) Study of the effect of DMSO concentration on the thickness of the PSS insulating barrier in PEDOT:PSS thin films. Synth Met 160:1501
81. Nardes AM, Kemerink M, de Kok MM, Vinken E, Maturova K, Janssen RAJ (2008) Conductivity, work function, and environmental stability of PEDOT:PSS thin films treated with sorbitol. Org Electron 9:727
82. Peng B, Guo X, Cui C, Zou Y, Pan C, Li Y (2011) Performance improvement of polymer solar cells by using a solvent-treated poly(3,4-ethylenedioxythiophene):poly (styrenesulfonate) buffer layer. Appl Phys Lett 98:243308
83. Xia Y, Sun K, Ouyang J (2012) Solution-processed metallic conducting polymer films as transparent electrode of optoelectronic devices. Adv Mater 24:2436

84. Ouyang J (2013) Solution-processed PEDOT:PSS films with conductivities as indium tin oxide through a treatment with mild and weak organic acids. ACS Appl Mater Interf 5:13082
85. http://CleviosTM.com/en/home/CleviosTM-homepage.aspx
86. http://www.agfa.com/sp/global/en/internet/main/solutions/orgacon_electronic_materials/index.jsp
87. Lang U, Müller E, Naujoks N, Dual J (2009) Microscopical investigations of PEDOT:PSS thin films. Adv Funct Mater 19:1215
88. Ouyang J, Guo TF, Yang Y, Higuchi H, Yoshioka M, Nagatsuka T (2002) High-performance, flexible polymer light-emitting diodes fabricated by a continuous polymer coating process. Adv Mater 14:915
89. Badre C, Marquant L, Alsayed AM, Hough LA (2012) Highly conductive poly(3,4-ethylenedioxythiophene):poly (styrenesulfonate) films using 1-ethyl-3-methylimidazolium tetracyanoborate ionic liquid. Adv Funct Mater 22:2723
90. Lipomi DJ, Lee JA, Vosgueritchian M, Tee BCK, Bolander JA, Bao Z (2012) Electronic properties of transparent conductive films of PEDOT:PSS on stretchable substrates. Chem Mater 24:373
91. Vosgueritchian M, Lipomi DJ, Bao Z (2012) Highly conductive and transparent PEDOT:PSS films with a fluorosurfactant for stretchable and flexible transparent electrodes. Adv Funct Mater 22:421
92. Lim FJ, Ananthanarayanan K, Luther J, Ho GW (2012) Influence of a novel fluorosurfactant modified PEDOT:PSS hole transport layer on the performance of inverted organic solar cells. J Mater Chem 22:25057
93. Yeo JS, Yun JM, Kim DY, Park S, Kim SS, Yoon MH, Kim TW, Na SI (2012) Significant vertical phase separation in solvent-vapor-annealed poly(3,4-ethylenedioxythiophene):poly (styrene sulfonate) composite films leading to better conductivity and work function for high-performance indium tin oxide-free optoelectronics. ACS Appl Mater Interf 4:2551
94. Marcus Y (1986) On enthalpies of hydration, ionization potentials, and the softness of ions. Thermochim Acta 104:389
95. Howden RM, McVay ED, Gleason KK (2013) CVD poly(3,4-ethylenedioxythiophene) conductivity and lifetime enhancement via acid rinse dopant exchang. J Mater Chem A 1:1334
96. Joo J, Long SM, Pouget JP, Oh EJ, MacDiarmid AG, Epstein AJ (1998) Charge transport of the mesoscopic metallic state in partially crystalline polyanilines. Phys Rev B 57:9567
97. Crispin X, Marciniak S, Osikowicz W, Zotti G, Denier van der Gon AW, Louwet F, Fahlman M, Groenendaal L, de Schryver F, Salaneck WR (2003) Conductivity, morphology, interfacial chemistry, and stability of poly(3,4-ethylene dioxythiophene)-poly(styrene sulfonate): a photoelectron spectroscopy study. J Polym Sci Polym Phys 41:2561
98. Takano T, Masunaga H, Fujiwara A, Okuzaki H, Sasaki T (2012) PEDOT nanocrystal in highly conductive PEDOT:PSS polymer films. Macromolecules 45:3859
99. Scott JC, Pfluger P, Krounbi MT, Street GB (1983) Electron-spin-resonance studies of pyrrole polymers: evidence for bipolarons. Phys Rev B 28:2140
100. Eom SH, Senthilarasu S, Uthirakumar P, Yoon SC, Lim J, Lee C, Lim HS, Lee J, Lee SH (2009) Polymer solar cells based on inkjet-printed PEDOT:PSS layer. Org Electron 10:536
101. Zhou Y, Fuentes-Hernandez C, Shim J, Meyer J, Giordano AJ, Li H, Winget P, Papadopoulos T, Cheun H, Kim J, Fenoll M, Dindar A, Haske W, Najafabadi E, Khan TM, Sojoudi H, Barlow S, Graham S, Brédas JL, Marder SR, Kahn A, Kippelen B (2012) A universal method to produce low–work function electrodes for organic electronics. Science 336:327
102. Zhou Y, Cheun H, Choi S, Potscavage WJ Jr, Fuentes-Hernandez C, Kippelen B (2010) Indium tin oxide-free and metal-free semitransparent organic solar cells. Appl Phys Lett 97:153304
103. Ahlswede E, Mühleisen W, Wahi MWNM, Hanisch J, Powalla M (2008) Highly efficient organic solar cells with printable low-cost transparent contacts. Appl Phys Lett 92:143307

104. Huang J, Li G, Yang Y (2008) A semi-transparent plastic solar cell fabricated by a lamination process. Adv Mater 20:415
105. Cai M, Ye Z, Xiao T, Liu R, Chen Y, Mayer RW, Biswas R, Ho KM, Shinar R, Shinar J (2012) Extremely efficient indium–tin-oxide-free green phosphorescent organic light-emitting diodes. Adv Mater 24:4337
106. Lin Y, Li Y, Zhan X (2012) Small molecule semiconductors for high-efficiency organic photovoltaics. Chem Soc Rev 41:4245
107. Li Y (2012) Molecular design of photovoltaic materials for polymer solar cells: toward suitable electronic energy levels and broad absorption. Acc Chem Res 45:723
108. Voigt MM, Mackenzie RCI, Yau CP, Atienzar P, Dane J, Keivanidis PE, Bradley DDC, Nelson J (2011) Gravure printing for three subsequent solar layers of inverted structures on flexible substrates. Sol Energy Mater Sol Cells 95:731
109. Colsmann A, Reinhard M, Kwon TH, Kayser C, Nickel F, Czolk J, Lemmer U, Clark N, Jasieniak J, Holmes AB, Jones D (2012) Inverted semi-transparent organic solar cells with spray coated, surfactant free polymer top-electrodes. Sol Energy Mater Sol Cells 98:118

MIX
Papier aus verantwortungsvollen Quellen
Paper from responsible sources
FSC® C105338

If you have any concerns about our products,
you can contact us on
ProductSafety@springernature.com

In case Publisher is established outside the EU,
the EU authorized representative is:
Springer Nature Customer Service Center GmbH
Europaplatz 3, 69115 Heidelberg, Germany

Printed by Libri Plureos GmbH
in Hamburg, Germany

MIX
Papier aus verantwortungsvollen Quellen
Paper from responsible sources
FSC® C105338

If you have any concerns about our products,
you can contact us on
ProductSafety@springernature.com

In case Publisher is established outside the EU,
the EU authorized representative is:
Springer Nature Customer Service Center GmbH
Europaplatz 3, 69115 Heidelberg, Germany

Printed by Libri Plureos GmbH
in Hamburg, Germany